Infectious Disease Ecology of Wild Birds

Infectious Disease Ecology of Wild Birds

EDITED BY

Jennifer C. Owen
Departments of Fisheries and Wildlife and Large Animal Clinical Sciences,
Michigan State University, USA

Dana M. Hawley
Department of Biological Sciences,
Virginia Tech, USA

Kathryn P. Huyvaert
Department of Fish, Wildlife, and Conservation Biology,
Colorado State University, USA

OXFORD
UNIVERSITY PRESS

OXFORD
UNIVERSITY PRESS

Great Clarendon Street, Oxford, OX2 6DP,
United Kingdom

Oxford University Press is a department of the University of Oxford.
It furthers the University's objective of excellence in research, scholarship,
and education by publishing worldwide. Oxford is a registered trade mark of
Oxford University Press in the UK and in certain other countries

© Oxford University Press 2021

The moral rights of the authors have been asserted

First Edition published in 2021

Impression: 1

Published in the United States of America by Oxford University Press
198 Madison Avenue, New York, NY 10016, United States of America

British Library Cataloguing in Publication Data

Data available

Library of Congress Control Number: 2021931163

ISBN 978–0–19–874624–9 (hbk.)
ISBN 978–0–19–874625–6 (pbk.)

DOI: 10.1093/oso/9780198746249.001.0001

Printed and bound by
CPI Group (UK) Ltd, Croydon, CR0 4YY

Foreword

In her book *The Coming Plague—Newly Emerging Diseases in a World Out of Balance*, Laurie Garrett (1994) describes a series of emerging human diseases and makes the point that 'Preparedness demands understanding. To comprehend the interactions between *Homo sapiens* and the vast and diverse microbial world, perspectives must be forged that meld many disparate fields.' She goes on to state that these include not only medicine, parasitology, entomology, and bacteriology, but also disciplines such as basic ecology and evolutionary biology, anticipating the One Health approach. While a fascinating book, the emphasis was on human health. Of course, diseases also emerge among non-human animals and plants. An advantage of studying emerging diseases in non-human systems is that it is easier to test hypotheses experimentally, and it may be possible to follow an epidemic over extended periods without intervention. Some of the most groundbreaking work in disease ecology today (and historically) is done in birds. Birds are both victims and reservoirs. What we learn from studies on birds, however, transcends taxonomic boundaries—hence, while this book may focus on birds, it is meant for any disease ecologist.

To really understand how parasites and pathogens emerge, thrive, evolve, and impact hosts, questions have to be asked over extended time periods and at multiple scales: individuals, populations, communities, and ecosystems (as nicely illustrated in the structure of this book), and across broad geographic scales. Here, birds play an important role because some data that need to be collected across large geographic regions can only be obtained in collaboration with community (citizen) scientists who love to watch birds and report their observations to scientists. The House Finch Disease Survey that we started in 1994, for example, not only allowed us to describe the expansion of an emerging infectious disease in great detail but also made it possible to describe differences in prevalence of conjunctivitis caused by *Mycoplasma gallisepticum* in wild bird species at a continental scale. As with many other bird studies, our study of mycoplasmal conjunctivitis has become long term, making it possible to detect effects of emerging infectious diseases on host and pathogen dynamics, while also studying evolutionary changes of both host and pathogen. Long-term studies also make it possible to detect effects of climate change on host–pathogen interactions, where bird studies play a big role (e.g., Fuller et al. 2012).

Disease ecology is special in various ways. By its integrative approach it requires collaboration of scientists with very different backgrounds, often from disciplines that ask very different questions and often speak different 'languages.' Nevertheless, the scientists involved must be willing to communicate constructively. While working on mycoplasmal conjunctivitis in house finches, for example, we had endless discussions to try to define virulence because veterinarians and mathematical modelers do not use that term in the same way; as is so often the case in ecology, it helps to define terms used up front so as to avoid confusion and to assure that everyone is on the same page, something the authors of the chapters in this volume address head on. Further, designing a project in avian infectious disease ecology requires extended conversations, because what interests a veterinarian may not interest an ecologist, and the data a modeler needs to make predictions may not be a priority for a microbiologist. The best way to understand such complex interactions among disciplines is by approaching infectious disease ecology from multiple perspectives and scales, as this book exemplifies. This text resource will help to train the next generation of infectious disease ecologists who can speak many languages, use many approaches, and, as a result, solve complex infectious disease ecology problems in avian systems and beyond.

However, to make these transdisciplinary collaborations possible so that work will truly advance our knowledge about infectious disease ecology, major funding is required. Until about 20 years ago, the problem in the US was that it was difficult to find a funding agency willing to support projects that study infectious disease in an ecological context. Over time, it was increasingly recognized that, without a firm foundation in disease ecology, our ability to understand disease dynamics across multiple scales was limited. Fortunately, through the collective vision of two scientists, Samuel Scheiner and Joshua Rosenthal, the National Science Foundation partnered with the National Institutes of Health to develop a program that was ultimately called the Ecology and Evolution of Infectious Diseases (EEID). The mission of the EEID program as it evolved over the subsequent decades was to support interdisciplinary research projects that combined to produce predictive understanding of infectious disease dynamics, with a focus on diseases with an environmental component (Scheiner and Rosenthal 2006). This program was a catalyst for disease ecology funding from more diverse sources. Collectively, the impact was beyond imagination. Not only do we have a vibrant annual EEID conference that alternates between North America and Europe, but this rapidly growing field has impacted so many young scientists that today we are looking at a book that can easily be used as a textbook to contribute to the training of disease ecologists around the world. It is noteworthy that chapter authors are mostly early- to mid-career scientists, women are prominently featured, and the chapters contribute important and diverse original insights, many of which emerged from interdisciplinary collaborations. Together, this illustrates the incredibly rapid growth and future potential of the exciting discipline that is avian infectious disease ecology.

André A. Dhondt
Member of the Academia Europaea,
Edwin H. Morgens Professor of Ornithology,
Cornell University, USA

References

Fuller, T., Bensch, S., Müller, I., et al. (2012). The ecology of emerging infectious diseases in migratory birds: an assessment of the role of climate change and priorities for future research. *Ecohealth*, 9, 80–88.

Garrett, L. (1994). *The Coming Plague: Newly Emerging Diseases in a World out of Balance.* Farrar, Straus and Giroux, New York.

Scheiner, S.M. and Rosenthal, J.P. (2006). Ecology of infectious disease: forging an alliance. *Ecohealth*, 3, 204–8.

Preface

Disease ecology is a rapidly growing discipline, which will undoubtedly only increase with the COVID-19 pandemic. With this growth comes a need for works that synthesize the core principles and concepts that underlie infectious disease dynamics from ecological and evolutionary perspectives. While there are many excellent disease ecology textbooks already in print, most focus on just one level of ecological hierarchy, such as the population, community, or ecosystem scales. When I was approached about writing a disease ecology book that focused on birds, I had been teaching a wildlife disease ecology course for 8 years. I structured the multiscale course by beginning at the organismal level with the host–parasite interaction and then incrementally scaling up to explore the unique properties that naturally emerge from these interactions within and among populations, communities, and ecosystems. No single textbook aligned with my course structure; rather, I pulled content from many books and papers for each unit. Hence, my motivation with this textbook was to provide a multiscale approach that I found missing in other books, using the lens of a well-studied taxonomic group. Like many ecologists, I am driven not only by scientific questions, which, in my case, are about the role of the environment in host–parasite interactions, but also by my fascination with a particular taxon—birds. Hence, it was such an exciting prospect to write this book; being able to immerse oneself in all facets of avian disease ecology seemed like a luxury, albeit an overwhelming one.

I began this journey solo. But, like many journeys, there were unexpected challenges en route that can force us to change course or adjust our approach. After I experienced a significant setback in life, it was clear that going solo on this textbook was not going to be possible; yet I wanted to see this project through. With that, I reached out to two amazing fellow disease ecologists and women, Dana Hawley and Kate Huyvaert, to join me as co-editors for this book. We are here today because they said *yes*. At the heart of scientific discovery and advancement is collaboration, communication, humility, and transparency. It is hard to imagine what this book would be without this collaboration—I know the book is better for it, as am I. And now this becomes *our* story.

We are proud of this book for many reasons—rather than a collection of independent chapters, each chapter builds upon another, which, in itself, demonstrates the transdisciplinary nature of what this field needs to be and clearly is becoming. Like the editors, each with our unique areas of expertise in disease ecology and avian biology, the authors' contributions span the breadth of ecological inquiry, taxonomic focus, and research approach. The selection of authors was deliberate—we looked for individuals whose careers were not just about a particular conceptual topic, but we also sought individuals for whom birds are at the heart of their research programs, to truly embrace the depth of understanding emerging at the interface of avian biology and disease ecology.

We are also proud of this book because it highlights a suite of early career and diverse scientists that are transforming the field and culture of disease ecology in much needed ways. Even in 2021, a scientific book with three women editors is notable, as is an author list that is largely female and includes numerous other forms of diversity, such as ethnicity, race, sexual orientation, and ability. But we also recognize that the field of disease ecology—much like many others in science—suffers from an overall lack of diversity, particularly with respect to race. We use this as a call to do more as educators and mentors to raise up, to encourage, and to empower those that continue to be marginalized and underserved in disease ecology and science more broadly.

We are scientists but we are also humans with lives and challenges that we each face—while writing this book we persisted through the loss of a parent, a debilitating car accident, death of beloved pets, a pandemic, Zoom fatigue, parenting/teaching young children with no childcare or school, and more. Hence, we could not have done this alone and we want to acknowledge all those that helped us get here today. Allie Shoffner worked behind the scenes to tackle the huge task of collecting and ensuring that all the many pieces of each chapter were ready and formatted properly for submission. We are also indebted to Ian Sherman, Charles Bath, and the publishing team at Oxford University Press for their support along this long journey, and our spouses and families (including the furry members) for their unwavering encouragement and support.

Through this endeavor, we grew individually and as a team, reminding ourselves that resiliency, personal and ecological, is a hallmark of a healthy, whole system. And, if we learned anything by crafting a book during the COVID-19 pandemic, it is that understanding and appreciating the importance of disease ecology is essential for the resilient world we aspire to live in and contribute to, both for humans and the feathered creatures that bring us and others so much joy.

Jen Owen, Dana Hawley, and Kate Huyvaert

Contents

List of Contributors   xiii

1 A Bird's Eye View of Avian Disease Ecology   **1**
Jennifer C. Owen, Dana M. Hawley, and Kathryn P. Huyvaert

2 The Nature of Host–Pathogen Interactions   **7**
Jennifer C. Owen, James S. Adelman, and Amberleigh E. Henschen

   2.1 Introduction   7
   2.2 Causation of disease   7
   2.3 Host–parasite interaction: infection timeline   12
   2.4 Developing pathogen-specific immunity   17
   2.5 Detecting and quantifying a host's infection status   18
   2.6 Population-level disease dynamics   22
   2.7 Summary   24

3 Ecoimmunology   **29**
Amberleigh E. Henschen and James S. Adelman

   3.1 Introduction   29
   3.2 Overview of the avian immune system   29
   3.3 Drivers of variation in immune defense   33
   3.4 Quantifying variation in immune defenses   38
   3.5 Bridging the first gap: do immune phenotypes predict infection
       status or host fitness?   39
   3.6 Bridging the second gap: do within-host measures inform
       population-level transmission dynamics?   44
   3.7 Conclusions   45

4 Behavior Shapes Infectious Disease Dynamics in Birds   **53**
Andrea K. Townsend and Dana M. Hawley

   4.1 Conceptual overview: the intersection of behavior and disease
       ecology   53
   4.2 Behaviors important to parasite exposure, resistance, and spread   54
   4.3 Synthesis and future directions   68

5   Host–Pathogen Evolution and Coevolution in Avian Systems   **77**
Camille Bonneaud

5.1 Introduction   77
5.2 Evolutionary responses of avian hosts to their pathogens   79
5.3 Evolutionary responses of pathogens to their avian hosts   85
5.4 Coevolution of avian hosts and their pathogens   89
5.5 Conclusions and implications   92

6   Fitness Effects of Parasite Infection in Birds   **99**
Jenny C. Dunn, Dana M. Hawley, Kathryn P. Huyvaert, and Jennifer C. Owen

6.1 Introduction   99
6.2 Effects of parasites on host survival   99
6.3 Effect of parasites on reproductive success   109
6.4 Context-dependence of fitness effects   115
6.5 Summary   115

7   Wild Bird Populations in the Face of Disease   **121**
Kathryn P. Huyvaert

7.1 Introduction   121
7.2 Tools for assessing effects of parasites on avian populations   121
7.3 Effects of parasites and pathogens on avian host populations   134
7.4 Lessons learned   140

8   Community-Level Interactions and Disease Dynamics   **145**
Karen D. McCoy

8.1 Introduction   145
8.2 Community structure and function   145
8.3 Community diversity and parasite/pathogen circulation   149
8.4 An example of Lyme borreliosis in seabird communities   156
8.5 Summary and future directions   162

9   Land Use Change and Avian Disease Dynamics   **171**
Maureen H. Murray and Sonia M. Hernandez

9.1 Introduction   171
9.2 Key changes associated with land use and influences on disease
     dynamics   172
9.3 Overall consequences of land use change for avian parasitism:
     urbanization as a case study   179
9.4 Synthesis: need for multidisciplinary approaches   182

10 Climate Change and Avian Disease   **189**
Richard J. Hall

10.1 Introduction   189
10.2 Infection consequences of changes in bird distribution and
      phenology   190

10.3 Changes in bird physiology, behavior, and response to infection   195
10.4 Effects of climate change on parasites with external
       transmission stages   198
10.5 Synthesis and future research directions   200

11 Pathogens from Wild Birds at the Wildlife–Agriculture Interface   207
Alan B. Franklin, Sarah N. Bevins, and Susan A. Shriner

11.1 Introduction   207
11.2 Avian use of agricultural operations   207
11.3 Host types and pathways for pathogen contamination by birds   210
11.4 Pathogen prevalence and transmission in wild birds   216
11.5 Contamination potential of agricultural operations by birds   220
11.6 Long-distance movements and pathogen introductions   220
11.7 Future directions   222

12 Pathogen Transmission at the Expanding Bird–Human Interface   229
Sarah A. Hamer and Gabriel L. Hamer

12.1 Introduction   229
12.2 Examples of zoonoses at the intersection of wild bird, domestic
       bird, and human populations   229
12.3 Ecological and evolutionary factors important for zoonotic
       emergence at the wild bird–human interface   232
12.4 Birds as vehicles for the movement of arthropod vectors and
       pathogens of human health conern   233
12.5 Reverse zoonoses (anthroponoses): examples of pathogen
       transmission from humans to birds   235
12.6 Parasite interactions in birds—implications for human health   236
12.7 Management strategies to minimize avian zoonoses   238
12.8 Summary   240

13 A Flight Path Forward for Avian Infectious Disease Ecology   245
Dana M. Hawley, Kathryn P. Huyvaert, and Jennifer C. Owen

13.1 Introduction   245
13.2 A flight path for avian disease ecology that crosses scales and
       disciplines   245
13.3 Leveraging existing strengths to promote innovation   248
13.4 A flight path forward for avian infectious disease ecology,
       management, and conservation   249
13.5 Conclusions   251

Index   255

List of Contributors

James S. Adelman Department of Biological Sciences, University of Memphis, USA

Sarah N. Bevins National Wildlife Research Center, United States Department of Agriculture, Wildlife Services, USA

Camille Bonneaud College of Life and Environmental Sciences, University of Exeter, UK

André A. Dhondt Cornell Lab of Ornithology and Department of Ecology and Evolutionary Biology, Cornell University, USA

Jenny C. Dunn School of Life Sciences, University of Lincoln, UK

Alan B. Franklin National Wildlife Research Center, United States Department of Agriculture, Wildlife Services, USA

Richard J. Hall Odum School of Ecology and Department of Infectious Diseases, University of Georgia, USA

Gabriel L. Hamer Department of Entomology, Texas A&M University, USA

Sarah A. Hamer Schubot Center for Avian Health, College of Veterinary Medicine & Biomedical Sciences, Texas A&M University, USA

Dana M. Hawley Department of Biological Sciences, Virginia Tech, USA

Amberleigh E. Henschen Department of Biological Sciences, University of Memphis, USA

Sonia M. Hernandez Warnell School of Forestry and Natural Resources and the Southeastern Cooperative Wildlife Disease Study, Department of Population Health, at the College of Veterinary Medicine, University of Georgia, USA

Kathryn P. Huyvaert Department of Fish, Wildlife, and Conservation Biology, Colorado State University, USA

Karen D. McCoy Centre for Research on the Ecology and Evolution of Diseases, MiVEGEC, CNRS-IRD-University of Montpellier, France

Maureen H. Murray Department of Conservation and Science, Lincoln Park Zoo, USA

Jennifer C. Owen Departments of Fisheries and Wildlife and Large Animal Clinical Sciences, Michigan State University, USA

Susan A. Shriner National Wildlife Research Center, United States Department of Agriculture, Wildlife Services, USA

Andrea K. Townsend Department of Biology, Hamilton College, USA

CHAPTER 1

A Bird's Eye View of Avian Disease Ecology

Jennifer C. Owen, Dana M. Hawley, and Kathryn P. Huyvaert

'Birds are not only birds but aviating zoological gardens'
(Shipley 1926).

Wild birds capture enormous human interest and joy—not just for the ornithologists and disease ecologists that study them but for all kinds of people around the globe. Birds function as national emblems, centers of cultural and spiritual rituals, and talisman symbols (Cocker 2014). Humans around the world also go to great lengths to care for and observe wild birds. Wild bird feeding ranks as one of the most popular forms of human–wildlife recreation in many countries (Cox and Gaston 2018; Jones 2018) and bird-based tourism is one of the fastest growing industries, with people making an estimated 3 million trips per year worldwide to watch birds (CREST 2014; CTO 2008). Wild game birds are an important source of subsistence and recreational hunting worldwide, and in the United States (US), revenue generated from the sale of duck stamps and hunting licenses has contributed hundreds of millions of dollars toward wetland conservation (Rubio-Cisneros et al. 2014; Shipley et al. 2019). Birds fascinate people of all ages worldwide; they come in a spectacular array of colors, shapes, and sizes and they possess remarkable adaptations that, more so than any other vertebrate, enable them to live in every habitat of the world. Finally, birds are unmatched in their ability to connect people to nature (Cox and Gaston 2016).

Not all the attention placed on birds is positive, however. Many wild birds are currently threatened by novel pathogens that pose conservation threats to their populations, some of which are already declining precipitously due to factors such as habitat degradation and loss, pollution, invasive species, and climate change (Rosenberg et al. 2019). Endemic Hawaiian honeycreepers have experienced dramatic population declines, and in some cases extinction, due to the combined impact of invasive bird and mammal species, the introduction of both the avian malaria parasite and its mosquito vector (Warner 1968), and, more recently, the threat of climate change (Paxton et al. 2016). Birds have also been linked to the spread of pathogens infectious to humans and domestic animals. In the last century, there have been significant increases in the numbers of emerging infectious diseases; further, approximately 75% are **zoonoses**, meaning that they originate in non-human animals but the parasites that cause them are transmissible to humans (Jones et al. 2008). Wild and domestic birds are the natural reservoir for many of these **zoonotic** pathogens of significant economic and public health importance (Reed et al. 2003). For example, West Nile virus (WNV), for which wild birds are the primary reservoir host, is the most geographically widespread arbovirus globally (Kramer et al. 2007). The virus invaded the US in 1999 and quickly spread across the country and between 1999 and 2018 led to over 50,000 human cases and 2,400 deaths (CDC 2020). The naïve avian population in North America also suffered, with notable mortality events and significant population declines in

Jennifer C. Owen, Dana M. Hawley, and Kathryn P. Huyvaert, *A Bird's Eye View of Avian Disease Ecology* In: *Infectious Disease Ecology of Wild Birds*.
Edited by: Jennifer C. Owen, Dana M. Hawley, and Kathryn P. Huyvaert, Oxford University Press. © Oxford University Press 2021.
DOI: 10.1093/oso/9780198746249.003.0001

the years following the invasion of WNV (LaDeau et al. 2007).

It is not surprising that bird-hosted pathogens continue to pose problems for bird conservation and global health—birds are the most diverse group of extant vertebrates and have evolved to utilize (exploit) every ecological niche on Earth, such that they have the propensity to serve as a host of pathogens in every part of the world. The diversity of birds is outmatched only by the diversity of the parasite fauna infecting them. Given the overwhelming diversity of both avian hosts and their parasites, we have only scratched the surface regarding the role that pathogens play in avian biology and the role that birds play in the maintenance and spread of zoonotic pathogens. For example, wild birds have been documented to harbor a suite of gamma and delta coronaviruses, but the zoonotic potential of avian coronaviruses remains entirely unknown (Wille and Holmes 2020). In addition to this understudied diversity, parasite–bird interactions are increasingly occurring in rapidly changing global environments—thus, their ecology is changing—and this shapes the complex ways by which parasites influence the interconnected health of birds, humans, and shared ecosystems. Parasite–bird interactions are both influenced by and have consequences for every level of ecological hierarchy, from the physiology, behavior, and evolution of individual hosts up to the complex biotic and abiotic interactions occurring within biological communities and ecosystems (Figure 1.1). Understanding these complex and multiscale interactions requires an inherently integrative approach.

Disease ecology is an interdisciplinary field that recognizes that the host–parasite interaction is shaped by the environment and can affect and be affected by the processes that occur across all levels of ecological organization (Hawley and Altizer 2011). The field of disease ecology is closely aligned, yet notably distinct from, the field of **epidemiology**, which studies the spatial and temporal patterns of disease in human populations and the likely factors that cause those patterns. Disease ecology focuses on understanding the complex ecological and evolutionary underpinnings of the host–parasite interactions of all biological taxa, allowing robust predictions about spatial and temporal patterns of pathogen transmission and disease risk to host and non-host populations (Kilpatrick and Altizer 2010). In this book, we focus on the dynamics of infectious diseases for wild avian hosts across different scales of biological organization—from within-host processes to landscape-level patterns (Figure 1.1). Because the infectious agents of wild birds can both influence and be affected by processes occurring in other taxa, we consider the role of domestic birds and non-avian hosts, including humans, in the dynamics of infectious diseases of wild birds.

Owen and colleagues begin in **Chapter 2** by discussing the basic principles of host–parasite interactions, as they apply more generally and specifically to birds. By introducing terms and providing definitions within the context of the broader field of disease ecology, the authors aim to provide a reference that is a starting point for anyone embarking on the study of infectious disease ecology. To better understand the population and community level dynamics of pathogen transmission, one must understand the timeline by which an individual host progresses through the key steps of infectious pathogen acquisition, clinical signs (where applicable), and spread. In fact, this timeline of infection at the level of the individual is one of the most critical aspects of being able to predict the magnitude of a disease outbreak and develop better control and prevention strategies.

One of the most important factors influencing a bird's ability to limit an invading pathogen and/or clinical disease expression is the type and effectiveness of the host's immune system. The field of ecoimmunology, or the study of immune variation, has been rapidly growing in the last two decades and much of the research in this field has been conducted on birds. In **Chapter 3**, Henschen and Adelman give a broad overview of the avian immune system and use the lens of ecoimmunology to explore some of the key factors that drive immunological variation within and among bird species. They then explore the complex and diverse relationships between immunological variation, host fitness, and infectious disease ecology, which, together, represent an intersection ripe for exciting progress in the coming decades.

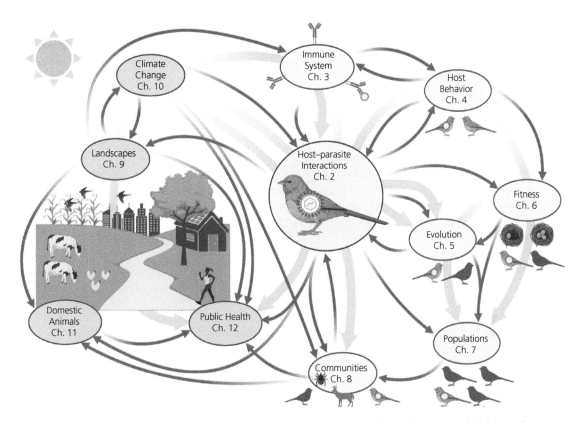

Figure 1.1 A conceptual framework for avian infectious disease ecology that is paralleled by the book's organization (with relevant chapter numbers in each topic oval). Central to this framework are interactions between avian hosts and their parasites (large gray circle in center), which directly connect to (gray arrows) as well as indirectly modify (green arrows) all other interactions. Disease ecology crosses levels of biological hierarchy and temporal and spatial scales (blue ovals), from the processes that occur in cells within an individual host's immune system, up through aspects of individuals (behavior, fitness), populations, and communities. Anthropogenic activities interact with these various biological processes to drive climate and landscape impacts on birds as well as the degree of overlap among wild birds, humans, domestic animals, and their shared parasites (green ovals). For visual clarity, only key linkages (arrows) are shown, but direct and indirect interactions likely exist between virtually all components of this conceptual framework, illustrating both the complexity and interconnectedness of infectious disease ecology. (Created with Biorender.com.)

A bird's behavior can act in concert with its immune system, in some cases acting as a first line of defense by preventing exposure and in other cases facilitating pathogen exposure and ongoing spread. In **Chapter 4**, Townsend and Hawley explore a suite of avian behaviors that have important consequences for infectious disease ecology. Birds are unique in their ability to readily transport pathogens over long distances via both daily foraging movements and annual migrations, with over half of the world's more than 10,000 species of birds classified as migratory. Townsend and Hawley emphasize the importance of understanding how a

host's infection status influences behaviors such as social interactions, foraging, and migration in order to understand the implications of many avian behaviors for ongoing disease spread.

Because of the importance of traits such as host immunity and behaviors for minimizing the fitness costs of pathogens, infectious agents are expected to exert strong selection on these components of avian biology. In **Chapter 5**, Bonneaud considers parasite-mediated evolution in birds, as well as antagonistic coevolution between birds and their parasites. Because birds are highly visible and well studied, they represent some of the most notable examples of the ways in which parasites exert strong selection on hosts and vice versa. Using a case study approach, Bonneaud discusses how some of these key avian examples have shed important light on the ability of hosts and pathogens to rapidly evolve in response to one another.

The extent to which avian hosts evolve in response to infectious agents will depend on the way in which parasites and pathogens affect the fitness of individual birds in the wild. While birds represent some of the most notable examples of mortality from pathogens, such as avian malaria causing acute mortality in endemic Hawaiian birds, the fitness effects of parasites on birds can be highly variable, in some cases constituting only subtle effects on survival and reproductive success. In **Chapter 6**, Dunn and colleagues discuss the ways that these fitness effects are typically measured in birds. They then consider the diverse sources of variation in fitness effects of parasites on hosts, including contributions of host and parasite variation, as well as the role of the environment.

The individual-level fitness effects of parasites on birds are increasingly well studied, but less is known about host population growth in light of parasitism and how and when parasites regulate wild bird populations. That said, the few examples in which a pathogen has been documented to regulate a wild animal population come from studies of wild birds, underscoring the importance of avian systems for our understanding of infectious disease ecology more broadly. In **Chapter 7**, Huyvaert introduces key concepts, important tools, and mathematical modeling approaches used to describe and deepen our understanding of the effects of parasites on bird population dynamics. At the heart of accurate estimates of the effects of parasites on avian populations are long-term studies—facilitated by bird banding—that employ quantitative approaches such as mark-recapture analyses to account for imperfect detection of individual animals or their infection status. Huyvaert ends by considering the potential for parasites to influence avian population dynamics by affecting population growth, regulation, and, in some cases, the likelihood of extinction.

Bird populations and the parasites they harbor are part of larger, more complex ecological communities. The attributes of these communities—such as species composition and richness and their structure and trophic networks—can greatly influence the way in which populations within the community interact with each other and their environment. In **Chapter 8**, McCoy explores community attributes and how the properties that emerge from these interactions influence avian infectious disease dynamics. For instance, an emergent property of the species diversity and composition present in a community is how it either amplifies or dilutes pathogen transmission. McCoy illustrates various community-level processes in disease ecology using long-term studies on polar seabird communities and Lyme disease bacteria. Attention to community-level interactions and the influence they have on disease dynamics is gaining warranted attention, particularly considering the ongoing global changes that have key downstream effects on community-level characteristics.

Human-led activities have altered landscapes for birds and the communities they reside in around the globe. In **Chapter 9**, Murray and Hernandez explore how changes in land and water use, including habitat loss, pollution, and supplemental resources, affect infectious disease dynamics in wild birds through several non-mutually exclusive mechanisms. For example, the intensification of agriculture and expansion of urban areas has led to degradation and loss of habitat, affecting abundance and composition of bird and vector populations and communities and, in some cases, the physiology of avian hosts and vectors in ways that alter their susceptibility or competence, respectively. Pollution—including toxicants, light, and noise—can reduce avian or vector population sizes and cause non-lethal effects that influence the capacity to resist infection. Finally, supplemental resources, whether intentional (i.e., bird feeding) or unintentional (e.g., landfills), can cause a suite of effects on disease dynamics by bringing avian hosts into close proximity and influencing their ability to resist infection. Understanding the emergent outcomes of anthropogenic land use for avian disease dynamics is an important area for future study in rapidly changing global landscapes.

One of the key environmental factors that interacts closely with land use and influences the population-level outcomes of disease is climate. In

Chapter 10, Hall discusses the consequences of a warming world on bird–parasite interactions and the diverse, and sometimes conflicting, ways in which climate change can influence avian disease dynamics to either augment or dampen pathogen transmission. Some of the most notable effects of climate change on bird–parasite interactions are the shifting, expanding, and shrinking ranges of many bird species concomitant with the changing distributions of the pathogens themselves and, for many, the vectors that transmit them. The environmental challenges linked to climate change can also affect the behavior and physiology of both hosts and vectors, which has consequences for the likelihood of infection and spread. Hall explores diverse examples of how climactic variables have directly or indirectly influenced the host organism, pathogen, and/or vector, to better predict what may happen in our uncertain future. What we do know for certain is that the effects of climate change on disease dynamics in birds is one of the largest knowledge gaps in avian disease ecology and where future research is sorely needed.

Understanding the role of factors such as land use and climate on avian pathogens has downstream effects for our own health and food security, because wild birds are a major source of pathogens that affect the health of domestic animals and humans. In **Chapter 11**, Franklin and colleagues discuss how wild birds, particularly granivorous species that frequent domestic farms due to the unintentional supplemental food provided by livestock operations, can affect pathogen transmission between wild birds and domestic livestock. Wild birds pose a risk to global food security by harboring pathogens such as food-borne zoonotic bacteria, including *Salmonella enterica* and *Campylobacter jejuni*, and viral pathogens such as virulent Newcastle Disease virus and the zoonotic highly pathogenic avian influenza viruses. Franklin and colleagues consider how the potential for pathogen transmission between wild birds and domestic animals, both avian and non-avian, is increasing with the intensification of agriculture and human encroachment on wildlife habitat. The ecological processes underlying the transmission and disease dynamics at the wild bird–agricultural interface are complex and tackling these problems requires a multiscale approach and expertise and tools that span disciplines.

Human health is also threatened by the emergence and reemergence of avian pathogens. Humans and birds engage in close contact across multiple interfaces, including interactions with wild birds via game-hunting, backyard bird feeding, or shared vectors; household or veterinary interactions with pet birds; and direct or indirect interactions with agricultural birds. In **Chapter 12**, Hamer and Hamer explore how these interfaces result in spillover events and ongoing transmission of pathogens between birds and humans. The circumstances of spillover events are often complex and driven by both ecological and evolutionary determinants of cross-species transmission events. Hamer and Hamer explore some of these determinants, as well as the role of migrating birds in the dispersal and establishment of zoonotic pathogens and their vectors in novel areas. Overall, this chapter highlights the increasing interconnectedness of avian health with that of humans and the environment and the need for better surveillance of pathogen transmission across the human–bird interface.

Where do we go from here? How do we predict where and when disease outbreaks will occur that negatively impact bird populations and/or where birds play a key role in disease dynamics? In **Chapter 13**, Hawley and colleagues provide a synthesis of the field of disease ecology in birds and a call for a flight path forward. The discipline of 'One Health' is a rapidly growing field that recognizes that the health of all living organisms—including birds and humans—is inherently interconnected and relies heavily on the health of the ecosystem. Hawley and colleagues highlight that the flight path forward for avian infectious disease ecology requires continued work that builds upon the One Health framework by embracing collaboration that crosses and transcends disciplines, harnessing emerging technologies and innovative approaches that leverage the unique characteristics of birds and expanding and merging across spatial and temporal scales of analysis. Approaches like those discussed in this volume are urgently needed because the challenges that face birds, humans, and the Earth we share are critical to a sustainable and healthy future for all living systems, including wild birds and the many people who value and cherish them.

Literature cited

CDC (2020). Final Cumulative Maps & Data for 1999–2018, https://www.cdc.gov/westnile/statsmaps/cummapsdata.html, accessed September 2020.

Cocker, M. (2014). *Birds and People*. Random House, London.

Cox, D.T. and Gaston, K.J. (2016). Urban bird feeding: connecting people with nature. *PLoS ONE*, 11, e0158717.

Cox, D.T. and Gaston, K.J. (2018). Human–nature interactions and the consequences and drivers of provisioning wildlife. *Philosophical Transactions of the Royal Society B: Biological Sciences*, 373, 20170092.

CREST (2014). *Market analysis of bird-based tourism: a focus on the US market to Latin America and the Caribbean, including fact sheets on the Bahamas, Belize, Guatemala and Paraguay*. CREST, Washington, DC.

CTO (2008). *Bird Watching*. St. Michael, Barbados.

Hawley, D.M. and Altizer, S.M. (2011). Disease ecology meets ecological immunology: understanding the links between organismal immunity and infection dynamics in natural populations. *Functional Ecology*, 25, 48–60.

Jones, D. (2018). *The Birds at my Table: Why we Feed Wild Birds and Why it Matters*. Cornell University Press, Ithaca, NY.

Jones, K.E., Patel, N.G., Levy, M.A., et al. (2008). Global trends in emerging infectious diseases. *Nature*, 451, 990–3.

Kilpatrick, A. and Altizer, S. (2010). Disease ecology. *Nature Education Knowledge*, 1, 408.

Kramer, L.D., Li, J., and Shi, P.Y. (2007). West Nile virus. *Lancet Neurology*, 6, 171–81.

LaDeau, S.L., Kilpatrick, A.M., and Marra, P.P. (2007). West Nile virus emergence and large-scale declines of North American bird populations. *Nature*, 447, 710–13.

Paxton, E.H., Camp, R.J., Gorresen, P.M., Crampton, L.H., Leonard, D.L., and VanderWerf, E.A. (2016). Collapsing avian community on a Hawaiian island. *Science Advances*, 2, e1600029.

Reed, K.D., Meece, J.K., Henkel, J.S., and Shukla, S.K. (2003). Birds, migration and emerging zoonoses: West Nile virus, Lyme disease, influenza A and enteropathogens. *Clinical Medicine and Research*, 1, 5–12.

Rosenberg, K.V., Dokter, A.M., Blancher, P.J., et al. (2019). Decline of the North American avifauna. *Science*, 366, 120–4.

Rubio-Cisneros, N.T., Aburto-Oropeza, O., Murray, J., Gonzalez-Abraham, C.E., Jackson, J., and Ezcurra, E. (2014). Transnational ecosystem services: the potential of habitat conservation for waterfowl through recreational hunting activities. *Human Dimensions of Wildlife*, 19, 1–16.

Shipley, A.E. (1926). Parasitism in evolution. *Science Progress in the Twentieth Century (1919–1933)*, 20, 637–61.

Shipley, N.J., Larson, L.R., Cooper, C.B., Dale, K., LeBaron, G., and Takekawa, J. (2019). Do birdwatchers buy the duck stamp? *Human Dimensions of Wildlife*, 24, 61–70.

Warner, R.E. (1968). The role of introduced diseases in the extinction of the endemic Hawaiian avifauna. *Condor*, 70, 101–20.

Wille, M. and Holmes, E.C. (2020). Wild birds as reservoirs for diverse and abundant gamma-and deltacoronaviruses. *FEMS Microbiology Reviews*, 44, 631–44.

CHAPTER 2

The Nature of Host–Pathogen Interactions

Jennifer C. Owen, James S. Adelman, and Amberleigh E. Henschen

'The complex effects of infectious disease at the scale of communities and ecosystems are fundamentally driven by the interaction between individual hosts and pathogens'

(Blaustein et al. 2012)

2.1 Introduction

The dynamics of infectious diseases are driven by the fundamental processes that mediate host–pathogen interactions. A basic understanding of the mechanisms underlying these interactions is essential for disease ecologists regardless of the scale of inquiry. In this chapter, we will summarize some of the terms and concepts commonly used in ecological studies of infectious disease across levels of organization and scales of inquiry, from the individual host organism to host populations and multispecies communities. By providing a brief introduction to epidemiological modeling, we will illustrate how the natural history of infection relates to population-level dynamics of infectious disease, a topic covered in more detail in Chapter 7. Further, the between-host processes discussed in the beginning of the chapter arise from the within-host processes between the pathogen and the host's immune system. Hence, we will finish the chapter with an overview of the pathogenesis of infection and the initial stages of the host immune response (the avian immune system is covered in more detail in Chapter 3). When applicable, we will highlight aspects of avian biology that are unique relative to other taxonomic groups. We do not mean this chapter to be exhaustive but, instead, to provide a common framework for readers approaching this topic

from unique backgrounds. Given the transdisciplinary nature of avian infectious disease ecology, many of the terms used have multiple meanings assigned to them that are taxon- or discipline-specific; hence, we will try to clarify some terms and provide definitions, many of which will be used throughout this book. In instances when definitions differ between chapters, they will be clarified accordingly.

2.2 Causation of disease

Disease is a condition when an organism's homeostasis has been disrupted or impaired and 'harm' is done. In this book, we focus on infectious diseases, which are disruptions to homeostatis specifically caused by infection with a pathogen or parasite (defined in Section 2.2.2). Importantly, disease is not always the outcome of infection, which we define as the condition when a pathogen or parasite has successfully invaded the host and begins to multiply (see Section 2.3.4). Instead, whether disease results in a given host will be a product of the interactions among a susceptible host, an infectious agent (pathogen or parasite) that has the capacity to cause disease in that host, and the environment that facilitates their interaction. These interactions are commonly illustrated by the epidemiological triangle in which each element (host, pathogen, and

Jennifer C. Owen, James S. Adelman, and Amberleigh E. Henschen, *The Nature of Host–Pathogen Interactions* In: *Infectious Disease Ecology of Wild Birds*. Edited by: Jennifer C. Owen, Dana M. Hawley, and Kathryn P. Huyvaert, Oxford University Press. © Oxford University Press 2021. DOI: 10.1093/oso/9780198746249.003.0002 environment) sits at one of the three vertices. Vectors, frequently absent from the classic triangle or incorporated into the environmental node, are an important component given the ubiquity of pathogens that require an arthropod vector (e.g., mosquito, tick) for biological transmission (see Section 2.2.3), and so we have depicted this epidemiological tetrad as a Venn diagram (Figure 2.1).

For infectious disease to occur, all elements—host, pathogen, environment, and competent vector (if applicable)—must be present and interacting (Figure 2.1). There are characteristics of each of the four elements that can influence the nature of interactions among them and occurrence of pathogen transmission and the potential for disease. Starting with the host, we will step through each element of the tetrad, defining important terms that will provide the reader with a foundation in eco-epidemiology.

2.2.1 Hosts

A host–parasite interaction requires a 'susceptible' host. The term **host** is used to classify any individual or population that interacts with the pathogen of interest. A susceptible host is an individual that is capable of being infected by a given pathogen at a given time. In many cases, such as for microparasites (see Section 2.2.2), susceptible hosts

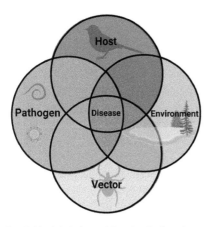

Figure 2.1 Epidemiological tetrad depicting the four elements, host, pathogen, environment, and vector, needed for occurrence of pathogen transmission and/or disease for vector-transmitted pathogens. Non-vector transmitted pathogens would be depicted with the host, pathogen, and environment triad. (Created with Biorender.com.)

are those that have never been exposed to a given pathogen in their lifetime; in other cases, susceptible hosts may have had prior exposure to a given pathogen or parasite but have since lost some or all protection conferred from initial infection (see Section 2.3) or otherwise remain susceptible, as is the case for macroparasites.

When pathogens infect multiple species, which describes most agents of infectious disease in birds, the word 'host' alone is not very meaningful (Haydon et al. 2002). In multihost–pathogen systems, species in a community have different epidemiological functions relative to their relationship with and potential contribution to the maintenance, amplification, and movement of the pathogen (Caron et al. 2015) within the system.

The host species in which parasites reach maturity and reproduce are called **definitive** hosts. This distinction is more relevant to macroparasites (see Section 2.2.2.1) that can pass through multiple hosts to complete their life cycle. **Intermediate hosts** are the obligatory species in which one or more of the parasite's life stages are completed. In some cases, a host may have a non-obligatory association with a parasite, and these are called mechanical, transport, or **paratenic hosts**. In the paratenic host, the parasite does not establish itself or undergo development/maturation; it only temporarily resides in (or on) the host until it is released into the environment, without any modification. While birds rarely serve as paratenic hosts, there are some examples in which birds can transport free-living stages of macroparasites or intermediate hosts harboring parasites on their plumage and/or legs (e.g., schistosome parasites; Huffman and Friend 2009). Additionally, hosts can ingest infective life stages and then release them into the environment. For instance, free-living waterbirds can serve as paratenic hosts of *Cryptosporidium parvum*, a waterborne protozoan pathogen, by ingesting the eggs and then eliminating them with the feces, without the parasite losing its viability and infectiousness (Graczyk et al. 1996). While the protozoa do not invade the bird's tissues, they can be dispersed during the bird's short- or long-distance flights within several days of being ingested.

Reservoir (maintenance) host is the most common host definition used in this volume; reservoir hosts are species within a community whose presence is required for the persistence of the pathogen and are the ultimate source of new infections. The word 'reservoir' alone also applies to an environmental source of a pathogen—such as water or food—when applicable. Not all species that can become infected can maintain the pathogen in the absence of a reservoir species; these are frequently called **dead-end** or **incidental hosts**. For example, eastern equine encephalitis virus (EEEV) is maintained in nature by an avian reservoir host, particularly passerine birds, and the ornithophilic (i.e., bird-loving) mosquito vector, *Culiseta melanura*. Non-avian vertebrates are dead-end hosts for EEEV because, while they can be fed upon by the vector, the virus does not replicate in these other taxa as it does in birds (Morens et al. 2019).

The inter- and intraspecific variation in host susceptibility to infection and capacity to serve as a reservoir can be substantial (Wilson et al. 2001). Host attributes that may influence their role in the epidemiological tetrad include demographic variables (age, sex; see Chapter 6), genetics, immunity (see Chapter 3), co-infection (see Chapter 12), historical exposure, life history (see Chapters 4, 5, and 6), and nutritional condition.

2.2.2 Pathogens and parasites

The etiological agents of infectious disease in wild birds include viruses, bacteria, fungi, protozoa, helminths, and endoparasitic arthropods, and, collectively, these are called pathogens and/or parasites. **Pathogen** is strictly defined as an agent that causes disease *in* the host organism (Casadevall and Pirofski 2002; Pirofski and Casadevall 2012). In an ecological context, a pathogen is one that causes a reduction in host fitness. **Parasites** are organisms that live in or on a host from which they obtain nourishment (Kennedy 1975). Thus, the term parasite also encompasses ectoparasitic arthropods such as ticks, fleas, and lice. In contrast, this classical definition of parasite would not include viruses, a quintessential pathogen, because they are not considered organisms. That said, how the terms are used and defined varies substantially across disciplines and scales of inquiry and has been the topic of several papers (Casadevall and Pirofski 2002;

Casadevall and Pirofski 2014; Méthot and Alizon 2014; Pirofski and Casadevall 2012). The word itself is less important than the meaning the authors assign to it. In this book, unless specified, the terms 'pathogen' and 'parasite' are used interchangeably to describe all infectious agents that cause disease, *can* cause disease, and/or disrupt the physiological state of a susceptible host.

2.2.2.1 Microparasites and macroparasites

In disease ecology it is common to categorize infectious agents, according to their unique biological features and life history, into two functional groups—microparasites and macroparasites—as described by Anderson and May (1991). Note that many pathogens do not fit under these discrete categories but rather can fall anywhere along a continuum between the two. **Microparasites** are microscopic or small-bodied and not visible to the human eye and include viruses, bacteria, fungi, and most protozoa. Microparasites have short generation times relative to their hosts and typically have direct replication or reproduction in one definitive host. For example, when bacteria invade a definitive host, they can rapidly produce many infectious progeny in the same individual host. Hosts invaded by microparasites typically either exhibit acute mortality from infection or, as occurs in most cases, mount an adaptive immune response that clears the pathogen from the host and protects the host from future reinfections; hence, microparasite infections are typically acute with rapid onset and short duration.

Macroparasites, such as helminths (e.g., trematodes, cestodes, acanthocephala), some protozoa, and ectoparasites such as ticks, biting midges, and mosquitoes (Han and Altizer 2013), have more complex life cycles and may require multiple host species to complete development and reach sexual maturity. While the term **infection** is used to describe the invasion of a microparasite into the host, for larger endo- and ectoparasites such as helminths and arthropods, colonization of the host is referred to as an **infestation**. Unlike microparasites, most macroparasites do not replicate and produce infectious progeny that stay in the definitive host. Instead, macroparasite progeny (eggs or larvae) leave the definitive host for further development;

typically, this occurs in another species (intermediate host) but may occur in the environment. The intensity of the macroparasite burden within or on a host, defined as the number of parasites per individual, is often highly variable, with a relatively small subset of hosts harboring large numbers of parasites and the majority of individuals harboring no or few parasites (Wilson et al. 2001; also see Chapter 7). Some macroparasites, such as helminths, elicit host antibody- and cell-mediated immune responses, but, given the large number of antigenic determinants, the generated antibodies are rarely protective. Mounting an immune response to a microparasite—like the protozoa that cause avian malaria—can also be detrimental to the host, in some cases more so than the parasite itself; hence, the host may tolerate chronic, low-level parasitemia (i.e., parasite load in blood) over its lifetime.

2.2.2.2 Routes of transmission

The ability of a pathogen or parasite to infect new susceptible hosts is paramount to its persistence. Transmission of a pathogen from one susceptible host to another can be categorized most broadly as horizontal versus vertical. **Horizontal transmission** is where the pathogen/parasite is passed between two hosts that are not in a parent–offspring relationship. In contrast, transmission between parents and offspring is called **vertical transmission**, which typically occurs when the reproductive organs of a female are infected and the pathogen is transferred to the egg and embryo. In birds, evidence of vertical transmission is scarce and primarily associated with domestic poultry (e.g., Cox et al. 2012; Zurfluh et al. 2014). However, there are some viruses that are known to be passed from an infectious female to the egg, including duck plague virus, aviadenoviruses, circoviruses, and retroviruses (Thomas et al. 2008).

Horizontal transmission between hosts can occur directly or indirectly (Wobeser 2013). **Direct transmission** is when the infection of a susceptible host occurs due to direct contact with an infectious host and exchange of infectious particles (e.g., respiratory droplets or aerosols, mucus, blood, feces, urine, tissue). Contact may arise from a variety of interactions, including predator/prey, scavenging, mating, aggression, or allopreening (see Chapter 4).

Additionally, direct transmission can occur through contact with pathogen-contaminated food, water, or fomites (infectious particles on inanimate objects, such as bird feeders; Dhondt et al. 2007). Fomite transmission is more likely with pathogens that can persist in an infectious state in the environment for long periods, such as Newcastle disease virus (NDV) (Brown and Bevins 2017), *Salmonella* bacteria (Winfield and Groisman 2003), and avian influenza viruses (AIVs) (Stallknecht et al. 1990), to name a few. Fecal–oral transmission is a fairly common route of infection, in which infected birds can shed pathogen in their feces into the environment (water, soil, nesting material) (Hartup and Kollias 1999) and susceptible birds may then acquire the pathogen from the environment through foraging and drinking (e.g., AIVs, NDV, *Pasteurella multocida*).

Indirect transmission is when the susceptible host is infected via an intermediate species, such as an intermediate host (see Section 2.2.1) or an arthropod vector (e.g., mosquito, tick, mite) (Wobeser 2013). Birds are commonly infected with arthropod-transmitted pathogens, such as avian poxvirus, *Borrelia burgdorferi*, and West Nile virus (WNV) among many others that affect birds. Many pathogens are transmitted via multiple routes; for example, avian poxvirus, a primarily vector-transmitted pathogen, can also be transmitted via direct contact with infectious individuals and fomites. Some transmission of WNV in American crows (*Corvus brachyrhynchos*) has been attributed to a fecal–oral route, presumably because they shed the virus in their feces (Dawson et al. 2007; Wheeler et al. 2014), which is also seen in other bird species and other arboviral pathogens (Nemeth et al. 2010; Owen et al. 2011). Overall, knowledge of transmission mode for a host–pathogen system is critical for developing and implementing strategies to contain or prevent disease outbreaks.

2.2.2.3 Virulence and pathogenicity

In addition to mode of transmission, two key characteristics of pathogens that influence their interaction with a susceptible host or vector are infectivity (i.e., its ability to invade and establish itself in a particular host) and the capacity to cause disease. The latter is frequently referred to as virulence and/or pathogenicity. However, the virulence and pathogenicity of a pathogen are not attributes of the pathogen alone but the emergent outcome of the host–pathogen interaction (Méthot and Alizon 2014) and the environment in which that interaction occurs. Moreover, the definitions of virulence and pathogenicity are often taxon- and discipline-specific (Thomas and Elkinton 2004). Nonetheless, virulence is often treated as an inherent pathogen trait by studies that hold host background and environment constant (e.g., Hawley et al. 2013).

Virulence is often defined as the severity of disease exhibited by a susceptible host in response to a particular dose of pathogen or, more broadly, as the degree to which the effects of a given pathogen reduce host fitness. Classically, pathogen virulence is measured under controlled, laboratory conditions using laboratory animal hosts and assumes that a pathogen's infectivity and host response are constant. Virulence is often quantified as the infectious dose (the number of pathogen units) that kills 50% of infected individuals, called the LD_{50} (lethal dose). Alternatively, if the pathogen does not directly kill the host in captivity or otherwise, the severity of disease may be scored (Leggett et al. 2012). Pathogen titers (their magnitude and duration) are also used in some cases to quantify virulence based on the idea that, when relevant, the onset of disease is positively correlated with titer. However, experimental quantifications of virulence, while lending important insight into host–pathogen interactions (see Hawley et al. 2013), are not feasible or even applicable for many wild avian systems.

A **pathogenic** microbe is one that can cause disease in the host organism. **Pathogenicity** describes the ability of a pathogen, of known virulence, to invade and establish itself in a host and to produce disease under natural conditions. In other words, pathogenicity is a product of the pathogen's virulence and infectivity for a particular host (Thomas and Elkinton 2004). The caveat of 'known virulence' limits the use of this term in wildlife studies, where there is likely to be substantial uncertainty in detecting infection or disease in a host population or in identifying the causative agent (see Section 2.4 and Chapter 7). Despite these limitations, these terms are an important way to describe host–parasite interactions, but their meaning in the context they are used must be clearly defined.

2.2.3 Vectors

An arthropod **vector** can infect a host with a pathogen through both mechanical and biological means. Mechanical transmission is when the vector is contaminated with a pathogen and incidentally transmits the pathogen during close association with a susceptible host. Most notable are avian poxviruses which can be transmitted to susceptible birds through contact with virus-contaminated mouthparts of biting arthropods. Biological transmission is when the pathogen is obligated to pass through the vector for development (Barreto et al. 2006). Some examples of bird-associated arthropod vectors and the pathogens they transmit biologically include biting midges (*Culicoides* spp.), vectors of many avian hematozoan parasites (Bennett et al. 1993; Svobodová et al. 2017), mosquitoes (primarily from the *Culicinae* subfamily), the known vectors of a suite of encephalitis viruses (including WNV and EEEV), and *Plasmodium* species (the protozoan agent of avian malaria).

Like hosts, there are numerous attributes of arthropod vectors that influence their ability to biologically transmit pathogens, many of which are vector-dependent. For pathogens that can be transmitted by multiple arthropod species, the relative competence of the vectors for the pathogen is important. Vectors vary in their susceptibility to infection and ability to support pathogen replication and dissemination, which, collectively, represent **vector competence**. Dissemination is when the virus is able to successfully invade the vector's midgut and salivary glands, a requirement for transmission. Additional factors that affect transmission events are the vectors' biting rates, survivorship, and extrinsic incubation periods, all of which are collectively called **vectorial capacity** (Garrett-Jones 1964). Variation in vectorial competence for WNV has received a lot of attention. WNV, which is maintained by avian definitive hosts, has been detected in 62 species of mosquito from 11 genera (Brault 2009), but transmission is primarily attributed to ornithophilic *Culex* species. However, *Culex* species are not the most competent vectors for WNV; instead, several species of *Aedes* mosquitoes (*Ae. albopictus, japonicus,* and *atropalpus*) have the highest competence (Turell et al. 2001; Turell et al.

2005); yet, they are not important in the WNV transmission cycle, given their feeding preference for non-avian vertebrate hosts.

The WNV system highlights that host preference is a critically important attribute of vectors that drives their likelihood of pathogen transmission. Arthropods (e.g., mosquitoes, midges, ticks) vary significantly in host feeding preferences. Some feed opportunistically based on vertebrate host availablity, while other species preferentially feed on certain vertebrate taxa, such as birds, mammals, or reptiles. Preferences also occur at the species level; for instance in the US, *Culex* mosquitoes, the primary vector for WNV, commonly feed on American robins (*Turdus migratorius*) in the northeast and midwest (Apperson et al. 2004; Hamer et al. 2008; Kilpatrick et al. 2006), house finches (*Haemorhous mexicanus*) in the west (Molaei et al. 2006), and northern mockingbirds (*Mimus polyglottos*) and northern cardinals (*Cardinalis cardinalis*) in the southern US (Apperson et al. 2004), disproportionate to the relative abundance of these bird species within the community.

2.2.4 Environment

The **environment** includes any extrinsic factor that promotes the survival of the host and pathogen (and the vector) and facilitates their interaction, thus providing opportunities for transmission. Both biotic and abiotic environmental features can modify disease dynamics, including community composition (Chapter 8), land use (Chapter 9), and climate variables (Chapter 10), to name a few. Effects of the abiotic environment are likely strongest for vectors and pathogens that have an external stage of their life cycle (see Chapter 10), where they are particularly vulnerable to conditions in the external environment. Nonetheless, abiotic factors can also alter the ability of a host to mount an adequate immune response (see Chapter 3) and thus can alter host–pathogen outcomes such as virulence. For example, the same dose and strain of *Mycoplasma gallisepticum* in house finches causes significantly less clinical disease when experimental infections occur at low ambient temperatures (Hawley et al. 2012). Overall, biotic and abiotic fea-

tures of the environment mediate and can dramatically alter all interactions among the players in the epidemiological tetrad; thus the variable contexts in which these interactions occur in nature are critical to consider.

2.3 Host–parasite interaction: infection timeline

Now that we have considered all of the components of the epidemiological tetrad, we consider the host–parasite interaction that occurs once successful infection of a susceptible host occurs (via contact with an infectious vector, host, or pathogen in the environment). We can depict a theoretical host–microparasite interaction (Figure 2.2A) to illustrate the progression of the susceptible host's state following infection with a pathogen, which is ultimately determined by a series of within-host processes (Figure 2.2C and see Section 2.3.4). We now step through the different host states for each timeline, starting with the disease state (Figure 2.2A—Disease States).

2.3.1 Progression of host's disease state

Once a host is infected by a pathogen, they enter what is called the **incubation period**, or the pre-disease state (Figure 2.2A—Disease States). The pathogen is in the early stages of replicating and has not caused significant damage to host tissue or host function; hence, the host's infection status is not yet apparent. This interval between infection and onset of disease may be a matter of hours or years and depends entirely on the attributes and interactions of the host, pathogen, and environment. Likewise, once the host enters the disease or clinical state, the severity of illness will vary according to the host–pathogen interaction and associated extrinsic and intrinsic modifiers that are explored in other chapters in this volume. A host that recovers and survives will move into a non-disease state. Additional information on the baseline responses for morbidity and mortality for a particular avian host–pathogen system can be found in two comprehensive books focused on avian pathogens and parasites (Atkinson et al. 2008; Thomas et al. 2008).

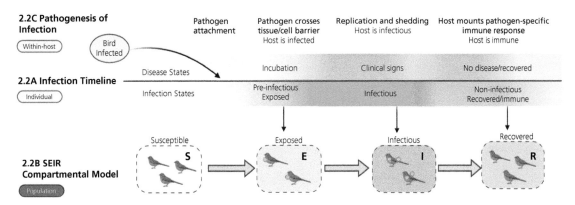

Figure 2.2 (A) Host's progression through different stages relative to disease (top of 2.2A) and infectiousness (bottom of 2.2A) for a hypothetical microparasite infection. Following infection, the discrete stages along the top delineate host's stage relative to when they exhibit clinical signs of infection. On the bottom, the states classify the host relative to when an infected host is infectious, that is, capable of transmitting pathogens to other susceptible hosts. (B) Structure of a hypothetical SEIR model for a population in which every individual is placed into different 'compartments' or assigned a 'state' (Susceptible, Exposed, Infectious, Recovered). Those state variables correspond to stages of infectiousness depicted in 2.2A. (C) Processes (chain/pathogenesis of infection) occurring within the host, such as pathogen binding, replication, and host immune response, following exposure to a pathogen determine where the host is along the timeline. (Created with Biorender.com.)

2.3.2 Progression of host's infectious state

While the above describes a host state with respect to clinical signs of infection (i.e., disease state), a host's state is also defined based on their ability to transmit the infectious agent to another susceptible host (Figure 2.2A—Infection States). After a pathogen successfully invades and infects the host, there may be a **latent** period when the host is infected but is not able to contribute to any new infections in other hosts, regardless of transmission route (Anderson and May 1991). This state is frequently referred to by mathematical modelers as the **exposed** state; however, with respect to defining infection states, we use the term exposure only to describe a host's initial contact with the pathogen. Another term to describe the latent state, when a host is not yet capable of transmitting the pathogen, is **pre-infectious**, proposed by Vynnycky and White (2010). Whether a host enters a pre-infectious state before becoming infectious varies with pathogen, host, and their interaction. Once the pathogen has multiplied in the host to the extent that it can lead to secondary infections, the individual is considered **infectious**. The concept of infectiousness is straightforward when we consider directly transmitted pathogens. A host is infectious when they

discharge any material (feces, saliva, mucus, or droplets) that contains an infectious dose of the pathogen. In vector-transmitted pathogens, infectiousness relates to the threshold of circulating pathogen needed to infect a blood-sucking arthropod (Lord et al. 2006), which varies with the vector's competence (see Section 2.2.3). If the host survives, they may transition from the infectious state to the **recovered** or immune (non-infectious) state, where they may remain for life or for some period of time before reverting to a susceptible state due to waning immunity (see Section 2.5).

2.3.3 Linking states of infection to compartmental models

These states of infectiousness described above can be used to determine the structure of compartmental 'SEIR' or 'SIR' epidemiological models (Figure 2.2B) commonly used to model transmission dynamics of microparasites (Anderson and May 1991; Kermack and McKendrick 1927) (see also Chapter 7). For example, in the hypothetical host–pathogen scenario depicted in Figure 2.2A, individuals would be assigned to a **compartment** (Figure 2.2B) based on where they are in relation to the infection timeline. An individual within the population would be assigned to a susceptible state ('S'), a latent/pre-infectious state (exposed; or 'E'), an infectious state ('I'), or a recovered/immune state ('R') (Figure 2.2B). How a population is compartmentalized varies with the different host–microparasite systems and their infection, disease, and immune outcomes.

A key piece of epidemiological information needed for developing and implementing effective strategies for managing disease outbreaks is how the host's disease and infectious states align. For instance, a host can be infectious, as depicted in Figure 2.2A, before they exhibit signs of disease, during which time they may be able to spread the pathogen through normal behaviors (e.g., movement, intraspecific interactions). This misalignment is fairly common; for instance during WNV infection, birds typically do not exhibit signs of infection until 3–4 days post-infection but are infectious to a biting mosquito within 48 hours of being infected. On the contrary, in house finches (*Haemorhous mexicanus*) infected with *M. gallisepticum*, birds are most infectious when they exhibit clinical signs of disease, although they can be mildly infectious while aclinical (Dhondt et al. 2008).

2.3.4 Within-host dynamics: pathogenesis

Organisms frequently come into contact with infectious agents, but few result in the host even becoming successfully infected. The host's trajectory along the infection timeline (Figure 2.2A) that we just laid out is ultimately determined by a series of within-host processes (Figure 2.2C), which include some or all of the following: (1) host contact and/or exposure to a pathogen/parasite; (2) colonization of the host target tissues by pathogen; (3) invasion of host tissue; (4) within-host replication and host immune response; and (5) shedding/release of pathogen and transmission to the environment or a naïve host (Baron 1996; Nash et al. 2000).

2.3.4.1 Exposure and colonization

A host's exposure/contact with a pathogen is mediated by a variety of factors, including host behavior and social structure (see Chapter 4), population size and density (see Chapter 7), and aspects of the abiotic environment (see Chapter 9). Following contact with a host, the pathogen must colonize the pathogen-specific site for entry, such as the mucous membranes of the respiratory tract, gastrointestinal tract, or conjunctiva; the skin; or the oviduct (Nash et al. 2000). Effective contact occurs when the pathogen has colonized the point of entry site.

2.3.4.2 Attachment and invasion

Following colonization, a pathogen must attach and/or cross host barriers to lead to an infection. The pathogen accomplishes this in three ways: (1) it *adheres* to the host tissues through its ability to recognize and bind to host receptors (seen predominantly with viral and bacterial pathogens); (2) it is injected into the host via the *deliberate penetration* of the host's skin by an arthropod vector; or (3) it *opportunistically invades* the host through breaks/lesions in the host's natural barriers (seen predominantly in fungal pathogens; e.g., *Aspergillus*).

Adherence to host target tissues requires the presence and availability of host receptors that the pathogen's adhesion proteins recognize. Whether a pathogen is a generalist or exhibits narrow tissue tropism (i.e., a pathogen's ability to infect a specific cell or location in the host) determines its **infectivity**, or its ability to invade and establish itself in the host (Thomas and Elkinton 2004). Pathogens with broad tissue tropism can adhere to more than one host receptor, which increases their odds of infecting the host. Other pathogens have narrow tissue tropism and may only recognize and bind to one host receptor. An example of this level of specificity is the tissue tropism exhibited by influenza type A viruses (IAVs; Box 2.1).

The second mode of entering the host is the deliberate penetration by an arthropod vector when it bites, probes, and/or takes a blood meal from the host (Styer et al. 2007). In addition to injecting the pathogen into the host, the arthropod secretes saliva, which can contain factors that promote virulence (Fontaine et al. 2011; Schneider et al. 2006). In fact, birds experimentally infected via an infectious mosquito develop higher viremias than birds inoculated with a needle (Styer et al. 2006).

Once the pathogen has invaded the host, the host can prevent its establishment through constitutive physical, microbiological, and physiological bar-

Box 2.1 Tissue tropism of IAVs

IAVs are the predominant type of influenza virus and, except for two distinct lineages that originate in bats, all IAVs have initially been isolated from birds (Mostafa et al. 2018). IAVs in birds originate almost entirely from aquatic birds, but they have evolved to adapt to various avian and non-avian hosts, including humans (Webster et al. 1992). IAVs are classified according to subtype, which is based on their surface proteins hemagglutinin (HA) and neuraminidase (NA). The virus's adhesion protein is the HA, which is responsible for recognition and binding to the host's receptors. Of the IAVs originally isolated from birds there are 16 recognized subtypes of HA (Spackman 2008), and each differs in their host recognition and binding specificity (Suzuki et al. 2000). IAVs adapted to humans preferentially bind to α2,6 linked sialic acid surface receptors, while avian-adapted IAVs (commonly written as avian influenza virus or AIV) bind to α2,3 linked sialic receptors (Suzuki 2005). This specificity is what limits highly pathogenic AIVs (HPAIV) from successfully spilling over and adapting to a human host, permitting human to human transmission. However, swine have both α2,3 and α2,6 linked sialic acid surface receptors and can serve as a mixing vessel where several IAVs can undergo reassortment and produce a highly pathogenic subtype that has binding affinity to the human α2,6 linked sialic acid surface receptors. The 1918 'Spanish' influenza virus (H1N1), which caused the 'mother of all pandemics' (Taubenberger and Morens 2006), killing over 20 million people, was determined by phylogenetic analyses to have diverged from a low-pathogenic AIV (LPAIV) in 1905 and then used swine as a mixing vessel ('intermediary host') prior to adapting to humans (Reid et al. 1999).

riers. The mucosal lining of the host respiratory and intestinal tract can trap foreign materials and expel them from the body via ciliary activity or phagocytic epithelial cells (Nash et al. 2000). Like other vertebrates, birds have a natural microbiota of commensal bacteria that can prevent establishment of pathogens (Kohl 2012). Another important barrier are antimicrobial peptides (AMPs) (e.g., defensins) that work by either destroying the microbe through membrane lysis or penetrating the microbe and disrupting metabolic function (Ganz 2003). One unique feature of AMPs in birds is their role in the immune protection of the developing embryo. The egg's albumin contains a high number of AMPs and the eggshell itself contains even more—as many as 520 identified AMPs have been identified in the eggshell (Wellman-Labadie et al. 2007).

If a host is successful at eliminating the pathogen at these early stages of the invasion, it is not considered to be currently or to have been previously infected. Furthermore, there would be no 'record' of that exposure; that is, a blood sample collected from the host would not contain pathogen or pathogen-specific antibodies.

2.3.4.3 Pathogen replication and activation of host immune response

The ability of a parasite or pathogen to replicate and cause damage within hosts will be determined largely by aspects of the host immune response, as well as the pathogen's ability to evade the host immune system, which is covered in detail in Chapter 3. Like other higher vertebrates, the avian immune system comprises two distinct, but highly integrated, arms of defense: an innate (non-specific) arm and an adaptive (specific) arm. The innate immune response is the first internal line of defense for eliminating or neutralizing invading pathogens. If the innate response is not successful in clearing an infection or does not recognize the pathogen, the adaptive immune response is induced, which has greater capacity to recognize and target foreign antigens (defined as any invading foreign substance that elicits an immune response, including pathogens). Antigens come in all sizes and structural complexities (consider a virus versus a helminth worm), and they are rarely presented to lymphocytes in their entirety. Each antigen has numerous antigenic determinants (epitopes) and each by themselves can be recognized by the antigen-binding receptors on the lymphocytes. Adaptive immune responses can take up to a week to activate, so innate responses are important in the days immediately following infection. Moreover, the innate immune system plays a key role in activating the adaptive immune response (Janeway et al. 2001), which, if successful, will lead to elimination of the pathogen and the host's recovery. If unsuccessful, the host may die or the pathogen may persist in an active or latent (non-circulating, sequestered in tissues) state.

The primary adaptive immune response to a novel antigen will typically generate antigen-specific antibodies within 4–7 days post infection, and then it may take an additional week or two for those antibody levels (titers) to peak. The production of the antigen-specific memory B and T cells (see Chapter 3) during the primary response are what constitute the host's long-term protection from becoming susceptible to reinfection with the same pathogen. The memory cells persist in the host and, upon subsequent infection with the same antigen, can mount a rapid and enhanced secondary immune response, preventing establishment of the pathogen within the host (Figure 2.3). See Section 2.4 for further discussion of pathogen-specific immunity.

2.3.4.4 Resistance and tolerance

The ability of a host's immune system to control or prevent pathogen replication is broadly defined as resistance, with resistant hosts able to clear or keep pathogen replication low within relevant tissues. In contrast, some hosts may respond to pathogen infection by exhibiting tolerance, a strategy that reduces the per-pathogen fitness impact of infection, often by limiting tissue damage (Medzhitov et al. 2012). Thus, tolerant hosts can harbor high parasite or pathogen loads but show limited associated tissue damage. For example, in red-winged blackbirds (*Agelaius phoeniceus*) infected with avian malarial parasites, higher levels of baseline corticosterone were associated with lower levels of anemia, potentially indicating tolerance of malaria infection via reductions in direct damage to red blood cells (Schoenle et al. 2018). While tolerance and resistance are often studied separately, they are not necessarily mutually exclusive (Bonneaud et al. 2019; Restif and Koella 2004) and may involve different balances among immune components and other physiological traits (Henschen and Adelman 2019; Medzhitov et al. 2012; Råberg et al. 2009).

2.3.4.5 Pathogen shedding/exit

In birds, the most common exit points for a pathogen include through the digestive tract, respiratory tract, skin, and blood. Where a pathogen or parasite exits the host frequently matches where within the host the pathogen/parasite colonizes and replicates. For instance, pathogens that replicate in the intestinal tract, such as LPAIV and enteric pathogens, will be shed in the feces. Pathogens associated with the respiratory tract are disseminated via fluid and

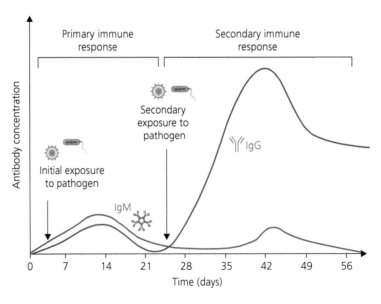

Figure 2.3 Primary and secondary immune responses. Initial exposure to a pathogen/antigen elicits a primary immune response characterized by increased production of membrane-bound immunoglobulin (IgM) (see Chapter 3) and a small amount of serum immunoglobulin G (IgG). When the individual is infected with the same or a similar pathogen, there is a secondary response accompanied by rapid production of IgG that is specific to the pathogen and able to neutralize the pathogen before it can replicate and/or harm the host. (Created with Biorender.com.)

mucosal secretions from the oral, nasal, and ocular cavities, such as *M. gallisepticum* and *P. multocida*. Other pathogens circulate in the host's blood and exit when the blood/skin barrier is penetrated either by a blood-sucking arthropod or injury. Some pathogens can exit the host via multiple routes. Where the pathogen exits the host is one determinant of the mode of transmission (see Section 2.2.2.2).

2.4 Developing pathogen-specific immunity

A host can actively or passively acquire pathogen-specific antibodies through natural or artificial routes. For active acquisition, a bird may become immune through natural exposure to a pathogen and mount a pathogen-specific immune response, as described above. Active immunity, analogous to that generated through natural exposure, can be produced via artificial routes such as vaccination with a pathogen or antigen (typically a live attenuated or inactivated strain of the pathogen) or pathogen-specific antibodies. These two artificial routes are routinely used in domestic poultry to protect large flocks from disease outbreaks. Our brief coverage here greatly understates the importance of all the recent advances that have been made in vaccine development and administration for domestic poultry (e.g., Kim and Samal 2019), which is critical for global food security. Use of artificial routes to generate immunity for a circulating pathogen is far less common in free-living birds (but see Ha et al. 2013), although in an experimental setting, vaccination is often used to simulate infection without using a live pathogen, allowing study of variation in immune responses under varying contexts (see Chapters 3 and 4). For passive acquisition of immunity, an immunologically naïve juvenile may have acquired antibodies through the passive transfer of maternally derived antibodies during yolk formation. This route of acquisition does not require any direct exposure of the juvenile bird to pathogen or antigen.

2.4.1 Maternal transfer of antibodies

The maternal transfer of antibodies is critical for protecting an immunologically naïve chick during the first few days or weeks of life (reviewed in Hasselquist et al. 2012). During egg development,

the female deposits a particular type of antibodies (termed immunoglobulin Y or IgY; the functional homolog to mammalian IgG) into the yolk sac, with smaller quantities of IgM and IgA deposited in the egg's albumen. The titer of maternal antibodies transferred and duration antibodies persist in newly hatched birds vary with the host–pathogen system, the female's historical pathogen exposure and associated immune response, and the female's condition at the time of egg-laying (Hasselquist et al. 2012).

On average, the chick's antibody levels positively correspond to the concentration of antibodies deposited in the yolk (Apanius 1998; Gasparini et al. 2002). Maternal antibody titers decline fairly quickly post-hatching, with loss of passive protection occurring within 14 days post-hatching and more often within 5 days of hatching (Hasselquist et al. 2012), although this may vary among bird species. To transfer antibodies during egg development, the female must increase synthesis of IgG while maintaining her own immunocompetence, which can have its own physiological costs (Hasselquist et al. 2012). Hence, the female's nutritional and energetic condition during egg development can affect the level of protection transferred to the chick (Karell et al. 2008).

Maternally derived antibodies provide protection against pathogens that the female may have been exposed to; hence chicks are likely to be protected from pathogens that are actively circulating in their natal population, increasing their chances of survival when their own immune system is still developing (Grindstaff et al. 2003). Studies have shown that females can also transfer antibodies generated from exposure to pathogens in previous seasons—providing evidence of persistence of pathogen-specific antibodies (Reid et al. 2006). Interestingly, there is evidence that the titer of antibodies deposited in the egg increases with egg-laying order, with the last chick to hatch having the highest titers and presumably higher protection from circulating pathogens (Hargitai et al. 2006), potentially mitigating any phenotypic consequences of asynchronous hatching.

The protection conferred to the chick by maternal antibodies is not well known due to the inherent difficulty of distinguishing between passively acquired maternal antibodies and the chick's endogenous production of pathogen-specific antibodies. Further, detectability of antibodies in itself may not be a robust metric of protection (see Section 2.5). For example, Nemeth and Bowen (2007) found that, while the apparent maternally acquired WNV-neutralizing antibodies in chickens (*Gallus gallus domesticus*) were undetectable by 3 weeks post-hatching, the chicks were resistant to experimental challenges with WNV up to 42 days post-hatching. Transgenerational effects on offspring's ability to mount effective immune responses to pathogens (e.g., van Dijk et al. 2014) is an open and needed area of research in avian disease ecology.

2.5 Detecting and quantifying a host's infection status

An understanding of these within-host processes and the host immune response is key for any disease investigation. For instance, identifying the individual's current state relative to the infection timeline is important to assess prevalence in the population (see Chapter 7). With respect to a particular pathogen or parasite, a bird may be actively infected, previously infected with protective immune memory (which is often associated with pathogen-specific antibodies), or neither, in which case the individual may have lost immunity over time or was never infected. Overall, a bird's infection status—also called its infection 'state'—can be assessed in a variety of ways, though all have their limitations.

Here we review the most common suite of methods used to assess a bird's infection status (Table 2.1). We do not intend this list of techniques or parasites to be exhaustive but rather to introduce some of the most utilized techniques for commonly studied avian pathogens and parasites. Many naturally occurring parasites are routinely quantified in avian hosts and we briefly discuss some examples in this section. Furthermore, we note that making conclusions about host fitness from such assessments alone is problematic (see Chapter 6). For example, tolerant hosts may have higher parasite loads but higher fitness than resistant hosts (Graham et al. 2011). This issue and ways to address it are also discussed in Chapter 3 (Section 3.5).

Despite the myriad assays/techniques for assessing an individual's infection status, there are frequently detection errors, which have consequences for determining prevalence of infected birds within a population or assessing historical exposure and population-level immunity (see Section 2.6 for more about herd immunity). **Prevalence** is defined as the proportion of the (total) population that is actively infected with the pathogen of interest or exhibiting disease. Detection issues can arise from a variety of factors related to sampling and assay characteristics (see Chapter 7 for more on prevalence and imperfect detection) or, more broadly, can occur if particular causative agents of disease are not part of active surveillance efforts.

2.5.1 Visual detection

The presence or absence of large ectoparasites (e.g., arthropods such as ticks, mites, and lice) can be determined in many cases by a visual inspection of avian hosts (Clayton and Walther 1997; Koop and Clayton 2013). Although these methods seem simple, and can be in some cases, correctly identifying and quantifying ectoparasites can be difficult for several reasons. In many cases, even a careful inspection can overlook many ectoparasites, which results in an underestimate of their abundance (see Chapter 7). For instance, birds are commonly infested with ticks but these ectoparasites may be difficult to detect when they are in larval or nymphal life stages or are located in difficult to examine regions (e.g., inside ears or eyelids; Hamer et al. 2012). Therefore, many researchers remove ectoparasites from hosts before identification and quantification; pesticides, aspirators, tweezers, and brushes can all be used to collect ectoparasites from hosts (Koop and Clayton 2013; Mori et al. 2019). After removal, these parasites can be more accurately quantified and identified using both microscopy and molecular detection (Clayton and Walther 1997; Miller et al. 2016). This is important as many species of ectoparasites look similar or identical and can prove impossible to differentiate without extensive expertise in arthropod identification (e.g., Miller et al. 2016). Identification of ticks, for example, may require a dissecting microscope to visualize mouth parts (Clayton and Walther 1997).

Table 2.1 Common methods for detecting avian parasites.

Methods	Benefits	Cautions	Reviews or examples
Visual detection			
Gross inspection	Simple and cost effective, can be done in field without invasive sampling	All ectoparasites are not easily visualized, may miss some individuals, many ectoparasites look similar, many parasites cause similar pathologies	Clayton and Walther 1997; Hamer et al. 2012; Mori et al. 2019; Wilcoxen et al. 2015
Microscopy			
Dissecting microscopes	Allows for greater visualization of morphological details of macroparasites	Many parasites may be morphologically similar, high level of expertise may be necessary to distinguish some species	Clayton and Walther 1997; Koop and Clayton 2013; Mori et al. 2019
Compound light microscopes	Samples can be easily collected in the field, cost effective	Many parasites may be morphologically similar, high level of expertise may be necessary to distinguish some species, low parasite loads may not be detected	Campbell 1995; DeGroote and Rodewald 2008; Dolnik et al. 2009; Godfrey et al. 1987; Krone et al. 2008; Svensson-Coelho et al. 2016; Valkiunas 2004
Immunological assays			
RIAs (radioimmunoassays) and ELISAs (enzyme-linked immunosorbent assays)	Can give infection history even if parasite has been cleared from the system, can be used with a small sample, efficient for a large number of individuals	May not give an accurate estimate of parasite load, assays may be very specific to hosts/parasites	Afanador-Villamizar et al. 2017; Grodio et al. 2009; Nakamura and Meireles 2015; Shriner et al. 2016
Molecular detection			
PCR (polymerase chain reaction)	Can detect a very small number of parasites, can be used to distinguish between different closely related species and lineages, more efficient than microscopy	Will detect DNA of both living and non-living parasites, more expensive than microscopy	Bell et al. 2015; Ciloglu et al. 2019; Dolnik et al. 2009; Miller et al. 2016; Nakamura and Meireles 2015; Nakamura et al. 2009; Pacheco et al. 2018; Ribeiro et al. 2005
Quantitative PCR	Can give an accurate estimate of parasite load	Will detect DNA of both living and non-living parasites, more expensive than microscopy	Knowles et al. 2010
-omics (e.g., 16S rRNA sequencing)	Can detect a large number of parasites at once	Data analysis can be challenging	Cerutti et al. 2018; Hamer et al. 2012; Kreisinger et al. 2018
Experimental manipulations			
Pesticides and medications, experimental infections	Allow for experimental manipulations of natural parasites	Some methods of removal may have toxic side-effects, parasites may not be entirely eradicated, immune memory may complicate interpretations of reinfection studies	Adelman et al. 2013; Hund et al. 2015; Knowles et al. 2010; Martínez-de la Puente et al. 2010; Merino et al. 2000

Visual detection can also be useful when parasite infections result in easily distinguishable pathology. Clinic signs (e.g., pathology) can be used to determine the presence or absence of certain types of parasitic infections and may even be detectable without capturing individuals. For example, conjunctivitis, pox lesions, unexplained feather loss, and cloacal swelling all indicate parasitic infections (Wilcoxen et al. 2015). However, caution must be taken when using these methods because many parasites cause similar clinical signs.

2.5.2 Microscopy

Microscopy can be used to detect and quantify parasites that are too small for visual detection. Using these methods, tissue samples are typically fixed to glass slides and then stained. These samples are then scanned using a light microscope for the presence of parasites (e.g., Campbell 1995; DeGroote and Rodewald 2008; Valkiunas 2004). Several common types of avian parasites can be detected using microscopy, including blood parasites (e.g., *Plasmodium* and *Haemoproteus*; Godfrey et al. 1987; Krone et al. 2008) and coccidial parasites (e.g., *Isospora* spp.; Brawner et al. 2000; Dolnik et al. 2009). Detecting these parasites using microscopy is useful in wild birds as the necessary samples (e.g., blood or excrement) are relatively non-invasive and easy to store. These methods can also be cost effective. However, examining slides for parasites is labor intensive and requires a high level of expertise to correctly identify parasites (Shirley et al. 2005). In addition, intricacies of parasite life cycles may result in infections being missed or parasite load being underestimated (Svensson-Coelho et al. 2016; Valkiunas 2004). These constraints have encouraged many researchers to rely on other methods, such as molecular detection, to both identify and quantify avian parasites (Hellgren et al. 2004; but see Jarvi et al. 2002).

2.5.3 Molecular detection

Molecular methods are generally used to detect parasite DNA or RNA (for certain viruses) within host samples. With these methods, even a small number of parasites can be detected. This is especially beneficial for parasites that maintain low-level infections within avian hosts. For example, using polymerase chain reaction (PCR) to detect blood parasite DNA within avian blood samples detects a higher prevalence of infection than blood smears examined with light microscopes (Ribeiro et al. 2005). However, care should be taken when interpreting these results as the DNA of both living and dead parasites will be detected using these

methods. Molecular methods can also be useful for correctly identifying which parasites individuals are infected with, especially when closely related parasites have a similar morphology. For example, in a study of tropical birds in Panama, researchers used a DNA barcoded reference library to determine the species identity of juvenile hard ticks (Ixodidae), as different species of these ticks look very similar to each other at this stage of development (Miller et al. 2016). The genera and lineages of blood parasites can also be identified by amplifying and sequencing portions of parasite DNA (Bell et al. 2015; Ciloglu et al. 2019; Pacheco et al. 2018), and similar methods have been developed to identify coccidian parasites (Dolnik et al. 2009; Nakamura et al. 2009; Saikia et al. 2017).

In addition to species identification, methods such as quantitative PCR can determine the amount of parasite DNA that is present in a sample and thus provide an estimate of parasite load (e.g., Knowles et al. 2010). Although these molecular methods can be more expensive than traditional methods of detection and quantification (e.g., light microscopy), they are much less laborious and can be more easily applied to many individuals at once. Given the frequent variation in parasite load and its implications for fitness (see Chapter 6), the ability to estimate variation in the number of parasites in a given individual is particularly important.

While non-viral pathogens are typically detected via DNA, viruses in birds commonly have an RNA genome and require different assays/reagents for detection. Detecting and estimating titer of viral RNA in tissue and blood samples is frequently done using real-time quantitative reverse transcription PCR (qRT-PCR) (Spackman 2020; Spackman and Lee 2020). Since real-time reverse transcription PCR (rRT-PCR) is only quantifying viral antigen and not necessarily live virus, additional assays, varying by pathogen type, are needed to validate/standardize RT-PCR output. For instance, the gold standard for quantifying WNV is through Vero cell plaque assays (Brien et al. 2013), which quantify live virus. Assays for IAV can also be done via plaque assay (Tobita et al. 1975) but are more commonly done using embryonated chicken eggs (for virus isolation) and hemagglutination assays for titers (see Swayne 1998).

Recently, molecular methods have been developed to detect the presence of more than one parasite at a time, even allowing for the characterization of a parasite's own suite of parasites. For example, 16S rRNA sequencing can be used to examine the entire microbiome of host samples (Killian 2020; Kreisinger et al. 2018). Furthermore, rRNA sequencing can be used to describe the microbiome of the ectoparasites found on hosts (Cerutti et al. 2018). This is particularly important for surveying ectoparasites such as ticks for potentially parasitic microbes, such as *Borrelia burgdorferi*, the causative agent of Lyme disease (Hamer et al. 2012). These techniques are also helpful in the detection of novel parasites and pathogens that more specific assays will miss.

2.5.4 Experimental manipulations of infection status

Manipulating parasite exposure and load can help researchers better understand why individuals differ in their response to pathogens and allows direct elucidation of the impact of parasites on hosts. However, if researchers want to use natural parasites of wild hosts for experimental challenges, they must ensure they are working only with naïve individuals, which can be challenging to find when disease prevalence is high. One way to achieve this is to find subsets of populations, such as juveniles, that are less likely to have been previously exposed to certain parasites or to identify populations that are naïve to particular parasites (e.g., Adelman et al. 2013; Staley et al. 2018). Alternatively, researchers can clear the infections of a subset of the population to compare naturally infected with experimentally uninfected birds. This technique has been used extensively in studies of avian malaria (e.g., Knowles et al. 2010; Martínez-de la Puente et al. 2010; Merino et al. 2000). In addition, after clearing infections, researchers can then experimentally reinfect hosts to determine the effects of a known parasite load. This is commonly done with nest ectoparasites. In these studies, nest ectoparasites are first killed (typically using pesticides or a heat gun) and then a known type and quantity of ectoparasites are introduced to nests to test their effects (Hund et al. 2015). Such studies, however, face complicated interpretation when the parasites of interest induce immune memory. In such cases, reinfection will reflect not just the impact of a known parasite load but also the differences in primary vs. secondary immune responses to the same parasite. Although they have their own potential drawbacks, these experimental methods can provide less messy studies for understanding the consequences of parasite infection in natural populations. Along with molecular methods that can detect and quantify a larger suite of parasites, these types of experimental studies may provide an important way forward for unraveling the complex relationships among immune responses, parasites, and host fitness outcomes in light of infection.

2.5.5 Immunological assays

Immunological assays are commonly used in wild birds and rely on antibodies to detect parasites (e.g., *Cryptosporidium*; Nakamura and Meireles 2015) or host antibodies specific to pathogens that illicit an immune response. In the latter, a host with pathogen-specific antibodies is **seropositive** for that pathogen, which suggests the individual was previously exposed to the pathogen (but see passive immunity in Section 2.4) and thus some degree of immune protection. Determining whether a bird is seropositive or not is the objective of serosurveillance efforts, where scientists are trying to determine seroprevalence in a population (i.e., proportion of population with pathogen-specific antibodies). Assays to assess enzyme-linked immunosorbent assays (ELISAs) are commonplace for detecting antibodies against specific parasites (particularly microparasites), often with commercially available kits—particularly for pathogens of high importance to health of domestic poultry, such as AIV and NDV (e.g., Cross et al. 2013). However, ELISAs are mainly used for determining whether a bird is seropositive or seronegative. A technique for detecting and quantifying antibody levels is through plaque-reduction neutralization tests (PRNTs), which are used to quantify serum levels of WNV-neutralizing antibodies (Calisher et al. 1989; Lindsey et al. 1976). Hemagglutination inhibition assays are another method for detecting and quantifying antibodies and have been found in some cases to be a more effective assay than ELISAs (e.g., Spackman et al. 2009).

Antibody assays are particularly useful when the infection history of an individual or population is of interest and when it is difficult to detect transient viremia or parasitemia in free-living birds. As with most approaches, however, understanding precisely how antibody titers relate to functional protection is not as simple as 'more is better.' For example, in house finches infected with *M. gallisepticum*, parasite-specific IgA and IgY titers tend to be highest among birds with the highest pathogen loads (Grodio et al. 2012). Further, for most pathogens, little is known about persistence of antibodies and a bird that is seronegative may have been infected and developed antibodies that waned over time (see Section 2.5).

2.5.5.1 Antibody persistence

How long antibodies persist following a host's initial exposure and their degree of protectiveness is critical information for developing accurate epidemiological models and the interpretation of serological data (see Chapter 7). In birds, we know little about the persistence of neutralizing antibodies and what we do know is from captive experiments or longitudinal studies that repeatedly sample the same bird in subsequent years. Repeated sampling, however, has limitations as it is hard to know if the presence of antibodies is due to persistence or from reexposure to the same pathogen, and detection issues may confound interpretation (see Chapter 7). Captive studies are rare, given the cost and feasibility of maintaining wild birds for enough time to document potential waning of antibody titers. In one example of such studies, naturally infected rock pigeons (*Columba livia*) were housed in captivity for over 15 months and maintained titers of WNV-neutralizing antibodies during this entire period (Gibbs et al. 2005). Nonetheless, looking at the rate of decline in antibody titers may not be predictive of the longevity of immune protection. For instance, Leighton and Heckert (2007) observed that Newcastle disease virus-infected double-crested cormorants (*Phalacrocorax auritus*) have peak antibody titers at three weeks post-infection, but titers initially wane quickly, the rate of which suggests that they would be undetectable by four months

post-infection (Leighton and Heckert 2007). Contrary to this prediction, a serosurveillance study of cormorants in the Great Lakes region of North America found that, in three years of sampling, 82–95% of adult cormorants were seropositive for NDV antibodies (Cross et al. 2013). Further, these high seroprevalences occurred in years when there was not an active NDV outbreak, suggesting the antibodies may persist from one breeding season to the next. Thus, quantifying the rate at which antibody levels wane is challenging.

Quantification of antibody persistence is only one part of the story and may not reflect waning immunity more broadly. Because antibodies may or may not be protective, and because the memory cells that produce protective antibodies can remain undetectable, it is critical to also measure protection by conducting challenge experimental infections after initial exposure. For example, in house finches, some degree of protection against *M. gallisepticum* infection remains for up to 14 months following initial exposure, but individual protection was negatively correlated with antibody presence, such that individuals that had detectable antibodies at the time of infection showed the highest rather than the lowest disease severity (Sydenstricker et al. 2005). Similarly, studies show that even when antibodies are not detected, there may still be residual protection. For instance, maternal antibodies transferred from WNV-infected chickens (*Gallus gallus domesticus*) persisted in chicks for up to 4 weeks post-hatching, but the chicks remained protected up to 42 days post-hatching (Nemeth and Bowen 2007). Overall, more studies are sorely needed to address this critical knowledge gap.

Finally, in some cases, a bird's infection state is not always linear. For example, some pathogens may leave circulation and reside in the host tissues in a state of latency, leading to later recrudescence when a host is stressed or immunocompromised (Gylfe et al. 2000; Owen et al. 2011).

2.6 Population-level disease dynamics

The relative proportions of individual infection states in a population at any point in time are critical for predicting the population-level dynamics of pathogens. The population-level occurrence of disease is defined by the magnitude, frequency, and extent of its occurrence in a particular geographic area (Barreto et al. 2006). A disease **epidemic** or outbreak is defined by periods when the number of infected individuals is greater than expected for a particular region for a specific period. The term **epizootic** is the (wild) animal equivalent of the human-based term, **epidemic**. While rarely used, the avian equivalent of epidemic/epizootic is **epornitic**. An example of an avian epizootic is the 2005 emergence of a protozoan parasite, *Trichomonas gallinae*, in wild passerine bird populations in Britain, which led to a large-scale mortality of greenfinches (*Chloris chloris*) and chaffinches (Robinson et al. 2010). Prior to 2005, the occurrence of disease in columbids (pigeons and doves) was at a relatively low and constant level, or **endemic**.

Predicting and tracking the intensity of an epidemic/epizootic is an integral part of all infectious disease investigations and is covered in more detail in Chapter 7; however, here we introduce a few key terms used to describe population-level disease dynamics, including basic and effective reproduction number (R) and herd immunity given their relevance to the infection timeline and SEIR modeling shown in Figure 2.2.

The **basic reproductive number**, R_0, is a key parameter and is the average number of individual hosts infected by one infectious individual in a completely susceptible host population (Anderson and May 1991). An R_0 greater than one indicates that the number of infected individuals will increase in the population exponentially (Figure 2.4), while R_0 of less than one will ultimately lead to the elimination of the pathogen. The value for R_0 is derived from a transmission parameter, beta (β; see Chapter 7), and duration that an individual is in the Infectious state (I; Figure 2.2B). The β term is the product of contact rate and probability of transmission upon contact; hence, it is the rate at which an individual leaves the Susceptible state (S; Figure 2.2B) and enters the Exposed (E) or Infectious (I) state. In short, the duration that an individual is in the Infectious state (Figure 2.2B) is one of the most important determinants of the severity and duration of a disease outbreak. Reducing R_0 can be achieved through interventions that break links

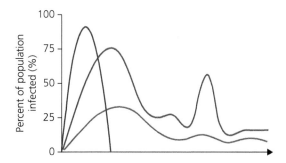

Figure 2.4 Epidemic curves can visually illustrate how number of actively infected or diseased individuals varies over time. Epidemic dynamics are modified by properties of the pathogen (i.e., R_0) that affect intensity of outbreak and host demography (i.e., birth and death rates) which influence the number of susceptible individuals (see Chapter 7). (Created with Biorender.com.)

between elements of the epidemiological tetrad discussed at the beginning of this chapter (see Section 2.2 and Figure 2.1).

The R_0 for a particular pathogen is not a fixed number; rather it varies as the proportion of susceptible to immune individuals changes within a population. For instance, through either vaccination or naturally derived immunity, the number of susceptible individuals an infectious individual encounters will decrease, ultimately reducing the number of secondary infections from one infectious individual, which is called the **effective reproduction number (R_e)**. As the number of immune individuals in a population increases, the R_e will eventually drop below 1, ending an outbreak. This population-level immunity at the point at which R_e has dropped below 1 is called **herd immunity**. Hence, the goal of vaccination efforts is to have enough of the population immune to drop the R_e below 1, with the threshold of population immunity needed varying with R_0 and population size. Whether a host is immune or not to a pathogen (see Sections 2.3.5 and 2.7.3) and how quickly that immunity wanes have important implications (both ecological and epidemiological) for host–parasite dynamics because they directly alter the proportion of susceptibles in the population (Boulinier and Staszewski 2008). Thus, understanding within-host dynamics (see Section 2.3.4) for a given pathogen is necessary for scaling up to population-level dynamics (Figure 2.2).

2.7 Summary

In this chapter, we aimed to provide a brief over-view of the within- and between-host processes that underlie avian infectious disease ecology and the way that these within- and between-processes par-allel one another. Ultimately, the host–(vector–) parasite interactions described here scale up to impact dynamics well beyond the host population—they underpin community-level interactions, eco-system processes, and the frequency of spillover events into other populations, such as domestic ani-mals and humans. In turn, top-down effects from higher-level events (e.g., land use changes that modify ecosystems and communities) can feed back to modify host–parasite interactions, highlighting the importance of multiscale interactions and approaches discussed throughout this book to the broader field of infectious disease ecology (see Chapter 13). Consequently, you will see many of the terms defined in this chapter used throughout this volume, but, depending on the context or the scale of the question, they may have slightly different interpretations, as defined in the respective chap-ters. Such variation in key terminology is, in large part, a consequence of the transdisciplinary and multiscaled approaches inherent in studying host–pathogen–vector–environment interactions. Ultimately, integrating these diverse approaches and perspectives allows a deeper understanding of the dynamic and complex role of infectious diseases in wild avian populations.

Literature cited

Adelman, J.S., Kirkpatrick, L., Grodio, J.L., and Hawley, D.M. (2013). House finch populations differ in early inflammatory signaling and pathogen tolerance at the peak of *Mycoplasma gallisepticum* infection. *The American Naturalist*, 181, 674–89.

Afanador-Villamizar, A. Gomez-Romero, C., Diaz, A., and Ruiz-Saenz, J. (2017). Avian influenza in Latin America: a systematic review of serological and molecular studies from 2000–2015. *PLoS ONE*, 12, e0179573.

Anderson, R.M. and May, R.M. (1991). *Infectious Disease of Humans*. Oxford University Press, Oxford.

Apanius, V. (1998). Ontogeny of immune function. In J.M. Starck and R.E. Ricklefs, eds. *Avian Growth and Development*, pp. 203–22. Oxford University Press, Oxford.

Apperson, C.S., Hassan, H.K., Harrison, B.A., et al. (2004). Host feeding patterns of established and potential mos-quito vectors of West Nile virus in the eastern United States. *Vector-Borne Zoonotic Diseases*, 4, 71–82.

Atkinson, C.T., Thomas, N.J., and Hunter, D.B. (2008). *Parasitic Diseases of Wild Birds*. Wiley-Blackwell, Ames, IA.

Baron, S. (ed.) (1996). *Medical Microbiology*, 4th edition. University of Texas Medical Branch at Galveston, Galveston, TX.

Barreto, M.L., Teixeira, M.G., and Carmo, E.H. (2006). Infectious diseases epidemiology. *Journal of Epidemiology and Community Health*, 60, 192–95.

Bell, J.A., Weckstein, J.D., Fecchio, A., and Tkach, V.V. (2015). A new real-time PCR protocol for detection of avian haemosporidians. *Parasites and Vectors*, 8, 383.

Bennett, G., Peirce, M., and Ashford, R. (1993). Avian hae-matozoa: mortality and pathogenicity. *Journal of Natural History*, 27, 993–1001.

Blaustein, A.R., Gervasi, S.S., Johnson, P.T., et al. (2012). Ecophysiology meets conservation: understanding the role of disease in amphibian population declines. *Philosophical Transactions of the Royal Society B: Biological Sciences*, 367, 1688–707.

Bonneaud, C., Tardy, L., Giraudeau, M., Hill, G.E., McGraw, K.J., and Wilson, A.J. (2019). Evolution of both host resistance and tolerance to an emerging bacterial pathogen. *Evolution Letters*, 3, 544–54.

Boulinier, T. and Staszewski, V. (2008). Maternal transfer of antibodies: raising immuno-ecology issues. *Trends in Ecology & Evolution*, 23, 282–88.

Brault, A.C. (2009). Changing patterns of West Nile virus transmission: altered vector competence and host sus-ceptibility. *Veterinary Research*, 40, 1–19.

Brawner, W.R., Hill, G.E., and Sundermann, C.A. (2000). Effects of coccidial and mycoplasmal infections on carotenoid-based plumage pigmentation in male house finches. *The Auk*, 117, 952–63.

Brien, J.D., Lazear, H.M., and Diamond, M.S. (2013). Propagation, quantification, detection, and storage of West Nile virus. *Current Protocols in Microbiology*, 31, 15D. 3.1–3.18.

Brown, V.R. and Bevins, S.N. (2017). A review of virulent Newcastle disease viruses in the United States and the role of wild birds in viral persistence and spread. *Veterinary Research*, 48, 1–15.

Calisher, C.H., Karabatsos, N., Dalrymple, J.M., et al. (1989). Antigenic relationships between flaviviruses as determined by cross-neutralization tests with poly-clonal antisera. *Journal of General Virology*, 70, 37–43.

Campbell, T.W. (1995). *Avian Hematology and Cytology*. Blackwell Publishing, Ames, IA.

Caron, A., Cappelle, J., Cumming, G.S., de Garine-Wichatitsky, M., and Gaidet, N. (2015). Bridge hosts, a missing link for disease ecology in multi-host systems. *Veterinary Research*, 46, 1–11.

Casadevall, A. and Pirofski, L.-A. (2002). What is a pathogen? *Annals of Medicine*, 34, 2–4.

Casadevall, A. and Pirofski, L. (2014). Ditch the term pathogen. *Nature*, 516, 165–66.

Cerutti, F., Modesto, P., Rizzo, F., et al. (2018). The microbiota of hematophagous ectoparasites collected from migratory birds. *PLoS ONE*, 13, e0202270.

Ciloglu, A., Ellis, V.A., Bernotiene, R., Valkiunas, G., and Bensch, S. (2019). A new one-step multiplex PCR assay for simultaneous detection and identification of avian haemosporidian parasites. *Parasitology Research*, 118, 191–201.

Clayton, D.H. and Walther, B.A. (1997). Collection and quantification of arthropod parasites of birds. In D.H. Clayton and J. Moore, eds. *Host–Parasite Evolution: General Principles and Avian Models*, pp. 419–40. Oxford University Press, Oxford.

Cox, N., Richardson, L., Maurer, J., et al. (2012). Evidence for horizontal and vertical transmission in *Campylobacter* passage from hen to her progeny. *Journal of Food Protection*, 75, 1896–902.

Cross, T.A., Arsnoe, D.M., Minnis, R.B., et al. (2013). Prevalence of avian paramyxovirus 1 and avian influenza virus in double-crested cormorants (*Phalacrocorax auritus*) in eastern North America. *Journal of Wildlife Diseases*, 49, 965–77.

Dawson, J.R., Stone, W.B., Ebel, G.D., et al. (2007). Crow deaths caused by West Nile virus during winter. *Emerging Infectious Diseases*, 13, 1912–14.

DeGroote, L.W. and Rodewald, P.G. (2008). An improved method for quantifying hematozoa by digital microscopy. *Journal of Wildlife Diseases*, 44, 446–50.

Dhondt, A.A., Dhondt, K.V., Hawley, D.M., and Jennelle, C.S. (2007). Experimental evidence for transmission of *Mycoplasma gallisepticum* in house finches by fomites. *Avian Pathology*, 36, 205–8.

Dhondt, A.A., Dhondt, K.V., and McCleery, B.V. (2008). Comparative infectiousness of three passerine bird species after experimental inoculation with *Mycoplasma gallisepticum*. *Avian Pathology*, 37, 635–40.

Dolnik, O.V., Palinauskas, V., and Bensch, S. (2009). Individual oocysts of *Isospora* (Apicomplexa: Coccidia) parasites from avian feces: from photo to sequence. *Journal of Parasitology*, 95, 169–75.

Fontaine, A., Diouf, I., Bakkali, N., et al. (2011). Implication of haematophagous arthropod salivary proteins in host–vector interactions. *Parasites & Vectors*, 4, 187.

Ganz, T. (2003). The role of antimicrobial peptides in innate immunity. *Integrative and Comparative Biology*, 43, 300–4.

Garrett-Jones, C. (1964). Prognosis for interruption of malaria transmission through assessment of the mosquito's vectorial capacity. *Nature*, 204, 1173–75.

Gasparini, J., McCoy, K.D., Tveraa, T., and Boulinier, T. (2002). Related concentrations of specific immunoglobulins against the Lyme disease agent *Borrelia burgdorferi* sensu lato in eggs, young and adults of the kittiwake (*Rissa tridactyla*). *Ecology Letters*, 5, 519–24.

Gibbs, S.E., Hoffman, D.M., Stark, L.M., et al. (2005). Persistence of antibodies to West Nile virus in naturally infected rock pigeons (*Columba livia*). *Clinical and Diagnostic Laboratory Immunology*, 12, 665–67.

Godfrey, R.D., Fedynich, A.M., and Pence, D.B. (1987). Quantification of hematozoa in blood smears. *Journal of Wildlife Diseases*, 23, 558–65.

Graczyk, T.K., Cranfield, M.R., Fayer, R., and Anderson, M.S. (1996). Viability and infectivity of *Cryptosporidium parvum* oocysts are retained upon intestinal passage through a refractory avian host. *Applied and Environmental Microbiology*, 62, 3234–37.

Graham, A.L., Shuker, D.M., Pollitt, L.C., Auld, S., Wilson, A.J., and Little, T.J. (2011). Fitness consequences of immune responses: strengthening the empirical framework for ecoimmunology. *Functional Ecology*, 25, 5–17.

Grindstaff, J.L., Brodie, E.D., and Ketterson, E.D. (2003). Immune function across generations: integrating mechanism and evolutionary process in maternal antibody transmission. *Proceedings of the Royal Society of London. Series B: Biological Sciences*, 270, 2309–19.

Grodio, J.L., Buckles, E.L., and Schat, K.A. (2009). Production of house finch (*Carpodacus mexicanus*) IgA specific anti-sera and its application in immunohistochemistry and in ELISA for detection of *Mycoplasma gallisepticum*-specific IgA. *Veterinary Immunology and Immunopathology*, 132, 288–94.

Grodio, J.L., Hawley, D.M., Osnas, E.E., et al. (2012). Pathogenicity and immunogenicity of three *Mycoplasma gallisepticum* isolates in house finches (*Carpodacus mexicanus*). *Veterinary Microbiology*, 155, 53–61.

Gylfe, A., Bergstrom, S., Lundstrom, J., and Olsen, B. (2000). Reactivation of *Borrelia* infection in birds. *Nature*, 403, 724–25.

Ha, H.J., Alley, M., Howe, L., and Gartrell, B. (2013). Evaluation of the pathogenicity of avipoxvirus strains isolated from wild birds in New Zealand and the efficacy of a fowlpox vaccine in passerines. *Veterinary Microbiology*, 165, 268–74.

Hamer, G.L., Kitron, U.D., Brawn, J.D., et al. (2008). *Culex pipiens* (Diptera: Culicidae): a bridge vector of West Nile virus to humans. *Journal of Medical Entomology*, 45, 125–8.

Hamer, S.A., Goldberg, T.L., Kitron, U.D., et al. (2012). Wild birds and urban ecology of ticks and tick-borne pathogens, Chicago, Illinois, USA, 2005–2010. *Emerging Infectious Diseases*, 18, 1589–95.

Han, B.A. and Altizer, S. (2013). Diseases, Conservation and. In S.A. Levin, ed. *Encyclopedia of Biodiversity*, 2nd edition, pp. 523–38. Academic Press, Cambridge, MA.

Hargitai, R., Prechl, J., and Török, J. (2006). Maternal immunoglobulin concentration in collared flycatcher (*Ficedula albicollis*) eggs in relation to parental quality and laying order. *Functional Ecology*, 20, 829–38.

Hartup, B.K. and Kollias, G.V. (1999). Field investigation of *Mycoplasma gallisepticum* infections in house finch (*Carpodacus mexicanus*) eggs and nestlings. *Avian Diseases*, 43, 572–6.

Hasselquist, D., Tobler, M., and Nilsson, J.-Å. (2012). *Maternal Modulation of Offspring Immune Function in Vertebrates*. Oxford University Press, New York.

Hawley, D.M., DuRant, S.E., Wilson, A.F., Adelman, J.S., and Hopkins, W.A. (2012). Additive metabolic costs of thermoregulation and pathogen infection. *Functional Ecology*, 26, 701–10.

Hawley, D.M., Osnas, E.E., Dobson, A.P., Hochachka, W.M., Ley, D.H., and Dhondt, A.A. (2013). Parallel patterns of increased virulence in a recently emerged wildlife pathogen. *PLoS Biology*, 11, e1001570.

Haydon, D.T., Cleaveland, S., Taylor, L.H., and Laurenson, M.K. (2002). Identifying reservoirs of infection: a conceptual and practical challenge. *Emerging Infectious Diseases*, 8, 1468–73.

Hellgren, O., Waldenstrom, J., and Bensch, S. (2004). A new PCR assay for simultaneous studies of *Leucocytozoon*, *Plasmodium*, and *Haemoproteus* from avian blood. *Journal of Parasitology*, 90, 797–802.

Henschen, A.E. and Adelman, J.S. (2019). What does tolerance mean for animal disease dynamics when pathology enhances transmission? *Integrative and Comparative Biology*, 59, 1220–30.

Huffman, J.E. and Fried, B. (2008). Schistosomes. In C.T. Atkinson, N.J. Thomas, and D.B. Hunter, eds. *Parasitic Diseases of Wild Birds*, pp. 246–60. Wiley-Blackwell, Ames, Iowa.

Hund, A.K., Blair, J.T., and Hund, F.W. (2015). A review of available methods and description of a new method for eliminating ectoparasites from bird nests. *Journal of Field Ornithology*, 86, 191–204.

Janeway, C.A., Travers, P., Walport, M., and Shlomchik, M.J. (2001). Infectious agents and how they cause disease. In *Immunobiology: The Immune System in Health and Disease*, 5th edition. Garland Science, New York.

Jarvi, S.I., Schultz, J.J., and Atkinson, C.T. (2002). PCR diagnostics underestimate the prevalence of avian malaria (*Plasmodium relictum*) in experimentally-infected passerines. *Journal of Parasitology*, 88, 153–8.

Karell, P., Kontiainen, P., Pietiäinen, H., Siitari, H., and Brommer, J. (2008). Maternal effects on offspring Igs and egg size in relation to natural and experimentally improved food supply. *Functional Ecology*, 22, 682–90.

Kennedy, C.R. (1975). *Ecological Animal Parasitology*. Blackwell Scientific Publications, Oxford.

Kermack, W.O. and McKendrick, A.G. (1927). A contribution to the mathematical theory of epidemics. *Proceedings of the Royal Society A*, 115, 700–21.

Killian, M.L. (2020). Hemagglutination assay for influenza virus. In E. Spackman, ed. *Animal Influenza Virus*, pp. 3–10. Humana Press, Totowa, NJ.

Kilpatrick, A.M., Daszak, P., Jones, M.J., Marra, P.P., and Kramer, L.D. (2006). Host heterogeneity dominates West Nile virus transmission. *Proceedings of the Royal Society B: Biological Sciences*, 273, 2327–33.

Kim, S.-H. and Samal, S.K. (2019). Innovation in Newcastle disease virus vectored avian influenza vaccines. *Viruses*, 11, 300.

Knowles, S.C.L., Palinauskas, V., and Sheldon, B.C. (2010). Chronic malaria infections increase family inequalities and reduce parental fitness: experimental evidence from a wild bird population. *Journal of Evolutionary Biology*, 23, 557–69.

Kohl, K.D. (2012). Diversity and function of the avian gut microbiota. *Journal of Comparative Physiology B*, 182, 591–602.

Koop, J.A.H. and Clayton, D.H. (2013). Evaluation of two methods for quantifying passeriform lice. *Journal of Field Ornithology*, 84, 210–15.

Kreisinger, J., Schmiedová, L., Petrželková, A., et al. (2018). Fecal microbiota associated with phytohaemagglutinin-induced immune response in nestlings of a passerine bird. *Ecology and Evolution*, 8, 9793–802.

Krone, O., Waldenström, J., Valkiūnas, G., et al. (2008). Haemosporidian blood parasites in European birds of prey and owls. *Journal of Parasitology*, 94, 709–16.

Leggett, H.C., Cornwallis, C.K., and West, S.A. (2012). Mechanisms of pathogenesis, infective dose and virulence in human parasites. *PLoS Pathogens*, 8, e1002512.

Leighton, F.A. and Heckert, R.A., eds. (2007). *Newcastle Disease and Related Avian Paramyxoviruses*, pp. 3–16. Blackwell Publishing, Ames, IA.

Lindsey, H.S., Calisher, C.H., and Mathews, J.H. (1976). Serum dilution neutralization test for California group virus identification and serology. *Journal of Clinical Microbiology*, 4, 503–10.

Lord, C.C., Rutledge, C.R., and Tabachnick, W.J. (2006). Relationships between host viremia and vector susceptibility for arboviruses. *Journal of Medical Entomology*, 43, 623.

Martínez-de la Puente, J., Merino, S., Tomás, G., et al. (2010). The blood parasite *Haemoproteus* reduces survival in a wild bird: a medication experiment. *Biology Letters*, 6, 663–5.

Medzhitov, R., Schneider, D.S., and Soares, M.P. (2012). Disease tolerance as a defense strategy. *Science*, 335, 936–41.

Merino, S., Moreno, J., José Sanz, J., and Arriero, E. (2000). Are avian blood parasites pathogenic in the wild? A medication experiment in blue tits (*Parus caeruleus*). *Proceedings of the Royal Society B: Biological Sciences*, 267, 2507–10.

Méthot, P.-O. and Alizon, S. (2014). What is a pathogen? Toward a process view of host–parasite interactions. *Virulence*, 5, 775–85.

Miller, M.J., Esser, H.J., Loaiza, J.R., et al. (2016). Molecular ecological insights into neotropical bird–tick interactions. *PLoS ONE*, 11, e0155989.

Molaei, G., Andreadis, T.G., Armstrong, P.M., Anderson, J.F., and Vossbrinck, C.R. (2006). Host feeding patterns of *Culex* mosquitoes and West Nile virus transmission, northeastern United States. *Emerging Infectious Diseases*, 12, 468.

Morens, D.M., Folkers, G.K., and Fauci, A.S. (2019). Eastern equine encephalitis virus—another emergent arbovirus in the United States. *New England Journal of Medicine*, 381, 1989–92.

Mori, E., Sala, J.P., Fattorini, N., Menchetti, M., Montalvo, T., and Senar, J.C. (2019). Ectoparasite sharing among native and invasive birds in a metropolitan area. *Parasitology Research*, 118, 399–409.

Mostafa, A., Abdelwhab, E.M., Mettenleiter, T.C., and Pleschka, S. (2018). Zoonotic potential of influenza A viruses: a comprehensive overview. *Viruses*, 10, 497.

Nakamura, A.A. and Meireles, M.V. (2015). *Cryptosporidium* infections in birds—a review. *Revista Brasileira de Parasitologia Veterinária*, 24, 253–67.

Nakamura, A.A., Simões, D.C., Antunes, R.G., da Silva, D.C., and Meireles, M.V. (2009). Molecular characterization of *Cryptosporidium* spp. from fecal samples of birds kept in captivity in Brazil. *Veterinary Parasitology*, 166, 47–51.

Nash, A.A., Mims, C.A., and Stephen, J. (2000). *Mims' Pathogenesis of Infectious Disease*. Academic Press, Cambridge, MA.

Nemeth, N.M. and Bowen, R.A. (2007). Dynamics of passive immunity to West Nile virus in domestic chickens (*Gallus gallus domesticus*). *American Journal of Tropical Medicine and Hygiene*, 76, 310–17.

Nemeth, N.M., Thomas, N.O., Orahood, D.S., Anderson, T.D., and Oesterle, P.T. (2010). Shedding and serologic responses following primary and secondary inoculation of house sparrows (*Passer domesticus*) and European starlings (*Sturnus vulgaris*) with low-pathogenicity avian influenza virus. *Avian Pathology*, 39, 411–18.

Owen, J.C., Moore, F.R., Williams, A.J., et al. (2011). Test of recrudescence hypothesis for overwintering of eastern equine encephalomyelitis virus in gray catbirds. *Journal of Medical Entomology*, 48, 896–903.

Pacheco, M.A., Cepeda, A.S., Bernotienė, R., et al. (2018). Primers targeting mitochondrial genes of avian haemosporidians: PCR detection and differential DNA amplification of parasites belonging to different genera. *International Journal for Parasitology*, 48, 657–70.

Pirofski, L.-a. and Casadevall, A. (2012). Q&A: what is a pathogen? A question that begs the point. *BMC Biology*, 10, 6.

Råberg, L., Graham, A.L., and Read, A.F. (2009). Decomposing health: tolerance and resistance to parasites in animals. *Philosophical Transactions of the Royal Society B: Biological Sciences*, 364, 37.

Reid, A.H., Fanning, T.G., Hultin, J.V., and Taubenberger, J.K. (1999). Origin and evolution of the 1918 'Spanish' influenza virus hemagglutinin gene. *Proceedings of the National Academy of Sciences of the United States of America*, 96, 1651–6.

Reid, J.M., Arcese, P., Keller, L.F., and Hasselquist, D. (2006). Long-term maternal effect on offspring immune response in song sparrows *Melospiza melodia*. *Biology Letters*, 2, 573–6.

Restif, O. and Koella, J.C. (2004). Concurrent evolution of resistance and tolerance to pathogens. *The American Naturalist*, 164, E90–102.

Ribeiro, S.F., Sebaio, F., Branquinho, F.C., Marini, M.A., Vago, A.R., and Braga, E.M. (2005). Avian malaria in Brazilian passerine birds: parasitism detected by nested PCR using DNA from stained blood smears. *Parasitology*, 130, 261–7.

Robinson, R.A., Lawson, B., Toms, M.P., et al. (2010). Emerging infectious disease leads to rapid population declines of common British birds. *PLoS ONE*, 5, e12215.

Saikia, M., Bhattacharjee, K., Sarmah, P., Tamuly, S., Dutta, B., and Konch, P. (2017). Pathology and molecular detection of coccidiosis in experimentally infected domestic pigeon. *Journal of Entomology and Zoology Studies*, 5, 1841–5.

Schneider, B.S., Soong, L., Girard, Y.A., Campbell, G., Mason, P., and Higgs, S. (2006). Potentiation of West Nile encephalitis by mosquito feeding. *Viral Immunology*, 19, 74–82.

Schoenle, L.A., Schoepf, I., Weinstein, N.M., Moore, I.T., and Bonier, F. (2018). Higher plasma corticosterone is associated with reduced costs of infection in red-winged blackbirds. *General and Comparative Endocrinology*, 256, 89–98.

Shirley, M.W., Smith, A.L., and Tomley, F.M. (2005). The biology of avian *Eimeria* with an emphasis on their control by vaccination. *Advances in Parasitology*, 60, 285–330.

Shriner, S.A., Root, J.J., Lutman, M.W., et al. (2016). Surveillance for highly pathogenic H5 avian influenza virus in synanthropic wildlife associated with poultry farms during an acute outbreak. *Scientific Reports*, 6, 36237.

Spackman, E. (2008). *Avian Influenza Virus*. Humana Press, Totowa, NJ.

Spackman, E. (2020). Avian influenza virus detection and quantitation by real-time RT-PCR. In E. Spackman, ed. *Animal Influenza Virus*, pp. 137–48. Humana Press, Totowa, NJ.

Spackman, E. and Lee, S.A. (2020). Avian influenza virus RNA extraction. In E. Spackman, ed. *Animal Influenza Virus*, pp. 123–35. Humana Press, Totowa, NJ.

Spackman, E., Pantin-Jackwood, M.J., Swayne, D.E., and Suarez, D.L. (2009). An evaluation of avian influenza diagnostic methods with domestic duck specimens. *Avian Diseases*, 53, 276–80.

Staley, M., Bonneaud, C., McGraw, K.J., Vleck, C.M., and Hill, G.E. (2018). Detection of *Mycoplasma gallisepticum* in house finches (*Haemorhous mexicanus*) from Arizona. *Avian Diseases*, 62, 14–17.

Stallknecht, D.E., Shane, S.M., Kearney, M.T., and Zwank, P.J. (1990). Persistence of avian influenza viruses in water. *Avian Diseases*, 34, 406–11.

Styer, L.M., Bernard, K.A., and Kramer, L.D. (2006). Enhanced early West Nile virus infection in young chickens infected by mosquito bite: effect of viral dose. *American Journal of Tropical Medicine and Hygiene*, 75, 337–45.

Styer, L.M., Kent, K.A., Albright, R.G., Bennett, C.J., Kramer, L.D., and Bernard, K.A. (2007). Mosquitoes inoculate high doses of West Nile virus as they probe and feed on live hosts. *PLoS Pathogens*, 3, 1262–70.

Suzuki, Y. (2005). Sialobiology of influenza: molecular mechanism of host range variation of influenza viruses. *Biological and Pharmaceutical Bulletin*, 28, 399–408.

Suzuki, Y., Ito, T., Suzuki, T., et al. (2000). Sialic acid species as a determinant of the host range of influenza A viruses. *Journal of Virology*, 74, 11825–31.

Svensson-Coelho, M., Silva, G.T., Santos, S.S., et al. (2016). Lower detection probability of avian *Plasmodium* in blood compared to other tissues. *Journal of Parasitology*, 102, 559–61.

Svobodová, M., Dolnik, O.V., Čepička, I., and Rádrová, J. (2017). Biting midges (Ceratopogonidae) as vectors of avian trypanosomes. *Parasites & Vectors*, 10, 1–9.

Swayne, D.E. (1998). *Laboratory Manual for the Isolation and Identification of Avian Pathogens*. American Association of Avian Pathologists, University of Pennsylvania, Philadelphia.

Sydenstricker, K.V., Dhondt, A.A., Ley, D.H., and Kollias, G.V. (2005). Re-exposure of captive house finches that recovered from *Mycoplasma gallisepticum* infection. *Journal of Wildlife Diseases*, 41, 326–33.

Taubenberger, J.K. and Morens, D.M. (2006). 1918 Influenza: the mother of all pandemics. *Revista Biomedica*, 17, 69–79.

Thomas, N.J., Hunter, D.B., and Atkinson, C.T. (2008). *Infectious Diseases of Wild Birds*. Blackwell Publishing, Ames, IA.

Thomas, S.R. and Elkinton, J.S. (2004). Pathogenicity and virulence. *Journal of Invertebrate Pathology*, 85, 146–51.

Tobita, K., Sugiura, A., Enomoto, C., and Furuyama, M. (1975). Plaque assay and primary isolation of influenza A viruses in an established line of canine kidney cells (MDCK) in the presence of trypsin. *Medical Microbiology and Immunology*, 162, 9–14.

Turell, M.J., O'Guinn, M., Dohm, D.J., and Jones, J.W. (2001). Vector competence of North American mosquitos (Diptera: Culicidae) for West Nile virus. *Journal of Medical Entomology*, 38, 130–4.

Turell, M.J., Dohm, D.J., Sardelis, M.R., Oguinn, M.L., Andreadis, T.G., and Blow, J.A. (2005). An update on the potential of North American mosquitoes (Diptera: Culicidae) to transmit West Nile virus. *Journal of Medical Entomology*, 42, 57–62.

Valkiunas, G. (2004). *Avian Malaria Parasites and Other Haemosporidia*. CRC Press, Boca Raton, FL.

van Dijk, J.G., Mateman, A.C., and Klaassen, M. (2014). Transfer of maternal antibodies against avian influenza virus in mallards (*Anas platyrhynchos*). *PLoS One*, 9, e112595.

Vynnycky, E. and White, R. (2010). *An Introduction to Infectious Disease Modelling*. Oxford University Press, Oxford.

Webster, R.G., Bean, W.J., Gorman, O.T., Chambers, T.M., and Kawaoka, Y. (1992). Evolution and ecology of influenza A viruses. *Microbiology and Molecular Biology Reviews*, 56, 152–79.

Wellman-Labadie, O., Picman, J., and Hincke, M. (2007). Avian antimicrobial proteins: structure, distribution and activity. *World's Poultry Science Journal*, 63, 421–38.

Wheeler, S.S., Woods, L.W., Boyce, W.M., et al. (2014). West Nile virus and non-West Nile virus mortality and coinfection of American crows (*Corvus brachyrhynchos*) in California. *Avian Diseases*, 58, 255–61.

Wilcoxen, T.E., Horn, D.J., Hogan, B.M., et al. (2015). Effects of bird-feeding activities on the health of wild birds. *Conservation Physiology*, 3, cov058.

Wilson, K., Bjornstad, O.N., Dobson, A.P., et al. (2001). Heterogeneities in macroparasite infections: patterns and processes. In P.J. Hudson, A.P. Rizzoli, B.T. Greenfell, J.A.P. Heesterbeek, and A.P. Dobson, eds. *The Ecology of Wildlife Diseases*, pp. 6–44. Oxford University Press, Oxford.

Winfield, M.D. and Groisman, E.A. (2003). Role of nonhost environments in the lifestyles of *Salmonella* and *Escherichia coli*. *Applied and Environmental Microbiology*, 69, 3687–94.

Wobeser, G.A. (2013). *Essentials of Disease in Wild Animals*. Blackwell Publishing, Ames, IA.

Zurfluh, K., Wang, J., Klumpp, J., Nüesch-Inderbinen, M., Fanning, S., and Stephan, R. (2014). Vertical transmission of highly similar blaCTX-M-1-harbouring IncI1 plasmids in *Escherichia coli* with different MLST types in the poultry production pyramid. *Frontiers in Microbiology*, 5, 519.

CHAPTER 3

Ecoimmunology

Amberleigh E. Henschen and James S. Adelman

3.1 Introduction

Avian immune responses show substantial variation among individuals, populations, and species, with important consequences for disease dynamics. Unlike traditional immunological studies in model systems, ecoimmunology aims to understand the sources of this variation in responses to exposure and infection (Schulenburg et al. 2009). Seminal research in the field set out to reveal factors that drive immune variation and how physiologically costly immune defenses trade-off against other critical organismal functions and life history traits (Brock et al. 2014). Early and ongoing studies have shown that such variation can arise from myriad sources, including innate characteristics of an individual such as age, current life history stage, or sex; physiological factors such as current stress levels or infection with other parasites; or environmental variables such as the availability of resources (Adelman et al. 2014a).

While these issues are still of great interest in the field (Brace et al. 2017), ecoimmunologists are also pursuing links with disease ecology by asking how variation in immune defenses can shape within- and between-host disease dynamics (Adelman 2015; Brock et al. 2014; Graham et al. 2011; Hawley and Altizer 2011). Throughout this chapter, we use the inclusive term 'parasite' to refer to all manner of infectious agents, from microparasites (e.g., viruses) to macroparasites (e.g., nematodes) and endoparasites to ectoparasites. At the individual level, mechanisms underlying both resistance, which kills invading parasites, and tolerance, which reduces fitness costs of infections without reducing parasite load, have been of keen interest to avian ecoimmunologists (Sorci 2013). Although less research has focused on the consequences of these within-host mechanisms for parasite transmission and prevalence, both empirical and theoretical studies in several avian systems are pushing the field in this direction (e.g., Adelman and Hawley 2017; Burgan et al. 2018; Burgan et al. 2019).

In this chapter, we first review avian immune defenses generally, then explore potential drivers of variation in these defenses. Following this, we briefly highlight common techniques that ecoimmunologists use to measure such variation, including some relative benefits and drawbacks of each method. Finally, we identify two critical knowledge gaps in linking ecoimmunology and disease ecology: (1) whether immune phenotypes predict infection outcomes or host fitness; and (2) whether within-host measures of immunity or host–parasite dynamics inform population-level transmission dynamics.

3.2 Overview of the avian immune system

This brief overview of immune defenses in birds is by no means exhaustive but rather provides a general introduction to some of the defenses discussed in this chapter and throughout this book (for additional detail see Schat et al. 2014). Like all immune

Amberleigh E. Henschen and James S. Adelman, *Ecoimmunology* In: *Infectious Disease Ecology of Wild Birds*. Edited by: Jennifer C. Owen, Dana M. Hawley, and Kathryn P. Huyvaert, Oxford University Press. © Oxford University Press 2021.
DOI: 10.1093/oso/9780198746249.003.0003

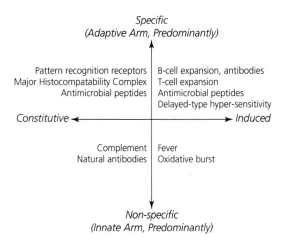

Specific
(Adaptive Arm, Predominantly)

Pattern recognition receptors | B-cell expansion, antibodies
Major Histocompatability Complex | T-cell expansion
Antimicrobial peptides | Antimicrobial peptides
| Delayed-type hyper-sensitivity

Constitutive ◄──────────────► Induced

Complement | Fever
Natural antibodies | Oxidative burst

Non-specific
(Innate Arm, Predominantly)

Figure 3.1 Functional immune protection results from a suite of defenses, conceptualized here along two complementary axes— constitutive vs. induced and specific vs. non-specific. In vertebrates, the adaptive arm of the immune system contributes most to specific defenses, whereas the innate arm of the immune system contributes most to non-specific defenses. Several commonly measured defenses in avian ecoimmunology appear in appropriate quadrants (some defenses are appropriate to quantify both constitutive and induced responses). (Adapted from Fig. 1 in Schmid-Hempel and Ebert. 2003. Trends in Ecology & Evolution 18: 27–32.)

systems, avian immune defenses are multifaceted, with numerous mechanisms working in concert to confer protection. One broad way to categorize the diverse defenses involves two complementary axes: constitutive vs. inducible and non-specific vs. specific (Schmid-Hempel and Ebert 2003; Figure 3.1). Constitutive defenses are continually present, regardless of current infection, whereas inducible defenses increase in abundance and/or activity during infection. Non-specific defenses protect against a diversity of invaders, often targeting common microbe-associated molecular patterns, whereas specific defenses are narrowly targeted, often against a small piece of one particular parasite. In birds and other vertebrates, non-specific defenses typically fall under the 'innate' arm of the immune system. In contrast, hyper-specific defenses, including memory of prior infections, typically fall under the 'adaptive' arm of the immune system. This distinction, however, is not totally binary: within both arms, some defenses are more or less specific than others.

3.2.1 The innate immune system

The innate immune system is typically the first line of defense after infection, with constitutive components continually at the ready. Such constitutive components include both soluble and cellular defenses. Among soluble defenses, myriad antimicrobial peptides (which can also be part of induced immune defenses) target a range of invaders, either directly lysing them or binding to their surfaces and activating other defenses (Juul-Madsen et al. 2014). Other commonly measured soluble defenses in ecoimmunology include natural antibodies and the complement system. Natural antibodies are a population of immunoglobulins (Ig) (see Section 3.2.2) that bind to a diversity of common invaders and circulate throughout the body (Carroll and Prodeus 1998; Härtle et al. 2014). These antibodies do not require prior exposure to specific invaders and are encoded directly in the germ line; thus they are not considered part of the adaptive immune system, as are other antibodies (Härtle et al. 2014). Once natural antibodies bind to an invader, they can activate the complement cascade, a group of constitutively expressed, circulating proteins that work in concert to bind an invader, facilitating further, induced immune responses, including phagocytosis and lysis, upregulation of inflammatory signals, and activation of the adaptive immune system (Juul-Madsen et al. 2014). Such links among components of the immune system highlight the integrative nature of a functional defense against infection.

The linkages rely heavily upon the cells of the innate immune system and their surface receptors. Collectively, these cells recognize parasites and activated soluble defenses, engulf or otherwise destroy invaders, and facilitate other innate and adaptive immune responses. Activating these cells requires a diversity of receptors (as is also true of cells in the adaptive arm of the immune system). Notably, pattern recognition receptors (PRRs), such as the Toll-like receptors (TLRs), bind to common microbial associated molecular patterns (MAMPS). Several distinct TLRs exist, activated by different classes of invaders, illustrating some degree of specificity within the innate system. For example, TLR4 responds to lipopolysaccharide (LPS), a highly conserved bacterial cell wall component and commonly used antigen in ecoimmunology (Juul-Madsen et al. 2014). After PRRs bind to MAMPs, subsequent signaling cascades result in the release of cytokines and chemokines, immune system signaling molecules that orchestrate further responses (Juul-Madsen et al. 2014; Kaiser and Stäheli 2014). Myriad cytokine and chemokine receptors allow immune cells to receive these signals (Kaiser and Stäheli 2014), enhancing processes like proliferation and recruitment to sites of infection. In addition, other receptors allow innate immune cells to interact with soluble factors, such as the Fc receptor, which allows binding of antibodies attached to invaders and facilitates phagocytosis (Juul-Madsen et al. 2014).

Principal cellular players of the innate immune system in birds include the phagocytes (heterophils and macrophages), natural killer cells, and the antigen presenting cells (dendritic cells and, again, macrophages) (Juul-Madsen et al. 2014; Kaspers and Kaiser 2014). Heterophils often serve as early cellular defenses against tissue infection, with the capacity to phagocytize invaders and lyse them intracellularly (Juul-Madsen et al. 2014). In addition, heterophils can release lytic enzymes, contributing to further immune activation and, in some cases, damage to the host's own tissues (Harmon 1998). Unlike their mammalian counterparts, neutrophils, heterophils rely heavily on non-oxidative lysis, producing fewer reactive oxygen species, which can be highly damaging to hosts (Harmon 1998; Juul-Madsen et al. 2014). Macrophages are also important phagocytes, though these cells tend to arrive at a site of infection after heterophils, remain active longer, and rely more heavily on oxidative burst to lyse invaders, creating more reactive oxygen species (Kaspers and Kaiser 2014). In contrast to the phagocytes, natural killer cells target intracellular parasites, lysing infected host cells (Rogers et al. 2008). Although most often described in the intestine, new cells with natural killer-like activity have recently been discovered elsewhere (Fenzl et al. 2017; Rogers et al. 2008). Finally, the antigen presenting cells, dendritic cells and macrophages, provide a crucial link to the adaptive arm of the immune system. Through major histocompatibility complex (MHC) molecules on their surfaces,

these cells display pieces of whatever they have engulfed, termed antigens, including both host and parasite-derived molecules. When specific lymphocytes (see below) recognize these presented antigens, this process becomes a critical step in activating the adaptive immune system (Kaspers and Kaiser 2014).

As a result of the above cellular and molecular responses, early innate responses frequently include what is known as the acute phase response (APR). The APR describes a suite of changes across the whole organism, including altered behavior, thermoregulation, and production of a suite of proteins known as acute phase proteins (Owen-Ashley and Wingfield 2007). Behavioral changes during the APR or sickness behaviors include lethargy and anorexia and arise in response to diverse types of infections. Although sickness behaviors may help conserve energy for use in other immune responses (Hart 1988), they can also trade off with other fitness enhancing behaviors like territory defense or mate-finding, making them especially interesting to ecoimmunologists (Adelman and Hawley 2017; Adelman and Martin 2009). Changes in thermoregulation, like fever or hypothermia, also result from a diversity of infections and can prove helpful in clearing parasites (Kluger et al. 1998). However, these responses incur significant energetic costs (Marais et al. 2011; Roe and Kinney 1965). Finally, acute phase proteins span a wide variety of functions, including directly disabling parasites (e.g., mannan-binding lectin) or mitigating damage induced during an infection (e.g., haptoglobin or PIT54) (Juul-Madsen et al. 2014; Matson et al. 2012). Commonly induced using LPS or other TLR agonists, the APR represents a highly integrative, whole-organism measure of the innate immune response.

3.2.2 The adaptive immune system

In contrast to innate defenses, adaptive immune defenses are generally highly specific and exhibit memory, such that secondary exposure to the same invader initiates a more targeted, rapid, and robust response. Although constitutive components and ongoing surveillance are important to initiating these responses, adaptive immune defenses are generally thought of as highly inducible. The key cellular players in these responses are the lymphocytes, B- and T- cells, named for their sites of origin/maturation—the bursa of Fabricius and thymus, respectively (Cooper et al. 1965; Glick et al. 1956). Broadly speaking, B-cells produce immunoglobulins, or antibodies, which bind extracellular invaders, a process referred to as antibody-mediated or humoral immunity. T-cells, in contrast, do not produce antibodies but consist of several subclasses that play distinct roles. First, cytotoxic or killer T-cells (T_c) induce apoptosis in cells infected with intracellular invaders, a process often referred to as cell-mediated immunity. Second, helper T-cells (T_h) serve to coordinate both antibody and cytotoxic responses. Finally, regulatory T-cells (T_{reg}) help rein in immune responses through cytokine signaling (Sompayrac 2016).

The adaptive immune system's ability to recognize highly specific antigens stems from the enormous diversity of individual B- and T-cell receptors. Each cell expresses copies of one unique receptor, capable of binding a very specific antigen. Although the exact mechanisms generating this diversity differ between mammals and birds, they rely on a modular genetic structure. Very briefly, the developing cell assembles different gene segments into unique combinations, leaving the mature cell with only one unique receptor gene out of a potential set of billions (Ratcliffe and Härtle 2014; Smith and Göbel 2014; Sompayrac 2016). In the case of B-cells, this receptor includes a membrane-bound immunoglobulin, with the same binding specificity as the antibodies that cell can secrete to disable invaders. The T-cell receptor, in contrast, does not correspond to any secreted protein but enables the cell to interact with cells presenting its specific antigen (Sompayrac 2016).

Such interactions between T-cell receptors and antigen presenting cells, most notably dendritic cells, are crucial in orchestrating subsequent responses. For instance, T-cell receptors on T_h cells allow these cells to interact with class II MHC molecules, which display antigen gathered from outside the cell. If that antigen–MHC complex matches the T_h's specific receptor, that T_h becomes activated and produces cytokines that facilitate B-cell and T_c proliferation and stimulates antibody production in B-cells that have also found their specific antigens. T_cs also undergo an activation process after finding

their matching antigens in class I MHC molecules on antigen presenting cells, which display antigens gathered from internal cellular processes. Following this activation and co-stimulation from T_h cytokines, T_cs can then interact with infected cells to trigger cell death (Sompayrac 2016).

In addition to these hyper-specific processes, another defining characteristic of the adaptive immune system is its capacity for memory. During an infection, activated B- and T-cells proliferate rapidly, producing not only more effector cells, which help tackle the immediate infection, but also long-lived memory cells. These cells reside in lymphoid tissues, like lymph nodes, and are capable of more rapid activation and proliferation after infection with the same prior invader (Sompayrac 2016).

Finally, another type of T-cell mediated response warrants discussion, as it has been frequently measured, and misinterpreted, in ecoimmunology (Martin et al. 2006). Delayed type hypersensitivity (DTH) involves stimulation of peripheral T_h cells, whose cytokines and chemokines then recruit innate cells, resulting in localized inflammation (Sompayrac 2016). Ecoimmunologists often induce DTH with antigens like phytohemagglutinin (PHA), which activates an array of T_h cells without the need for antigen presenting cells (Martin et al. 2006). However, numerous papers refer to such responses simply as cell-mediated immunity. Although DTH does involve T-cells, this response should not be confused with T_c responses. Rather, it is an example of integration across the immune system, with lymphocytes activating innate cells.

3.2.3 An integrated response to infection

Even through an overview as brief as this one, the immune system emerges as a highly complex array of diverse defenses. One natural conclusion from this diversity is that mounting an effective immune defense requires tight orchestration: a given type of response rarely occurs in isolation but rather depends on prior responses and helps facilitate subsequent ones. Importantly for ecoimmunology and disease ecology, much of the signaling that underlies this orchestration, for example cytokines, also helps integrate the immune system with other physiological and behavioral responses (Demas and

Carlton 2015). As such, understanding an immune response can help determine not only how rapidly an animal might clear an infection but also how such responses may trade off with other critical biology functions or impact behaviors critical to the spread of infectious disease (Adelman and Hawley 2017; Hawley et al. 2011; Henschen and Adelman 2019; Martin et al. 2019). In addition, the diverse interactions among immune components present a logistical challenge to ecoimmunologists: which defenses to measure and when? In the sections that follow, we synthesize current thinking around these conceptual links and practical considerations.

3.3 Drivers of variation in immune defense

Ecoimmunologists seek to understand the abiotic and biotic drivers of natural variation in immune defenses and quantify these defenses. Given the diverse and multifaceted nature of immune systems, it is not surprising that their development and use can be costly in terms of energy, other nutrients, and time (Brace et al. 2017; Klasing 2004; Sheldon and Verhulst 1996). Moreover, immune responses interact with diverse physiological systems, which can respond to myriad factors such as sex, age, life history stage, condition, and environmental factors (Cohen et al. 2012). As such, trade-offs between the immune system and other functions are expected and common (Brace et al. 2017). Understanding such trade-offs among immune responses, parasite load, and other life history functions is extremely important for revealing how immunity is related to overall fitness in wild birds. Here, we review some of the major relationships among immune system functioning and both host-level and environmental factors. It should be noted that each of these factors may simultaneously impact, or be impacted by, immune responses. Thus, studies that aim to understand the relationship between one of these factors and immune responses should control or account for differences in other factors.

3.3.1 Sex

In theory, differential selective pressures between males and females should drive variation in

investment in many physiological processes, including immune defense. Indeed, sex differences in immune responses and parasite load are well documented in avian species. In general, parasite prevalence is often lower and immune responses are often stronger in females than in males (Hasselquist 2007). For example, the prevalence of malaria is lower in female than male tawny pipits (*Anthus campestris*) and bacterial killing ability in the blood of female house sparrows (*Passer domesticus*) is higher during the breeding season than males (Calero-Riestra and Garcia 2016; Pap et al. 2010). Recently, ecoimmunologists have begun unraveling the mechanistic and evolutionary drivers of sex-based immune variation (Roved et al. 2017).

Mechanistically, differences in levels of steroid hormones between the sexes can have an impact on immune responses. This is because high levels of testosterone can be immunosuppressive (Casto et al. 2001; Folstad and Karter 1992), although the relationship between testosterone and immune responses is highly variable between species and studies (Foo et al. 2017). Estrogen and progesterone also affect immune responses, although these hormones can enhance some immune responses and inhibit others (Foo et al. 2017; Roved et al. 2017).

Mating systems may also impact the optimal investment in immune response for both males and females and, thus, may exaggerate or diminish sex-based immune differences. In polygynous systems (e.g., leks), where male fitness is heavily skewed between individuals, males may be selected to invest more in reproduction, including increasing levels of testosterone, at the expense of immune responses. In contrast, there may be little difference between male and female immune responses in monogamous systems (Zuk et al. 1990). Comparisons of polygynous and monogamous systems suggest that mating systems, indeed, can have a strong effect on sexual dimorphism in immune responses (Hasselquist 2007). This may be partially due to differences in testosterone production during the breeding season; while testosterone production remains high during the breeding season in males of polygynous species, males of monogamous species experience a gradual decline of testosterone production after pair formation (Wingfield et al. 1990).

3.3.2 Age

Much of the work investigating the effects of age on immune responses in avian systems has compared juvenile with adult individuals. As in other animal taxa, juvenile immune responses are often undeveloped and juveniles rely, in part, on maternal antibodies for protection from infection (Figure 3.2). Great tit (*Parus major*) nestlings, for example, have complement systems that are not activated until after the nestling stage, although they have similar levels of natural antibodies compared to adult birds (De Coster et al. 2010). Similarly, levels of natural antibodies and complement systems continue to develop through fledging in house sparrows (Killpack et al. 2013; Figure 3.2). Acquired immunity also continues to develop over the life of avian hosts and may explain some of the age-biased mortality during disease outbreaks (Hill et al. 2016). For example, the prevalence of antibodies to the causative agent of Newcastle disease (avian paramyxovirus serotype 1, APMV-1) is much lower, and mortality much higher, in double-crested cormorant (*Phalacrocorax auritus*) chicks compared to older birds (Cross et al. 2013; Kuiken et al. 1998). Given these differences in acquired immunity and parasite loads, morbidity and mortality are often higher in juveniles than adults.

Infection during juvenile stages can also have long-term fitness consequences. Barn swallow (*Hirundo rustica*) nestlings challenged with injections of LPS produced lower quality feathers, which can decrease flying efficiency (Romano et al. 2011). Interestingly, in this study, female barn swallow nestlings showed more pronounced effects of LPS injections than did males. In contrast, nestling male tawny pipits infected with avian malaria gained mass more slowly than uninfected male nestlings, while infection status did not affect mass gain in females (Calero-Riestra and Garcia 2016). These studies suggest that there may be important interactions between age and sex on the fitness costs of infection in young birds (see also Chapter 6).

Given the documented fitness costs to early infection, muted immune responses in juvenile birds may be necessary for proper growth and development. Some types of immune responses, such as the APR, may impact growth directly by reducing energy intake (due to anorexia) and shunting available energy away from growth (Lochmiller and Deerenberg 2000). Humoral immunity may also trade off with growth: in Leach's storm-petrels (*Oceanodroma leucorhoa*), nestling growth rate decreases as the rate of natural antibody production increases (Mauck et al. 2005). Furthermore, in domestic poultry, selection for rapid growth results in individuals that respond less effectively to immune challenges (van der Most et al. 2011). Thus, there may be strong trade-offs between growth and immune responses in many avian systems during juvenile stages.

3.3.3 Life history stage

Over the course of a year, birds engage in many costly endeavors, including breeding, migration, molt, and overwinter survival. Trade-offs may therefore arise between the costly activities in each of these life history stages and immune responses. Actively breeding birds, for example, will often have dampened immune responses and increased parasite loads (Knowles et al. 2009). For instance, female great tits have lower levels of natural antibodies and dampened complement activation while they are breeding (De Coster et al. 2010). Such lowered protection mechanisms may lead to increased parasite loads during the breeding season, due to either new infections or relapses of chronic infections (e.g., with avian blood parasites; Valkiunas 2004). Likewise, immune challenges during the breeding season can result in reduced investment in current reproduction (Wegmann et al. 2015). However, immune challenges do not always reduce breeding effort. For example, immune activation may trigger older individuals to increase investment in current reproduction, consistent with 'terminal investment' in the face of reduced survival probability (Bowers et al. 2012; Clutton-Brock 1984; Velando et al. 2014; see also Chapter 6).

As with reproduction, immune responses are likely to be lower during migration due to the high costs of both migration and the immune system (Owen and Moore 2006). In mallards (*Anas platyrhynchos*) infected with avian influenza virus, for example, local migrants have higher natural antibody levels than distant migrants (van Dijk et al.

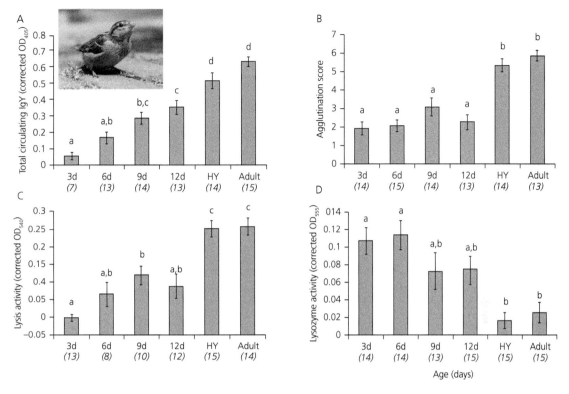

Figure 3.2 As in other taxa, immune function develops across the juvenile stage in wild birds. For example, in free-living house sparrows (*Passer domesticus*) in Wisconsin, USA, circulating IgY antibodies (A) increased throughout the nestling period (3, 6, 9, and 12 days post hatch) and continued to increase after fledging (i.e., hatch-year (HY) birds). Similarly, hemagglutination score (B) and complement-mediated hemolysis (C) activity increased to adult levels only after birds fledged. In contrast, lysozyme activity (D) decreased throughout the nestling period and tended to be lowest in fledgling and adult birds, which may indicate declining levels of maternal lysozyme. Letters indicate groups that are not significantly different (based on Tukey's post hoc tests) and sample sizes are shown in parentheses. (Photo credit: iStock.com/Agami stock. Figure first appeared in Killpack et al. 2013. Reprinted with permission from John Wiley & Sons, Inc.)

2015; Figure 3.3A). Migratory common blackbirds (*Turdus merula*) also show some reductions in innate immune function during the migratory season compared with resident blackbirds (Eikenaar and Hegemann 2016; Figure 3.3B–D). In contrast, skylarks (*Alauda arvensis*) with higher levels of an acute phase protein, haptoglobin, were more likely to migrate from breeding territories to separate wintering grounds during the winter months. This pattern could suggest that either high quality individuals (with strong immune responses) are more likely to migrate or migrating individuals temporarily invest more in immune responses, potentially to fight parasites encountered during migration (Hegemann et al. 2015). Indeed, migratory birds may host a wider diversity of parasites (Leung and

Koprivnikar 2016), potentially due to the immuno-compromising nature of migration or to the exposure to a greater diversity of environments.

Winter is also a challenging time for many bird species, as decreased day length shortens foraging times and decreased temperatures increase energy needs. Thus, wintering conditions may have large effects on avian immune responses and parasite infections, during both the overwintering period and subsequent migratory and breeding seasons (see also Chapter 8). Although winter may pose several challenges, the winter immunoenhancement hypothesis suggests that increased levels of an immunity-enhancing hormone (i.e., melatonin) may mitigate the negative effects of winter on immune function (Nelson and Demas 1996). However, this

Figure 3.3 Birds may have lower immune responses during migration, potentially due to energetic trade-offs. (A) Levels of natural antibodies were higher in short-distance as opposed to long-distance migratory mallards (*Anas platyrhynchos*; left photo) infected with avian influenza virus, although there were no differences between resident and distant migratory individuals (Eikenaar and Hegemann 2016). (B–D) Migratory common blackbirds (*Turdus merula*; right photo) have lower bacterial killing activity (BKA) and haptoglobin-like activity than resident blackbirds, although levels of total immunoglobulins did not differ between these two groups (van Dijk et al. 2015). (Photo credits: mallard, iStock.com/steveo73; blackbird, Wikimedia Commons, photo by Andreas Trepte, reprinted under the Creative Commons Attribution-Share Alike 2.5 Generic license.)

hypothesis is not well supported in birds (Buehler et al. 2009b; O'Neal 2013). In red knots (*Calidris canutus*), for example, relative shifts in immune function during winter vary, depending on the component of the immune system measured (Buehler et al. 2008). Interestingly, temperature did not have a large effect on immune function in this study. Indeed, captive studies suggest that cold temperatures do not necessarily inhibit immune function but that the costs of cold stress and immune challenges are additive (Hawley et al. 2012; King and Swanson 2013). Although this work has begun unraveling the effects of winter conditions on immune function, relatively few studies have measured the immune function of free-living birds during winter months. As such, much work is still needed to uncover how the immune system responds to the combined facets of winter life (e.g., temperature, day length, food availability, changes to social structure, absence of reproductive pressures).

Although current life history stages play a role in immune response, some immune responses do not appear to vary during the avian life cycle. For example, skylarks maintain a consistent acute phase response to LPS injections throughout breeding, migration, and overwintering stages (Hegemann et al. 2013). As the authors suggest in this paper, this may not be surprising as the fitness consequences of limiting certain immune responses may outweigh the potential benefits. A species' overall life history (e.g., fast vs. slow living species) may also impact the strategies used by avian hosts faced with immune challenges. For example, during the breeding season, short-lived species may invest more in current reproduction, at the cost of adequate immune responses and self-maintenance. In contrast, long-lived species may decrease current investment in reproduction in favor of survival and future reproductive endeavors (Pap et al. 2015).

3.3.4 Environmental stressors and physiological stress responses

Like other animals, birds must respond to constant environmental and physiological stressors. Here we discuss how a few environmental (e.g., food availability and anthropogenic factors) and physiological (e.g., free radicals) stressors interact with immune responses. We also discuss a well-studied physiological mechanism that animals use to cope with stressors (the release of glucocorticoid hormones) and how this stress response is related to the immune system.

3.3.4.1 Environmental stressors: food availability

Both the quantity and quality of available food may affect the ability of individuals to mount an effective immune response. General food restriction can inhibit both innate (e.g., house sparrow) (Killpack et al. 2015) and cell-mediated (e.g., yellow-legged gull, *Larus cachinnans*) (Alonso-Alvarez and Tella 2001) immune responses in birds. The restriction of particular nutrients (e.g., proteins), even under otherwise normal nutritional conditions, may also inhibit immune responses (Gonzalez et al. 1999). However, when food resources are limited, individuals may still be able to mount strong immune responses if they are able to suppress food-limiting sickness behaviors (e.g., anorexia and lethargy) to ensure adequate nutritional intake, as demonstrated in red knots (Buehler et al. 2009a). Thus, as with other drivers of variation, food availability and nutritional status may have a complex relationship with the strength of immune responses.

3.3.4.2 Environmental stressors: anthropogenic factors

Anthropogenic impacts on the environment and climate can potentially affect avian immune systems in several ways (see also Chapters 9 and 10). Changes in climate can cause changes in plant and insect phenology, resulting in a mismatch between avian breeding seasons and the peak availability of food (Sanz et al. 2003), which may inhibit immune responses in nestling birds (see Section 3.3.4.1). Population declines driven by habitat fragmentation and loss can reduce immune gene diversity and immunocompetence of avian populations (Bateson et al. 2016). Pollution, especially heavy metal pollution, can also have an immunosuppressive effect in birds (Hawley et al. 2009; Snoeijs et al. 2004). For example, levels of lead in mallards are negatively related to responses to PHA injection, although the opposite is true of antibody production (Vallverdú-Coll

et al. 2015). Finally, urbanization could have strong effects on immune responses of wild birds (see Chapter 9). For example, because levels of potentially immunosuppressive pollutants (e.g., lead) can be higher in urban environments, urbanization may inhibit some immune responses (Bichet et al. 2013). However, there is also evidence that birds in more urbanized areas have stronger DTH responses (e.g., Barbados bullfinches, *Loxigilla barbadensis*) (Audet et al. 2016). Thus, the effects of urbanization on avian immune systems, as with many other anthropogenic factors, are not clear at this time and warrant further study.

3.3.4.3 Physiological stressors: oxidative stress

Oxidative stress results when harmful oxidants are not fully mitigated by protective antioxidants and can result in oxidative damage to cells and tissues (Monaghan et al. 2009). Oxidative stress and damage are known results of parasite infection (Asghar et al. 2015; Delhaye et al. 2016) and immune responses (Torres and Velando 2007; van de Crommenacker et al. 2010). However, how an individual's current level of oxidative stress or their ability to resist oxidative stress affects immune responses has not been widely tested in wild birds. One study of white-browed sparrow weavers (*Plocepasser mahali*) demonstrated that individuals with stronger baseline antioxidant protection (i.e., superoxide dismutase activity) had weaker DTH responses than individuals with lower baseline antioxidant protection. However, because antioxidant protection may be elevated in response to oxidative stress, it is unclear if these individuals showed reduced immune responses because they were in a state of oxidative stress or because superoxide dismutase activity itself can inhibit inflammatory responses (Cram et al. 2015). Further studies are necessary to determine if oxidative status prior to immune activation enhances or inhibits immune responses and how this affects the outcomes of infections.

3.3.4.4 Physiological stressors: endocrine-mediated stress response

Animals, including birds, cope with stressors (e.g., predators and inclement weather) using a set of physiological and behavioral modifications known as the stress response (Romero 2004). Stress responses are partially mediated by corticosterone (CORT), the major glucocorticoid hormone in birds. Increases in CORT help individuals cope with stressors by releasing stored energy and modulating behaviors (Sapolsky et al. 2000). Thus, CORT is commonly used by avian researchers to quantify an individual's current level of stress. However, the relationship between CORT levels and the strength of immune responses in wild birds is not fully understood. Temporary (acute) or low-level increases in CORT may enhance innate immunity in some species (Vagasi et al. 2018) but inhibit it in others (Chin et al. 2013; Merrill et al. 2012). Long-term (e.g., chronic) increases in CORT can also decrease cell-mediated (e.g., DTH response and lymphocyte counts) and antibody responses, but this may not be the case across all sexes, populations, or species (Martin et al. 2005; Nazar and Marin 2011). Furthermore, increases in CORT can have long-term effects on immune responses, which may be difficult to take into consideration in many studies of wild birds. For example, juvenile song sparrows (*Melospiza melodia*) treated with simulated chronic stress, experimentally elevated CORT, showed less-pronounced DTH responses than control birds and for months after treatment. However, this long-term effect of CORT was sex-dependent and present only for certain components of the immune system (Schmidt et al. 2015). Together, these studies demonstrate the complexity of the relationship between CORT levels and immune responses. Indeed, some studies have begun to recognize CORT as an immunomodulatory hormone that can simultaneously inhibit some aspects of the immune system, while enhancing others (Martin 2009; Sapolsky et al. 2000). Given this complexity, taking an individual's stress 'status' into account should help reveal how environmental and physiological stressors alter immune phenotypes and responses to infection.

3.3.4.5 Relationships among individual and environmental factors and immune responses

While researchers have made much headway into understanding how various factors affect immune responses, the results of these studies can often be difficult to interpret for a variety of reasons. Trade-offs between the immune system and other functions

may be masked by other factors, such as behavioral changes of immune-challenged individuals (Nord et al. 2014) or their mates (Vermeulen et al. 2016). Furthermore, some relationships may only be apparent under challenging environmental conditions (Brzęk and Konarzewski 2007). The effects of some individual or environmental factors may also change, or even reverse, the relationship between immune responses and other organismal traits (Nettle et al. 2016; Figure 3.4). Future studies should take care to control individual and environmental variation and examine the relationships between these factors and immune responses under a variety of environmental conditions (e.g., across years or under different experimental conditions) when possible. Such a thorough understanding of the effects of individual traits and environmental factors on immune response will also be crucial for understanding parasite transmission and, thus, predicting and preventing the spread of parasites (see Sections 3.5 and 3.6 for further discussion).

3.4 Quantifying variation in immune defenses

The broad diversity of immune defenses, coupled with the myriad ways in which individual and environmental traits can alter these defenses, highlight two considerable challenges for ecoimmunologists: which defenses to measure and how to do so (Boughton et al. 2011; Brock et al. 2014; Demas et al. 2011). In general, the current ecoimmunology toolkit includes a variety of endpoints to measure, ranging from whole-organism responses (e.g., fever, sickness behaviors) to specific host molecules (e.g., antimicrobial peptides, antibodies) or nucleic acids related to those molecules (e.g., mRNA). In the interest of space, we provide only a brief overview of several such techniques in tabular form (Table 3.1). For both a broader survey and more in-depth discussions of specific techniques, we recommend several reviews (e.g., Adelman et al. 2014b; Boughton et al. 2011; Demas et al. 2011; Graham et al. 2011; Martin et al. 2006; Zylberberg 2019).

While numerous studies have successfully used the techniques outlined in Table 3.1 to reveal eco-

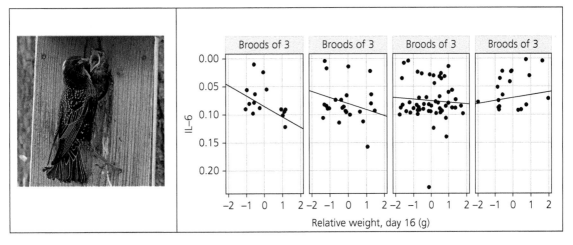

Figure 3.4 Relationship between relative nestling weight and development of a nestling's immune system on day 16 after hatching, measured by plasma levels of interleukin-6 (IL-6), is regulated by natural brood size in free-living European starlings (*Sturnus vulgaris*). In smaller broods, where nestling competition is lower, heavier nestlings have reduced levels of IL-6 when compared to lighter nestlings, suggesting a trade-off between growth and immune system development. However, in larger broods (where nestling competition is stronger), heavier nestlings tended to have higher levels of IL-6 (although this relationship was not statistically significant), suggesting there is no trade-off between growth and immune system development. Relative nestling weight was determined by the following formula: (nestling weight − mean weight of nestlings in the brood) / SD of weights in the brood. Plasma levels of IL-6 were reciprocally transformed and then multiplied by −1, so higher numbers indicate higher levels of IL-6. (Photo credit: iStock.com/Toby Photos. Figure from Nettle et al. 2016 [open access].)

logical drivers of variation in avian immune defenses, the diversity of immune assays available for wild birds still lags behind that available for domesticated species (Boughton et al. 2011; Brock et al. 2014; Demas et al. 2011; Schat et al. 2014). However, next-generation sequencing and '-omics' technologies present one promising path forward. Pairing such approaches with data on functional protection, for example during experimental infection, will likely identify novel genes and proteins that predict differences in resistance or tolerance among individuals, populations, or species. As these tools become increasingly affordable, ecoimmunology may be poised for a 'leapfrog' moment, wherein we jump to measuring not only immune-related genes but also the proteins that carry out actual immune defense and signaling—all without the need for species-specific, single-molecule assays. As we discuss in further detail in Section 3.5, these new approaches provide excellent potential to address a major question that still looms over the field of ecoimmunology: which metrics of immune variation are most salient to functional protection

and overall host fitness (Adamo 2004; Brock et al. 2014; Graham et al. 2011; Viney et al. 2005).

3.5 Bridging the first gap: do immune phenotypes predict infection status or host fitness?

Understanding links among immune phenotypes, infection status, and fitness represents an important gap at the intersection of ecoimmunology and disease ecology. Although ecoimmunology seeks to understand both the causes of natural variation in immune system function and the consequences of that variation, it can be difficult to know what individual immune responses mean for overall host fitness. One important step in revealing these fitness consequences is to link immune measures directly to parasitic infections (Graham et al. 2011; Horrocks et al. 2011a). Such links can help determine whether a given immune response predicts functional protection in the form of resistance (e.g., a higher magnitude response leads to lower parasite loads).

Table 3.1 Commonly used techniques in avian ecoimmunology that quantify immune variation among individuals and species, though none is without its drawbacks—see some of the major drawbacks for each approach below; for consideration of the broader drawbacks that apply to all ecoimmunology techniques see Section 3.5 and several past reviews (Adamo 2004; Brock et al. 2014; Graham et al. 2011; Viney et al. 2005).

Measurement	Type of defense	Benefits	Cautions	Reviews and/or examples
Whole-organism or whole-tissue approaches				
Delayed-type hyper-sensitivity (e.g., via phytohemmaglutinin [PHA] injection)	Innate/adaptive, induced	Relatively easy and cheap; metric integrates a range of signaling and cellular responses.	Challenging to inject and measure accurately; does not measure general T-cell activity, as was often assumed in early ecoimmunology papers.	Martin et al. 2006
Fever and sickness behaviors (e.g., lipopolysaccharide [LPS] injection)	Innate, induced	Equipment (e.g., thermocouples, video recording, GPS, radio telemetry) is not species-specific; a wide range of parasites induce these responses.	Requires multiple captures; cost can be high depending on technology used; often constrained to captivity; temperature can change rapidly in response to capture stress.	Adelman et al. 2014b
Leukocyte counts	Innate/adaptive, induced/constitutive	Relatively easy and cheap; possible with small, single blood samples; easily coupled with surveillance for blood parasites.	Time-intensive visual inspection of blood smears required for most species; cell morphology can vary across species.	Campbell 1995; Owen 2011
Bacterial killing (whole blood or plasma)	Innate, constitutive	Relatively easy and cheap; integrates multiple defenses in one functional readout.	Whole blood must be brought to lab rapidly; plasma-only assay integrates fewer defenses; assays may not be repeatable within individuals. Unknown how well killing by blood translates to defense within other tissues.	Liebl and Martin 2009; Millet et al. 2007
Acute phase proteins (e.g., haptoglobin) in blood or plasma	Innate, induced	Easy to measure alongside other immune functions, variable across individuals.	Relationships between some proteins (e.g., haptoglobin) and other immune functions may change between seasons and only be repeatable over the short term; assay kits can be costly.	Hegemann et al. 2015; Matson et al. 2012
Cellular approaches				
T-cell proliferation assays (e.g., induced via PHA and concanavalin A)	Innate/adaptive, induced	Can assess multiple cell-types' responses to multiple stimuli.	In vitro, requires some specialized equipment, potential rapid transport to lab.	Palacios et al. 2007
Phagocytosis	Innate, induced	Clear metric of a widespread, common defense in near real-time.	In vitro, requires rapid transport to lab.	Millet et al. 2007
Specific effector or signaling molecule approaches				
Natural antibody titers	Innate, constitutive	Relatively cheap and easy; performed on single blood samples.	Accurately interpreting the readout requires considerable practice.	Matson et al. 2005

Measure	Immune classification	Advantages	Limitations	References
Complement activity	Innate, constitutive	Relatively cheap and easy; performed on single blood samples.	Multiple pathways can activate complement; this only measures one.	Greives et al. 2006; Matson et al. 2005
Total immunoglobulin (Ig) levels	Innate/adaptive, constitutive/ induced	Highly integrative—measures all Ig, not just parasite-specific or induced; performed on single blood samples.	Detection from different families or species may differ because of reagent reactivity, rather than Ig differences.	Fassbinder-Orth et al. 2016
Antigen- or parasite-specific antibody titers (e.g., antibody response to sheep red blood cell [SRBC] challenge; often measured via enzyme-linked immunosorbent assays)	Specific, induced	Relevance to a particular parasite is often clearer than in other assays.	Requires antigen- or parasite-specific reagents, which can be expensive or difficult to obtain; antibody type (e.g., IgA vs. IgY) may not be most salient defense against parasite or strain of interest.	Casagrande et al. 2012; Deerenberg et al. 1997; Grodio et al. 2012
Oxidative burst	Innate, induced	A defense that is broadly effective against parasites but has potential for self-harm; commercially available kits.	Best coupled with metrics of oxidative damage to assess costs to hosts; may not reflect intracellular free radicals.	Sild and Horak 2010
Antimicrobial peptides	Innate, constitutive/ induced	Relatively straightforward assays; several commercially available kits have been adapted for use in wild birds.	Peptide-specific reagents required; numerous peptides exist, so choosing one or two risks an incomplete picture; variation is extensive and can be difficult to interpret.	Horrocks et al. 2011b; Matson et al. 2012; Millet et al. 2007
Polymerase chain reaction (PCR)/traditional sequencing approaches				
Cytokine, pattern recognition receptor, and other mRNA expression	Innate/adaptive, constitutive/ induced	Isolating mRNA and developing necessary primers for quantitative PCR has become easier and affordable; can quantify expression of myriad genes in non-model species.	Some tissues (e.g., blood) can yield low or inconsistent expression; choice of 'housekeeping' genes can be difficult; some questions require terminal experiments; non-trivial costs, especially for equipment.	Fassbinder-Orth 2014; Martin et al. 2014; Vinkler et al. 2018
Immune gene diversity; allele identity	Innate/adaptive, constitutive/ induced	Can examine how allele number and ID relate to disease outcomes; can be performed on relatively small tissue samples.	Requires a lab with standard molecular equipment; some primer sets will be too specific for cross-species work.	Hawley and Fleischer 2012
Next-generation sequencing/"-omics" approaches (see Zylberberg 2019)				
Immune gene diversity	Innate/adaptive, constitutive/ induced	With major sequencing projects underway (e.g., B10K project, Avian Phylogenomics Consortium), sequence for myriad immune genes and homologs is readily available.	Assembly and annotation of genomes is lagging behind sequencing, meaning some diversity may be missed when using such datasets.	Eöry et al. 2015; Velova et al. 2018; Zhang et al. 2014

(Continued)

Table 3.1 Continued

Measurement	Type of defense	Benefits	Cautions	Reviews and/or examples
RNA-seq (gene expression)	Innate/adaptive, constitutive/induced	Can generate entire transcriptome of expressed genes in non-model species; can do so in different states of infection and/or under different environmental conditions.	For species without a sequenced genome, de novo transcriptome assembly is possible, but results vary with techniques used; still costly (though becoming more affordable).	Fassbinder-Orth 2014; Zylberberg 2019
Metagenomics (microbiota links with functional/immune defense)	Innate/adaptive, constitutive/induced	Assessing disruptions to microbial symbionts under various conditions can predict how animals respond to parasites.	Costs are still high, though coming down; expertise in bioinformatics required; some sequencing platforms have had trouble with bleed between samples.	Knutie 2018; Thomason et al. 2017
Proteomics	Innate/adaptive, constitutive/induced	Techniques like tandem mass spectrometry (TMS) can create vast datasets of peptide concentrations, allowing us to move beyond gene expression to ask questions about actual effector proteins.	Equipment requires massive capital and specific expertise; as with genomic or transcriptomic data, the proteome requires assembly and annotation; no studies to date in wild birds.	TMS generally: Castellana et al. 2008. Chicken proteomics: Haq et al. 2010

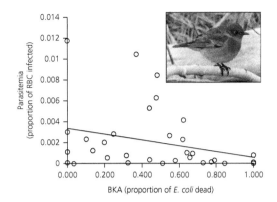

Figure 3.5 Among western bluebirds (*Sialia mexicana*) infected with blood parasites (*Haemoproteus* and *Plasmodium* lineages), individuals with higher levels of plasma bacterial killing ability (BKA) showed a lower proportion of infected red blood cells (RBC) than those with lower bacterial killing ability. (Photo credit: bluebird, Creative Commons). Figure from Jacobs et al. 2015. Reprinted by permission from Springer Nature, (Behavioral Ecology and Sociobiology) by Jacobs, A.C. et al. © 2015.)

However, simple negative correlations between immune responses and parasite load are unlikely to reveal all relevant links to host fitness (Graham et al. 2011). Specifically, because immune responses are physiologically costly and often carry the risk of self-harm, or immunopathology, overall host fitness may actually be highest at intermediate levels of immune defense (Graham et al. 2005; Viney et al. 2005). For example, Råberg and Stjernman (2003) found that blue tits (*Cyanistes caeruleus*) that mounted intermediate antibody responses against a diphtheria vaccine were more likely to survive over winter. In that study, however, birds mounted responses against novel antigens, not against parasites they were likely to encounter (and would need to resist) in the wild. So, while this work illustrates that intermediate responses can correlate with fitness, a full exploration of links among immune system variation, functional defense, and host fitness would require measuring resistance to (and tolerance against) relevant parasites as well (Graham et al. 2011).

Many studies have tested the relationships between immune measures and natural parasitic infections in the wild (Biard et al. 2015; Chakarov et al. 2017; Jacobs et al. 2015; Vaziri et al. 2019; Figure 3.5). This approach can be highly relevant to the

potential for parasite transmission (see Section 3.6) and is often more feasible than experimental manipulations, especially when hosts are difficult to obtain or are protected or when a community of avian hosts, rather than a single host species, is of interest. For example, higher levels of natural antibodies were associated with a lower ectoparasite load in Galápagos hawks (*Buteo galapagoensis*) (Whiteman et al. 2006). Measures of immunity may also be useful for predicting offspring parasite load if parasites are vertically transmitted. For example, female birds with higher levels of natural antibodies transmit a lower bacterial parasite load onto eggs than females with lower natural antibody titers (Soler et al. 2011). While such studies can reveal important links between immune variation, parasite resistance, and the potential for parasite transmission, they cannot fully predict how these links translate into fitness.

Moreover, when studies find no or contradictory relationships between immune measures and parasite load, several interpretations are possible (Kreisinger et al. 2018; Owen et al. 2013). These results may reflect the lack of a relationship between a specific immune trait and functional defense against a given parasite. As not all components of the immune system will respond to (or offer protection from) a given type of parasite or a given stage of infection, it is important to carefully consider which immune system measures may be most relevant. True relationships between immune responses and parasite load may also be masked in the wild. This may be partially due to common temporal lags from peak parasite load to peak immune response. This is likely the case when measuring aspects of inducible immunity, particularly those like antibody responses that take some time to activate. For example, antibody levels peak during West Nile virus infection after parasite loads have already begun to decrease (VanDalen et al. 2013). Furthermore, parasites outside of focal infections likely influence immune responses (e.g., Vaziri et al. 2019) and it is not always possible to control for these confounding effects on the immune system. In addition, because antibody titers wane over time and at different rates for different infections (Härtle et al. 2014), accurately assessing a host's previous exposure to a given parasite is challenging. Overall, revealing correlations between immune responses and parasite burdens can be very difficult in wild birds, particularly when collecting samples at a single time-point and/or when the precise timing of infection is unknown (but see Pepin et al. 2017 for a creative way of utilizing such time lags in population surveillance).

Immune responses and parasite load may also be uncoupled when hosts respond to infection with tolerance, rather than resistance. As tolerance mechanisms can decrease fitness costs of infection without directly decreasing parasite load, high parasite loads do not necessarily reveal a failure of the immune system to protect host fitness. Tolerance mechanisms may be particularly relevant for systems in which much of the cost of infection stems from immunopathology, rather than direct damage from parasites themselves. For instance, in house finches (*Haemorhous mexicanus*), infections with the emerging bacterial parasite *Mycoplasma gallisepticum* result in a suite of inflammatory responses, which carry high risks of immunopathology (Adelman et al. 2013; Graham et al. 2005; Vinkler et al. 2018). In this system, birds from a population naïve to *M. gallisepticum* showed more pro-inflammatory signaling and greater pathology than did birds from a population where *M. gallisepticum* has been endemic, despite similar parasite loads in birds from both populations (Adelman et al. 2013; Bonneaud et al. 2019). This suggests that tolerance mechanisms, which may reduce certain immune responses like inflammation, could occlude correlations between immune responses and parasite load.

Studies of tolerance also highlight that measuring both immune responses and parasite loads may not give a full picture of the fitness costs of infection (Graham et al. 2011). Although directly measuring fitness can be challenging, incorporating one or more proxies of fitness can improve our interpretation of such correlations between immune responses and parasite loads. Recently, Burgan and colleagues (2018) applied this approach in house sparrows (*Passer domesticus*) experimentally infected with West Nile virus. That study reported parasite load, immune response, and fitness-related measures (e.g., body mass, breast muscle scores, activity level, and flight performance) from all individuals. Results suggested that single immune measures (i.e., pro- or anti-inflammatory cytokine

levels) could predict both tolerance and resistance (in one case, in opposite directions). Experimental infection studies like this one, which incorporate diverse immune metrics and fitness proxies, may thus improve our interpretation of the relationships between immune responses and parasite load in wild birds.

3.6 Bridging the second gap: do within-host measures inform population-level transmission dynamics?

Long before the origins of ecoimmunology, Bartholomew (1966) wrote that every biological 'level finds its explanations of mechanism in the levels below, and its significances in the levels above it.' Applied to ecoimmunology, this suggests that without inquiry across levels—for example, from molecule to organism or from organism to population—we will likely miss the causes and consequences of variation in immune responses. While technological advances like RNA-seq will help unravel the mechanistic causes of immune variation, revealing the consequences of this variation for parasite spread remains one of the biggest knowledge gaps in ecoimmunology. Uncovering these organism-to-population links is critical if ecoimmunology is to make the most meaningful possible contributions to avian disease ecology.

To address this knowledge gap, ecoimmunologists have begun to incorporate the idea of (vertebrate) host competence, or the propensity of one host to cause infections in others. This term, borrowed from disease ecology and medical entomology, has underpinnings that date back at least to MacDonald's theoretical work on malaria vectors (e.g., Komar et al. 1999; Lord et al. 2006; Macdonald 1952; Martin et al. 2016; VanderWaal and Ezenwa 2016). The concept is also closely related to the individual reproductive number, V, defined as the number of new infections caused by a given infected host (Lloyd-Smith et al. 2005). VanderWaal and Ezenwa (2016) used a simple equation to highlight how variation in aspects of individual infections contributes to variation in V and thus host competence. Briefly, they defined V as:

$$V = (\beta_c \cdot \beta_p) \cdot infectious\,period \qquad (3.1)$$

where $(\beta_c \cdot \beta_p)$ represents the overall parasite transmission rate (often written as β) as the product of two components: the contact rate (β_c, or how often an infected host contacts naïve hosts or vectors) and transmissibility (β_p, or how likely an new infection occurs given contact) (Dobson 1995). In theory, immune or behavioral responses that alter any component of Equation 3.1 could impact parasite spread at the population level (see also Chapter 7). Empirically identifying such responses and linking them directly to V, however, remains a significant challenge.

To date, ecoimmunologists have approached this challenge using two main types of experiments: those that identify proxies for infectiousness and those that directly test transmission to naïve individuals. In the former category, Burgan and colleagues (2018) used an estimate of the infectious load of West Nile virus (Komar et al. 2003) to approximate both transmissibility and the duration of the infectious period in experimentally infected house sparrows. Moreover, they showed that both of these metrics were lower for individuals that had higher expression of the interferon-γ gene (a proinflammatory cytokine that upregulates myriad early immune defenses). In addition to measuring such proxies, experimental studies can directly relate individual host differences to successful parasite transmission. For example, experimentally infected house finches that spent more time on bird feeders, likely fomites for some avian parasites (Dhondt et al. 2007; Wilcoxen et al. 2015), transmitted *M. gallisepticum* to flockmates more rapidly than infected birds that spent less time on feeders (Adelman et al. 2015). In addition to directly assessing overall transmission (i.e., $\beta_c \cdot \beta_p$ from Equation 3.1), this study showed that both high- and low-feeding birds had similar parasite loads and outward pathology. This suggests that variation in responses that impact contact rate β_c alone, such as sickness behaviors, can have direct consequences for transmission dynamics in this avian disease system.

Although focusing on host competence has yielded important insights into how individual immune responses may help predict transmission, it ignores a critical component of the transmission process: the susceptibility of uninfected individuals. In discussing Equation 3.1, we have thus far focused on the infected individual's contribution to β, but naïve hosts also influence transmissibility.

For example, behavior toward infected individuals can have marked effects on β_c (see Chapter 6 for details). In addition, naïve hosts' abilities to resist infection will affect transmissibility via β_p. For example, Tolf and colleagues (2013) showed that when mallards were housed in a common, captive environment with frequent exposure to wild, free-living conspecifics, they showed marked temporal and interindividual variation in rates of infection with avian influenza viruses. Although their study was small (10 birds), it highlighted at least two important patterns. First, as individuals aged, they were less likely to become infected (likely due to prior exposure and immune memory). Second, both antibody production and viral prevalence varied across the annual cycle in these birds, with individuals showing active infection shortly after their lowest antibody titers. Revealing what such variation truly means for infection dynamics will require further experimental transmission studies. In particular, future studies will need to quantify or manipulate immune variation among susceptible individuals, while minimizing such variation among initially infected individuals.

Pepin and colleagues (2017) devised a different, though related, application that links individual immune responses with population-level epidemics. Using surveillance data on individual antibody titers from a population of snow geese, they estimated temporal fluctuations in population-level force of infection (the rate at which susceptible animals become infected). While their results highlighted a highly practical application of quantifying variation in immune responses, they also revealed important impacts of individual-level variation in making such predictions. Specifically, with substantial interindividual variation in antibody decay rates over time, their models tended to overestimate the magnitude of the force of infection and miss its temporal peak. Such results highlight the need to further characterize interindividual variation in immune responses generally—and particularly as such variation pertains to parasite transmission dynamics.

3.7 Conclusions

Although the field of ecoimmunology emerged largely independently of disease ecology, recent advances in techniques and growing interest in linking these two disciplines leave us poised to make important contributions to understanding how avian parasites spread and evolve in wild populations. In this chapter, we have identified two major areas in which ecoimmunology will help drive avian disease ecology forward in the coming years: (1) linking immune defenses directly to individual fitness and parasite burdens; and (2) revealing how differences in those individual immune phenotypes impact parasite transmission. In particular, we argue that advances in next-generation sequencing techniques and proteomics may enable a period of rapid discovery in avian ecoimmunology. These advances will have ramifications not only for disease ecology but also for One Health pursuits, like uncovering immune mechanisms that allow certain wildlife species to serve as effective (or ineffective) reservoirs of zoonotic disease (e.g., Smith et al. 2015; see also Chapter 12). Using such next-generation tools, ecoimmunologists could potentially evaluate genetic, transcriptomic, or even proteomic predictors of tolerance, resistance, and competence across species in both controlled experiments and field surveillance. More generally, these new tools will enable broader basic research in non-model avian species that effectively tests how and why immune phenotypes vary in the wild and how this variation predicts individual disease outcomes, fitness, and the transmission of parasites across hosts.

Literature cited

Adamo, S.A. (2004). How should behavioural ecologists interpret measurements of immunity? *Animal Behaviour*, 68, 1443–9.

Adelman, J.S. (2015). Immune systems: linking organisms, populations, and evolution through disease. In L.B. Martin, C.K. Ghalambor, and H.A. Woods, eds. *Integrative Organismal Biology*, pp. 169–85. Wiley-Blackwell, Hoboken.

Adelman, J.S. and Hawley, D.M. (2017). Tolerance of infection: a role for animal behavior, potential immune mechanisms, and consequences for parasite transmission. *Hormones and Behavior*, 88, 79–86.

Adelman, J.S. and Martin, L.B. (2009). Vertebrate sickness behaviors: adaptive and integrated neuroendocrine immune responses. *Integrative and Comparative Biology*, 49, 202–14.

Adelman, J.S., Kirkpatrick, L., Grodio, J.L., and Hawley, D.M. (2013). House finch populations differ in early inflammatory signaling and pathogen tolerance at the peak of *Mycoplasma gallisepticum* infection. *American Naturalist*, 181, 674–89.

Adelman, J.S., Ardia, D.R., and Schat, K.A. (2014a). Ecoimmunology. In K.A. Schat, B. Kaspers and P. Kaiser, eds. *Avian Immunology*, 2nd ed., pp. 391–411. Elsevier, Amsterdam.

Adelman, J.S., Moyers, S.C., and Hawley, D.M. (2014b). Using remote biomonitoring to understand heterogeneity in immune-responses and disease-dynamics in small, free-living animals. *Integrative and Comparative Biology*, 54, 377–86.

Adelman, J.S., Moyers, S.C., Farine, D.R., and Hawley, D.M. (2015). Feeder use predicts both acquisition and transmission of a contagious pathogen in a North American songbird. *Proceedings of the Royal Society B: Biological Sciences*, 282, 20151429.

Alonso-Alvarez, C. and Tella, J.L. (2001). Effects of experimental food restriction and body-mass changes on the avian T-cell-mediated immune response. *Canadian Journal of Zoology*, 79, 101–5.

Asghar, M., Hasselquist, D., Hansson, B., Zehtindjiev, P., Westerdahl, H., and Bensch, S. (2015). Hidden costs of infection: chronic malaria accelerates telomere degradation and senescence in wild birds. *Science*, 347, 436–8.

Audet, J.-N., Ducatez, S., and Lefebvre, L. (2016). The town bird and the country bird: problem solving and immunocompetence vary with urbanization. *Behavioral Ecology*, 27, 637–44.

Bartholomew, G.A. (1966). Interaction of physiology and behaviour under natural conditions. In R.I. Bowman, ed. *The Galápagos: Proceedings of the Symposia of the Galápagos International Project*, pp. 39–45. University of California Press, Berkeley.

Bateson, Z.W., Hammerly, S.C., Johnson, J.A., Morrow, M.E., Whittingham, L.A., and Dunn, P.O. (2016). Specific alleles at immune genes, rather than genome-wide heterozygosity, are related to immunity and survival in the critically endangered Attwater's prairie-chicken. *Molecular Ecology*, 25, 4730–44.

Biard, C., Monceau, K., Motreuil, S., and Moreau, J. (2015). Interpreting immunological indices: the importance of taking parasite community into account. An example in blackbirds *Turdus merula*. *Methods in Ecology and Evolution*, 6, 960–72.

Bichet, C., Scheifler, R., Cœurdassier, M., Julliard, R., Sorci, G., and Loiseau, C. (2013). Urbanization, trace metal pollution, and malaria prevalence in the house sparrow. *PLoS ONE*, 8, e53866.

Bonneaud, C., Tardy, L., Giraudeau, M., Hill, G.E., McGraw, K.J., and Wilson, A.J. (2019). Evolution of both host resistance and tolerance to an emerging bacterial pathogen. *Evolution Letters*, 3, 544–54.

Boughton, R.K., Joop, G., and Armitage, S.A.O. (2011). Outdoor immunology: methodological considerations for ecologists. *Functional Ecology*, 25, 81–100.

Bowers, E.K., Smith, R.A., Hodges, C.J., Zimmerman, L.M., Thompson, C.F., and Sakaluk, S.K. (2012). Sex-biased terminal investment in offspring induced by maternal immune challenge in the house wren (*Troglodytes aedon*). *Proceedings of the Royal Society B: Biological Sciences*, 279, 2891–8.

Brace, A.J., Lajeunesse, M.J., Ardia, D.R., et al. (2017). Costs of immune responses are related to host body size and lifespan. *Journal of Experimental Zoology Part A: Ecological and Integrative Physiology*, 327, 254–61.

Brock, P.M., Murdock, C.C., and Martin, L.B. (2014). The history of ecoimmunology and its integration with disease ecology. *Integrative and Comparative Biology*, 54, 353–62.

Brzęk, P. and Konarzewski, M. (2007). Relationship between avian growth rate and immune response depends on food availability. *Journal of Experimental Biology*, 210, 2361.

Buehler, D.M., Piersma, T., Matson, K.D., and Tieleman, B.I. (2008). Seasonal redistribution of immune function in a migrant shorebird: annual-cycle effects override adjustments to thermal regime. *The American Naturalist*, 172, 783–96.

Buehler, D. M., Encinas-Viso, F., Petit, M., Vézina, F., Tieleman, B.I., and Piersma, T. (2009a). Limited access to food and physiological trade-offs in a long-distance migrant shorebird, II: constitutive immune function and the acute-phase response. *Physiological and Biochemical Zoology*, 82, 561–71.

Buehler, D.M., Koolhaas, A., Van't Hof, T.J., et al. (2009b). No evidence for melatonin-linked immunoenhancement over the annual cycle of an avian species. *Journal of Comparative Physiology A—Neuroethology Sensory Neural and Behavioral Physiology*, 195, 445–51.

Burgan, S.C., Gervasi, S.S., and Martin, L.B. (2018). Parasite tolerance and host competence in avian host defense to West Nile virus. *EcoHealth*, 15, 360–71.

Burgan, S.C., Gervasi, S.S., Johnson, L.R., and Martin, L.B. (2019). How individual variation in host tolerance affects competence to transmit parasites. *Physiological and Biochemical Zoology*, 92, 49–57.

Calero-Riestra, M. and Garcia, J.T. (2016). Sex-dependent differences in avian malaria prevalence and consequences of infections on nestling growth and adult condition in the tawny pipit, *Anthus campestris*. *Malaria Journal*, 15, 178.

Campbell, T.W. (1995). *Avian Hematology and Cytology*. Blackwell Publishing, Ames, IA.

Carroll, M.C. and Prodeus, A.P. (1998). Linkages of innate and adaptive immunity. *Current Opinion in Immunology*, 10, 36–40.

Casagrande, S., Costantini, D., and Groothuis, T.G.G. (2012). Interaction between sexual steroids and immune response in affecting oxidative status of birds. *Comparative Biochemistry and Physiology A—Molecular & Integrative Physiology*, 163, 296–301.

Castellana, N.E., Payne, S.H., Shen, Z.X., Stanke, M., Bafna, V., and Briggs, S.P. (2008). Discovery and revision of *Arabidopsis* genes by proteogenomics. *Proceedings of the National Academy of Sciences of the United States of America*, 105, 21034–8.

Casto, J.M., Nolan, V., and Ketterson, E.D. (2001). Steroid hormones and immune function: experimental studies in wild and captive dark-eyed juncos (*Junco hyemalis*). *The American Naturalist*, 157, 408–20.

Chakarov, N., Pauli, M., and Krüger, O. (2017). Immune responses link parasite genetic diversity, prevalence and plumage morphs in common buzzards. *Evolutionary Ecology*, 31, 51–62.

Chin, E.H., Quinn, J.S., and Burness, G. (2013). Acute stress during ontogeny suppresses innate, but not acquired immunity in a semi-precocial bird (*Larus delawarensis*). *General and Comparative Endocrinology*, 193, 185–92.

Clutton-Brock, T.H. (1984). Reproductive effort and terminal investment in iteroparous animals. *The American Naturalist*, 123, 212–29.

Cohen, A.A., Martin, L.B., Wingfield, J.C., McWilliams, S.R., and Dunne, J.A. (2012). Physiological regulatory networks: ecological roles and evolutionary constraints. *Trends in Ecology & Evolution*, 27, 428–35.

Cooper, M.D., Peterson, R.D., and Good, R.A. (1965). Delineation of the thymic and bursal lymphoid systems in the chicken. *Nature*, 205, 143–6.

Cram, D.L., Blount, J.D., York, J.E., and Young, A.J. (2015). Immune response in a wild bird is predicted by oxidative status, but does not cause oxidative stress. *PLoS ONE*, 10, e0122421.

Cross, T.A., Arsnoe, D.M., Minnis, R.B., et al. (2013). Prevalence of avian paramyxovirus 1 and avian influenza in double-crested cormorants (*Phalacrocorax auritus*) in eastern North America. *Journal of Wildlife Diseases*, 49, 965–77.

De Coster, G., De Neve, L., Martín-Gálvez, D., Therry, L., and Lens, L. (2010). Variation in innate immunity in relation to ectoparasite load, age and season: a field experiment in great tits (*Parus major*). *Journal of Experimental Biology*, 213, 3012–18.

Deerenberg, C., Apanius, V., Daan, S., and Bos, N. (1997). Reproductive effort decreases antibody responsiveness. *Proceedings of the Royal Society B: Biological Sciences*, 264, 1021–9.

Delhaye, J., Jenkins, T., and Christe, P. (2016). *Plasmodium* infection and oxidative status in breeding great tits, *Parus major*. *Malaria Journal*, 15, 531.

Demas, G.E. and Carlton, E.D. (2015). Ecoimmunology for psychoneuroimmunologists: considering context in neuroendocrine–immune–behavior interactions. *Brain Behavior and Immunity*, 44, 9–16.

Demas, G.E., Zysling, D.A., Beechler, B.R., Muehlenbein, M.P., and French, S.S. (2011). Beyond phytohaemagglutinin: assessing vertebrate immune function across ecological contexts. *Journal of Animal Ecology*, 80, 710–30.

Dhondt, A.A., Dhondt, K.V., Hawley, D.M., and Jennelle, C.S. (2007). Experimental evidence for transmission of *Mycoplasma gallisepticum* in house finches by fomites. *Avian Pathology*, 36, 205–8.

Dobson, A. (1995). The ecology and epidemiology of Rinderpest virus in Serengeti and Ngorongoro Conservation Area. In A.R.E. Sinclair and P. Arcese, eds. *Serengeti II. Dynamics, management, and conservation of an ecosystem*, pp. 485–505. University of Chicago Press, Chicago.

Eikenaar, C. and Hegemann, A. (2016). Migratory common blackbirds have lower innate immune function during autumn migration than resident conspecifics. *Biology Letters*, 12, 20160078.

Eöry, L., Gilbert, M.T.P., Li, C., et al. (2015). Avianbase: a community resource for bird genomics. *Genome Biology*, 16.

Fassbinder-Orth, C.A. (2014). Methods for quantifying gene expression in ecoimmunology: from qPCR to RNA-Seq. *Integrative and Comparative Biology*, 54, 396–406.

Fassbinder-Orth, C.A., Wilcoxen, T.E., Tran, T., et al. (2016). Immunoglobulin detection in wild birds: effectiveness of three secondary anti-avian IgY antibodies in direct ELISAs in 41 avian species. *Methods in Ecology and Evolution*, 7, 1174–81.

Fenzl, L., Gobel, T.W., and Neulen, M.L. (2017). Gamma(delta) T cells represent a major spontaneously cytotoxic cell population in the chicken. *Developmental and Comparative Immunology*, 73, 175–83.

Folstad, I. and Karter, A.J. (1992). Parasites, bright males, and the immunocompetence handicap. *The American Naturalist*, 139, 603–22.

Foo, Y.Z., Nakagawa, S., Rhodes, G., and Simmons, L.W. (2017). The effects of sex hormones on immune function: a meta-analysis. *Biological Reviews*, 92, 551–71.

Glick, B., Chang, T.S., and Jaap, R.G. (1956). The bursa of Fabricius and antibody production. *Poultry Science*, 35, 224–5.

Gonzalez, G., Sorci, G., and de Lope, F. (1999). Seasonal variation in the relationship between cellular immune response and badge size in male house sparrows (*Passer domesticus*). *Behavioral Ecology and Sociobiology*, 46, 117–22.

Graham, A.L., Allen, J.E., and Read, A.F. (2005). Evolutionary causes and consequences of immunopathology. *Annual Review of Ecology, Evolution, and Systematics*, 36, 373–97.

Graham, A.L., Shuker, D.M., Pollitt, L.C., Auld, S., Wilson, A.J., and Little, T.J. (2011). Fitness consequences of immune responses: strengthening the empirical framework for ecoimmunology. *Functional Ecology*, 25, 5–17.

Greives, T.J., McGlothlin, J.W., Jawor, J.M., Demas, G.E., and Ketterson, E.D. (2006). Testosterone and innate immune function inversely covary in a wild population of breeding dark-eyed Juncos (*Junco hyemalis*). *Functional Ecology*, 20, 812–18.

Grodio, J.L., Hawley, D.M., Osnas, E.E., et al. (2012). Pathogenicity and immunogenicity of three *Mycoplasma gallisepticum* isolates in house finches (*Carpodacus mexicanus*). *Veterinary Microbiology*, 155, 53–61.

Haq, K., Brisbin, J.T., Thanthrige-Don, N., Heidari, M., and Sharif, S. (2010). Transcriptome and proteome profiling of host responses to Marek's disease virus in chickens. *Veterinary Immunology and Immunopathology*, 138, 292–302.

Harmon, B.G. (1998). Avian heterophils in inflammation and disease resistance. *Poultry Science*, 77, 972–7.

Hart, B.L. (1988). Biological basis of the behavior of sick animals. *Neuroscience and Biobehavioral Reviews*, 12, 123–37.

Härtle, S., Magor, K.E., Göbel, T.W., Davison, F., and Kaspers, B. (2014). Structure and evolution of avian immunoglobulins. In K.A. Schat, B. Kaspers, and P. Kaiser, eds. *Avian Immunology*, 2nd ed., pp. 103–20. Elsevier, Amsterdam.

Hasselquist, D. (2007). Comparative immunoecology in birds: hypotheses and tests. *Journal of Ornithology*, 148, 571–82.

Hawley, D.M. and Altizer, S.M. (2011). Disease ecology meets ecological immunology: understanding the links between organismal immunity and infection dynamics in natural populations. *Functional Ecology*, 25, 48–60.

Hawley, D.M. and Fleischer, R.C. (2012). Contrasting epidemic histories reveal pathogen-mediated balancing selection on class II MHC diversity in a wild songbird. *PLoS ONE*, 7, e30222.

Hawley, D.M., Hallinger, K.K., and Cristol, D.A. (2009). Compromised immune competence in free-living tree swallows exposed to mercury. *Ecotoxicology*, 18, 499–503.

Hawley, D.M., Etienne, R.S., Ezenwa, V.O., and Jolles, A.E. (2011). Does animal behavior underlie covariation between hosts' exposure to infectious agents and susceptibility to infection? Implications for disease dynamics. *Integrative and Comparative Biology*, 51, 528–39.

Hawley, D.M., DuRant, S.E., Wilson, A.F., Adelman, J.S., and Hopkins, W.A. (2012). Additive metabolic costs of thermoregulation and pathogen infection. *Functional Ecology*, 26, 701–10.

Hegemann, A., Matson, K.D., Versteegh, M.A., Villegas, A., and Tieleman, B.I. (2013). Immune response to an endotoxin challenge involves multiple immune parameters and is consistent among the annual-cycle stages of a free-living temperate zone bird. *Journal of Experimental Biology*, 216, 2573–80.

Hegemann, A., Marra, P.P., and Tieleman, B.I. (2015). Causes and consequences of partial migration in a passerine bird. *The American Naturalist*, 186, 531–46.

Henschen, A.E. and Adelman, J.S. (2019). What does tolerance mean for animal disease dynamics when pathology enhances transmission? *Integrative and Comparative Biology*, 59, 1220–30.

Hill, S.C., Manvell, R.J., Schulenburg, B., et al. (2016). Antibody responses to avian influenza viruses in wild birds broaden with age. *Proceedings of the Royal Society B: Biological Sciences*, 283, 20162159.

Horrocks, N.P.C., Matson, K.D., and Tieleman, B.I. (2011a). Pathogen pressure puts immune defense into perspective. *Integrative and Comparative Biology*, 51, 563–76.

Horrocks, N.P.C., Tieleman, B.I., and Matson, K.D. (2011b). A simple assay for measurement of ovotransferrin—a marker of inflammation and infection in birds. *Methods in Ecology and Evolution*, 2, 518–26.

Jacobs, A.C., Fair, J.M., and Zuk, M. (2015). Parasite infection, but not immune response, influences paternity in western bluebirds. *Behavioral Ecology and Sociobiology*, 69, 193–203.

Juul-Madsen, H.R., Viertlböeck, B., Härtle, S., Smith, A.L., and Göbel, T.W. (2014). Innate immune responses. In K.A. Schat, B. Kaspers, and P. Kaiser, eds. *Avian Immunology*, 2nd ed., pp. 121–47. Elsevier, Amsterdam.

Kaiser, P. and Stäheli, P. (2014). Avian cytokines and chemokines. In K.A. Schat, B. Kaspers, and P. Kaiser, eds. *Avian Immunology*, 2nd ed., pp. 189–204. Elsevier, Amsterdam.

Kaspers, B. and Kaiser, P. (2014). Avian antigen-presenting cells. In K.A. Schat, B. Kaspers, and P. Kaiser, eds. *Avian Immunology*, 2nd ed., pp. 169–88. Elsevier, Amsterdam.

Killpack, T.L., Oguchi, Y., and Karasov, W.H. (2013). Ontogenetic patterns of constitutive immune parameters in altricial house sparrows. *Journal of Avian Biology*, 44, 513–20.

Killpack, T.L., Carrel, E., and Karasov, W.H. (2015). Impacts of short-term food restriction on immune development in altricial house sparrow nestlings. *Physiological and Biochemical Zoology*, 88, 195–207.

King, M.O. and Swanson, D.L. (2013). Activation of the immune system incurs energetic costs but has no effect on the thermogenic performance of house sparrows during acute cold challenge. *Journal of Experimental Biology*, 216, 2097–102.

Klasing, K.C. (2004). The costs of immunity. *Acta Zoologica Sinica*, 50, 961–9.

Kluger, M.J., Kozak, W., Conn, C.A., Leon, L.R., and Soszynski, D. (1998). Role of fever in disease. *Annals of the New York Academy of Sciences*, 856, 224–33.

Knowles, S.C.L., Nakagawa, S., and Sheldon, B.C. (2009). Elevated reproductive effort increases blood parasitaemia and decreases immune function in birds: a meta-regression approach. *Functional Ecology*, 23, 405–15.

Knutie, S.A. (2018). Relationships among introduced parasites, host defenses, and gut microbiota of Galapagos birds. *Ecosphere*, 9, e02286.

Komar, N., Dohm, D.J., Turell, M.J., and Spielman, A. (1999). Eastern equine encephalitis virus in birds: Relative competence of European starlings (*Sturnus vulgaris*). *American Journal of Tropical Medicine and Hygiene*, 60, 387–91.

Komar, N., Langevin, S., Hinten, S., et al. (2003). Experimental infection of North American birds with the New York 1999 strain of West Nile virus. *Emerging Infectious Diseases*, 9, 311–22.

Kreisinger, J., Schmiedová, L., Petrželková, A., et al. (2018). Fecal microbiota associated with phytohaemagglutinin-induced immune response in nestlings of a passerine bird. *Ecology and Evolution*, 8, 9793–802.

Kuiken, T., Heckert, R.A., Riva, J., Leighton, F.A., and Wobeser, G. (1998). Excretion of pathogenic Newcastle disease virus by double-crested cormorants (*Phalacrocorax auritus*) in absence of mortality or clinical signs of disease. *Avian Pathology*, 27, 541–6.

Leung, T.L. and Koprivnikar, J. (2016). Nematode parasite diversity in birds: the role of host ecology, life history and migration. *Journal of Animal Ecology*, 85, 1471–80.

Liebl, A.L. and Martin, L.B. (2009). Simple quantification of blood and plasma antimicrobial capacity using spectrophotometry. *Functional Ecology*, 23, 1091–6.

Lloyd-Smith, J.O., Schreiber, S.J., Kopp, P.E., and Getz, W.M. (2005). Superspreading and the effect of individual variation on disease emergence. *Nature*, 438, 355–9.

Lochmiller, R.L. and Deerenberg, C. (2000). Trade-offs in evolutionary immunology: just what is the cost of immunity? *Oikos*, 88, 87–98.

Lord, C.C., Rutledge, C.R., and Tabachnick, W.J. (2006). Relationships between host viremia and vector susceptibility for arboviruses. *Journal of Medical Entomology*, 43, 623–30.

Macdonald, G. (1952). The analysis of equilibrium in malaria. *Tropical Diseases Bulletin*, 49, 813–29.

Marais, M., Maloney, S.K., and Gray, D.A. (2011). The metabolic cost of fever in Pekin ducks. *Journal of Thermal Biology*, 36, 116–20.

Martin, L.B. (2009). Stress and immunity in wild vertebrates: timing is everything. *General and Comparative Endocrinology*, 163, 70–76.

Martin, L.B., Gilliam, J., Han, P., Lee, K., and Wikelski, M. (2005). Corticosterone suppresses cutaneous immune function in temperate but not tropical house sparrows, *Passer domesticus*. *General and Comparative Endocrinology*, 140, 126–35.

Martin, L.B., Han, P., Lewittes, J., Kuhlman, J.R., Klasing, K.C., and Wikelski, M. (2006). Phytohemagglutinin-induced skin swelling in birds: histological support for a classic immunoecological technique. *Functional Ecology*, 20, 290–99.

Martin, L.B., Coon, C.A.C., Liebl, A.L., and Schrey, A.W. (2014). Surveillance for microbes and range expansion in house sparrows. *Proceedings of the Royal Society B: Biological Sciences*, 281, 20132690.

Martin, L.B., Burgan, S.C., Adelman, J.S., and Gervasi, S.S. (2016). Host competence: an organismal trait to integrate immunology and epidemiology. *Integrative and Comparative Biology*, 56, 1225–37.

Martin, L.B., Addison, B., Bean, A.G.D., et al. (2019). Extreme competence: keystone hosts of infections. *Trends in Ecology & Evolution*, 34, 303–14.

Matson, K.D., Ricklefs, R.E., and Klasing, K.C. (2005). A hemolysis-hemagglutination assay for characterizing constitutive innate humoral immunity in wild and domestic birds. *Developmental and Comparative Immunology*, 29, 275–86.

Matson, K.D., Horrocks, N.P.C., Versteegh, M.A., and Tieleman, B.I. (2012). Baseline haptoglobin concentrations are repeatable and predictive of certain aspects of a subsequent experimentally-induced inflammatory response. *Comparative Biochemistry and Physiology A–Molecular & Integrative Physiology*, 162, 7–15.

Mauck, R.A., Matson, K.D., Philipsborn, J., and Ricklefs, R.E. (2005). Increase in the constitutive innate humoral immune system in Leach's storm-petrel (*Oceanodroma leucorhoa*) chicks is negatively correlated with growth rate. *Functional Ecology*, 19, 1001–7.

Merrill, L., Angelier, F., O'Loghlen, A.L., Rothstein, S.I., and Wingfield, J.C. (2012). Sex-specific variation in brown-headed cowbird immunity following acute stress: a mechanistic approach. *Oecologia*, 170, 25–38.

Millet, S., Bennett, J., Lee, K.A., Hau, M., and Klasing, K.C. (2007). Quantifying and comparing constitutive immunity across avian species. *Developmental and Comparative Immunology*, 31, 188–201.

Monaghan, P., Metcalfe, N.B., and Torres, R. (2009). Oxidative stress as a mediator of life history trade-offs: mechanisms, measurements and interpretation. *Ecology Letters*, 12, 75–92.

Nazar, F. and Marin, R. (2011). Chronic stress and environmental enrichment as opposite factors affecting the immune response in Japanese quail (*Coturnix coturnix japonica*). *Stress*, 14, 166–73.

Nelson, R.J. and Demas, G.E. (1996). Seasonal changes in immune function. *Quarterly Review of Biology*, 71, 511–48.

Nettle, D., Andrews, C., Reichert, S., et al. (2016). Brood size moderates associations between relative size, telomere length, and immune development in European starling nestlings. *Ecology and Evolution*, 6, 8138–48.

Nord, A., Sköld-Chiriac, S., Hasselquist, D., and Nilsson, J.Å. (2014). A tradeoff between perceived predation risk and energy conservation revealed by an immune challenge experiment. *Oikos*, 123, 1091–100.

O'Neal, D.M. (2013). Eco-endo-immunology across avian life history stages. *General and Comparative Endocrinology*, 190, 105–11.

Owen-Ashley, N.T. and Wingfield, J.C. (2007). Acute phase responses of passerine birds: characterization and seasonal variation. *Journal of Ornithology*, 148, S583–91.

Owen, J.C. (2011). Collecting, processing, and storing avian blood: a review. *Journal of Field Ornithology*, 82, 339–54.

Owen, J.C. and Moore, F.R. (2006). Seasonal differences in immunological condition of three species of thrushes. *The Condor*, 108, 389–98.

Owen, J.C., Cornelius, E.A., Arsnoe, D.A., and Garvin, M.C. (2013). Leukocyte response to eastern equine encephalomyelitis virus in a wild passerine bird. *Avian Diseases*, 57, 744–9.

Palacios, M.G., Cunnick, J.E., Winkler, D.W., and Vleck, C.M. (2007). Immunosenescence in some but not all immune components in a free-living vertebrate, the tree swallow. *Proceedings of the Royal Society B: Biological Sciences*, 274, 951–7.

Pap, P.L., Czirják, G.Á., Vágási, C.I., Barta, Z., and Hasselquist, D. (2010). Sexual dimorphism in immune function changes during the annual cycle in house sparrows. *Naturwissenschaften*, 97, 891–901.

Pap, P.L., Vágási, C.I., Vincze, O., Osváth, G., Veres-Szászka, J., and Czirják, G.Á. (2015). Physiological pace of life: the link between constitutive immunity, developmental period, and metabolic rate in European birds. *Oecologia*, 177, 147–58.

Pepin, K.M., Kay, S.L., Golas, B.D., et al. (2017). Inferring infection hazard in wildlife populations by linking data across individual and population scales. *Ecology Letters*, 20, 275–92.

Råberg, L. and Stjernman, M. (2003). Natural selection on immune responsiveness in blue tits *Parus caeruleus*. *Evolution*, 57, 1670–78.

Ratcliffe, M.J. and Härtle, S. (2014). B cells, the bursa of Fabricius and the generation of antibody repertoires. In K.A. Schat, B. Kaspers, and P. Kaiser, eds. *Avian Immunology*, 2nd ed., pp. 65–90. Elsevier, Amsterdam.

Roe, C.F. and Kinney, J.M. (1965). Caloric equivalent of fever II. Influence of major trauma. *Annals of Surgery*, 161, 140–48.

Rogers, S.L., Viertlboeck, B.C., Gobel, T.W., and Kaufman, J. (2008). Avian NK activities, cells and receptors. *Seminars in Immunology*, 20, 353–60.

Romano, A., Rubolini, D., Caprioli, M., Boncoraglio, G., Ambrosini, R., and Saino, N. (2011). Sex-related effects of an immune challenge on growth and begging behavior of barn swallow nestlings. *PLoS ONE*, 6, e22805.

Romero, L.M. (2004). Physiological stress in ecology: lessons from biomedical research. *Trends in Ecology & Evolution*, 19, 249–55.

Roved, J., Westerdahl, H., and Hasselquist, D. (2017). Sex differences in immune responses: hormonal effects, antagonistic selection, and evolutionary consequences. *Hormones and Behavior*, 88, 95–105.

Sanz, J.J., Potti, J., Moreno, J., Merino, S., and Frías, O. (2003). Climate change and fitness components of a migratory bird breeding in the Mediterranean region. *Global Change Biology*, 9, 461–72.

Sapolsky, R.M., Romero, L.M., and Munck, A.U. (2000). How do glucocorticoids influence stress responses? Integrating permissive, suppressive, stimulatory, and preparative actions. *Endocrine Reviews*, 21, 55–89.

Schat, K.A., Kaspers, B., and Kaiser, P. (eds) (2014). *Avian Immunology*, 2nd ed. Elsevier, Amsterdam.

Schmid-Hempel, P. and Ebert, D. (2003). On the evolutionary ecology of specific immune defence. *Trends in Ecology & Evolution*, 18, 27–32.

Schmidt, K.L., Kubli, S.P., MacDougall-Shackleton, E.A., and MacDougall-Shackleton, S.A. (2015). Early-life stress has sex-specific effects on immune function in adult song sparrows. *Physiological and Biochemical Zoology*, 88, 183–94.

Schulenburg, H., Kurtz, J., Moret, Y., and Siva-Jothy, M.T. (2009). Ecological immunology. *Philosophical Transactions of the Royal Society B: Biological Sciences*, 364, 3–14.

Sheldon, B.C. and Verhulst, S. (1996). Ecological immunology: costly parasite defences and trade-offs in evolutionary ecology. *Trends in Ecology & Evolution*, 11, 317–21.

Sild, E. and Horak, P. (2010). Assessment of oxidative burst in avian whole blood samples: validation and application of a chemiluminescence method based on Pholasin. *Behavioral Ecology and Sociobiology*, 64, 2065–76.

Smith, A.L. and Göbel, T.W. (2014). Avian T cells: antigen recognition and lineages. In K.A. Schat, B. Kaspers, and P. Kaiser, eds. *Avian Immunology*, 2nd ed., pp. 91–102. Elsevier, Amsterdam.

Smith, J., Smith, N., Yu, L., et al. (2015). A comparative analysis of host responses to avian influenza infection in ducks and chickens highlights a role for the interferon-induced transmembrane proteins in viral resistance. *BMC Genomics*, 16, 574.

Snoeijs, T., Dauwe, T., Pinxten, R., Vandesande, F., and Eens, M. (2004). Heavy metal exposure affects the humoral immune response in a free-living small songbird, the great tit (*Parus major*). *Archives of Environmental Contamination and Toxicology*, 46, 399–404.

Soler, J.J., Peralta-Sánchez, J.M., Flensted-Jensen, E., Martín-Platero, A.M., and Møller, A.P. (2011). Innate humoural immunity is related to eggshell bacterial load of European birds: a comparative analysis. *Naturwissenschaften*, 98, 807.

Sompayrac, L. (2016). *How the Immune System Works*, 5th ed. Wiley-Blackwell, Oxford.

Sorci, G. (2013). Immunity, resistance and tolerance in bird–parasite interactions. *Parasite Immunology*, 35, 350–61.

Thomason, C.A., Leon, A., Kirkpatrick, L.T., Belden, L.K., and Hawley, D.M. (2017). Eye of the finch: characterization of the ocular microbiome of house finches in relation to mycoplasmal conjunctivitis. *Environmental Microbiology*, 19, 1439–49.

Tolf, C., Latorre-Margalef, N., Wille, M., et al. (2013). Individual variation in influenza A virus infection histories and long-term immune responses in mallards. *PLoS ONE*, 8, e61201.

Torres, R. and Velando, A. (2007). Male reproductive senescence: the price of immune-induced oxidative damage on sexual attractiveness in the blue-footed booby. *Journal of Animal Ecology*, 76, 1161–8.

Vagasi, C.I., Pătraș, L., Pap, P.L., et al. (2018). Experimental increase in baseline corticosterone level reduces oxidative damage and enhances innate immune response. *PLoS ONE*, 13, e0192701.

Valkiunas, G. (2004). *Avian Malaria Parasites and Other Haemosporidia*. CRC Press, Boca Raton, FL.

Vallverdú-Coll, N., López-Antia, A., Martinez-Haro, M., Ortiz-Santaliestra, M.E., and Mateo, R. (2015). Altered immune response in mallard ducklings exposed to lead through maternal transfer in the wild. *Environmental Pollution*, 205, 350–56.

van de Crommenacker, J., Horrocks, N.P.C., Versteegh, M.A., Komdeur, J., Tieleman, B.I., and Matson, K.D. (2010). Effects of immune supplementation and immune challenge on oxidative status and physiology in a model bird: implications for ecologists. *Journal of Experimental Biology*, 213, 3527.

van der Most, P.J., de Jong, B., Parmentier, H.K., and Verhulst, S. (2011). Trade-off between growth and immune function: a meta-analysis of selection experiments. *Functional Ecology*, 25, 74–80.

van Dijk, J.G., Fouchier, R.A., Klaassen, M., and Matson, K.D. (2015). Minor differences in body condition and immune status between avian influenza virus-infected and noninfected mallards: a sign of coevolution? *Ecology and Evolution*, 5, 436–49.

VanDalen, K.K., Hall, J.S., Clark, L., McLean, R.G., and Smeraski, C. (2013). West Nile virus infection in American robins: new insights on dose response. *PLoS ONE*, 8, e68537.

VanderWaal, K.L., and Ezenwa, V.O. (2016). Heterogeneity in pathogen transmission: mechanisms and methodology. *Functional Ecology*, 30, 1606–22.

Vaziri, G.J., Muñoz, S.A., Martinsen, E.S., and Adelman, J.S. (2019). Gut parasite levels are associated with severity of response to immune challenge in a wild songbird. *Journal of Wildlife Diseases*, 55, 64–73.

Velando, A., Beamonte-Barrientos, R., and Torres, R. (2014). Enhanced male coloration after immune challenge increases reproductive potential. *Journal of Evolutionary Biology*, 27, 1582–9.

Velova, H., Gutowska-Ding, M.W., Burt, D.W., and Vinkler, M. (2018). Toll-like receptor evolution in birds: gene duplication, pseudogenization, and diversifying selection. *Molecular Biology and Evolution*, 35, 2170–84.

Vermeulen, A., Eens, M., Zaid, E., and Müller, W. (2016). Baseline innate immunity does not affect the response to an immune challenge in female great tits (*Parus major*). *Behavioral Ecology and Sociobiology*, 70, 585–92.

Viney, M.E., Riley, E.M., and Buchanan, K.L. (2005). Optimal immune responses: immunocompetence revisited. *Trends in Ecology & Evolution*, 20, 665–9.

Vinkler, M., Leon, A.E., Kirkpatrick, L., Dalloul, R.A., and Hawley, D.M. (2018). Differing house finch cytokine expression responses to original and evolved isolates of *Mycoplasma gallisepticum*. *Frontiers in Immunology*, 9, 13.

Wegmann, M., Voegeli, B., and Richner, H. (2015). Physiological responses to increased brood size and ectoparasite infestation: adult great tits favour self-maintenance. *Physiology & Behavior*, 141, 127–34.

Whiteman, N.K., Matson, K.D., Bollmer, J.L., and Parker, P.G. (2006). Disease ecology in the Galápagos hawk (*Buteo galapagoensis*): host genetic diversity, parasite load and natural antibodies. *Proceedings of the Royal Society B: Biological Sciences*, 273, 797–804.

Wilcoxen, T.E., Horn, D.J., Hogan, B.M., et al. (2015). Effects of bird-feeding activities on the health of wild birds. *Conservation Physiology*, 3, cov058.

Wingfield, J.C., Hegner, R.E., Dufty Jr, A.M., and Ball, G.F. (1990). The 'challenge hypothesis': theoretical implications for patterns of testosterone secretion, mating systems, and breeding strategies. *The American Naturalist*, 136, 829–46.

Zhang, G.J., Jarvis, E.D., and Gilbert, M.T.P. (2014). A flock of genomes. *Science*, 346, 1308–9.

Zuk, M., Thornhill, R., Ligon, J.D., and Johnson, K. (1990). Parasites and mate choice in red jungle fowl. *American Zoologist*, 30, 235–44.

Zylberberg, M. (2019). Next-generation ecological immunology. *Physiological and Biochemical Zoology*, 92, 177–88.

CHAPTER 4

Behavior Shapes Infectious Disease Dynamics in Birds

Andrea K. Townsend and Dana M. Hawley

4.1 Conceptual overview: the intersection of behavior and disease ecology

Animal behavior can both influence and be influenced by parasites or pathogens. Because the diverse and bidirectional interactions between parasites and host behavior can have important consequences for host–parasite evolution and infectious disease dynamics (Ezenwa et al. 2016a), it is important to consider how avian behavior influences each step in the parasite transmission cycle (Figure 4.1). First, avian behavior can influence the degree of exposure that hosts have to parasites or their vectors (Bush and Clayton 2018; Figure 4.1A). Second, some behaviors and their associated physiological mediators can alter the ability of a host to resist parasites once exposed ('resistance'), as well as the degree of parasitemia reached (Hawley et al. 2011; Figure 4.1B). At the same time, the parasite load experienced by an infected host often causes changes in the behavior of animals (dashed arrow, Figure 4.1B), whether due to pathology, host immune responses, or manipulation of the host by a parasite (e.g., Hart 1988; Poulin 2013). These changes in behavior during infection, together with behavioral changes that vectors or healthy hosts might make in response to cues from infected hosts, can influence rates of contact of infected hosts with vectors or healthy conspecifics

(e.g., Cornet et al. 2013; Stroeymeyt et al. 2018). Thus, behaviors during infection, in combination with host parasite load (Figure 4.1B), contribute significantly to the third and final step of the parasite transmission process: the probability of successful transmission of the parasite from that host to another one (Figure 4.1C).

Avian behaviors have the potential to either augment or suppress components of the transmission process at each of these steps (Table 4.1). In some cases, the same behavior (e.g., preening of feathers to remove ectoparasites) can simultaneously

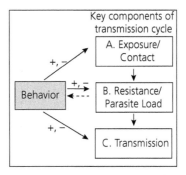

Figure 4.1 Behavior Can Augment (+) or reduce (−) the three key components of the pathogen transmission cycle: (A) risk of exposure to or contact with pathogens, (B) ability to resist or minimize infection once exposed, and (C) likelihood of transmitting pathogens to others. Likewise, current infection (B) can alter host behavior (dashed arrow) as well as the behaviors of vectors and healthy conspecifics with which infected hosts come into contact.

Andrea K. Townsend and Dana M. Hawley, *Behavior Shapes Infectious Disease Dynamics in Birds* In: *Infectious Disease Ecology of Wild Birds.*
Edited by: Jennifer C. Owen, Dana M. Hawley, and Kathryn P. Huyvaert, Oxford University Press. © Oxford University Press 2021.
DOI: 10.1093/oso/9780198746249.003.0004

Table 4.1 Some avian behaviors that augment or reduce the key components of the transmission cycle.

Key components	Behaviors that can augment	Behaviors that can reduce
1. Exposure/contact	Preening/allopreening; foraging behaviors (e.g., ground feeding); migration; boldness; use of high-risk food or sites; mating; group living	Preening/allopreening; behaviors to deter aerial parasites (e.g., foot stomping); sickness behaviors; avoidance of high-risk food, sites, or mates; 'migratory escape;' shyness; 'encounter-dilution' via group living
2. Resistance	Sickness behaviors; self-medication; dominance/elevated body condition; preference for high-quality/genetically dissimilar partners; social facilitation; immune priming	Migratory stress; dominance/stress; inbreeding; terminal investment
3. Transmission	Allopreening; masking sickness behaviors; pathogen transport during migration; boldness; mating; gregariousness	Sickness behavior/social isolation; shyness; sedentariness; 'migratory culling'

augment exposure to one type of parasite while reducing exposure to others. It is critical to understand the diverse and dynamic ways that behavior interacts with each component of the transmission cycle in order to predict how avian hosts should evolve in response to parasites and how variation in host behavior will scale up to influence the likelihood or severity of disease epidemics.

The intersection of behavior and infectious disease in birds is an exciting and rapidly growing field (Bush and Clayton 2018). Given the inherent breadth and recent increase in work on this topic, we do not attempt an exhaustive review. Instead, we focus on major categories of behavior important for the likelihood of contact with pathogens or parasites (terms we use interchangeably throughout this chapter), resistance to infection or degree of parasitemia reached once exposed, and the likelihood of spreading (Table 4.1). Throughout, we highlight and integrate areas in which recent advances have been made or for which more data are needed in avian systems, emphasizing directions for future research at the intersection of avian behavior and disease ecology.

4.2 Behaviors important to parasite exposure, resistance, and spread

In this section, we begin by discussing the behaviors that birds may use to deter or avoid parasites. Because complete avoidance of parasites is impossible, however, we next discuss the immediate effects of infection on behavior. We then focus on three broad classes of behaviors (foraging and movement, personality, and social interactions) that can interact with all three steps of the transmission process and thus both influence and respond to pathogens. For brevity, we do not discuss parasite manipulation of host behavior here; instead, we refer the interested reader to other excellent sources on that topic (e.g., Klein 2003; Moore 2002).

4.2.1 Antiparasite behaviors

Birds engage in several antiparasite behaviors that can reduce their own exposure to parasites and affect the extent to which they transmit parasites to others (Table 4.1). Here, we focus on behaviors that deter ectoparasites (e.g., lice, fleas, ticks, and flies), which can have direct negative effects on the health, reproductive success, and survival of birds (Moller 1990; Richner et al. 1993). Ectoparasites can also have indirect negative consequences for birds by vectoring endoparasites (e.g., bacteria, viruses, protists); therefore, behaviors that repel ectoparasites may also reduce individual risk of vector-borne pathogens.

The effectiveness of many behaviors with putative antiparasite functions in birds is variable or uncertain (Table 4.2) and can come at the cost of energy expenditure or opportunities lost (e.g., social, vigilance, or foraging opportunities), which can be particularly challenging for birds already under pathogen stress (Hart 1990). Moreover, some antiparasite behaviors reduce exposure to certain parasites but increase exposure to others. Avian

Table 4.2 Feather maintenance behaviors hypothesized to serve antiparasite functions in birds. Direct parasite removal via preening and allopreening are discussed in the text.

Behavior	Description	Effectiveness?	Example sources
Use of preen gland oil	Oil (secreted from uropygial gland) spread over plumage with bill during preening	Antimicrobial and ectoparasite-repelling effects *in vitro* but not *in situ*	Moreno-Rueda 2017
Anting	Ants allowed to swarm over bird or are actively crushed and rubbed over feathers, coating them with defensive compounds	Does not appear to reduce lice, mites, or feather-degrading fungi and bacteria	Revis and Waller 2004
Sunning	Feathers spread in direct sunlight	Heat and UV light may inhibit, desiccate, or dislodge ectoparasites; may inhibit growth of feather-degrading bacteria	Bush and Clayton 2018; Saranathan and Burtt 2007
Dust-bathing	Dust ruffled through feathers	Sand treated with kaolin clay or diatomaceous earth reduced louse burdens on domestic hens; no effect of bathing with untreated sand	Martin and Mullens 2012; Vezzoli et al. 2015

Figure 4.2 Preening is an effective method of ectoparasite removal in birds. (Photo credit: J. Owen.)

antiparasite behaviors have been reviewed in-depth recently (Bush and Clayton 2018); therefore, we limit our detailed discussion to three types of antiparasite behaviors (preening, deterrence of aerial parasites, and nest 'self-medication') that are relatively well studied, with emphasis on their relevance to parasite transmission, positive and negative.

4.2.1.1 Preening

Preening (Figure 4.2) may be the most important and effective method by which birds rid and protect themselves from ectoparasites, as made clear when preening is reduced or prevented (reviewed in Bush and Clayton 2018). Indeed, avian beak morphology may be specifically adapted for removing ectoparasites: when the small overhang of the culmen is removed from their upper mandibles, for example,

rock pigeons (*Columba livia*) have significantly more lice and feather damage than control birds but the same feeding efficiency (Clayton et al. 2005). Further, in Darwin's finches, all three studied species had very similar relative mandibular overhang lengths despite divergent selection on overall beak size, suggesting strong stabilizing selection for a beak component known to be critical for parasite control (Villa et al. 2018).

Preening can be costly in terms of time and energy, however. A meta-analysis found that birds spend an average of 9.2% of their daylight time budget on 'maintenance behaviors,' primarily preening (Cotgreave and Clayton 1994). For incubating king penguins (*Apternodytes patagonicus*), energy expenditure during behaviors like preening and scratching is, on average, 1.24 times higher than the resting metabolic rate (Viblanc et al. 2011). Although abstaining from preening could therefore save energy, 'scruffy' feathers lose insulation efficiency, thereby increasing the cost of immune function as well as increasing ectoparasite burden (Hart 1988). However, birds also appear to adjust how much they invest in preening behaviors depending on their current parasite load, indicating that preening is an inducible defense in birds (Villa et al. 2016a).

Ectoparasite burden can also be reduced through allopreening (mutual preening), through which ectoparasites are removed from areas that are difficult to reach by self-preening. Paired penguins that engaged in allopreening, for example, were less likely to be infested with ticks and had lower average tick burdens than unpaired penguins (Brooke 1985). Likewise, allopreening was 17-fold more effective than self-preening in rock pigeons, possibly because most allopreening was directed at the head and neck, where ectoparasites tend to congregate (Villa et al. 2016b). However, allopreening may induce costs by increasing transmission of endoparasites directly or via fecal–oral routes (see Section 4.2.6). Both self-preening and allopreening may also facilitate transmission of parasites with trophic transmission routes, such as helminths that use lice and fleas as intermediate hosts and birds as definitive hosts (reviewed in Bush and Clayton 2018).

4.2.1.2 Deterrence of aerial parasites

Birds can respond to attack by aerial ectoparasites with a combination of foot stomping, head shaking, and body fluffing, although the effectiveness of these movements as vector deterrents can vary. For example, the proportion of successful blood meals among *Culex* mosquitoes decreased with frequency of parasite deterrence movements for young chickens but not house sparrows (*Passer domesticus*); this ineffectiveness might contribute to the importance of house sparrows as amplification hosts for West Nile virus (Darbro and Harrington 2007).

Although deterrence movements might reduce the direct effects of ectoparasites on individual hosts (i.e., by reducing blood loss), they may also increase transmission of endoparasites (i.e., internal parasites) by interrupting vectors mid-meal, forcing them to feed on multiple hosts to get a full blood meal and increasing the chance of cross-infection among hosts (Darbro and Harrington 2007). In addition, current endoparasitic infection can make hosts more attractive to mosquitoes (Cornet et al. 2013) and more lethargic (as demonstrated by house finches [*Haemorhous mexicanus*] infected with *Mycoplasma gallisepticum*; Darbro et al. 2007), thereby promoting vector-borne parasite transmission. Lethargy and suppression of antiparasite behaviors due to current infection could also promote secondary infection of a diseased host by other parasites, increasing the likelihood of co-infection.

4.2.1.3 Nest 'self-medication'

Some birds use plant material in nest construction, apparently to repel parasites from their nests.

European starlings (*Sturnus vulgaris*), for example, preferred plants with higher concentrations of volatile compounds than expected by availability, and the volatiles of preferred plants had microbiocidal properties and inhibited the growth of lice and hematophagous mites (Clark and Mason 1985; Clark and Mason 1988). Likewise, mixtures of aromatic herbs deterred *Culex* mosquitoes from Corsican blue tit (*Parus caeruleus ogliastrae*) nestlings or at least masked their chemical cues from this vector (Lafuma et al. 2001). Urban birds (e.g., house sparrows and house finches) sometimes incorporate cigarette butts in their nest, and correlational data suggest that the nicotine in the butts repels nest ectoparasites (Suarez-Rodriguez et al. 2013; Figure 4.3); moreover, house finches add more cigarette material to their nests when ectoparasite load is experimentally increased (Suarez-Rodriguez and Garcia 2017). However, cigarettes in the nest are associated with increased genotoxicity for both nestlings and breeding adults (Suarez-Rodriguez et al. 2017). Thus, much like preening behaviors, nest 'self-medication' can be associated with both costs and benefits, the balance of which are likely to vary as parasite pressures change.

4.2.2 Sickness behaviors

The behaviors considered in Section 4.2.1 help to reduce avian contact with parasites, but when successful infection occurs, a suite of behavioral changes often ensues. Sickness behaviors are changes in behavior that can occur immediately after infection as part of a host's early immune response (Hart 1988). These behaviors—which include lethargy, inactivity, a hunched posture, ptiloerection, and disinterest in food, water, sex, and other social interactions (Hart 1988; Hennessy et al. 2014; Figure 4.4)—can affect disease dynamics in several ways (Table 4.1). First, because sickness behaviors are a component of vertebrate immune responses, they can alter a host's resistance to infection by promoting recovery and reducing the exposure of infected individuals to additional stressors and pathogens. Second, by restricting host movement and activity, sickness behaviors can alter the encounter rates of infectious hosts with susceptible conspecifics or vectors. Third, because they can also

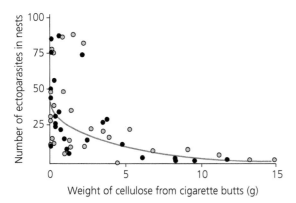

Figure 4.3 Number of ectoparasites in nests of house finches (gray circles) and house sparrows (black circles) declines with increasing quantity of cigarette butts in the nest. (Figure redrawn with permission from Suarez-Rodriguez et al. 2013.)

Figure 4.4 Male house finch exhibiting sickness behaviors (e.g., lethargy, inactivity, ptiloerection) associated with mycoplasmal conjunctivitis. (Photo credit: L. Pinkham/Project Feederwatch.)

signal a host's infection status, sickness behaviors may trigger social avoidance (or facilitation) by healthy conspecifics. However, when the social costs of exhibiting sickness behaviors are too high, they can be suppressed or masked (Lopes et al. 2012), which can augment the potential for transmission (Adelman and Hawley 2017). In this section, we describe the manifestations and putative functions of sickness behaviors in birds, highlight context-specific suppression of sickness behaviors, and discuss how sickness behaviors could scale up to affect transmission.

4.2.2.1 Manifestations and putative function of sickness behaviors in birds

Sickness behaviors are exhibited as part of the acute phase and are mediated by the release of pro-inflammatory cytokines (e.g., interleukin-1 [IL-1] and tumor necrosis factor α) from innate immune cells activated during infection (Konsman et al. 2002). Sickness behaviors are widespread among animals (Hart 1988; Kazlauskas et al. 2016) and are now generally hypothesized to be adaptations that promote recovery from infection (Aubert 1999; Hart 1988; Hennessy et al. 2014). Energy-conserving behaviors like lethargy, social withdrawal, and loss of interest in surroundings, for example, could conserve fuel for immune responses, as observed among zebra finches (*Taeniopygia guttata*) in which immune defenses increased with time spent resting (Lopes et al. 2014). Reduced interest in food or water could also be beneficial if it reduces dangerous foraging activity during the acute phase response, at which time the host is particularly vulnerable to predators (as seen in house finches infected with *M. gallisepticum*; Adelman et al. 2017).

Although sickness behaviors are usually associated with social isolation, they can sometimes increase social interactions among conspecifics (Hennessey et al. 2014). For example, healthy house finch males strongly preferred to feed next to conspecifics infected with *M. gallisepticum*, possibly because diseased birds exhibited sickness behaviors and were thus less competitive (Bouwman and Hawley 2010). In contrast, however, Zylberberg et al. (2012) found that male house finches avoided associating with conspecific males injected with bacterial lipopolysaccharide (LPS), which activates the immune system and triggers sickness behaviors. The responses of birds to conspecifics exhibiting sickness behaviors appear to be variable, although the reasons for this variation are not clear.

4.2.2.2 Context-specific sickness behaviors

Hart (1988) proposed that sickness behaviors represent an all-out effort to recover from a potentially life-threatening infection without regard to other costs. These costs can include loss of social status, reduced parental care, and foregone reproductive opportunities. However, sickness behaviors can be masked or suppressed (at least partially or temporarily) if the infection is secondary in urgency to social pressures or other motivational states (Aubert 1999; Hennessy et al. 2014; Lopes et al. 2012). For example, highly social male zebra finches are less active and lose mass after LPS injection when in isolation but exhibit no visible sickness behaviors in a colony (Lopes et al. 2012), potentially so they do not lose status in a social setting. Likewise, LPS injections decreased territorial aggression for male northwestern song sparrows (*Melospiza melodia morphna*) during the non-breeding season but not in the early breeding season, when the risk for loss of offspring was higher (Owen-Ashley and Wingfield 2006). Birds also may suppress sickness behaviors during migration, although not without costs. Western sandpipers (*Calidris mauri*), for example, did not reduce their flight duration in a wind tunnel after LPS injection, although flying during an acute phase response appeared to suppress their immune response (Nebel et al. 2013; see also Section 4.2.3.2). Context-dependent suppression of sickness behavior can also vary between the sexes (Yirmiya et al. 1995), though work in avian systems on this topic is lacking.

In some cases, the degree to which sickness behaviors can be suppressed appears condition dependent. In European starlings, for example, control males reduced their song rate after injection with LPS; however, males supplemented with carotenoids (a dietary micronutrient with immunomodulatory effects) did not reduce their song rate, suggesting that birds can downregulate the expression of sickness behaviors when they have the appropriate resources (Casagrande et al. 2015). The extent to which sickness behaviors can be suppressed is also likely to depend on the intensity of cytokine response and the virulence of the pathogen (Hennessy et al. 2014).

4.2.2.3 Sickness behaviors and transmission

Sickness behaviors are likely to influence the degree to which an individual contributes to ongoing transmission. The lethargy associated with sickness behaviors may reduce antivector defenses (e.g., Darbro et al. 2007) and thus augment the ability of an individual to transmit a vector-borne pathogen. In contrast, lethargy and social withdrawal are likely to reduce contact with conspecifics and thus suppress the spread of directly transmitted pathogens. Although this has not been explicitly examined in birds, Lopes et al. (2016) used LPS injections to show that mice exhibiting sickness behaviors have reduced connectivity within social groups. When disease outbreaks were then simulated over the empirically quantified social networks, the observed behavioral changes during mock infection resulted in highly attenuated disease outbreaks. The social isolation of individuals expressing sickness behaviors will be even stronger if healthy conspecifics use sickness behaviors as a cue to avoid contagious groupmates. Given the evidence that healthy birds do sometimes avoid conspecifics expressing sickness behaviors (Zylberberg et al. 2012) and that sickness behavior expression can be context-dependent, it will be critical to understand the behavior of both sick and healthy birds in social groups during infection to predict how these behaviors will influence disease dynamics.

4.2.3 Foraging and movement behaviors

Foraging behavior and movements, both local and long-distance, can affect all three steps of the pathogen transmission process (Table 4.1). We first focus on the relationship between foraging behavior and pathogen exposure risk. We then discuss the links between migratory behaviors and pathogen exposure and resistance. We end this section by considering how the effects of infection on movement behaviors in birds can impact pathogen transmission.

4.2.3.1 Foraging behavior and exposure risk

Diet and foraging microhabitat can have a strong impact on the risk of exposure to pathogens that are acquired either directly from food or via fecal–oral or fomite transmission (reviewed in Benskin et al. 2009; Figure 4.5). For example, raptors, scavengers, and ground-feeding birds appear to be at high risk for acquiring *Campylobacter jejuni* (Waldenstrom et al. 2002), and ground-feeding birds are most likely to harbor high levels of intestinal coccidia (Dolnik et al. 2010). Ground-feeding birds are also more likely to acquire ticks from grass or leaf litter (e.g., Eisen et al. 2004; Marsot et al. 2012) and thus may be at high risk for picking up tick-associated

Figure 4.5 Exposure to pathogens and parasites can be increased through (A) dabbling and (B) ground foraging. These American crows (*Corvus brachyrhynchos*) have poxviral dermatitis lesions on their right hallux ('IL;' left) or bill ('WT;' right). (Photo credits: (A) inkknife_2000/CC BY-SA; (B) A. Townsend.)

endoparasites. For example, Comstedt et al. (2006) found that risk of infestation with ticks carrying *Borrelia burgdorferi* (the causative agent of Lyme borreliosis) was five times higher among ground-feeding birds relative to birds foraging on other substrates. Bark feeding has also been associated with elevated rates of tick infestation in birds in both northwestern California (Newman et al. 2015) and the Atlantic Forest of Brazil (Marini et al. 1996).

Specific foraging methods may increase exposure to pathogens. Amongst waterfowl, dabbling ducks are at particularly high risk for avian influenza A viruses (AIVs) (Munster and Fouchier 2009). This may at least partly relate to their feeding in shallow or surface waters, which harbor higher concentrations of particulates that facilitate virus binding and viability (Bitton 1980). Within dabbling ducks, differences in the density of lamellae (the comb-like structure on the sides of their bill) among species is associated with variation in AIV prevalence, suggesting a possible role for specific foraging tactics in the likelihood of AIV transmission (Hill et al. 2010). However, intrinsic differences in viral receptivity may play a more important role in among-species differences in AIV prevalence than foraging behavior (Gaidet et al. 2012). Individual variation in foraging behavior within a species also can lead to differences in the likelihood of acquiring a pathogen. For example, free-living Bewick's swans

(*Cygnus columbianus*) that preferentially foraged in aquatic habitats harbored a greater risk of AIV infection (Hoye et al. 2012), and the amount of time that free-living house finches spent per day on bird feeders, a known fomite of infection for *M. gallisepticum*, was associated with higher likelihood of infection (Adelman et al. 2015).

A suite of animal taxa have been shown to avoid foraging in or using habitats associated with high pathogen risk. For example, mammalian herbivores including sheep (Hutchings et al. 1998) and wild ungulates (Ezenwa 2004) avoid grazing in areas with fresh feces. Little is known about avoidance of contaminated habitats or prey by birds. However, birds do appear to have the ability to recognize parasite risk in non-foraging contexts. For example, great tits (*Parus major*) forced to choose between nest boxes that were either infested with ectoparasites or sterilized largely chose the parasite-free nests (Oppliger et al. 1994), and cliff swallows (*Petrochelidon pyrrhonota*) from fumigated (parasite-free) colonies showed higher breeding site fidelity than those from unfumigated colonies (Brown et al. 2017). Furthermore, there is some evidence that birds may avoid heavily parasitized prey: oystercatchers (*Haematopus ostralegus*) that serve as the final host of a trematode parasite prefer intermediate-sized molluscs over larger ones that contained more calories but are also more heavily parasitized (Norris 1999). Research is sorely needed to determine to what extent birds actively avoid foraging areas or prey that pose a high risk of parasite or pathogen contamination and whether the degree of avoidance varies with the cost of a given parasite infection (i.e., its virulence).

It also remains largely unknown whether birds forage on particular food types as a form of self-medication, which would help to augment resistance once exposed. Work suggests that great bustards (*Otis tarda*) ingest cantharidin-containing blister beetles, which provide some protection against gastrointestinal bacteria (Bravo et al. 2014). Furthermore, this behavior appears strongest in males, potentially contributing to sexual selection in this species (see also Section 4.2.5.1). However, it remains unknown whether the degree of ingestion of these beetles depends on variation in infection status or risk (deRoode et al. 2013), which would help confirm the use of these food items for true self-medication in birds.

4.2.3.2 Migration and disease dynamics

Birds are highly mobile, covering broad spatial areas through their daily, seasonal, and annual movements, with important consequences for pathogen spread (Table 4.1). Considerable research has been focused on understanding how long-distance migratory movements in birds and other taxa alter infectious disease dynamics (Altizer et al. 2011). There are several potential effects of migration on parasite and pathogen dynamics for wild animals, including birds (see also Chapter 10). First, migratory birds may have greater exposure to parasites because they spend substantial amounts of time in distinct habitats with distinct parasite communities. For example, waterfowl that migrate longer distances show higher levels of blood parasite richness and prevalence (Figuerola and Green 2000). Alternatively, migration may reduce exposure by allowing individuals to escape some of their parasites ('migratory escape;' Loehle 1995) and, at the population level, may reduce prevalence and ongoing transmission by culling infected individuals that cannot effectively withstand the migratory journey ('migratory culling;' Bradley and Altizer 2005). While evidence for migratory escape has primarily been found in ungulates (Altizer et al. 2011), there is some evidence that migratory culling may occur in birds: migratory

Bewick's swans infected with low-pathogenic AIV showed delayed migration initiation and reduced travel distance (van Gils et al. 2007). In contrast, migratory mallards (*Anas platyrhynchos*) naturally infected with AIV did not differ in either the speed or distance of migration (Latorre-Margalef et al. 2009), suggesting that the effects of migratory culling are likely to be species-specific.

Migration can also alter host resistance to pathogens by suppressing immune function. Three species of North American thrushes, for example, show seasonal changes in innate immune function that are consistent with the hypothesis that migration can suppress innate immunity (Owen and Moore 2006). Likewise, the photoperiodic stimulation of migratory restlessness was associated with reactivation of latent *Borrelia* infection in redwing thrushes (*Turdus iliacus*; Gylfe et al. 2000). In contrast, however, migratory status did not appear to influence viremia of West Nile virus in two neotropical migrant songbird species (Owen et al. 2006) and immune suppression associated with migration was not detected in red knots (Buehler et al. 2008; Hasselquist et al. 2007).

Finally, migratory birds could transport pathogens into new populations, as suggested by the movement of highly pathogenic AIV H5N1 virus along migratory flyways (Tian et al. 2015). Non-migratory spatial movements are also likely to play a role in exposure to and spatial spread of pathogens in birds. For example, bird movement between colonies of cliff swallows predicts the prevalence of Buggy Creek virus in insect vectors (Brown et al. 2007) and post-nesting dispersal of year-round and summer resident birds could contribute to the late summer movement of West Nile virus in California (Reisen et al. 2004). Overall, however, relatively little is known about the importance of short-distance movements for pathogen exposure and transmission in birds.

4.2.3.3 Effects of disease on movement and foraging

Like most interactions between parasites and host behavior, the relationship between movement and infection is bidirectional, with infection status both responding to and, in some cases, directly altering movement behaviors. Effects of infection on migratory distance or propensity can alter a host's potential contribution to landscape-level spread by reducing traveling distances, as occurs in Bewick's swans (van Gils et al. 2007). However, in gray catbirds (*Dumatella carolensis*) experimentally infected with West Nile virus, infectious individuals showed similar degrees of migratory restlessness to those of controls (Owen et al. 2006). Although Swainson's thrushes (*Catharus ustulatus*) in the same study showed average reductions in migratory activity during the viremic period, approximately half of these viremic individuals maintained their migratory activity during the infectious period, indicating substantial intraspecific variation in response to infection (Owen et al. 2006).

For non-migratory movements, the anorexia and lethargy associated with acute pathogen infections may reduce daily foraging movements and home ranges. For example, juvenile apapane (*Himatione sanguinea*) experimentally infected with blood parasites that cause avian malaria devoted significantly less time to all locomotory activities including flight, hopping, and foraging (Yorinks and Atkinson 2000). Given that avian malarial parasites are vector-borne, this reduction in movement may augment local transmission by decreasing antivector defense behaviors (see Section 4.2.2.3) but ultimately reduce spatial spread across the landscape. Finally, free-living house finches with visible mycoplasmal conjunctivitis spent more time sitting on bird feeders than healthy conspecifics but moved less frequently between bird feeders and nearby perches (Hawley et al. 2007a), suggesting that infectious individuals may deposit *M. gallisepticum* onto bird feeders within a relatively limited spatial area. While less studied than acute infections, chronic infection with parasites and pathogens may also result in changes in foraging and movement in free-living birds. Feather ectoparasites, for example, can influence locomotory performance and foraging, as was seen in male barn swallows (*Hirundo rustica*), where the number of holes produced in the feathers by lice was associated with an increased percentage time flapping during level flight (Barbosa et al. 2002).

4.2.4 Personality

Animals often exhibit consistent differences in correlated sets of behaviors across contexts, referred to as behavioral syndromes or personalities (Sih et al. 2004; Stamps 2007). Some of the behaviors typically considered to constitute personality in wild animals include boldness (defined as willingness to engage in risky behaviors), exploration, aggressiveness, and sociability (Sih et al. 2004). Because these differences in personality can affect the extent to which individuals interact with conspecifics, heterospecifics, and the environment, which are all relevant for pathogen spread, there is growing interest in exploring the intersection between animal personality and disease (Barber and Dingemanse 2010; Kortet et al. 2010; Figure 4.6). Furthermore, personality traits are often associated with neuroendocrine differences relevant for immune responses and pathogen resistance (Koolhaas 2008). Thus, differences in animal personality are particularly interesting in the context of infectious disease because they can covary with all three steps of the transmission process (Table 4.1). Here, we explore the specific components of personality that are most likely to impact pathogen exposure, resistance, and spread and briefly review the current evidence and open questions at the intersection of personality and infectious disease dynamics.

4.2.4.1 Personality and exposure

Most work to date on the role of personality in the context of parasitism has focused on potential links between personality traits and the likelihood of exposure to pathogens, although this topic has received little attention in birds. For ectoparasites or pathogens that are transmitted via the environment (i.e., fecal–oral or fomite transmission), willingness to explore novel environments or objects may be associated with increased risk of parasite exposure, as supported by correlative evidence from Siberian chipmunks (*Eutamias sibiricus*; Boyer et al. 2010). Although no studies of birds have explicitly examined whether exploration behavior predicts pathogen exposure, free-ranging great tits that are more exploratory responded to changes in food availability by visiting new potential food sources at greater spatial distances from their original feeding sites (van Overveld and Matthysen 2010), which may increase their degree of exposure to environmentally transmitted pathogens that are spatially variable.

Figure 4.6 Aspects of personality such as sociability, boldness, and exploratoriness can influence parasite exposure, resistance, and transmission. (Photo credit: Magnus Johansson, CC BY-SA 2.0, https://commons.wikimedia.org/w/index.php?curid=42071585.)

For directly transmitted parasites that are spread via close interactions with conspecifics, bold individuals and/or those who are most sociable may be at highest risk for parasite or pathogen exposure. Two studies of mammalian species found higher seroprevalence of pathogens that are largely spread via aggressive interactions amongst bolder individuals (Dizney and Dearing 2013; Natoli et al. 2005). Sociability, often measured using social network degree, has also been found to correlate with boldness and/or exploration in some taxa, including several species of birds (Aplin et al. 2013; Moyers et al. 2018; Snijders et al. 2014). Thus, for any pathogens transmitted via social interactions, personality traits associated with boldness or sociability may increase the risk of parasite or pathogen exposure.

4.2.4.2 Personality and resistance

Because personality traits often covary with components of the immune system and/or its physiological mediators (reviewed in Lopes 2017), personality traits may also play a role in the ability to resist a pathogen once exposed. Determining whether patterns of pathogen infection in nature arise from differences in the degree of exposure or host resistance once exposed is extremely challenging, and thus immune assay studies are often used to isolate covariation between personality traits and immune traits likely to influence resistance to pathogens once exposed. These studies reveal the

presence of often strong covariation between immune system components and personality traits across a range of taxa, including birds (Lopes 2017). Although personality-mediated covariation between behavior and immunity can arise via several indirect mechanisms, there may also be adaptive reasons why personality traits covary with immune responses (Lopes 2017). For example, the risk-of-parasitism hypothesis argues that boldness or risk-taking behaviors, which may augment pathogen exposure, should be correlated with stronger immune defenses (Jacques-Hamilton et al. 2017). In support of this hypothesis, more exploratory house finches harbor stronger innate immune defenses (Zylberberg et al. 2014) and collared flycatchers (*Ficedula albicollis*) with stronger risk-taking behavior (measured as flight initiation distance) had higher levels of allelic diversity at sites in the major histocompatibility complex (MHC) (Garamszegi et al. 2015). In contrast, more exploratory superb fairy wrens (*Malurus cyaneus*) had lower levels of natural antibodies (Jacques-Hamilton et al. 2017), which the authors argue may instead provide support for the pace-of-life hypothesis, whereby individuals that seek to maximize survival and thus mount stronger immune responses are also expected to be more risk-averse than individuals that prioritize current reproduction.

While these hypotheses provide adaptive explanations for immune–personality covariation, much of the covariation between immune responses and personality may arise simply from the dependence of both types of traits on neuroendocrine physiology (Koolhaas 2008). For example, greenfinches (*Chloris chloris*) that showed calmer coping styles (measured as tail feather damage) when brought into captivity also mounted stronger antibody responses to a novel antigen (Sild et al. 2011), which the authors hypothesize was a result of differences in stress reactivity. In fact, numerous studies have documented relationships between personality traits and stress physiology in birds, though the directionality of those relationships is highly variable (e.g., Atwell et al. 2012; Baugh et al. 2013; Moyers et al. 2018). Overall, more work is needed to determine why personality traits often covary with immune responses and/or stress physiology and the extent to which this translates to pathogen resistance.

Given the potential links among host personality, risk of exposure, and resistance, pathogens may represent strong selective forces on personality traits in bird populations (Kortet et al. 2010), but this area remains largely unexplored. Personality traits appear to be heritable in the avian taxa studied to date (e.g., Dingemanse et al. 2002; Korsten et al. 2010) and thus should be subject to selection. Future studies should examine the extent to which personality profiles in avian populations and species represent the outcome of selection from parasites and pathogens.

4.2.4.3 Personality and infection spread

Infection state might alter personality traits relevant for pathogen transmission. Acute infections that cause sickness behaviors are predicted to decrease the expression of certain personality traits such as boldness and sociability, for example, which would likely dampen spread for directly transmitted parasites (Ezenwa et al. 2016a). However, infection may in some cases exacerbate differences in individual personality traits via state-dependent feedbacks (Barber and Dingemanse 2010; Sih et al. 2015), which could augment pathogen spread. For example, a bold individual with an inherently higher risk of becoming infected due to differences in foraging behavior or aggression may then need to forage even more heavily to account for the increased energetic demands of parasitism (Sih et al. 2015), thus leading to increased aggressive interactions and a higher likelihood of pathogen spread once infected. Finally, parasites that manipulate the behavior of their hosts may be under strong selection pressure to augment particular components of host personality that facilitate their spread, which could potentially decouple behavioral syndromes (Barber and Dingemanse 2010; Poulin 2013).

Regardless of the underlying mechanism, bold individuals could act as 'superspreaders' by contributing disproportionately to pathogen spread (Lloyd-Smith et al. 2005) in any system where boldness is both relevant to transmission and maintained or even augmented during infection. There has been very little work on how infection state alters personality in birds, and experimental work is sorely needed to determine causality. However, a study of free-living great tits found that females

infected with avian malarial parasites were more exploratory than uninfected females (Dunn et al. 2011). Overall, more information is needed on how individual personality changes after infection to elucidate how personality might influence ongoing pathogen spread.

4.2.5 Mate choice and sexual selection

Mating and reproduction are essential and often taxing functions that may augment all three components of the transmission process (Table 4.1). However, birds can modify their reproductive decisions in ways that decrease pathogen risk. The relationship between parasites and sexual selection in birds has been the subject of intense scrutiny since Hamilton and Zuk (1982) proposed that female birds use the bright plumage of males to assess their genetically based pathogen resistance. Therefore, we do not attempt a comprehensive review of this topic but briefly highlight some of the ways in which avian mating or reproductive behavior might affect pathogen exposure, resistance, and transmission and some areas of inquiry that are poorly understood in birds.

4.2.5.1 Reproductive strategies and exposure

The close contact necessitated by mating and parental care can expose birds and their offspring to directly transmitted pathogens (Figure 4.7). An experimental transmission study of zebra finches, for example, found that a single mating event resulted in successful *Bacillus licheniformis* transfer to almost 90% of female birds by their mating partners (Kulkarni and Heeb 2007). In contrast, female to male transmission was significantly lower, suggesting that the infection risks associated with mating in birds may differ between the sexes. Risks associated with parasite exposure are likely to be important selective forces for the evolution of mating strategies in birds. For example, extra-pair copulations (EPCs) are a common strategy that both male and female birds employ to increase reproductive success, but this strategy may be associated with increased exposure to and transmission of sexually transmitted pathogens (Sheldon 1993). On the other hand, transfer of beneficial microbes may also occur during EPCs and may, in some cases,

Figure 4.7 Copulating tree swallows. Pathogens with direct transmission routes can be transferred during mating and parental care. (Photo credit: commons.wikimedia.org.)

outweigh the risks or costs of exposure to pathogenic bacteria (Lombardo et al. 1999). To date, the study of sexually transmitted pathogens in birds is still in its infancy, and further study is needed in birds to determine to what extent EPCs augment exposure to parasites or beneficial microbes.

The high risk of pathogen exposure associated with mating behaviors such as copulation and parental care suggests that there should be strong selection for avoidance of individuals exhibiting sickness behaviors or other signs of poor health in the context of mate choice (Hennessy et al. 2014). Mating success does appear to be higher for males with lower parasite burdens in some polygynous avian taxa (reviewed in Hart 1990), although direct evidence for avoidance of infected partners in birds remains rare. In great bustards, females intensively inspect the cloaca of males for cleanliness prior to mating (Hidalgo de Trucios and Carranza 1991), which work suggests may be an adaption to reduce risk of sexually transmitted pathogens (Bravo et al. 2014). Furthermore, sexually selected traits, which are often correlated with mating success in birds, have been associated with current infection status (Thompson et al. 1997), suggesting that female birds may use sexually selected traits to minimize direct risks of infection from their partners. For example, male common greenfinches with larger yellow tail patches had lower viremia and a faster clearance rate after experimental infection with Sindbis virus

(Lindström and Lundström 2000), suggesting that this sexually selected trait is indicative of parasite resistance in this species. There may also be direct fitness benefits to pairing with parasite-free partners if they are more likely to provide parental care than parasitized partners.

4.2.5.2 Reproductive strategies and resistance

In addition to reducing direct exposure to pathogens by choosing healthy mates, an individual's offspring might benefit by inheriting genes conferring pathogen resistance. The 'good genes hypothesis' posits that individuals will choose partners with characteristics that indicate 'outstanding…health and vigor,' which might increase the resistance of their offspring to pathogens (Hamilton and Zuk 1982). Choosing genetically dissimilar partners—and thereby producing less homozygous offspring—is another dimension of the 'good genes hypothesis' with particular relevance to pathogen transmission. Many studies have documented links between homozygosity and pathogens in birds and other taxa. For example, free-living American crows (*Corvus brachyrhynchos*) that died with signs of infectious disease (e.g., poxvirus, West Nile virus, and bacterial or fungal infections) were less heterozygous than crows that survived or died from other causes (Townsend et al. 2009). Polymorphism at the MHC may be particularly critical to pathogen recognition and resistance (Zelano and Edwards 2002), and preference for MHC-dissimilar partners has been noted across a range of taxa, including birds (e.g., Freeman-Gallant et al. 2003; Juola and Dearborn 2012). Although the cues that birds use to identify MHC-dissimilar partners are not usually identified, work suggests that at least some species (e.g., blue petrel; *Halobaena caerulea*) can distinguish the degree of MHC dissimilarity based on odor cues (Leclaire et al. 2017).

Life history trade-offs associated with reproduction can also have important consequences for pathogen resistance. For example, the 'immunocompetence handicap hypothesis' posits that the high testosterone loads associated with male mating strategies come at a cost in terms of immune defense (Folstad and Karter 1992; see Chapter 3 for further details). In support of this hypothesis, male red junglefowl (*Gallus gallus*) with longer comb lengths (a sexually selected trait) had both higher testosterone levels and fewer circulating lymphocytes (Zuk et al. 1995). However, overall support for the immunocompetence handicap hypothesis in birds and other taxa has been mixed (e.g., Roberts et al. 2004). Investment in parental care is another reproductive behavior shown to result in life history trade-offs by reducing the ability of birds to mount immune responses against novel antigens and increasing parasitemia (e.g., Knowles et al. 2009). Thus, investment in sexually selected ornaments and parental care may have important consequences for the ability of hosts to fight off pathogens once infected and to influence the degree of parasitemia that hosts reach.

4.2.5.3 Reproductive strategies and pathogen spread

There are several ways in which reproductive strategies of birds can influence pathogen spread once infected. One mechanism by which birds could augment pathogen transmission (in this case, to the subsequent generation) is through greater investment in reproduction when their chances of survival are lower (e.g., the 'terminal investment hypothesis;' Clutton-Brock 1984). Consistent with this idea, some birds invest more heavily in reproduction after experimental infection. House sparrows injected with Newcastle disease virus vaccine, for example, were more likely to lay a replacement clutch and produced larger nestlings (Bonneaud et al. 2004), and senescent male blue-footed boobies (*Sula nebouxii*) had higher reproductive success after injection with LPS (Velando et al. 2006). In addition to the risks of transmission for offspring, increased reproductive effort in times of pathogen stress could reduce an individual's probability or pace of recovery (see Section 4.2.2.1). Thus, many birds reduce rather than increase their reproductive effort during infection (e.g., Marzal et al. 2005), potentially because the energetic costs of infection and reproduction are too difficult to bear simultaneously. These reductions in reproductive investment would be predicted to decrease the potential for pathogen spread to the next generation.

Reproductive behavior could also increase pathogen transmission in socially monogamous or cooperatively breeding species, if mating partners or group members facilitate rather than avoid sick family members. Facilitation could be favored by selection if it increases the likelihood of recovery and if kin selection or the need for help with parental care, territorial defense, or other direct benefits outweighs the potential costs of pathogen transmission (Hart 1990). Behavioral facilitation of sick group members is common, for example, in the social hymenoptera, although the extent to which they receive special care or are quarantined, killed, or removed from the colony varies with context (e.g., Biganski et al. 2018). The extent to which 'caring for' (or agonism toward) sick family members occurs in birds, however, has received little attention.

4.2.6 Social behavior

Elevated disease risk has long been proposed as a major cost of living in groups (Alexander 1974). Pathogen exposure and transmission is predicted to be higher in groups (Table 4.1) because hosts are in close proximity. However, the evidence that infectious disease increases with group size is mixed, in part because transmission route may influence the way in which sociality affects risk (e.g., Cote and Poulin 1995; Rifkin et al. 2012; Schmid-Hempel 2017). Moreover, social organisms can behave in ways that reduce pathogen transmission or resistance within the social group, and thus the likelihood of exposure or the negative effects of some pathogens may be ameliorated by sociality (Ezenwa et al. 2016b). Here, we review how parasites with distinct transmission routes affect social behavior, the way in which dominance status can interact with all three components of the transmission process, 'collective defense' behaviors through which social animals might decrease pathogen transmission, and the potential for (and limitations on) pathogens to drive changes in social behaviors or degree of sociality.

4.2.6.1 Transmission routes and social behavior

Directly transmitted pathogens are predicted to be elevated in gregarious animals because of higher local densities of and more interactions among hosts (Rifkin et al. 2012). The prevalence of *M. gallisepticum*, for example, tends to increase with mean flock size in house finches (Altizer et al. 2004). In contrast, the burden of mobile ectoparasites and, potentially, the vector-borne endoparasites that they carry, may decrease with group size. Specifically, the 'encounter-dilution effect' (Mooring and Hart 1992) posits that individual risk will be lower in larger groups because there will be fewer parasites per host, as long as (1) the number of ectoparasites does not scale proportionately with group size and (2) each ectoparasite does not bite more hosts when in larger host groups. Encounter-dilution benefits may be particularly high for individuals in the center of the group (i.e., the geometry for the selfish herd; Hamilton 1971). Among American robins (*Turdus migratorius*), for example, risk for West Nile virus may be lower in larger flocks: estimated *per capita* mosquito bite rates for robins were lower within roosts, and the risk of West Nile virus exposure (i.e., the rate of seroconversion) was lower for sentinel house sparrows within the roosts than in non-roost sites (Krebs et al. 2014). Overall, however, support for a negative correlation between group size and level of ectoparasitism in birds and other taxa is mixed. Although one meta-analysis reported negative associations between ectoparasitism level and group size across taxa (Cote and Poulin 1995), another found a positive overall association between group size and infection risk, regardless of transmission route (Rifkin et al. 2012). However, most of the avian studies used in the latter analysis were focused on nest ectoparasites, which generally tend to increase in intensity with colony size and against which the encounter-dilution effect (originally proposed for volant parasites) may be less applicable. Patterns of risk with group size may therefore be different for ectoparasites that infest nests and aerial ectoparasites that feed on adults.

4.2.6.2 Dominance behavior

Some individuals within social groups may be more likely than others to experience the costs or benefits of sociality with respect to parasite risk. Dominance status is one key source of individual variation in the costs and benefits of group living. Much like personality (see Section 4.2.4), which is often correlated with dominance status in birds (e.g., David et al. 2011), dominance status can act to alter exposure to pathogens and physiological resistance once exposed (Fairbanks and Hawley 2012). For example,

because dominance status is associated with greater access to resources such as food or mates, dominant individuals may have higher exposure to parasites and pathogens (e.g., Halvorsen 1986) or may be able to mount stronger immune responses (e.g., Zuk and Johnsen 2000) due to better body condition (e.g., Piper and Wiley 1990; Senar et al. 2000) and/or lower levels of stress hormones (Goymann and Wingfield 2004). In contrast, dominant individuals may have lower resistance to parasites if they face trade-offs between maintaining dominance and mounting adequate immune responses due to pleiotropic effects of hormones (e.g., Muehlenbein and Watts 2010; Poiani et al. 2000) or resource-based mechanisms.

Relatively few studies have directly examined patterns of parasite load in relation to dominance status in birds. However, several studies have identified plumage characteristics associated with both dominance and parasite load. For example, in wild turkeys (*Meleagris gallopavo*), dominant males have longer snood lengths (Buchholz 1997) and snood length is negatively correlated with coccidian parasite load (Buchholtz 1995). Conversely, a meta-analysis of the relationship between rank and parasitism across vertebrates found that dominant males, on average, harbor higher parasite loads than subordinate males (Habig et al. 2018), supporting the hypothesis that, for many vertebrate taxa, dominant individuals may have greater exposure and/or lower resistance to parasites. However, the majority of bird studies included in the meta-analysis were of territorial male birds during the breeding season, which is likely not relevant for understanding dominance status in birds that winter or breed in social groups. Thus, further work is needed in birds to understand how dominance status relates to patterns of parasitism in the wild.

Because of the inherent difficulty in determining whether rank-related differences in parasite loads stem from differences in exposure versus resistance, experimental infection studies are useful in isolating effects of rank on resistance to pathogens once exposed. These studies, however, have yielded variable or contradictory results, suggesting that dominance status in birds can be associated with either reduced or enhanced resistance to pathogens and the overall effects of dominance may thus be system-dependent. Lindström (2004), for example, experimentally infected greenfinches with Sindbis virus and found that dominant finches had higher viremia earlier in infection but cleared infection faster than subordinate animals. Hawley et al. (2007b) experimentally infected group-living house finches with *M. gallisepticum* and found that dominant males showed significantly lower symptom severity and proportion of time symptomatic. In contrast, socially dominant canaries (*Serinus canaria domestica*) experimentally infected with *Plasmodium relictum* (one causative agent of avian malaria) showed higher parasitemia and disease severity than did their subordinate counterparts and this pattern remained regardless of the degree of competition for food (Larcombe et al. 2013).

Effects of dominance rank on pathogen resistance may be inconsistent across studies because the physiological effects of dominance are likely to be strongly influenced by the type of social system examined, which will alter the relative costs and benefits of dominance (Goymann and Wingfield 2004). Social status may also have differential effects on components of the immune system, leading to differential rank-related patterns of resistance depending on the relevant immune mechanisms for infection with a particular pathogen. An intriguing study of rhesus macaques (*Macaca mulatta*) found that changes in social rank were associated with polarized immune responses to antigens, potentially making dominant animals more susceptible to some types of pathogens but subordinate animals more susceptible to others (Snyder-Mackler et al. 2016). Consistent with this possibility, dominant male house finches mounted stronger inflammation responses to phytohemagglutinin (see Chapter 3) but weaker antibody titers in response to sheep red blood cells (Hawley et al. 2007b). As the immunological reagents available for studies in birds improve, future studies should explore whether social rank is associated with differential immune profiles, as this may help to explain discrepant patterns in rank-related resistance across systems.

An individual's infection status may also alter the ability to obtain high dominance within a group (Freeland 1981). For example, experimental infections with intestinal nematodes caused female red junglefowl to obtain lower dominance status in new social settings than their uninfected counterparts (Zuk et al. 1998). However, there is less evidence that dominance status, once established, changes when an individual becomes infected (Freeland 1981). In experimental infections of flock-housed greenfinches, infection with Sindbis virus had no impact on dominance status (Lindström 2004); likewise, there was no effect of LPS injection on dominance status in flock-housed, black-capped chickadees (*Poecile atricapillus*; Stewart and Greives 2016). However, dominant house sparrows with acute coccidiosis were more likely to lose rank (Dolnik and Hoi 2010) and dominant (but not subordinate) house finches injected with LPS showed reduced levels of aggression toward their flockmates (Moyers et al. 2015), suggesting that changes in social behavior with infection status may, in some cases, be rank-dependent. Further study is needed on how dominance status or aggressive interactions change with infection status in birds, as these changes are likely to impact further pathogen spread for directly transmitted pathogens.

4.2.6.3 Collective defense behaviors

Collective defense behaviors, also referred to as 'social immunity', are collective or cooperative actions to avoid, control, or eliminate infection (Cremer et al. 2018; Schmid-Hempel 2017). These actions, which include hygienic behaviors and reducing contact with diseased group members, can benefit both the individual host and/or the social group (Schmid-Hempel 2017) by reducing contact and augmenting resistance. Although best studied in the social insects, some social behaviors exhibited by birds would fall under the umbrella of collective defense behaviors, including self-medication of the nest and the reduction of ectoparasite burden through allopreening (introduced in Section 4.2.1.1). Allopreening and other forms of close contact may also have pathogen costs, however. In the social hymenoptera, for example, allopreening, allofeeding, and contact-based communication are thought to be major pathways for fecal–oral pathogen transmission (Biganski et al. 2018). Likewise, allopreening in American crows (Figure 4.8), combined with their communal roosting behavior and propensity to forage in contaminated areas, might all contribute

Figure 4.8 Allopreening (seen in these American crows) can reduce ectoparasite burden while facilitating spread of parasites with fecal–oral transmission routes. (Photo credit: Kevin McGowan, Cornell Lab of Ornithology.)

to the high prevalence of food-borne pathogens in this species (Taff et al. 2016).

Conversely, however, allopreening and other forms of close contact with infected conspecifics may lead to self-vaccination (Ezenwa et al. 2016b). In the social insects, this 'immune priming' triggered by contact with low doses of an antigen could provide resistance to later encounters with that pathogen (Schmid-Hempel 2017). Immune priming (albeit not in a social context) has been documented in rock pigeons: upon secondary experimental exposure, pigeons that had been previously primed via exposure to hippoboscid flies mounted a higher antifly IgY antibody response than naïve pigeons (Waite et al. 2014). The relative benefits of behaviors like allopreening could change with the pathogens that are prevalent in each environment (e.g., ectoparasites vs. parasites with fecal–oral transmission routes) and the extent to which contact leads to infection or promotes resistance.

Reducing contact with diseased group members is another example of collective defense behavior in social taxa. Although avoidance of infected group members has been documented in diverse types of taxa, including mammals (Poirotte et al. 2017), fish (Stephenson et al. 2018), and lobsters (Behringer et al. 2006), the extent to which birds avoid infected conspecifics is variable and not well understood (see Section 4.2.2.1 for an example of attraction to diseased conspecifics in birds). No studies in birds to date have documented avoidance of conspecifics actively infected with a parasite or pathogen, but Zylberberg et al. (2012) found that male house finches avoided associating with conspecific males injected with LPS and thus displaying visible signs of sickness. Intriguingly, avoidance of visibly sick conspecifics was most prominent amongst house finch males with lower levels of two innate immune parameters, suggesting that avoidance behavior may be context-dependent within species. Further work in birds is needed to determine whether and under what conditions pathogens that cause visible lesions or sickness behaviors result in active avoidance by other group members.

4.3 Synthesis and future directions

As highlighted throughout this chapter, behavior can influence every aspect of the parasite transmission process, yet many of the behaviors that might affect or be affected by parasites have not been well studied in birds. Moreover, even behaviors with seemingly straightforward relationships with parasite risk can have subtle complexities. Preening and allopreening, for example, clearly function to reduce ectoparasite load but may simultaneously increase transmission of parasites with trophic or fecal–oral transmission routes. The extent to which any antiparasite behavior is favored by selection could therefore vary with factors such as the prevalence, mode of transmission, and virulence of the pathogens within a community at a given time, which may themselves be in a constant state of ecological and evolutionary flux. Nevertheless, if the selection pressure of a pathogen is strong enough, behavioral evolution may occur.

Emerging infectious diseases, which can have devastating effects on avian populations (e.g., West Nile virus; LaDeau et al. 2007), may be particularly likely to exert strong selection pressure on behavior, although data supporting this idea are scarce. Two species of Hawaiian honeycreeper appear to have evolved daily altitudinal migrations in response to avian malaria, whereby both species move downslope to forage during the day when mosquito activity is low and move back to upper elevations that are largely mosquito-free to roost (van Riper et al. 1986). Further, roosting observations made by Warner (1968) and van Riper et al. (1986), almost 20 years apart, suggest that Hawaiian honeycreepers have begun to roost with their extremities tucked in, presumably to prevent mosquito access. However, it remains unknown whether these separate observations are, in fact, an example of behavioral evolution. Further study of whether and how recently emerged pathogens in birds have resulted in behavioral evolution is an exciting area for future exploration.

An important challenge at the intersection of animal behavior and disease ecology is understanding how differences in individual behavior scale up to impact disease dynamics at the group or population level (Barron et al. 2015). Although studies of how individual behavior influences epidemic dynamics are rare in birds, Adelman et al. (2015) showed that experimental flock epidemics of the avian bacterial pathogen *M. gallisepticum* started by house finches that spend the longest amount of time per day on bird feeders are both faster and stronger than epidemics started by birds that spend low amounts of time on bird feeders (Adelman et al. 2013). This work suggests that house finches that spend ample time on bird feeders may act as 'superspreaders' of *M. gallisepticum* by contributing disproportionately to transmission. Studies of other avian systems should examine whether personality traits, foraging or movement behaviors, or social behaviors predict 'superspreading,' and how such behaviors change during infection, when behavior is most relevant for ongoing spread. In addition, the way in which healthy individuals behave toward sick individuals is critical for predicting how host behavior will scale up to influence the dynamics of directly transmitted pathogens. Studies in avian systems are beginning to examine how sickness behavior expression and the social interactions of healthy and sick birds are influenced by both social context and the infection status of flockmates (Lopes et al. 2012; Moyers et al. 2015; Vaziri et al. 2019). Future work should build on this to understand whether there are particular contexts (i.e., breeding season, group size) in which the expression of sickness behaviors and conspecific responses to visibly sick flockmates scale up to exacerbate or dampen disease outbreaks in birds.

Overall, the intersection of animal behavior and disease ecology in birds is an exciting one that is still arguably in its infancy, in part because of the difficulties of monitoring disease and determining causation in wild populations. Thus, future work at the intersection of behavior and disease in birds needs to balance the benefits of using controlled experiments (which are not possible for all host–pathogen systems and often must occur in captivity) with the study of unmanipulated behavior in free-living animals. By integrating both types of approaches, studies in birds will continue to uncover the diverse ways in which their behavior is intimately and fundamentally connected to the ecology and evolution of their parasites.

Literature cited

Adelman, J. and Hawley, D. (2017). Tolerance of infection: a role for animal behavior, potential immune mechanisms, and consequences for parasite transmission. *Hormones and Behavior*, 88, 79–86.

Adelman, J.S., Moyers, S.C., Farine, D.R., and Hawley, D.M. (2015). Feeder use predicts both acquisition and transmission of a contagious pathogen in a North American songbird. *Proceedings of the Royal Society B: Biological Sciences*, 282, 20151429.

Adelman, J.S., Mayer, C., and Hawley, D.M. (2017). Infection reduces anti-predator behaviors in house finches. *Journal of Avian Biology*, 48, 519–28.

Alexander, R.D. (1974). The evolution of social behavior. *Annual Review of Ecology and Systematics*, 5, 325–83.

Altizer, S., Hochachka, W.M., and Dhondt, A.A. (2004). Seasonal dynamics of mycoplasmal conjunctivitis in eastern North American house finches. *Journal of Animal Ecology*, 73, 309–22.

Altizer, S., Bartel, R., and Han, B.A. (2011). Animal migration and infectious disease risk. *Science*, 331, 296–302.

Aplin, L.M., Farine, D.R., Morand-Ferron, J., Cole, E.F., Cockburn, A., and Sheldon, B.C. (2013). Individual personalities predict social behaviour in wild networks of great tits (*Parus major*). *Ecology Letters*, 16, 1365–72.

Atwell, J.W., Cardoso, G.C., Whittaker, D.J., Campbell-Nelson, S., Robertson, K.W., and Ketterson, E.D. (2012). Boldness behavior and stress physiology in a novel urban environment suggest rapid correlated evolutionary adaptation. *Behavioral Ecology*, 23, 960–9.

Aubert, A. (1999). Sickness and behaviour in animals: a motivational perspective. *Neuroscience and Biobehavioral Reviews*, 23, 1029–36.

Barber, I. and Dingemanse, N.J. (2010). Parasitism and the evolutionary ecology of animal personality. *Philosophical Transactions of the Royal Society of London B*, 365, 4077–88.

Barbosa, A., Merino, S., Lope, F., and Møller, A.P. (2002). Effects of feather lice on flight behavior of male barn swallows (*Hirundo rustica*). *The Auk*, 119, 213–16.

Barron, D.G., Gervasi, S.S., Pruitt, J.M., and Martin, L.B. (2015). Behavioral competence: how host behaviors can interact to influence parasite transmission risk. *Current Opinion in Behavioral Sciences*, 6, 35–40.

Baugh, A.T., van Oers, K., Naguib, M., and Hau, M. (2013). Initial reactivity and magnitude of the acute stress response associated with personality in wild great tits (*Parus major*). *General and Comparative Endocrinology*, 189, 96–104.

Behringer, D.C., Butler, M.J., and Shields, J.D. (2006). Avoidance of disease by social lobsters. *Nature*, 441, 421.

Benskin, C.M.H., Wilson, K., Jones, K., and Hartley, I.R. (2009). Bacterial pathogens in wild birds: a review of the frequency and effects of infection. *Biological Reviews*, 84, 349–73.

Biganski, S., Kurze, C., Muller, M.Y., and Moritz, R.F.A. (2018). Social response of healthy honeybees towards *Nosema ceranae*-infected workers: care or kill? *Apidologie*, 49, 325–34.

Bitton, G. (1980). *Introduction to Environmental Virology*. John Wiley & Sons, New York.

Bonneaud, C., Mazuc, J., Chastel, O., Westerdahl, H., and Sorci, G. (2004). Terminal investment induced by immune challenge and fitness traits associated with major histocompatibility complex in the house sparrow. *Evolution*, 58, 2823–30.

Bouwman, K.M. and Hawley, D.M. (2010). Sickness behaviour acting as an evolutionary trap? Male house finches preferentially feed near diseased conspecifics. *Biology Letters*, 6, 462–5.

Boyer, N., Réale, D., Marmet, J., Pisanu, B., and Chapuis, J.-L. (2010). Personality, space use and tick load in an introduced population of Siberian chipmunks *Tamias sibiricus*. *Journal of Animal Ecology*, 79, 538–47.

Bradley, C.A. and Altizer, S. (2005). Parasites hinder monarch butterfly flight: implications for disease spread in migratory hosts. *Ecology Letters*, 8, 290–300.

Bravo, C., Bautista, L.M., García-Paris, M., Blanco, B., and Alonso, J.C. (2014). Males of a strongly polygynous species consume more poisonous food than females. *PLoS ONE*, 9, e111057.

Brooke, M.D. (1985). The effect of allopreening on tick burdens of molting eudyptid penguins. *The Auk*, 102, 893–5.

Brown, C.R., Brown, M.B., Moore, A.T., and Komar, N. (2007). Bird movement predicts Buggy Creek virus infection in insect vectors. *Vector-Borne and Zoonotic Diseases*, 7, 304–14.

Brown, C.R., Roche, E.A., and Brown, M.B. (2017). Why come back home? Breeding-site fidelity varies with group size and parasite load in a colonial bird. *Animal Behaviour*, 132, 167–80.

Buchholz, R. (1995). Female choice, parasite load and male ornamentation in wild turkeys. *Animal Behaviour*, 50, 929–43.

Buchholz, R. (1997). Male dominance and variation in fleshy head ornamentation in wild turkeys. *Journal of Avian Biology*, 28, 223–30.

Buehler, D.M., Piersma, T., Matson, K., and Tieleman, B.I. (2008). Seasonal redistribution of immune function in a migrant shorebird: annual-cycle effects override adjustments to thermal regime. *The American Naturalist*, 172, 783–96.

Bush, S.E. and Clayton, D.H. (2018). Anti-parasite behaviour of birds. *Philosophical Transactions of the Royal Society B: Biological Sciences*, 373, 13.

Casagrande, S., Pinxten, R., Zaid, E., and Eens, M. (2015). Birds receiving extra carotenoids keep singing during the sickness phase induced by inflammation. *Behavioral Ecology and Sociobiology*, 69, 1029–37.

Clark, L. and Mason, J.R. (1985). Use of nest material as insecticidal and anti-pathogenic agents by the European starling. *Oecologia*, 67, 169–76.

Clark, L. and Mason, J.R. (1988). Effect of biologically active plants used as nest material and the derived benefit to starling nestlings. *Oecologia*, 77, 174–80.

Clayton, D.H., Moyer, B.R., Bush, S.E., et al. (2005). Adaptive significance of avian beak morphology for ectoparasite control. *Proceedings of the Royal Society B: Biological Sciences*, 272, 811–17.

Clutton-Brock, T.H. (1984). Reproductive effort and terminal investment in iteroparous animals. *The American Naturalist*, 123, 212–29.

Comstedt, P., Bergström, S., Olsen, B., et al. (2006). Migratory passerine birds as reservoirs of Lyme borreliosis in Europe. *Emerging Infectious Disease Journal*, 12, 1087.

Cornet, S., Nicot, A., Rivero, A., and Gandon, S. (2013). Malaria infection increases bird attractiveness to uninfected mosquitoes. *Ecology Letters*, 16, 323–9.

Cote, I.M. and Poulin, R. (1995). Parasitism and group-size in social animals—a metaanalysis. *Behavioral Ecology*, 6, 159–65.

Cotgreave, P. and Clayton, D.H. (1994). Comparative analysis of time spent grooming by birds in relation to parasite load. *Behaviour*, 131, 171–87.

Cremer, S., Pull, C.D., and Furst, M.A. (2018). Social immunity: emergence and evolution of colony-level disease protection. In M.R. Berenbaum (ed.) *Annual Review of Entomology, Vol. 63*. Annual Reviews, Palo Alto.

Darbro, J.M. and Harrington, L.C. (2007). Avian defensive behavior and blood-feeding success of the West Nile vector mosquito, *Culex pipiens*. *Behavioral Ecology*, 18, 750–57.

Darbro, J.M., Dhondt, A.A., Verrneylen, F.M., and Harrington, L.C. (2007). *Mycoplasma gallisepticum* infection in house finches (*Carpodacus mexicanus*) affects mosquito blood feeding patterns. *American Journal of Tropical Medicine and Hygiene*, 77, 488–94.

David, M., Auclair, Y., and Cézilly, F. (2011). Personality predicts social dominance in female zebra finches, *Taeniopygia guttata*, in a feeding context. *Animal Behaviour*, 81, 219–24.

de Roode, J.C., Lefèvre, T., and Hunter, M.D. (2013). Self-medication in animals. *Science*, 340, 150–51.

Dingemanse, N., Both, C., Drent, P., Van Oers, K., and Van Noordwijk, A. (2002). Repeatability and heritability of exploratory behaviour in great tits from the wild. *Animal Behaviour*, 64, 929–38.

Dizney, L. and Dearing, M.D. (2013). The role of behavioural heterogeneity on infection patterns: implications for pathogen transmission. *Animal Behaviour*, 86, 911–16.

Dolnik, O.V. and Hoi, H. (2010). Honest signaling, dominance hierarchies and body condition in house sparrows *Passer domesticus* (Aves: Passeriformes) during acute coccidiosis. *Biological Journal of the Linnean Society*, 99, 718–26.

Dolnik, O.V., Dolnik, V.R., and Bairlein, F. (2010). The effect of host foraging ecology on the prevalence and intensity of coccidian infection in wild passerine birds. *Ardea*, 98, 97–103.

Dunn, J.C., Cole, E.F., and Quinn, J.L. (2011). Personality and parasites: sex-dependent associations between avian malaria infection and multiple behavioural traits. *Behavioral Ecology and Sociobiology*, 65, 1459–71.

Eisen, L., Eisen, R.J., and Lane, R.S. (2004). The roles of birds, lizards, and rodents as hosts for the western black-legged tick *Ixodes pacificus*. *Journal of Vector Ecology*, 29, 295–308.

Ezenwa, V.O. (2004). Selective defecation and selective foraging: anti-parasite behavior in wild ungulates? *Ethology*, 110, 851–62.

Ezenwa, V.O., Archie, E.A., Craft, M., et al. (2016a). Host–behavior–parasite feedbacks: an essential link between animal behavior and disease ecology. *Proceedings of the Royal Society B: Biological Sciences*, 283, 20153078.

Ezenwa, V.O., Ghai, R.R., McKay, A.F., and Williams, A.E. (2016b). Group living and pathogen infection revisited. *Current Opinion in Behavioral Sciences*, 12, 66–72.

Fairbanks, B.M. and Hawley, D.M. (2011). Interactions between host social behavior, physiology, and disease susceptibility: the role of dominance status and social context. In G.E. Demas and R.J. Nelson, eds. *Ecoimmunology*, pp. 440–67. Oxford University Press, Oxford.

Figuerola, J. and Green, A.J. (2000). Haematozoan parasites and migratory behaviour in waterfowl. *Evolutionary Ecology*, 14, 143–53.

Folstad, I. and Karter, A.J. (1992). Parasites, bright males, and the immunocompetence handicap. *The American Naturalist*, 139, 603–22.

Freeland, W.J. (1981). Parasitism and behavioral dominance among male mice. *Science*, 213, 461–2.

Freeman-Gallant, C.R., Meguerdichian, M., Wheelwright, N.T., and Sollecito, S.V. (2003). Social pairing and female mating fidelity predicted by restriction fragment length polymorphism similarity at the major histocompatibility complex in a songbird. *Molecular Ecology*, 12, 3077–83.

Gaidet, N., Caron, A., Cappelle, J., et al. (2012). Understanding the ecological drivers of avian influenza virus infection in wildfowl: a continental-scale study across Africa. *Proceedings of the Royal Society B: Biological Sciences*, 279, 1131–41.

Garamszegi, L.Z., Zagalska-Neubauer, M., Canal, D., et al. (2015). Malaria parasites, immune challenge, MHC variability, and predator avoidance in a passerine bird. *Behavioral Ecology*, 26, 1292–302.

Goymann, W. and Wingfield, J.C. (2004). Allostatic load, social status and stress hormones: the costs of social status matter. *Animal Behaviour*, 67, 591–602.

Gylfe, A., Bergström, S., Lundström, J., and Olsen, B. (2000). Reactivation of *Borrelia* infection in birds. *Nature*, 403, 724.

Habig, B., Doellman, M.M., Woods, K., Olansen, J., and Archie, E.A. (2018). Social status and parasitism in male and female vertebrates: a meta-analysis. *Scientific Reports*, 8, 3629.

Halvorsen, O. (1986). On the relationship between social status of host and risk of parasitic infection. *Oikos*, 47, 71–4.

Hamilton, W.D. (1971). Geometry for the selfish herd. *Journal of Theoretical Biology*, 31, 295–311.

Hamilton, W.D. and Zuk, M. (1982). Heritable true fitness and bright birds: a role for parasites? *Science*, 218, 384–7.

Hart, B.L. (1988). Biological basis of the behavior of sick animals. *Neuroscience and Biobehavioral Reviews*, 12, 123–37.

Hart, B.L. (1990). Behavioral adaptations to pathogens and parasites—five strategies. *Neuroscience and Biobehavioral Reviews*, 14, 273–94.

Hasselquist, D., Lindström, A., Jenni-Eiermann, S., Koolhaas, A., and Piersma, T. (2007). Long flights do not influence immune responses of a long-distance migrant bird: a wind-tunnel experiment. *Journal of Experimental Biology*, 210, 1123–31.

Hawley, D.M., Davis, A.K., and Dhondt, A.A. (2007a). Transmission-relevant behaviours shift with pathogen infection in wild house finches (*Carpodacus mexicanus*). *Canadian Journal of Zoology*, 85, 752–7.

Hawley, D.M., Jennelle, C., Sydenstricker, K., and Dhondt, A. (2007b). Pathogen resistance and immunocompetence covary with social status in house finches (*Carpodacus mexicanus*). *Functional Ecology*, 21, 520–27.

Hawley, D.M., Etienne, R.S., Ezenwa, V.O., and Jolles, A.E. (2011). Does animal behavior underlie covariation between hosts' exposure to infectious agents and susceptibility to infection? Implications for disease dynamics. *Integrative and Comparative Biology*, 51, 528–39.

Hennessy, M.B., Deak, T., and Schiml, P.A. (2014). Sociality and sickness: have cytokines evolved to serve social functions beyond times of pathogen exposure? *Brain, Behavior, and Immunity*, 37, 15–20.

Hildalgo de Trucios, S.J. and Carranza, J. (1991). Timing, structure, and functions of the courtship display in male great bustard. *Ornis Scandinavica*, 22, 360–66.

Hill, N., Takekawa, J.Y., Cardona, C.J., et al. (2010). Waterfowl ecology and avian influenza in California: do host traits inform us about viral occurrence? *Avian Diseases*, 54, 426–32.

Hoye, B.J., Fouchier, R.A.M., and Klaassen, M. (2012). Host behaviour and physiology underpin individual variation in avian influenza virus infection in migratory Bewick's swans. *Proceedings of the Royal Society B: Biological Sciences*, 279, 529–34.

Hutchings, M.R., Kyriazakis, I., Anderson, D.H., Gordon, I.J., and Coop, R.L. (1998). Behavioural strategies used by parasitized and non-parasitized sheep to avoid ingestion of gastro-intestinal nematodes associated with faeces. *Animal Science*, 67, 97–106.

Jacques-Hamilton, R., Hall, M.L., Buttemer, W.A., et al. (2017). Personality and innate immune defenses in a wild bird: evidence for the pace-of-life hypothesis. *Hormones and Behavior*, 88, 31–40.

Juola, F.A. and Dearborn, D.C. (2012). Sequence-based evidence for major histocompatibility complex-disassortative mating in a colonial seabird. *Proceedings of the Royal Society B: Biological Sciences*, 279, 153–62.

Kazlauskas, N., Klappenbach, M., Depino, A.M., and Locatelli, F.F. (2016). Sickness behavior in honey bees. *Frontiers in Physiology*, 7, 10.

Klein, S.L. (2003). Parasite manipulation of the proximate mechanisms that mediate social behavior in vertebrates. *Physiology & Behavior*, 79, 441–9.

Knowles, S.C.L., Nakagawa, S., and Sheldon, B. C. (2009). Elevated reproductive effort increases blood parasitaemia and decreases immune function in birds: a meta-regression approach. *Functional Ecology*, 23, 405–15.

Konsman, J.P., Parnet, P., and Dantzer, R. (2002). Cytokine-induced sickness behaviour: mechanisms and implications. *Trends in Neurosciences*, 25, 154–9.

Koolhaas J.M (2008). Coping style and immunity in animals: making sense of individual variation. *Brain, Behavior, and Immunity*, 22, 662–7.

Korsten, P., Mueller, J.C., Hermannstädter, C., et al. (2010). Association between DRD4 gene polymorphism and personality variation in great tits: a test across four wild populations. *Molecular Ecology*, 19, 832–43.

Kortet, R., Hedrick, A.V., and Vainikka, A. (2010). Parasitism, predation and the evolution of animal personalities. *Ecology Letters*, 13, 1449–58.

Krebs, B.L., Anderson, T.K., Goldberg, T.L., et al. (2014). Host group formation decreases exposure to vector-borne disease: a field experiment in a 'hotspot' of West Nile virus transmission. *Proceedings of the Royal Society B: Biological Sciences*, 281, 7.

Kulkarni, S. and Heeb, P. (2007). Social and sexual behaviours aid transmission of bacteria in birds. *Behavioural Processes*, 74, 88–92.

Ladeau, S.L., Kilpatrick, A.M., and Marra, P.P. (2007). West Nile virus emergence and large-scale declines of North American bird populations. *Nature*, 447, 710–13.

Lafuma, L., Lambrechts, M.M., and Raymond, M. (2001). Aromatic plants in bird nests as a protection against blood-sucking flying insects? *Behavioural Processes*, 56, 113–20.

Larcombe, S., Bichet, C., Cornet, S., Faivre, B., and Sorci, G. (2013). Food availability and competition do not modulate the costs of *Plasmodium* infection in dominant male canaries. *Experimental Parasitology*, 135, 708–14.

Latorre-Margalef, N., Gunnarsson, G., Munster, V., et al. (2009). Effects of influenza virus A infection on migrating mallard ducks. *Proceedings of the Royal Society B: Biological Sciences*, 276, 20081501.

Leclaire, S., Strandh, M., Mardon, J., Westerdahl, H., and Bonadonna, F. (2017). Odour-based discrimination of similarity at the major histocompatibility complex in birds. *Proceedings of the Royal Society B: Biological Sciences*, 284, 5.

Lindström, K.M. (2004). Social status in relation to Sindbis virus infection clearance in greenfinches. *Behavioral Ecology and Sociobiology*, 55, 236–41.

Lindström, K. and Lundström, J. (2000). Male greenfinches (*Carduelis chloris*) with brighter ornaments have higher virus infection clearance rate. *Behavioral Ecology and Sociobiology*, 48, 44–51.

Lloyd-Smith, J., Schreiber, S., Kopp, P., and Getz, W. (2005). Superspreading and the effect of individual variation on disease emergence. *Nature*, 438, 355–9.

Loehle, C. (1995). Social barriers to pathogen transmission in wild animal populations. *Ecology*, 76, 326–35.

Lombardo, M.P., Thorpe, P.A., and Power, H.W. (1999). The beneficial sexually transmitted microbe hypothesis of avian copulation. *Behavioral Ecology*, 10, 333–7.

Lopes, P.C. (2017). Why are behavioral and immune traits linked? *Hormones and Behavior*, 88, 52–9.

Lopes, P.C., Adelman, J., Wingfield, J. C., and Bentley, G.E. (2012). Social context modulates sickness behavior. *Behavioral Ecology and Sociobiology*, 66, 1421–8.

Lopes, P.C., Springthorpe, D., and Bentley, G.E. (2014). Increased activity correlates with reduced ability to mount immune defenses to endotoxin in zebra finches. *Journal of Experimental Zoology Part A—Ecological Genetics and Physiology*, 321, 422–31.

Lopes, P.C., Block, P., and König, B. (2016). Infection-induced behavioural changes reduce connectivity and the potential for disease spread in wild mouse contact networks. *Scientific Reports*, 6, 31790.

Marini, M.A., Reinert, B.L, Bornschein, M.R., Pinto, J.C., and Pichorim, M.A. (1996). Ecological correlates of ectoparasitism on Atlantic Forest birds, Brazil. *Ararajuba*, 4, 93–102.

Marsot, M., Henry, P.-Y., Vourc'h, G., et al. (2012). Which forest bird species are the main hosts of the tick, *Ixodes ricinus*, the vector of *Borrelia burgdorferi* sensu lato, during the breeding season? *International Journal for Parasitology*, 42, 781–8.

Martin, C.D. and Mullens, B.A. (2012). Housing and dust-bathing effects on northern fowl mites (*Ornithonyssus sylviarum*) and chicken body lice (*Menacanthus stramineus*) on hens. *Medical and Veterinary Entomology*, 26, 323–33.

Marzal, A., De Lope, F., Navarro, C., and Moller, A.P. (2005). Malarial parasites decrease reproductive success: an experimental study in a passerine bird. *Oecologia*, 142, 541–5.

Moller, A.P. (1990). Effects of parasitism by a hematophagous mite on reproduction in the barn swallow. *Ecology*, 71, 2345–57.

Moore, J. (2002). *Parasites and the Behavior of Animals*. Oxford University Press, Oxford.

Mooring, M.S. and Hart, B.L. (1992). Animal grouping for protection from parasites—selfish herd and encounter-dilution effects. *Behaviour*, 123, 173–93.

Moreno-Rueda, G. (2017). Preen oil and bird fitness: a critical review of the evidence. *Biological Reviews*, 92, 2131–43.

Moyers, S.C., Kosarski, K.B., Adelman, J.S., and Hawley, D.M. (2015). Interactions between social behavior and the acute phase response in house finches. *Behaviour* 15: 2039–58.

Moyers, S.C., Adelman, J.S., Moore, I., Farine, D., and Hawley, D.M. (2018). Exploratory behavior is linked to stress physiology and social network centrality in free-living house finches (*Haemorhous mexicanus*). *Hormones and Behavior*, 102, 105–13.

Muehlenbein, M.P. and Watts, D.P. (2010). The costs of dominance: testosterone, cortisol and intestinal parasites in wild male chimpanzees. *Biopsychosocial Medicine*, 4, 21.

Munster, V. J. and Fouchier, R.A.M. (2009). Avian influenza virus: of virus and bird ecology. *Vaccine*, 27, 6340–44.

Natoli, E., Say, L., Cafazzo, S., Bonanni, R., Schmid, M., and Pontier, D. (2005). Bold attitude makes male urban feral domestic cats more vulnerable to feline immunodeficiency virus. *Neuroscience & Biobehavioral Reviews*, 29, 151–7.

Nebel, S., Buehler, D.M., Macmillan, A., and Guglielmo, C.G. (2013). Flight performance of western sandpipers, *Calidris mauri*, remains uncompromised when mounting an acute phase immune response. *Journal of Experimental Biology*, 216, 2752–9.

Newman, E.A., Eisen, L., Eisen, R.J., et al. (2015). *Borrelia burgdorferi* sensu lato spirochetes in wild birds in northwestern California: associations with ecological factors, bird behavior and tick infestation. *PLoS ONE*, 10, e0118146.

Norris, K. (1999). A trade-off between energy intake and exposure to parasites in oystercatchers feeding on a bivalve mollusc. *Proceedings of the Royal Society B: Biological Sciences*, 266, 1703–09.

Oppliger, A., Richner, H., and Christe, P. (1994). Effect of an ectoparasite on lay date, nest-site choice, desertion, and hatching success in the great tit (*Parus major*). *Behavioral Ecology*, 5, 130–34.

Owen, J.C. and Moore, F.R. (2006). Seasonal differences in immunological condition of three species of thrushes. *The Condor*, 108, 389–98.

Owen, J., Moore, F., Panella, N., et al. (2006). Migrating birds as dispersal vehicles for West Nile virus. *EcoHealth*, 3, 79–85.

Owen-Ashley, N.T. and Wingfield, J.C. (2006). Seasonal modulation of sickness behavior in free-living northwestern song sparrows (*Melospiza melodia morphna*). *Journal of Experimental Biology*, 209, 3062–70.

Piper, W. and Wiley, R. (1990). The relationship between social dominance, subcutaneous fat, and annual survival in wintering white-throated sparrows (*Zonotrichia albicollis*). *Behavioral Ecology and Sociobiology*, 26, 201–8.

Poiani, A., Goldsmith, A.R., and Evans, M.R. (2000). Ectoparasites of house sparrows (*Passer domesticus*): an experimental test of the immunocompetence handicap hypothesis and a new model. *Behavioral Ecology and Sociobiology*, 47, 230–42.

Poirotte, C., Massol, F., Herbert, A., et al. (2017). Mandrills use olfaction to socially avoid parasitized conspecifics. *Science Advances*, 3, 8.

Poulin, R. (2013). Parasite manipulation of host personality and behavioural syndromes. *Journal of Experimental Biology*, 216, 18–26.

Reisen, W., Lothrop, H., Chiles, R., et al. (2004). West Nile virus in California. *Emerging Infectious Diseases*, 10, 1369–78.

Revis, H.C. and Waller, D.A. (2004). Bactericidal and fungicidal activity of ant chemicals on feather parasites: an evaluation of anting behavior as a method of self-medication in songbirds. *The Auk*, 121, 1262–8.

Richner, H., Oppliger, A., and Christe, P. (1993). Effect of an ectoparasite on reproduction in great tits. *Journal of Animal Ecology*, 62, 703–10.

Rifkin, J.L., Nunn, C.L., and Garamszegi, L.Z. (2012). Do animals living in larger groups experience greater parasitism? A meta-analysis. *The American Naturalist*, 180, 70–82.

Roberts, M.L., Buchanan, K.L., and Evans, M.R. (2004). Testing the immunocompetence handicap hypothesis: a review of the evidence. *Animal Behaviour*, 68, 227–39.

Saranathan, V. and Burtt, E.H. (2007). Sunlight on feathers inhibits feather-degrading bacteria. *Wilson Journal of Ornithology*, 119, 239–45.

Schmid-Hempel, P. (2017). Parasites and their social hosts. *Trends in Parasitology*, 33, 453–62.

Senar, J.C., Polo, V., Uribe, F., and Camerino, M. (2000). Status signalling, metabolic rate and body mass in the siskin: the cost of being a subordinate. *Animal Behaviour*, 59, 103–10.

Sheldon, B.C. (1993). Sexually transmitted disease in birds: occurrence and evolutionary significance. *Philosophical Transactions of the Royal Society B: Biological Sciences*, 339, 491–7.

Sih, A., Bell, A., and Johnson, J.C. (2004). Behavioral syndromes: an ecological and evolutionary overview. *Trends in Ecology and Evolution*, 19, 372–8.

Sih, A., Mathot, K.J., Moiron, M., Montiglio, P.O., Wolf, M., and Dingemanse, N.J. (2015). Animal personality and state-behaviour feedbacks: a review and guide for empiricists. *Trends in Ecology and Evolution*, 30, 50–60.

Sild, E., Sepp, T., and Hõrak, P. (2011). Behavioural trait covaries with immune responsiveness in a wild passerine. *Brain, Behavior, and Immunity*, 25, 1349–54.

Snijders, L., van Rooij, E.P., Burt, J.M., Hinde, C.A., van Oers, K., and Naguib, M. (2014). Social networking in territorial great tits: slow explorers have the least central social network positions. *Animal Behaviour*, 98, 95–102.

Snyder-Mackler, N., Sanz, J., Kohn, J.N., et al. (2016). Social status alters immune regulation and response to infection in macaques. *Science*, 354, 1041–5.

Stamps, J.A. (2007). Growth-mortality tradeoffs and 'personality traits' in animals. *Ecology Letters*, 10, 355–63.

Stephenson, J.F., Perkins, S.E., and Cable, J. (2018). Transmission risk predicts avoidance of infected conspecifics in Trinidadian guppies. *Journal of Animal Ecology*, 87, 1525–33.

Stewart, E.C. and Greives, T.J. (2016). Short-term immune challenge does not influence social dominance behaviour in top-ranked black-chapped chickadees. *Animal Behaviour*, 120, 77–82.

Stroeymeyt, N., Grasse, A.V., Crespi, A., Mersch, D.P., Cremer, S., and Keller, L. (2018). Social network plasticity decreases disease transmission in a eusocial insect. *Science*, 362, 941–5.

Suarez-Rodriguez, M. and Garcia, C.M. (2017). An experimental demonstration that house finches add cigarette butts in response to ectoparasites. *Journal of Avian Biology*, 48, 1316–21.

Suarez-Rodriguez, M., Lopez-Rull, I., and Garcia, C.M. (2013). Incorporation of cigarette butts into nests reduces nest ectoparasite load in urban birds: new ingredients for an old recipe? *Biology Letters*, 9.

Suarez-Rodriguez, M., Montero-Montoya, R., and Garcia, C.M. (2017). Anthropogenic nest materials may increase breeding costs for urban birds. *Frontiers in Ecology and Evolution*, 5, 10.

Taff, C.C., Weis, A.M., Wheeler, S., et al. (2016). Influence of host ecology and behavior on *Campylobacter jejuni* prevalence and environmental contamination risk in a synanthropic wild bird species. *Applied and Environmental Microbiology*, 82, 4811–20.

Thompson, C.W., Hillgarth, N., Leu, M., and Mcclure, H.E. (1997). High parasite load in house finches (*Carpodacus mexicanus*) is correlated with reduced expression of a sexually selected trait. *The American Naturalist*, 149, 270–94.

Tian, H.Y., Zhou, S., Dong, L., et al. (2015). Avian influenza H5N1 viral and bird migration networks in Asia. *Proceedings of the National Academy of Sciences of the United States of America*, 112, 172–7.

Townsend, A.K., Clark, A.B., Mcgowan, K.J., Buckles, E.L., Miller, A.D., and Lovette, I.J. (2009). Disease-mediated inbreeding depression in a large, open population of cooperative crows. *Proceedings of the Royal Society B: Biological Sciences*, 276, 2057–64.

van Gils, J.A., Munster, V.J., Radersma, R., Liefhebber, D., Fouchier, R.A.M., and Klaassen, M. (2007). Hampered foraging and migratory performance in swans infected with low-pathogenic avian influenza A virus. *PLoS ONE*, 2, e184.

van Overveld, T. and Mattysen, E. (2010). Personality predicts spatial responses to food manipulations in free-ranging great tits (*Parus major*). *Biology Letters*, 6, 187–90.

van Riper, C. III, van Riper, S.G., Goff, M.L., and Laird, M. (1986). The epizootiology and ecological significance of malaria in Hawaiian land birds. *Ecological Monographs*, 56, 327–44.

Vaziri, G.J., Johny, M.M., Caragea, P.C., and Adelman, J.S. (2019). Social context affects thermoregulation but not activity level during avian immune response. *Behavioral Ecology*, 30, 383–92.

Velando, A., Drummond, H., and Torres, R. (2006). Senescent birds redouble reproductive effort when ill: confirmation of the terminal investment hypothesis. *Proceedings of the Royal Society B: Biological Sciences*, 273, 1443–8.

Vezzoli, G., Mullens, B.A., and Mench, J.A. (2015). Dustbathing behavior: do ectoparasites matter? *Applied Animal Behaviour Science*, 169, 93–9.

Viblanc, V.A., Mathien, A., Saraux, C., Viera, V.M., and Groscolas, R. (2011). It costs to be clean and fit: energetics of comfort behavior in breeding-fasting penguins. *PLoS ONE*, 6, 10.

Villa, S.M., Campbell, H.E., Bush, S.E., and Clayton, D.H. (2016a). Does anti-parasite behavior improve with experience? An experimental test of the priming hypothesis. *Behavioral Ecology*, 27, 1167–71.

Villa, S.M., Goodman, G.B., Ruff, J.S., and Clayton, D.H. (2016b). Does allopreening control avian ectoparasites? *Biology Letters*, 12, 4.

Villa, S.M., Koop, J.A.H., Le Bohec, C., and Clayton, D.H. (2018). Beak of the pinch: anti-parasite traits are similar among Darwin's finch species. *Evolutionary Ecology*, 32, 443–52.

Waite, J.L., Henry, A.R., Owen, J.P., and Clayton, D.H. (2014). An experimental test of the effects of behavioral and immunological defenses against vectors: do they interact to protect birds from blood parasites? *Parasites & Vectors*, 7, 1–11.

Waldenström, J., Broman, T., Carlssoon, I., et al. (2002). Prevalence of *Campylobacter jejuni*, *Campylobacter lari*, and *Campylobacter coli* in different ecological guilds and taxa of migrating birds. *Applied and Environmental Microbiology*, 68, 5911–17.

Warner, R.E. (1968). The role of introduced diseases in the extinction of the endemic Hawaiian avifauna. *Condor*, 70, 101–20.

Yirmiya, R., Avitsur, R., Donchin, O., and Cohen, E. (1995). Interleukin-1 inhibits sexual behavior in female but not in male rats. *Brain Behavior and Immunity*, 9, 220–33.

Yorinks, N. and Atkinson, C.T. (2000). Effects of malaria on activity budgets of experimentally infected juvenile apapane (*Himatione sanguinea*). *The Auk*, 117, 731–8.

Zelano, B. and Edwards, S.V. (2002). An MHC component to kin recognition and mate choice in birds: predictions, progress, and prospects. *The American Naturalist*, 160, S225–37.

Zuk, M. and Johnsen, T.S. (2000). Social environment and immunity in male red jungle fowl. *Behavioral Ecology*, 11, 146–53.

Zuk, M., Johnsen, T.S., and Maclarty, T. (1995). Endocrine–immune interactions, ornaments and mate choice in red jungle fowl. *Proceedings of the Royal Society B: Biological Sciences*, 260, 205–10.

Zuk, M., Kim, T., Robinson, S.I., and Johnsen, T.S. (1998). Parasites influence social rank and morphology, but not mate choice, in female red junglefowl, *Gallus gallus*. *Animal Behaviour*, 56, 493–9.

Zylberberg, M., Klasing, K.C., and Hahn, T.P. (2012). House finches (*Carpodacus mexicanus*) balance investment in behavioural and immunological defences against pathogens. *Biology Letters*, 9, 20120856.

Zylberberg, M., Klasing, K.C., and Hahn, T.P. (2014). In house finches, *Haemorhous mexicanus*, risk takers invest more in innate immune function. *Animal Behaviour*, 89, 115–22.

CHAPTER 5

Host–Pathogen Evolution and Coevolution in Avian Systems

Camille Bonneaud

5.1 Introduction

Antagonistic interactions between hosts and their pathogens can generate selection pressures that shape the evolutionary trajectories of both interacting species (Stearns 1999; Thompson 1994; Woolhouse et al. 2002). The impacts of pathogens on their hosts can be so strong that they can lead to host population declines and even extinctions (see Chapter 7). In response, hosts can evolve a range of counter-strategies that prevent infection or minimize its costs, with critical repercussions for pathogen persistence and transmission. Such reciprocal selection pressures on hosts and pathogens are expected to give rise to patterns of adaptations and counter-adaptations—the hallmarks of coevolution (Gaba and Ebert 2009; Thompson 2005). While there is accumulating evidence for adaptive evolution and coevolution of hosts and pathogens in laboratory settings (e.g., Buckling and Rainey 2002; Eizaguirre et al. 2012), demonstrating such processes in free-ranging populations, particularly of vertebrate hosts, remains challenging.

The paucity of studies demonstrating adaptive host–pathogen evolution and coevolution in free-ranging populations of vertebrates stems largely from the difficulties of demonstrating adaptive change in both antagonists in direct response to each other (Box 5.1). Consequently, studies typically contrast the traits of one antagonist across space in the presence or absence of the other (Nuismer 2006;

Thompson 2005) or infer adaptive change from genetic data (e.g., Di Giallonardo et al. 2016). Population and genetic differences can, however, be shaped by other unidentified antagonists, environmental factors (e.g., climate, food, predators), or even drift (Gaba and Ebert 2009). Recent outbreaks of novel infectious pathogens in birds have provided unique opportunities to complement such approaches with more direct observations of adaptations and counter-adaptations of interacting hosts and pathogens over multiple generations and in real-time.

In this chapter, I summarize the different adaptive strategies that avian hosts and their pathogens can evolve in response to each other. I then present emerging empirical evidence in support of the evolution and coevolution of such strategies in the wild, largely adopting a case study approach to best illustrate each point. In adopting a case study approach, I will be inevitably selective, and so I caution readers against concluding that adaptive evolution and coevolution is the predominant outcome of host–pathogen interactions—it might be, but owing to the challenges of demonstrating each, it would be premature to conclude one way or the other. To guide future tests of host–pathogen evolution and coevolution in avian systems, I summarize the necessary supporting criteria (Box 5.1) and outline the strengths and weaknesses of candidate experimental approaches that allow us to shed light on the evolutionary processes at

Camille Bonneaud, *Host–Pathogen Evolution and Coevolution in Avian Systems* In: *Infectious Disease Ecology of Wild Birds.*
Edited by: Jennifer C. Owen, Dana M. Hawley, and Kathryn P. Huyvaert, Oxford University Press. © Oxford University Press 2021.
DOI: 10.1093/oso/9780198746249.003.0005

Box 5.1 Schematic of core requirements for demonstrating evolution and coevolution

Evolutionary theory is often used to help understand and predict host–pathogen interactions, but testing the degrees to which such interactions are shaped by adaptive evolution and coevolution is challenging. At its core, demonstrating evolution by natural selection requires three steps (Boxed Figure 5.1). **Step I**: a specific environmental change (A to B), such as a disease outbreak, is identified. **Step II**: this change positively selects for one genotype over another (e.g., stars over circles). **Step III**: this fitness difference generates a change in frequencies of specific genotypes (i.e., stars increase). Note: because demonstrating the precise genes under selection in Steps II and III is challenging, and because insights into evolution and coevolution are often more tangible at the phenotypic level, many studies focus on phenotypes. In such cases, one must assume that phenotypic variation is underpinned by genetic variation and that phenotypic plasticity does not confound evidence of phenotypic evolution (**Step III′**)—a problem that can be mitigated by investigating associated changes in genes of functional relevance (e.g., immune genes) (**Step III″**).

Omitting any steps also requires assumptions to be highlighted. For example, genomic approaches tend to bypass all steps and analyze evidence for molecular evolution by natural selection (square 'Genotypes Environment B'), resulting in assumptions about both the selective drivers and phenotypic consequences. Studies testing Step II (correlation between genetic [or phenotypic] variants and fitness),

without demonstrating the precise selective driver in isolation of numerous confounding selective forces (Step I), must make assumptions about the driving selection pressures. Finally, confirming Steps I–III′ but not III or III″ must caveat that phenotypic change might be driven by plasticity rather than allelic change (see above).

Coevolution is the process of adaptation and counter-adaptation to the reciprocal selection pressures imposed by interacting species (Gaba and Ebert 2009; Thompson 2005; Woolhouse et al. 2002). Demonstrating coevolution is even more challenging not only because it requires demonstrating evolution by natural selection in both host and pathogen (i.e., Steps I–III) but also because the evolutionary change in each *must*, in turn, represent the demonstrated environmental change (Step I) for the other. Coevolution *cannot* be concluded solely on either (1) hosts and pathogens have evolved over time (since they need not have done so in response to each other); or (2) one antagonist has evolved in response to the other (since the reverse must also hold).

Thus, we need to be careful about our use of the term evolution, let alone coevolution. That said, with standing genetic variation for a heritable trait and a strong selection pressure, adaptive evolution can be extremely rapid (e.g., within one generation for Darwin's finches; Grant and Grant 2006; Lamichhaney et al. 2015), whereas evolution requiring *de novo* mutations will take longer (e.g., Tenaillon et al. 2016).

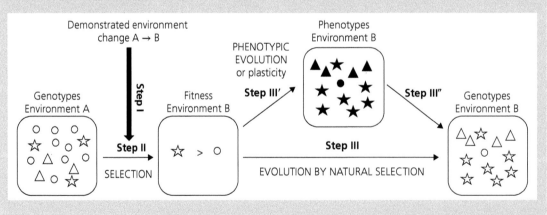

Boxed Figure 5.1 Core requirements for demonstrating evolution.

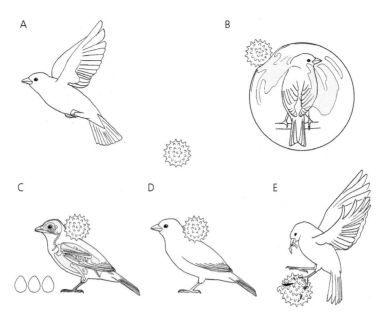

Figure 5.1 Host responses to infection. First, birds can prevent infection by evolving either (A) avoidance behaviors that minimize contact with pathogens or (B) mechanisms that prevent infection. Second, birds can reduce costs of being infected by evolving (C) life history changes, such as increasing pace of life, (D) tolerance to harboring the pathogen, or (E) resistance costs fighting infection through immune defenses (illustrated as a bird 'attacking' a pathogen). (Illustrations by Emma Wood.)

a population level. Finally, I discuss the broader implications of studying the evolution and coevolution of birds and their pathogens.

5.2 Evolutionary responses of avian hosts to their pathogens

In response to infection, hosts may evolve a range of different strategies that either prevent infection or reduce its cost (Figure 5.1). Infection prevention includes behavioral avoidance strategies that limit contact with the pathogen (see Chapter 4) or cellular/molecular modifications that inhibit its establishment. By contrast, cost reduction could occur through life history changes, the evolution of tolerance, or the evolution of resistance via immune defense and clearance (see also Chapter 3). While interspecific differences suggest that infection prevention and life history strategies may play a role in shaping patterns of infection prevalence and severity across avian species, evidence for their evolution in response to infection is still largely lacking, including in birds. In contrast,

evidence for the evolution of both resistance and tolerance is accumulating, with implications for our understanding of the drivers, costs, and epidemiological consequences of these avian responses.

5.2.1 Preventing infection

One efficient way of dealing with pathogens is by not getting infected in the first place. This can be achieved by avoiding contact with the pathogen or, following contact, by preventing its establishment. There is growing evidence that birds exhibit behavioral strategies that minimize contact with parasitized or sick individuals (see Chapter 4). For example, house finches (*Haemorhous mexicanus*) were found to associate more with control conspecifics than with immune-challenged ones (Zylberberg et al. 2013). Interestingly, finches that displayed the greatest levels of avoidance tended to be those with lower levels of immune defenses. Further work is required to test any causality underlying this relationship and to determine whether the evolution of behavioral responses to infection can limit (or slow) the evolution of cost reduction responses (e.g., immune resistance). Either way, while we might expect avoidance behaviors to evolve, investigative studies are lacking.

Interspecific differences in susceptibility to becoming infected suggest that it should be possible to evolve molecular/cellular changes that limit pathogen establishment following contact. Such processes may, for instance, explain the variation in infection outcome (pathogen load, clinical severity) and transmission potential observed among 25 bird species experimentally infected with West Nile virus (WNV) (Komar et al. 2003; see also Chapter 2). In some cases, preventing infection can be achieved through mutations that modify the host cell surface receptors used by pathogens to attach and invade the host environment (Imai et al. 2012). Because such mutations can reduce normal cell function and therefore be costly in the absence of infection, I predict that, if such modifications did evolve, they would do so primarily *de novo* under intense pathogen-driven selection. The evolution of host responses through *de novo* mutation(s) will be slower than when evolving from standing genetic variation (Tenaillon et al. 2016). It is therefore perhaps not surprising that we are still lacking evidence for the evolution of infection prevention in natural avian populations, even in response to particularly virulent pathogens.

5.2.2 Reducing the costs of infection

For hosts that are unable to prevent infection, selection may instead favor the evolution of responses that mitigate or minimize the cost of infection through three complementary ways.

5.2.2.1 Life history changes

Whenever the external environment reduces the survival or the future reproductive opportunities of individuals, selection should favor those with a faster 'pace of life,' exemplified by reduced ages at first reproduction and increased investment in current versus future offspring (Agnew et al. 2000). Indeed, a faster pace of life could evolve when pathogens impair future host reproduction through castration or death (Ohlberger et al. 2011) or when the probability of becoming infected increases with host age (Johnson et al. 2012). For example, patterns of seroprevalence in geese suggest that the likelihood of exposure to avian influenza viruses

increases with age (Ely et al. 2013). Life history adjustments may, at least partly, compensate for the consequences of infection. Yet, to my knowledge, no study has tested for such life history changes in response to infection in birds whilst controlling for concurrent changes expected to arise as a result of reduced competition following the death of infected individuals.

5.2.2.2 Resistance and tolerance

In contrast, there is growing evidence that birds can evolve to minimize the costs of infection through two other strategies: resistance and tolerance. Resistance is defined here as the ability to clear an infection through an immune response, such that resistant individuals display lower pathogen loads than non-resistant ones (Boots and Bowers 2004; Janeway 2005). While there are diverse ways of defining tolerance, mechanisms of tolerance are generally considered to be those that reduce disease severity by mitigating the somatic damage caused by the infection (Medzhitov et al. 2012; Råberg et al. 2009; Råberg 2014). Thus, tolerant hosts are typically defined as those that experience a lower fitness cost of infection at a given parasite burden (Råberg et al. 2009). That resistance and/or tolerance can evolve in avian hosts in response to infection is based on two primary types of evidence: (1) positive selection and local adaptation at immune genes; and (2) phenotypic changes in host responses following emerging infectious outbreaks, which I present in the form of two case studies.

Evidence from avian immune genes

One approach to testing for the evolution of resistance (or tolerance) strategies in response to pathogens is to test for evidence of selection using genomic approaches. When a gene is under positive selection, it should display an elevated non-synonymous/synonymous substitution ratio (dN/dS) relative to expectations under neutral evolution (Kimura 1983; McDonald and Kreitman 1991). Genome-wide scans in humans and birds have found that genes related to the immune system and/or that are upregulated in response to infection show evidence of positive selection, supporting a role for pathogens as significant selective forces (Fumagalli et al. 2011; Shultz and Sackton 2019).

Such findings at the level of the genome are consistent with studies of the evolutionary dynamics of specific avian immune genes. In particular, genes of the major histocompatibility complex (*Mhc*) and Toll-like receptors (*Tlr*) display evidence of positive selection across bird species (Burri et al. 2010; Edwards et al. 2000; Grueber et al. 2014). The *Mhc* is a group of highly variable genes encoding receptors on cell surfaces that play an important role in immune responses, with the binding and presentation of intracellular or extracellular foreign peptides by MHC class I or II molecules, respectively, leading to T-cell activation (Janeway 2005). When considering 16 bird species, Alcaide et al. (2013) found strong evidence of positive selection acting on the third exon of *Mhc* class I genes, which encodes the polymorphic sites of the peptide-binding region, with the strength of positive selection in songbirds being about double that of non-passerines. TLR proteins, on the other hand, bind structurally conserved molecular patterns of pathogens, leading to innate immune responses and the development of antigen-specific acquired immunity (see also Chapter 3). Investigation of the variation at *Tlr* genes across seven bird species revealed that, while avian *Tlr* genes appear dominated by stabilizing selection, codons that were likely to be situated at species-specific ligand-binding sites displayed patterns of positive selection (Alcaide and Edwards 2011). Such evidence of positive selection in extracellular, ligand-binding domains of avian *Tlr*s was upheld in a comparison of 63 bird species covering all major avian clades (Velova et al. 2018). These studies suggest that pathogens contribute to the diversity of avian immune genes and elucidate potential targets of selection at the genetic level.

Studies of *Mhc* genes also shed light on the potential fitness consequences of genetic variation for responses against pathogens. Westerdahl et al. (2005) found that great reed warblers (*Acrocephalus arundinaceus*) harboring either a large number of *Mhc* class I alleles or the specific *Mhc* class I allele B4b were more likely to be infected with the *Plasmodium* lineage GRW2. While the former result is consistent with a high *Mhc* allelic diversity providing protection through immune defense against a high antigenic repertoire (thereby increasing survival during infection), the latter suggests a role of B4b in the specific immune response against

GRW2. A similar result was found in a study of great tits (*Parus major*) which combined *Mhc* alleles into functional groups according to the physico-chemical amino acid properties of their antigen binding sites (Sepil et al. 2013). Individuals infected with *Plasmodium circumflexum* were more likely to harbor *Mhc* supertype 6, with this supertype also found at a higher frequency in older-aged birds in high-risk areas. In contrast, however, great tits infected with *P. relictum* were less likely to harbor *Mhc* supertype 17, suggesting a role in resistance to becoming infected (Sepil et al. 2013). Effects on infection prevention versus immune resistance may be elucidated by jointly considering infection status (infected or not) and intensity (pathogen load/parasitemia). In doing so, Westerdahl et al. (2013) showed that blue tits (*Cyanistes caeruleus*) with the *Mhc* class I allele 242 had a reduced intensity of infection but not a reduced probability of infection with the blood parasite *Haemoproteus majoris*.

Finally, *Mhc* genes also show patterns of variation consistent with local adaptation (Thompson 2005). For example, two distinct populations of house sparrows (*Passer domesticus*) displayed population-specific *Mhc* alleles associated with differences in the probability of being infected with the *Plasmodium* lineage SGS1 (Bonneaud et al. 2006). Subsequent work on 13 house sparrow populations across France showed stronger among-population patterns of isolation by distance for *Mhc* class I genes than for neutral microsatellite markers (Loiseau et al. 2009). In addition, *Mhc* alleles were associated with an increased or reduced probability of infection with *P. relictum* in a population-specific way, with some alleles displaying opposite effects in different populations. While comparative approaches can be powerful means of investigating the effect of a known environmental difference, populations can differ in an array of confounding ways, from historical differences in pathogen exposure, through differences in microbiome and diet, to differences in weather and predation pressure. For these reasons, extreme caution is required when drawing conclusions about causality, and common-garden experiments are ideally required to minimize confounding effects.

Together these studies show that avian immune genes are under selection and that variation at avian immune genes can be associated with different infection outcomes (Box 5.1). As such, these studies contribute to our understanding of how host resistance through immune defense can evolve in response to pathogen-driven selection. What remains to be demonstrated, however, are the specific selective drivers (i.e., specific pathogens) of immune gene evolution and the phenotypic outcomes. Below, I detail two case studies that have largely adopted a phenotypic approach using novel pathogens to forward our understanding of the effect of pathogens on the evolution of host resistance and tolerance.

Evidence of phenotypic evolution

Case study 1: Avian malaria in endemic Hawaiian birds

The invasion of the Hawaiian archipelago by the avian malaria parasite *P. relictum* in the 1820s is one of the best known examples of the strong selection pressure that emerging pathogens can impose on avian hosts, as well as of the ability of those hosts to evolve rapidly to reduce the costs of infection. A series of field and laboratory studies by Van Riper et al. (1986), conducted between 1977 and 1980, showed that the endemic bird species most severely affected by *P. relictum* were those found in mid-elevational forests, most likely due to a reduced abundance of birds at low elevations and of mosquito vectors at high elevations. Following experimental inoculation, endemic bird species displayed a higher quantity of parasites in the blood (parasitemia) than exotic introduced ones. Endemic species, nevertheless, differed inherently in their ability to deal with the infection, with Hawai'i 'amakihi (*Hemignathus virens*) and 'apapane (*Himatione sanguinea*) suffering moderate mortality and Laysan finches (*Telespyza cantans*) displaying 100% mortality along with the highest parasitemia (Van Riper et al. 1986).

Since these early experiments, there has been growing evidence that native Hawaiian birds have evolved ways of dealing with malaria. First, there is a circumstantial suggestion that some species were able to evolve behaviors to limit mosquito exposure (Van Riper et al. 1986); see Chapter 4. Second, there is empirical support for the evolution of a response at the physiological/immunological level. For example, in the last few decades, endemic species such as the 'apapane (Atkinson et al. 2005) and Oahu 'amakihi (*Hemignathus chloris*) have been found to persist at low elevation (Shehata et al. 2001), while the Hawai'i 'amakihi has been expanding into low-elevation sites of high malaria prevalence (Spiegel et al. 2006; et al. 2005). To test the hypothesis that Hawai'i 'amakihi were evolving adaptively to *P. relictum*, Atkinson et al. (2013) conducted experimental inoculations of 'amakihi from low and high elevations. There was no difference between high and low elevation birds in the likelihood of becoming infected or in peak parasitemia. However, high elevation 'amakihi displayed a more rapid and more extreme drop in food consumption, greater body mass loss, and potentially greater mortality (5 of 10 and 2 of 12 birds died from the high and low elevation groups, respectively) over the course of the infection. The lowered clinical severity of low altitude 'amakihi relative to high elevation ones, despite equivalent parasite burden, is consistent with the evolution of tolerance to infection, with low elevation birds having evolved the ability to mitigate the pathological effects of acute malaria infection.

As in humans, differences in the ability to deal with malaria infections in birds are, however, likely to result from variation in both tolerance and resistance (Hill 1991). In support of a role for immune defense in t.he response to *Plasmodium* infection, Hawai'i 'amakihi that survived an initial infection were found to be more resistant to subsequent challenges, indicating a capacity for acquired immunity (Atkinson et al. 2001). Furthermore, a recent comparative genomics study revealed that low and high elevation populations of Hawai'i 'amakihi displayed significant differences at immune-related genes (Cassin-Sackett et al. 2019). Indeed, among the genes found to be under selection in low elevation populations were those of *Mhc*, *Tlr*, interferons, and tumor necrosis factor, as well as genes encoding proteins known to interact with T-cells. Together, these findings are consistent with the evolution of both tolerance and resistance mechanisms in Hawai'i 'amakihi in response to avian malaria. Whether other species of Hawaiian honeycreepers have similarly evolved to deal with *Plasmodium* infection remains to be determined.

Case study 2: Mycoplasmal conjunctivitis in house finches

The first sightings of house finches with conjunctivitis occurred in the suburbs of Washington DC, USA, in 1994 (Dhondt et al. 1998). Suspecting the onset of an epidemic and with the aim of monitoring its spread, Dhondt and colleagues quickly capitalized on a Cornell University-based volunteer project using backyard birdwatchers across North America to monitor signs of disease in birds at feeders (Dhondt et al. 2005). By September 1994, this surveillance program allowed tracking of the disease as it moved northwards into Canada and then west and southwards, reaching North Dakota, Nebraska, Kansas, and Texas by 1997 (Dhondt et al. 1998). In 1996, captures of birds allowed for estimates of infection prevalence which were close to 50% in New Jersey (Dhondt et al. 1998) and 60% in Alabama (Nolan et al. 1998), with many infected house finches dying of starvation or predation, presumably as a result of conjunctivitis-induced blindness and lethargy (Adelman et al. 2017; Kollias et al. 2004). The eastern population of house finches, which until then had been growing since the release of a small number of individuals (originally of western stock) near New York in the 1940s, suffered measurable population declines due to the death of millions and which only stabilized after several years (Hochachka and Dhondt 2000).

The etiological agent of infection was confirmed through culture and polymerase chain reaction (PCR) to be the bacterium *Mycoplasma gallisepticum*, a pathogen known for causing chronic respiratory disease in chickens and infectious sinusitis in turkeys (Yoder 1991). This minute, wall-less bacterium is an obligate pathogen that can persist within or outside host cells and displays distinctive abilities to evade and manipulate host immune systems (Razin et al. 1998; Vogl et al. 2008), which are critical to successful infection and transmission (Staley and Bonneaud 2015; Szczepanek et al. 2010). *M. gallisepticum* induces a misdirected inflammatory response that facilitates the invasion of the mucosal surfaces of the conjunctiva and upper respiratory tract, as well as allowing transmission through ocular secretions transferred directly or left on inert surfaces acting as fomites (Gaunson et al. 2000). Infection persistence depends on the ability of *M. gallisepticum* to evade and suppress more pathogen-specific immune components known to otherwise play a role in its control (e.g., Naylor et al. 1992).

Given the devastating impact of this outbreak, it is not surprising that house finches appear to have evolved rapidly in response (Gates et al. 2018). Evidence that genetic resistance had spread from standing genetic variation within only 12 years of initial outbreak was found in a study conducted in 2007 (Bonneaud et al. 2011). Following experimental inoculation with a 2007 *M. gallisepticum* isolate, house finches from pathogen-exposed populations displayed significantly lower bacterial loads than those from unexposed populations in a common-garden set-up. These results cannot easily be explained by population differences other than the history of pathogen exposure, because finches from exposed and unexposed populations had displayed equivalent gene expression in response to *M. gallisepticum* in 2000 (Bonneaud et al. 2011), when they still exhibited similar levels of susceptibility to infection (Farmer et al. 2002). By contrast, functionally relevant transcriptional responses were found to differ between birds from exposed and unexposed populations in 2007, suggesting that exposed populations had, by then, evolved the ability to resist pathogen-induced immunosuppression and mount a protective cell-mediated immune response (Bonneaud et al. 2012a). In addition, none of the individual birds used in this experiment had themselves previously been exposed to the pathogen and all were maintained in comparable conditions with *ad libitum* food for three months before the onset of the experiment, further reducing the capacity for phenotypic plasticity to influence these results (Bonneaud et al. 2012b; Box 5.1).

Nevertheless, using a comparable experimental approach, evidence also emerged that house finches from pathogen-exposed populations were more tolerant to infection than finches from unexposed populations (Adelman et al. 2013). Following inoculation with a bacterial isolate collected at the start of the epidemic (i.e., in 1994), finches from exposed populations in 2010 had similar peak pathogen load but reduced clinical signs (conjunctival swelling) relative to finches from unexposed populations. Although the studies of Bonneaud et al. (2011) and Adelman et al. (2013) suggested differing

Box 5.2 Experimental approaches for testing pathogen trait differences and evolution

Two experimental approaches have largely been adopted for investigating pathogen trait differences and evolution in vertebrate hosts, for which ethical and logistical limitations operate. In the first approach (Boxed Figure 5.2A), a limited number of pathogen isolates are obtained from just a few time-points across the host–pathogen coevolutionary interaction (here n = 3 isolates from three time-points), and each isolate is inoculated into several hosts (here n = 5 hosts/isolate; n total = 15 hosts). Such experimental design controls for among-host variation in responses to infection and thereby allows the characterization of among-isolate differences in pathogen traits such as virulence. However, caution is required for interpreting evolutionary changes because the single pathogen isolate used from a given time-point may be drawn anywhere from the natural distribution of pathogen trait values and thus may not be broadly representative.

In the second approach (Boxed Figure 5.2B), a larger number of isolates are obtained from more time-points across the coevolutionary interaction (here n = 16 isolates sampled from eight time-points), and each isolate is inoculated into a limited number of hosts. In each approach, a comparable number of host birds are used for ethical reasons, but here (Boxed Figure 5.2B), the eight pairs of isolates are inoculated into a single bird each. Such an experimental design is more robust for characterizing the trajectory of evolutionary change because sixteen isolates will better represent the natural distribution of pathogen trait values than three isolates. However, this approach is not suitable for testing among-isolate differences, since these will be confounded by among-host variation in resistance/tolerance. Evolutionary inferences need to be made more cautiously using the former approach, while inferences about isolate differences need to be cautioned in the latter.

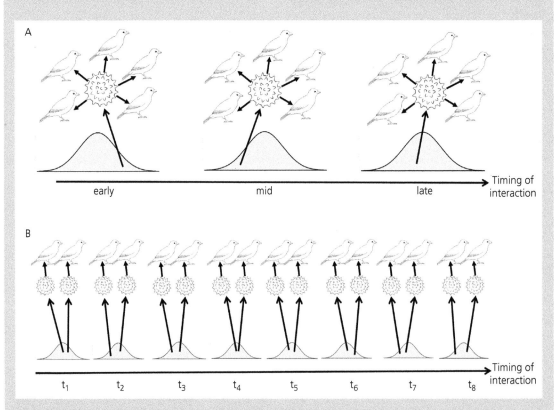

Boxed Figure 5.2 Visual depiction of two experimental approaches used to investigate pathogen trait differences and evolution in vertebrate hosts, where 't' refers to a given time-point in the host–pathogen interaction. (Illustrations by Emma Wood.) From Bensch et al. 2000; published in Proceedings of the Royal Society B-Biological Sciences, volume 267, Figure 2, pg 1586.

evolutionary solutions to infection with *M. gallisepticum*, both were limited by the inoculation of a single isolate into multiple hosts (Box 5.2).

More compelling evidence to suggest that genetic resistance and tolerance have both evolved in house finches, and in tandem, came from a larger-scale infection experiment using 55 *M. gallisepticum* isolates collected from the initial outbreak (1994) and over the subsequent 20 years (until 2015; Bonneaud et al. 2019a). Using a large number of isolates collected throughout the epidemic reduces the possibility that results were confounded by the use of a single outlying isolate from a given time period (Box 5.2). Following inoculation, finches from pathogen-exposed and unexposed populations, again in a common-garden set-up, displayed comparable peak pathogen loads around eight days post-inoculation. However, finches from the former group thereafter cleared the pathogen, consistent with previous evidence of resistance operating through cell-mediated immune processes (Bonneaud et al. 2012a). Furthermore, finches from exposed populations exhibited lower clinical signs for a given pathogen load, suggesting that tolerance had evolved to minimize damage (Bonneaud et al. 2019a). Such damage limitation is consistent with the lower inflammatory cytokine signaling and fever of finches from exposed populations relative to those from unexposed populations that were detected previously (Adelman et al. 2013). Together, these findings highlight the combined importance of immune resistance and tolerance mechanisms in the evolutionary responses of avian hosts to emerging pathogens. Further work is required to demonstrate the heritable genetic basis of phenotypic evolution in house finches.

5.3 Evolutionary responses of pathogens to their avian hosts

The continuously changing environment of the host is expected to represent a significant selection pressure for the pathogen but arguably never more so than when the host is novel (e.g., after a host shift). Several outbreaks of novel infections in birds have furthered our understanding of the molecular evolution of avian pathogens in the wild. For instance,

studies of avian influenza and WNV have yielded insight into the broad molecular changes that these pathogens undergo as they are transmitted between host species and/or within a host population over time (Mann et al. 2013; Taubenberger and Morens 2017; Venkatesh et al. 2018). As a result, we now have a sound understanding of how avian pathogens can evolve at the molecular level. In contrast, however, our empirical knowledge of the extent to which this evolution is adaptive and of its phenotypic consequences is more limited (Boxes 5.1 and 5.2). A thorough understanding of how pathogens evolve adaptively in avian hosts will require that we bridge molecular and phenotypic approaches, but such studies are still largely lacking.

5.3.1 Molecular evolution

In the last ten years, access to high-throughput deep sequencing technologies has enabled the acquisition of large sets of sequence data from infectious pathogens of birds. This has yielded an understanding that ranges from the description of phylogenetic relationships between pathogen isolates and/or documentation of molecular changes in pathogens over time (e.g., avian poxviruses [Gyuranecz et al. 2013] and *M. gallisepticum* [Delaney et al. 2012]), to an in-depth knowledge of the molecular basis of disease emergence (e.g., avian influenza viruses [Venkatesh et al. 2018]) and evolution (e.g., St. Louis encephalitis virus and WNV [Auguste et al. 2009; Di Giallonardo et al. 2016]). The extent to which the molecular changes observed are adaptive and what their selective context and phenotypic consequences are, however, remain to be investigated to fully understand how and why these pathogens are evolving. Best exemplifying the level of molecular evolution known in an emerging avian pathogen is the outbreak of WNV.

5.3.1.1 West Nile virus (WNV)

The large-scale outbreak of flaviviral encephalitis caused by WNV in North America in 1999 has become a classic example of how an exotic infectious pathogen can spread and evolve in naïve hosts. WNV is an RNA virus, naturally transmitted in an enzootic cycle between mosquitoes (primarily *Culex* spp.) that act as vector and birds (primarily

Passeriformes) that act as amplifying hosts (Pesko and Ebel 2012). In 1999, WNV emerged in New York and gave rise to a highly virulent outbreak (Hayes 2001), which resulted in large die-offs of wild birds, as well as potentially lethal infections in humans (CDC 1999). For example, disease was manifest in over 90% of American crows (*Corvus brachyrhynchos*) infected, with the population size of this species decreasing by 45% as a result (LaDeau et al. 2007; Yaremych et al. 2004). Since its emergence, the virus has been found in close to 300 bird species, although a few species are thought to play a critical role in its transmission. These include species with high pathogen loads (viremias) such as the American crow (McLean et al. 2001) and species with moderate pathogen loads that are strongly preferred by mosquito vectors such as the American robin (*Turdus migratorius*) (Kilpatrick et al. 2006), as well as vagrants and migrants which are likely to have played a role in spreading the virus to and within the Americas (May et al. 2011).

The high surveillance efforts involved in this outbreak have generated an unusual understanding of the molecular evolution of WNV. The genotype emerging in New York (referred to as NY99) was found to be most closely related to viral isolates obtained in Israel in 1998 and Hungary in 2003 (Jia et al. 1999; Zehender et al. 2011). Very little viral diversification was observed initially, with samples taken in the first two years following the outbreak being remarkably conserved genetically (Anderson et al. 2001; Ebel et al. 2001; Lanciotti et al. 1999). However, by 2003, NY99 had been displaced by a derived strain (WN02), which contained an amino acid substitution in an envelope protein likely responsible for a more efficient transmission through mosquitoes (i.e., through shortening the incubation period; Beasley et al. 2003; Davis et al. 2005). A phylogenetic analysis of 696 WNV isolates obtained from birds and mosquitoes in the US between 1999 and 2012 revealed a star-like tree topology, characteristic of an explosive expansion from outbreak, and little spatial structure, particularly for bird isolates (Di Giallonardo et al. 2016). This is consistent with the rapid spread of WNV across the US (i.e., within four years) and suggests that the virus encountered no major geographic barriers as it expanded through its current range. In

addition, non-synonymous/synonymous substitution ratios (dN/dS) measured in the 696 isolates were found to be low overall, suggesting weak positive selection and that the evolution of WNV was mainly driven by purifying selection (Di Giallonardo et al. 2016).

Although RNA-viruses generally exhibit high mutation rates that allow their rapid adaptation to novel environmental conditions (Cleaveland et al. 2007; Woolhouse et al. 2005), they also display surprisingly low interhost pairwise sequence dissimilarity and relatively slow rates of long-term evolution (Holmes 2009; Jenkins et al. 2002). Two hypotheses may explain this pattern in WNV. The first hypothesis states that contrasting selection pressures experienced in birds and mosquitoes constrain viral evolution (Ciota and Kramer 2010; Holmes 2003). In contrast, the 'selective sieve' hypothesis proposes that intense purifying selection in birds reduces the genetic diversity generated in vectors (Grubaugh and Ebel 2016; Jerzak et al. 2008). Support for the latter was found in a study of the intrahost population dynamics of WNV, which showed that genetic diversity was greater in viruses that had been repeatedly grown by serial passage through mosquitoes than in those that had been passaged through chicken. In addition, chicken-passaged WNV displayed increased fitness (estimated as the proportion of viral particles in a competition against a reference strain) and a lower endpoint genetic diversity than WNV either passaged through mosquitoes or cycled through chickens and mosquitoes. This suggests that WNV was under purifying selection in birds but not in mosquitoes and that specialization on avian hosts may be constrained by transmission through mosquito vectors (Deardorff et al. 2011).

Experimental evolution of WNV in natural avian hosts indicates that such increased competitive ability and associated reduced genetic diversity may, in fact, be avian species-specific. To test such a hypothesis, Grubaugh et al. (2015) serially passaged WNV through one of three host species that typically exhibit varying levels of viremia and mortality following infection: American crows (high viremia/high mortality: Brault et al. 2004), house sparrows (high viremia/intermediate mortality: Langevin et al. 2005), and American robins (intermediate viremia/low mortality: VanDalen et al. 2013). All passaged viruses displayed increased replicative ability in comparison with the unpassaged virus, although fitness gains were lowest in crows and highest in robins. Furthermore, the number of non-synonymous mutations (dN) did not increase during passage, while the number of synonymous ones (dS) did, indicating that WNV was evolving primarily through purifying selection in all three host species. However, the number of intrahost single nucleotide variants at high frequency was greatest in the robin-passaged WNV populations and similar in the sparrow- and crow-passaged ones, consistent with selection being strongest in the less susceptible host species. Finally, passaged WNV populations were found to have accumulated the most deleterious mutations in crows and the least in robins. Taken together, these results suggest that host-driven selection is more intense in the less susceptible host species, which likely limits the accumulation of deleterious mutations in the pathogen population (Grubaugh et al. 2015).

Despite the potentially strong selection pressures exerted by some avian host species, there is little evidence that adaptive mutations that improve transmission in birds have arisen and increased in frequency in WNV since the initial 1999 outbreak. Brault and colleagues (2007) were able to recreate the virulence phenotype of the NY99 strain in American crows from a low virulence Kenyan strain by introducing an amino acid substitution in the NS3 helicase at the only site across the NS3 gene found to have been under positive (rather than purifying) selection. This suggests that mutation at this site may have facilitated the WNV outbreak in North American birds. However, there is no further evidence of adaptation to avian hosts. The 696 WNV isolates obtained between 1999 and 2012 revealed that evidence of positive selection was consistently found more often in non-structural genes than in structural ones, suggesting that WNV evolution is unlikely to have been driven by host immunity (Di Giallonardo et al. 2016). One explanation for the lack of adaptive evolution to birds is that the highly generalist nature of WNV precludes any one host species imposing sufficiently strong selection to sustain the spread of an adaptive variant. Further work is required to test whether the

more commonly infected bird species are capable of evolving resistance and how this might impact the evolution of WNV.

5.3.2 Adaptive phenotypic responses of pathogens

The 'virulence-transmission trade-off hypothesis' provides a theoretical framework for predicting adaptive evolution in pathogens (Alizon et al. 2009; Alizon and Michalakis 2015; Anderson and May 1982; Ewald 1983). This hypothesis assumes that transmission to secondary hosts, the key metric of pathogen fitness (see Chapter 2), requires exploitation of the primary host, which incidentally hastens host death, thereby decreasing the duration of the infection and the time window for transmission. As a consequence, theory predicts that selection will push pathogens to optimize the trade-off between transmission rate and infection duration, hence favoring pathogens of intermediate virulence. While there is accumulating evidence for the assumed positive association between virulence and transmission rate (reviewed in Alizon et al. 2009), few studies have demonstrated the predicted fitness advantage for pathogens of intermediate virulence (e.g., de Roode et al. 2008; Jensen et al. 2006; Read et al. 2015).

Under the trade-off hypothesis, any environmental factor that impacts the duration of infection or transmission rate will, in turn, have consequences for pathogen virulence evolution. Predicting adaptive evolution in pathogens therefore requires evaluating the pathogen's evolutionarily stable strategy under a given set of changing environmental conditions (Lion and Metz 2018). One environmental factor that is likely to be critical in shaping pathogen fitness is the host response to infection (Barclay et al. 2012; Gandon and Michalakis 2000; Porco et al. 2005). Theoretical models show that preventing infection, mounting effective immune responses, and hindering transmission can modify virulence evolution (Gandon et al. 2001; Gandon et al. 2003; Miller et al. 2006). For instance, preventing infection or pathogen transmission is predicted to select for the evolution of lower virulence, because both limit transmission opportunities and therefore tip the balance in favor of longer infections. In contrast, host responses that reduce pathogen growth through immune clearance are predicted to select for higher pathogen virulence. The reasons for this increase are twofold. First, immune clearance will decrease infection duration and therefore favor pathogens that transmit faster. Second, because resistant hosts should die more slowly (if at all) from the infection, virulence can increase because the costs of high virulence that limit increases in transmission rate are alleviated. While a wealth of theoretical models of pathogen adaptive evolution exist, empirical tests in natural populations, including of birds, remain rare.

5.3.2.1 Empirical evidence in birds

Support for a role of host immune responses in driving adaptive pathogen virulence evolution comes from studies conducted in two systems. In the first, Read et al. (2015) investigated the impact of an incomplete immune response (i.e., one that does not completely clear infection) mounted by chickens on the rate of transmission of Marek's disease virus (MDV). To do so, they vaccinated a subset of chickens with a 'leaky' MDV vaccine that protected from infection-induced mortality without preventing viral replication and transmission and left others unvaccinated. Following subsequent inoculation with highly lethal strains of virus, they found that transmission to uninfected chicken sentinels occurred successfully in vaccinated groups only, with all unvaccinated chickens dying before transmission (Read et al. 2015). By alleviating the cost of virulence (i.e., through a decrease in host mortality rate), leaky vaccines and the incomplete immune responses they produce can therefore favor more virulent pathogen strains, with implications for the design of effective vaccination programs.

A second system for which there is growing evidence that immunity can lead to increased virulence evolution is the epidemic of *M. gallisepticum* in house finches. In an experiment designed to test the direction of *M. gallisepticum* virulence evolution from epidemic outbreak, Hawley et al. (2013) compared eye lesion severity among isolates collected between 1994 and 2008 during the initial eastern US epidemic and between 2006 and 2010 during the

later western US epidemic. In both areas, virulence was found to be greater in the latter isolates. Further clarification of the trajectory of virulence evolution over the 20 years of the epidemic was provided by another study in which 55 isolates collected before (at outbreak in 1994–1996), during (1998–2007), and after (2007–2015) the spread of genetic resistance were inoculated into susceptible house finches from unexposed populations (Tardy et al. 2019; Box 5.2). This study showed that virulence has increased linearly over this period. This result is consistent with the hypothesis that virulence evolution arose, at least in part, because genetically resistant finches from disease-exposed populations are better able to clear the infection (Bonneaud et al. 2019a). Thus, by decreasing the duration of the infection through pathogen clearance, host immunity would have selected for the evolution of higher virulence.

Such impact of host resistance on virulence evolution is likely amplified by the fact that, upon secondary exposure to *M. gallisepticum*, house finch immune responses are less effective against high virulence pathogen isolates. Indeed, Fleming-Davies et al. (2018) showed that, while high virulence isolates caused higher disease-induced mortality, house finches that survived first infection and subsequently got reinfected mounted secondary immune responses that cleared low virulence infections but not high virulence ones. Further, the pre-existing immunity generated by the first infection with *M. gallisepticum* prevented the most virulent isolates from causing host mortality, thereby favoring the evolution of higher virulence. How host genetic resistance and partial immune protection acquired over the lifetime of an individual have worked together to shape pathogen virulence evolution in this system remains to be determined. However, one hypothesis is that, by driving the initial increases in virulence, partial immunity during reinfections intensified pathogen-driven selection for the evolution of resistance, which in turn would have further increased selection for higher pathogen virulence.

Additional work conducted on this system suggests that the change in virulence undergone by *M. gallisepticum* had consequences for its transmission rate and infection duration (the inverse of host recovery rate), in line with assumptions of the virulence-transmission trade-off hypothesis. Williams et al. (2014) investigated transmission dynamics within groups of house finches, where one bird per group was inoculated with one of three *M. gallisepticum* isolates that varied in virulence. As assumed, a negative association was found between transmission and host recovery rates, and a positive one was found between transmission rate and virulence among isolates. In contrast, however, in a separate study no support was found for the commonly held expectation that pathogen replication rate (and hence pathogen load) underlies virulence when considering variation across 55 *M. gallisepticum* isolates (Tardy et al. 2019). This suggests that changes in pathogen virulence need not necessarily be driven by pathogen replication rate and hence by pathogen load. Further work is required in *M. gallisepticum* to identify the mechanism underlying virulence evolution, as well as whether pathogen fitness is maximized for intermediate levels of virulence in this system, as predicted by the trade-off hypothesis.

5.4 Coevolution of avian hosts and their pathogens

While there is accumulating evidence that hosts can evolve in response to pathogens and that pathogens can evolve in response to hosts, studies that demonstrate their coevolution remain comparatively rare. The main reason for this is that, beyond the challenges of showing adaptive change in each interacting species, there is the added difficulty of doing so in a selective environment (represented by each antagonist) that is continuously changing. This is because, by continuously evolving in response to each other, each species counters and therefore obscures the adaptive evolution of the other, thereby creating the overall impression that there is no change in the interaction over time. Coevolution is, as a result, most often inferred from the duration and specificity of a host–pathogen interaction over long (macroevolutionary) timeframes or from spatial variation across landscapes and patterns of local adaptation. Less often, host–pathogen coevolution is demonstrated using specific experimental designs to test for co-adaptation over time (Gaba and Ebert

2009). While there are logistical challenges in carrying out tests of temporal changes, particularly in vertebrate hosts, such approaches further elucidate the process of host–pathogen coevolution and should therefore be preferred, whenever possible (Box 5.1).

5.4.1 Macroevolutionary studies

The interaction of avian hosts and their pathogens can be investigated over long macroevolutionary timescales by comparing the match between their respective phylogenetic trees (Brooks 1988; Clayton et al. 2003). A poor match between trees is indicative of host shifting, a process by which the jump of a pathogen from one host species to another is followed by specialization on the novel host and loss of infectivity of the original donor host (Agosta et al. 2010; Giraud et al. 2010). In contrast, high congruence between trees is thought to reflect high host specificity and the co-speciation of hosts and their pathogens. Although co-speciation is highly suggestive of coevolutionary interaction, it need not be equivalent. Indeed, co-speciation events can also be seen when pathogens track the speciation of their hosts, either neutrally as a result of isolation or adaptively by specializing on the novel host species (Clayton et al. 2003; Moran and Baumann 1994). Regardless, the duration of the association between a host and a pathogen and the degree of specificity of their interaction will shape our level of confidence that they are likely to have coevolved.

Phylogenetic analyses conducted on avian blood parasites and their hosts have yielded important insights into their interactions over long periods of time. For instance, Bensch et al. (2000) found a poor match between the phylogenetic trees of *Haemoproteus* and *Plasmodium* parasites and their 12 passerine host species, suggesting that host shifts have occurred repeatedly over time (Figure 5.2). Subsequent studies conducted with a larger number of species have provided further support for this pattern (Ricklefs and Fallon 2002; Ricklefs et al. 2004). Ricklefs et al. (2014) analyzed 181 genetic (mitochondrial DNA) lineages of *Plasmodium* and *Haemoproteus* spp. and found that such host-shifts often occurred across large host taxonomic distances, with parasite species most likely forming in

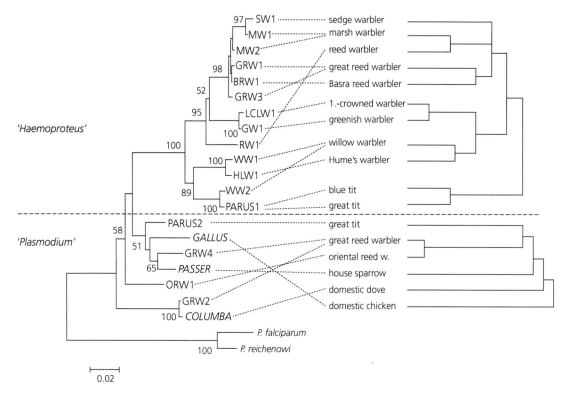

Figure 5.2 Phylogenetic tree of partial cytochrome b gene of *Plasmodium* and *Haemoproteus* parasites (left) and of the 12 avian host species that they were collected from (right; connected to their parasites via dotted lines). (Republished with permission of Royal Society Journal, from Bensch et al. 2000.)

allopatry and subsequently spreading before secondarily coming back into contact. Such high frequency of host shifts may make for very dynamic coevolutionary interactions between birds and their malaria parasites, with each having been repeatedly confronted with a novel antagonist, which would likely have led to bouts of rapid evolution in both.

5.4.2 Microevolutionary studies

5.4.2.1 Spatial adaptation experiments

Evidence of coevolution can be gained by studying spatial variation in host–pathogen interactions. The geographic mosaic of coevolution theory posits that interactions between species will vary geographically across landscapes, with consequences for the selection pressures experienced by each species (Thompson 1994; Thompson 2005). This will result in a 'geographic mosaic' in which interactions will

be reciprocal in some sites (coevolutionary hotspots) and weak or even absent in others (coevolutionary coldspots). Spatial patterns of coevolution should be maintained by new mutations, gene flow, and random genetic drift, which will all generate variation in the range of traits (i.e., defenses and counter-defenses in the case of hosts and pathogens) that are available for selection to act upon within each region (Thompson 2005). While a growing number of studies have investigated population-specific variation at key (immune) genes (see Section 5.2.2.2), few have so far taken advantage of spatial variation in avian host–pathogen interactions to test key predictions regarding their coevolutionary interactions.

One prediction that can be tested experimentally is that hosts should be best adapted and therefore more resistant to their own local (sympatric) pathogens, while pathogens should be better able to infect and therefore have higher fitness in sympatric hosts

(Blanquart et al. 2013; Ebert 1994; Lively et al. 2004). In accordance with the hypothesis that local hosts are more resistant to their sympatric parasites, local white-crowned sparrows (*Zonotrichia leucophrys*) were found to be less severely infected by *Haemoproteus* blood parasites than non-local ones (MacDougall-Shackleton et al. 2002). Experimental tests of local adaptation can be conducted by exposing hosts from different sites to pathogens obtained in sympatry and allopatry (Kalbe and Kurtz 2006; Oppliger et al. 1999). Such a reciprocal experimental design has been used, for example, in great tits and their *Plasmodium* parasites: uninfected juveniles that had been reciprocally transplanted between two sites tended to be more likely to be infected with allopatric parasites (Jenkins et al. 2015). This is consistent with an adaptation of hosts to their local parasites but not of parasites to their local hosts. Such lack of support for the adaptation of *Plasmodium* to their local host, however, may be explained by the fact that *Plasmodium* parasites are broadly considered to be generalists (Bensch et al. 2009), which should make them less likely to display local host adaptation (Gandon 2002).

Similar results were found in another study in which adult song sparrows (*Melospiza melodia*) were inoculated with sympatric or allopatric lineages of *Plasmodium* that were sampled at a distance of 440 m (Sarquis-Adamson and MacDougall-Shackleton 2016). However, although infection risk was lower in birds exposed to sympatric relative to allopatric parasites, this effect was found to be unlikely to be mediated by differences at *Mhc* class I or II genes (Slade et al. 2017). Indeed, there was no evidence of population genetic differentiation at *Mhc* loci, suggesting that other genes and/or ecological factors may drive this pattern of local adaptation. Further work is required to determine the scale of local adaptation of different bird species and their pathogens, particularly given that movements such as migration may greatly complicate any spatial pattern of host–pathogen associations (Pulgarin-R et al. 2019).

5.4.2.2 Temporal adaptation experiments

Temporal studies, in which populations of hosts and pathogens are monitored over multiple generations, present an advantage over spatial studies in that measures of change over time are less likely to be confounded by genetic and/or ecological effects (Gaba and Ebert 2009). Indeed, assumptions of equivalent gene flow and biotic/abiotic habitat are more likely to hold over time than space, such that phenotypic and genetic change over time can be linked more confidently to evolutionary processes (including selection) rather than to ecological ones (i.e., environmental variation). One way of determining the phenotypic changes of hosts and their pathogens over time, as well as their adaptive consequences, is to use a time-shift experimental design. In this case, each antagonist is confronted with antagonists from the past, the present, and the future, thereby allowing phenotypic metrics and their consequences for success to be measured in each antagonist, controlling for any counter-response of the other (Decaestecker et al. 2007; Gaba and Ebert 2009). However, it is not possible to conduct this type of experiment in vertebrate systems, since vertebrates cannot be maintained in an arrested state. As a result, such time-shift approaches have instead been primarily conducted on bacterial and invertebrate hosts (Decaestecker et al. 2007).

A time-shift type of experimental design can nevertheless be adopted in a vertebrate system when there is access to recently separated host populations that differ in their exposure to a specific pathogen. This approach was used in house finches and the bacterial pathogen *M. gallisepticum* (Bonneaud et al. 2018). House finches were caught from Alabama populations that had been exposed to the pathogen since initial emergence (i.e., 'adapted' populations) and from Arizona populations that were unexposed to the pathogen at the time of the study (i.e., 'ancestral' populations). After verifying that none of the finches caught had been exposed to the pathogen over the course of their lives, finches were then inoculated with 1 of 55 bacterial isolates collected over the course of the epidemic (see Section 5.3.2). The probability that finches became infected increased with the year of isolate collection but did not differ between host populations. Peak clinical signs were also found to increase with year of isolate collection but, in contrast to infection probability, were higher in finches from ancestral populations. Thus, phenotypic measures indicate the evolution of increasing pathogen virulence and increasing host resistance over time.

Crucially, these phenotypic changes were found to have measurable consequences for metrics of fitness in both host and pathogen. Finches from adapted populations were three times less likely to develop lethal clinical signs of conjunctivitis than finches from ancestral populations, with the divergence in clinical severity occurring when finches were inoculated with isolates collected after the spread of host resistance. Furthermore, the probability that isolates successfully established an infection in the primary host and, before inducing death, transmitted to an uninfected sentinel was highest when recent isolates were inoculated in hosts from ancestral populations and lowest when early isolates were inoculated in hosts from adapted populations. These results provide compelling evidence for antagonistic coevolution between house finches and their *M. gallisepticum* bacterial pathogen, with finches displaying increased resistance and *M. gallisepticum* increased virulence in response to each other over time (Bonneaud et al. 2018).

5.5 Conclusions and implications

In conclusion, there is convincing evidence that avian hosts and their pathogens can evolve by natural selection in response to one another and can even antagonistically coevolve. Nevertheless, this evidence is derived from a variety of different empirical approaches applied across a range of host–pathogen systems. As a result, our understanding of the adaptive evolution of avian hosts and their pathogens over the course of their coevolutionary interactions remains incomplete. Bridging this gap in knowledge will require adopting lines of methodological inquiry from a range of disciplines (including behavioral ecology, epidemiology, molecular and cellular biology, genomics and quantitative genetics) to weave together a broad understanding that will scale from the precise underlying genetic and phenotypic drivers to the whole organismal phenotypic outcome and fitness benefits of evolutionary change and will do so in both antagonists. If anything, the complexity involved in putting together all of the pieces of the puzzle should

remind us that rigorously demonstrating adaptive evolution and coevolution is challenging and cannot be concluded hastily (Box 5.1).

How birds and their pathogens evolve responses and counter-responses to each other has, at least, three important implications. First, the ability of birds to respond adaptively to pathogens will shape epidemiological dynamics and thereby host population persistence. We have seen with the emergence of avian malaria parasites in Hawaii that an endemic bird species that initially might have appeared to be doomed may, in fact, persist due to its evolved ability to resist and tolerate infection. One implication is that individuals displaying such adaptation (or its associated genetic make-up) could be preferentially targeted in reintroduction programs, particularly given the predicted spread of *P. relictum* to higher elevations as a result of climate change (Benning et al. 2002). Second, the mechanisms by which birds and their pathogens coevolve will have repercussions in terms of the trajectory of virulence evolution. This will be particularly important when designing disease control programs, because the use of inadequate control measures may push the pathogen down an evolutionary path that the host may find even more difficult to counter. Third, how birds and their pathogens coevolve will have repercussions for the host specificity of pathogens and thereby their ability to infect and even emerge in other host species (Bonneaud et al. 2019b; Simmonds et al. 2018), which may be particularly important for domestic animal and human hosts. For instance, if host evolutionary responses push pathogens into evolving host species-specific counter-responses that lead to host specialization, then artificial control measures may, in contrast, select for pathogens that are more generalists. While human intervention may, nevertheless, in some instances be necessary to prevent host extinction, an understanding of how birds and their pathogens coevolve may ensure that any negative consequences of intervention are minimized.

Literature cited

Adelman, J.S., Kirkpatrick, L., Grodio, J.L., and Hawley, D.M. (2013). House finch populations differ in early inflammatory signaling and pathogen tolerance at the peak of *Mycoplasma gallisepticum* infection. *The American Naturalist*, 181, 674–89.

Adelman, J.S., Mayer, C., and Hawley, D.H. (2017). Infection reduces anti-predator behaviors in house finches. *Journal of Avian Biology*, 48, 519–28.

Agnew, P., Koella, J.C., and Michalakis, Y. (2000). Host life history responses to parasitism. *Microbes and Infection*, 2, 891–6.

Agosta, S.J., Janz, N., and Brooks, D.R. (2010). How specialists can be generalists: resolving the 'parasite paradox' and implications for emerging infectious disease. *Zoologia*, 27, 151–62.

Alcaide, M. and Edwards, S.V. (2011). Molecular evolution of the toll-like receptor multigene family in birds. *Molecular Biology and Evolution*, 28, 1703–15.

Alcaide, M., Liu, M., and Edwards, S.V. (2013). Major histocompatibility complex class I evolution in songbirds: universal primers, rapid evolution and base compositional shifts in exon 3. *PeerJ*, 1, e86.

Alizon, S. and Michalakis, Y. (2015). Adaptive virulence evolution: the good old fitness-based approach. *Trends in Ecology & Evolution*, 30, 248–54.

Alizon, S., Hurford, A., Mideo, N. and Van Baalen, M. (2009). Virulence evolution and the trade-off hypothesis: history, current state of affairs and the future. *Journal of Evolutionary Biology*, 22, 245–59.

Anderson, J.F., Vossbrinck, C.R., Andreadis, T.G., Iton, A., Beckwith, W.H., and Mayo, D.R. (2001). A phylogenetic approach to following West Nile virus in Connecticut. *Proceedings of the National Academy of Sciences of the United States of America*, 98, 12885–9.

Anderson, R.M. and May, R.M. (1982). Coevolution of hosts and parasites. *Parasitology*, 85, 411–26.

Atkinson, C.T., Dusek, R.J., and Lease, J.K. (2001). Serological responses and immunity to superinfection with avian malaria in experimentally-infected Hawaii Amakihi. *Journal of Wildlife Diseases*, 37, 20–7.

Atkinson, C.T., Lease, J.K., Dusek, R.J., and Samuel, M.D. (2005). Prevalence of pox-like lesions and malaria in forest bird communities on leeward Mauna Loa Volcano, Hawaii. *Condor*, 107, 537–46.

Atkinson, C.T., Saili, K.S., Utzurrum, R.B., and Jarvi, S.I. (2013). Experimental evidence for evolved tolerance to avian malaria in a wild population of low elevation Hawai'i 'Amakihi (*Hemignathus virens*). *Ecohealth*, 10, 366–75.

Auguste, A.J., Pybus, O.G., and Carrington, C.V.F. (2009). Evolution and dispersal of St. Louis encephalitis virus in the Americas. *Infection Genetics and Evolution*, 9, 709–15.

Barclay, V.C., Sim, D., Chan, B.H.K., et al. (2012). The evolutionary consequences of blood-stage vaccination on the rodent malaria *Plasmodium chabaudi*. *PLoS Biology*, 10, 11.

Beasley, D.W.C., Davis, C.T., Guzman, H., et al. (2003). Limited evolution of West Nile virus has occurred during its southwesterly spread in the United States. *Virology*, 309, 190–5.

Benning, T.L., LaPointe, D., Atkinson, C.T., and Vitousek, P.M. (2002). Interactions of climate change with biological invasions and land use in the Hawaiian Islands: modeling the fate of endemic birds using a geographic information system. *Proceedings of the National Academy of Sciences of the United States of America*, 99, 14246–9.

Bensch, S., Stjernman, M., Hasselquist, D., et al. (2000). Host specificity in avian blood parasites: a study of *Plasmodium* and *Haemoproteus* mitochondrial DNA amplified from birds. *Proceedings of the Royal Society B: Biological Sciences*, 267, 1583–9.

Bensch, S., Hellgren, O. and Perez-Tris, J. (2009). MalAvi: a public database of malaria parasites and related haemosporidians in avian hosts based on mitochondrial cytochrome b lineages. *Molecular Ecology Resources*, 9, 1353–8.

Blanquart, F., Kaltz, O., Nuismer, S.L., and Gandon, S. (2013). A practical guide to measuring local adaptation. *Ecology Letters*, 16, 1195–205.

Bonneaud, C., Perez-Tris, J., Federici, P., Chastel, O., and Sorci, G. (2006). Major histocompatibility alleles associated with local resistance to malaria in a passerine. *Evolution*, 60, 383–9.

Bonneaud, C., Balenger, S., Russell, A.F., Zhang, J., Hill, G.E., and Edwards, S.V. (2011). Rapid evolution of disease resistance is accompanied by functional changes in gene expression in a wild bird. *Proceedings of the National Academy of Sciences of the United States of America*, 108, 7866–71.

Bonneaud, C., Balenger, S. L., Zhang, J., Edwards, S.V., and Hill, G.E. (2012a). Innate immunity and the evolution of resistance to an emerging infectious disease in a wild bird. *Molecular Ecology*, 21, 2628–39.

Bonneaud, C., Balenger, S.L., Hill, G.E., and Russell, A.F. (2012b). Experimental evidence for distinct costs of pathogenesis and immunity against a natural pathogen in a wild bird. *Molecular Ecology*, 21, 4787–96.

Bonneaud, C., Giraudeau, M., Tardy, L., Staley, M., Hill, G.E., and McGraw, K.J. (2018). Rapid antagonistic coevolution in an emerging pathogen and its vertebrate host. *Current Biology*, 28, 2978–83.

Bonneaud, C., Tardy, L., Giraudeau, M., Hill, G.E., McGraw, K.J., and Wilson, A.J. (2019a). Evolution of both host resistance and tolerance to an emerging bacterial pathogen. *Evolution Letters*, 3, 544–54.

Bonneaud, C., Weinert, L.A., and Kuijper, B. (2019b). Understanding the emergence of bacterial pathogens.

Philosophical Transactions of the Royal Society B: Biological Sciences, 374, 20180328.

Boots, M. and Bowers, R.G. (2004). The evolution of resistance through costly acquired immunity. *Proceedings of the Royal Society B: Biological Sciences*, 271, 715–23.

Brault, A.C., Langevin, S.A., Bowen, R.A., et al. (2004). Differential virulence of West Nile strains for American crows. *Emerging Infectious Diseases*, 10, 2161–8.

Brault, A.C., Huang, C.Y.H., Langevin, S.A., et al. (2007). A single positively selected West Nile viral mutation confers increased virogenesis in American crows. *Nature Genetics*, 39, 1162–6.

Brooks, D.R. (1988). Macroevolutionary comparisons of host and parasite phylogenies. *Annual Review of Ecology and Systematics*, 19, 235–59.

Buckling, A. and Rainey, P.B. (2002). Antagonistic coevolution between a bacterium and a bacteriophage. *Proceedings of the Royal Society B: Biological Sciences*, 269, 931–6.

Burri, R., Salamin, N., Studer, R.A., Roulin, A., and Fumagalli, L. (2010). Adaptive divergence of ancient gene duplicates in the avian MHC class II beta. *Molecular Biology and Evolution*, 27, 2360–74.

Cassin-Sackett, L., Callicrate, T.E., and Fleischer, R.C. (2019). Parallel evolution of gene classes, but not genes: evidence from Hawaiʻian honeycreeper populations exposed to avian malaria. *Molecular Ecology*, 28, 568–83.

CDC (1999). Outbreak of West Nile-like viral encephalitis—New York, 1999. *Morbidity and Mortality Weekly Report*, 48, 845–9.

Ciota, A.T. and Kramer, L.D. (2010). Insights into arbovirus evolution and adaptation from experimental studies. *Viruses*, 2, 2594–617.

Clayton, D.H., Bush, S.E., Goates, B.M., and Johnson, K.P. (2003). Host defense reinforces host–parasite cospeciation. *Proceedings of the National Academy of Sciences of the United States of America*, 100, 15694–9.

Cleaveland, S., Haydon, D.T., and Taylor, L. (2007). Overviews of pathogen emergence: which pathogens emerge, when and why? *Current Topics in Microbiology and Immunology*, 315, 85–111.

Davis, C.T., Ebel, G.D., Lanciotti, R.S., et al. (2005). Phylogenetic analysis of North American West Nile virus isolates, 2001–2004: evidence for the emergence of a dominant genotype. *Virology*, 342, 252–65.

de Roode, J.C., Yates, A.J., and Altizer, S. (2008). Virulence-transmission trade-offs and population divergence in virulence in a naturally occuring butterfly parasite. *Proceedings of the National Academy of Sciences of the United States of America*, 105, 7489–94.

Deardorff, E.R., Fitzpatrick, K. A., Jerzak, G.V.S., Shi, P.Y., Kramer, L.D., and Ebel, G.D. (2011). West Nile virus experimental evolution in vivo and the trade-off hypothesis. *PLoS Pathogens*, 7, e1002335.

Decaestecker, E., Gaba, S., Raeymaekers, J.A.M., et al. (2007). Host–parasite 'Red Queen' dynamics archived in pond sediment. *Nature*, 450, 870–3.

Delaney, N.F., Balenger, S., Bonneaud, C., et al. (2012). Ultrafast evolution and loss of CRISPRs following a host shift in a novel wildlife pathogen, *Mycoplasma gallisepticum*. *PLoS Genetics*, 8, e1002511.

Dhondt, A.A., Tessaglia, D.L., and Slothower, R.L. (1998). Epidemic mycoplasmal conjunctivitis in house finches from eastern North America. *Journal of Wildlife Diseases*, 34, 265–80.

Dhondt, A.A., Altizer, S., Cooch, E.G., et al. (2005). Dynamics of a novel pathogen in an avian host: Mycoplasmal conjunctivitis in house finches. *Acta Tropica*, 94, 77–93.

Di Giallonardo, F., Geoghegan, J.L., Docherty, D.E., et al. (2016). Fluid spatial dynamics of West Nile virus in the United States: rapid spread in a permissive host environment. *Journal of Virology*, 90, 862–72.

Ebel, G.D., Dupuis, A.P., Ngo, K., et al. (2001). Partial genetic characterization of West Nile virus strains, New York State, 2000. *Emerging Infectious Diseases*, 7, 650–3.

Ebert, D. (1994). Virulence and local adaptation of a horizontally transmitted parasite. *Science*, 265, 1084–6.

Edwards, S.V., Gasper, J., Garrigan, D., Martindale, D., and Koop, B.F. (2000). A 39-kb sequence around a blackbird Mhc class II gene: ghost of selection past and songbird genome architecture. *Molecular Biology and Evolution*, 17, 1384–95.

Eizaguirre, C., Lenz, T.L., Kalbe, M., and Milinski, M. (2012). Rapid and adaptive evolution of MHC genes under parasite selection in experimental vertebrate populations. *Nature Communications*, 3, 6.

Ely, C.R., Hall, J.S., Schmutz, J.A., et al. (2013). Evidence that life history characteristics of wild birds influence infection and exposure to influenza A viruses. *PLoS ONE*, 8, 11.

Ewald, P.W. (1983). Host–parasite relations, vectors, and the evolution of disease severity. *Annual Review of Ecology and Systematics*, 14, 465–85.

Farmer, K.L., Hill, G.E., and Roberts, S.R. (2002). Susceptibility of a naive population of house finches to *Mycoplasma gallisepticum*. *Journal of Wildlife Diseases*, 38, 282–6.

Fleming-Davies, A.E., Williams, P.D., Dhondt, A.A., et al. (2018). Incomplete host immunity favors the evolution of virulence in an emergent pathogen. *Science*, 359, 1030–3.

Fumagalli, M., Sironi, M., Pozzoli, U., Ferrer-Admettla, A., Pattini, L., and Nielsen, R. (2011). Signatures of environmental genetic adaptation pinpoint pathogens as the main selective pressure through human evolution. *PLoS Genetics*, 7, e1002355.

Gaba, S. and Ebert, D. (2009). Time-shift experiments as a tool to study antagonistic coevolution. *Trends in Ecology & Evolution*, 24, 226–32.

Gandon, S. (2002). Local adaptation and the geometry of host–parasite coevolution. *Ecology Letters*, 5, 246–56.

Gandon, S. and Michalakis, Y. (2000). Evolution of parasite virulence against qualitative or quantitative host resistance. *Proceedings of the Royal Society B: Biological Sciences*, 267, 985–90.

Gandon, S., Mackinnon, M.J., Nee, S., and Read, A.F. (2001). Imperfect vaccines and the evolution of pathogen virulence. *Nature*, 414, 751–6.

Gandon, S., Mackinnon, M., Nee, S., and Read, A. (2003). Imperfect vaccination: some epidemiological and evolutionary consequences. *Proceedings of the Royal Society B: Biological Sciences*, 270, 1129–36.

Gates, D.E., Valletta, J.J., Bonneaud, C., and Recker, M. (2018). Quantitative host resistance drives the evolution of increased virulence in an emerging pathogen. *Journal of Evolutionary Biology*, 31, 1704–14.

Gaunson, J.E., Philip, C.J., Whithear, K.G., and Browning, G.F. (2000). Lymphocytic infiltration in the chicken trachea in response to *Mycoplasma gallisepticum* infection. *Microbiology*, 146, 1223–9.

Giraud, T., Gladieux, P., and Gavrilets, S. (2010). Linking the emergence of fungal plant diseases with ecological speciation. *Trends in Ecology & Evolution*, 25, 387–95.

Grant, P.R. and Grant, B.R. (2006). Evolution of character displacement in Darwin's finches. *Science*, 313, 224–26.

Grubaugh, N.D. and Ebel, G.D. (2016). Dynamics of West Nile virus evolution in mosquito vectors. *Current Opinion in Virology*, 21, 132–8.

Grubaugh, N.D., Smith, D.R., Brackney, D.E., et al. (2015). Experimental evolution of an RNA virus in wild birds: evidence for host-dependent impacts on population structure and competitive fitness. *PLoS Pathogens*, 11, 19.

Grueber, C.E., Wallis, G.P., and Jamieson, I.G. (2014). Episodic positive selection in the evolution of avian Toll-like receptor innate immunity genes. *PLoS ONE*, 9, e89632.

Gyuranecz, M., Foster, J.T., Dan, A., et al. (2013). Worldwide phylogenetic relationship of avian poxviruses. *Journal of Virology*, 87, 4938–51.

Hawley, D.M., Osnas, E.E., Dobson, A.P., Hochachka, W.M., Ley, D.H. and Dhondt, A.A. (2013). Parallel patterns of increased virulence in a recently emerged wildlife pathogen. *PLoS Biology*, 11, 11.

Hayes, C.G. (2001). West Nile virus: Uganda, 1937, to New York city, 1999. *Annals of the New York Academy of Sciences*, 951, 25–37.

Hill, A.V.S. (1991). HLA associations with malaria in Africa: some implications for MHC evolution. In J. Klein and D. Klein, eds. *Molecular Evolution of the Major Histocompatibility Complex*, pp. 403–20. Springer, Berlin.

Hochachka, W.M. and Dhondt, A.A. (2000). Density-dependent decline of host abundance resulting from a new infectious disease. *Proceedings of the National Academy of Sciences of the United States of America*, 97, 5303–6.

Holmes, E.C. (2003). Error thresholds and the constraints to RNA virus evolution. *Trends in Microbiology*, 11, 543–6.

Holmes, E.C. (2009). *The Evolution and Emergence of RNA Viruses*. Oxford University Press, New York.

Imai, M., Watanabe, T., Hatta, M., et al. (2012). Experimental adaptation of an influenza H5 HA confers respiratory droplet transmission to a reassortant H5 HA/H1N1 virus in ferrets. *Nature*, 486, 420.

Janeway, C. (2005). *Immunobiology: The Immune System in Health and Disease*. Garland Science, New York.

Jenkins, G.M., Rambaut, A., Pybus, O.G., and Holmes, E.C. (2002). Rates of molecular evolution in RNA viruses: a quantitative phylogenetic analysis. *Journal of Molecular Evolution*, 54, 156–65.

Jenkins, T., Delhaye, J., and Christe, P. (2015). Testing local adaptation in a natural great tit–malaria system: an experimental approach. *PLoS ONE*, 10, e0141391.

Jensen, K.H., Little, T., Skorping, A., and Ebert, D. (2006). Empirical support for optimal virulence in a castrating parasite. *PLoS Biology*, 4, 1265–9.

Jerzak, G.V.S., Brown, I., Shi, P.Y., Kramer, L.D., and Ebel, G.D. (2008). Genetic diversity and purifying selection in West Nile virus populations are maintained during host switching. *Virology*, 374, 256–60.

Jia, X.Y., Briese, T., Jordan, I., et al. (1999). Genetic analysis of West Nile New York 1999 encephalitis virus. *Lancet*, 354, 1971–2.

Johnson, P.T.J., Rohr, J.R., Hoverman, J.T., Kellermanns, E., Bowerman, J., and Lunde, K.B. (2012). Living fast and dying of infection: host life history drives interspecific variation in infection and disease risk. *Ecology Letters*, 15, 235–42.

Kalbe, M. and Kurtz, J. (2006). Local differences in immunocompetence reflect resistance of sticklebacks against the eye fluke *Diplostomum pseudospathaceum*. *Parasitology*, 132, 105–16.

Kilpatrick, A.M., Kramer, L.D., Jones, M.J., Marra, P.P., and Daszak, P. (2006). West Nile virus epidemics in North America are driven by shifts in mosquito feeding behavior. *PLoS Biology*, 4, 606–10.

Kimura, M.T. (1983). *The Neutral Theory of Molecular Evolution*. Cambridge University Press, Cambridge.

Kollias, G.V., Sydenstricker, K.V., Kollias, H.W., et al. (2004). Experimental infection of house finches with *Mycoplasma gallisepticum*. *Journal of Wildlife Diseases*, 40, 79–86.

Komar, N., Langevin, S., Hinten, S., et al. (2003). Experimental infection of North American birds with the New York 1999 strain of West Nile virus. *Emerging Infectious Diseases*, 9, 311–22.

LaDeau, S.L., Kilpatrick, A.M., and Marra, P.P. (2007). West Nile virus emergence and large-scale declines of North American bird populations. *Nature*, 447, 710–3.

Lanciotti, R.S., Roehrig, J.T., Deubel, V., et al. (1999). Origin of the West Nile virus responsible for an outbreak of encephalitis in the northeastern United States. *Science*, 286, 2333–7.

Langevin, S.A., Brault, A.C., Panella, N.A., Bowen, R.A., and Komar, N. (2005). Variation in virulence of West Nile virus strains for house sparrows (*Passer domesticus*). *American Journal of Tropical Medicine and Hygiene*, 72, 99–102.

Lamichhaney, S., Berglund, J., Almen, M.S., et al. (2015). Evolution of Darwin's finches and their beaks revealed by genome sequencing. *Nature*, 518, 371–75.

Lion, S. and Metz, J.A.J. (2018). Beyond R-0 maximisation: on pathogen evolution and environmental dimensions. *Trends in Ecology & Evolution*, 33, 458–73.

Lively, C.M., Dybdahl, M.F., Jokela, J., Osnas, E.E., and Delph, L.F. (2004). Host sex and local adaptation by parasites in a snail–trematode interaction. *The American Naturalist*, 164, S6–18.

Loiseau, C., Richard, M., Garnier, S., et al. (2009). Diversifying selection on MHC class I in the house sparrow (*Passer domesticus*). *Molecular Ecology*, 18, 1331–40.

MacDougall-Shackleton, E.A., Derryberry, E.P., and Hahn, T.P. (2002). Nonlocal male mountain white-crowned sparrows have lower paternity and higher parasite loads than males singing local dialect. *Behavioral Ecology*, 13, 682–9.

Mann, B.R., McMullen, A.R., Swetnam, D.M., and Barrett, A.D.T. (2013). Molecular epidemiology and evolution of West Nile virus in North America. *International Journal of Environmental Research and Public Health*, 10, 5111–29.

May, F.J., Davis, C.T., Tesh, R.B., and Barrett, A.D.T. (2011). Phylogeography of West Nile virus: from the cradle of evolution in Africa to Eurasia, Australia, and the Americas. *Journal of Virology*, 85, 2964–74.

McDonald, J.H. and Kreitman, M. (1991). Adaptive protein evolution at the adh locus in Drosophila. *Nature*, 351, 652–4.

McLean, R.G., Ubico, S.R., Docherty, D.E., Hansen, W.R., Sileo, L., and McNamara, T.S. (2001). West Nile virus transmission and ecology in birds. *Annals of the New York Academy of Sciences*, 951, 54–7.

Medzhitov, R., Schneider, D.S., and Soares, M.P. (2012). Disease tolerance as a defense strategy. *Science*, 335, 936–41.

Miller, M.R., White, A., and Boots, M. (2006). The evolution of parasites in response to tolerance in their hosts: the good, the bad, and apparent commensalism. *Evolution*, 60, 945–56.

Moran, N. and Baumann, P. (1994). Phylogenetics of cytoplasmically inherited microorganisms of arthropods. *Trends in Ecology & Evolution*, 9, 15–20.

Naylor, C.J., Alankari, A.R., Alafaleq, A.I., Bradbury, J.M., and Jones, R.C. (1992). Exacerbation of *Mycoplasma gallisepticum* infection in turkeys by rhinotracheitis virus. *Avian Pathology*, 21, 295–305.

Nolan, P.M., Hill, G.E., and Stoehr, A.M. (1998). Sex, size, and plumage redness predict house finch survival in an epidemic. *Proceedings of the Royal Society B: Biological Sciences*, 265, 961–5.

Nuismer, S.L. (2006). Parasite local adaptation in a geographic mosaic. *Evolution*, 60, 24–30.

Ohlberger, J., Langangen, O., Edeline, E., et al. (2011). Pathogen-induced rapid evolution in a vertebrate life-history trait. *Proceedings of the Royal Society B: Biological Sciences*, 278, 35–41.

Oppliger, A., Vernet, R., and Baez, M. (1999). Parasite local maladaptation in the Canarian lizard *Gallotia galloti* (Reptilia : Lacertidae) parasitized by haemogregarian blood parasite. *Journal of Evolutionary Biology*, 12, 951–5.

Pesko, K.N. and Ebel, G.D. (2012). West Nile virus population genetics and evolution. *Infection Genetics and Evolution*, 12, 181–90.

Porco, T.C., Lloyd-Smith, J.O., Gross, K.L., and Galvani, A.P. (2005). The effect of treatment on pathogen virulence. *Journal of Theoretical Biology*, 233, 91–102.

Pulgarin-R, P.C., Gomez, C., Bayly, N.J., et al. (2019). Migratory birds as vehicles for parasite dispersal? Infection by avian haemosporidians over the year and throughout the range of a long-distance migrant. *Journal of Biogeography*, 46, 83–96.

Råberg, L. (2014). How to Live with the enemy: understanding tolerance to parasites. *PLoS Biology*, 12, 9.

Råberg, L., Graham, A.L., and Read, A.F. (2009). Decomposing health: tolerance and resistance to parasites in animals. *Philosophical Transactions of the Royal Society B: Biological Sciences*, 364, 37–49.

Razin, S., Yogev, D., and Naot, Y. (1998). Molecular biology and pathogenicity of mycoplasmas. *Microbiology and Molecular Biology Reviews*, 62, 1094–156.

Read, A.F., Baigent, S.J., Powers, C., et al. (2015). Imperfect vaccination can enhance the transmission of highly virulent pathogens. *PLoS Biology*, 13, 18.

Ricklefs, R.E. and Fallon, S.M. (2002). Diversification and host switching in avian malaria parasites. *Proceedings of the Royal Society B: Biological Sciences*, 269, 885–92.

Ricklefs, R.E., Fallon, S.M., and Bermingham, E. (2004). Evolutionary relationships, cospeciation, and host switching in avian malaria parasites. *Systematic Biology*, 53, 111–19.

Ricklefs, R.E., Outlaw, D.C., Svensson-Coelho, M., Medeiros, M.C.I., Ellis, V.A., and Latta, S. (2014). Species formation by host shifting in avian malaria parasites. *Proceedings of the National Academy of Sciences of the United States of America*, 111, 14816–21.

Sarquis-Adamson, Y. and MacDougall-Shackleton, E.A. (2016). Song sparrows *Melospiza melodia* have a home-field advantage in defending against sympatric malarial parasites. *Royal Society Open Science*, 3, 160216.

Sepil, I., Lachish, S., Hinks, A.E., and Sheldon, B.C. (2013). Mhc supertypes confer both qualitative and quantitative resistance to avian malaria infections in a wild bird population. *Proceedings of the Royal Society B: Biological Sciences*, 280, 20130134.

Shehata, C.L., Freed, L.A., and Cann, R.L. (2001). Changes in native and introduced bird populations on O'ahu: infectious diseases and species replacement. *Studies in Avian Biology*, 22, 264–73.

Shultz, A.J. and Sackton, T.B. (2019). Immune genes are hotspots of shared positive selection across birds and mammals. *Elife*, 8, e41815.

Simmonds, P., Aiewsakun, P., and Katzourakis, A. (2018). Prisoners of war—host adaptation and its constraints on virus evolution. *Nature Reviews Microbiology*, 17, 321–8.

Slade, J.W.G., Sarquis-Adamson, Y., Gloor, G.B., Lachance, M.A., and MacDougall-Shackleton, E.A. (2017). Population differences at mhc do not explain enhanced resistance of song sparrows to local parasites. *Journal of Heredity*, 108, 127–34.

Spiegel, C.S., Hart, P.J., Woodworth, B.L., Tweed, E.J., and LeBrun, J. (2006). Distribution and abundance of native forest birds in low-elevation areas on Hawai'i island: evidence of range expansion. *Bird Conservation International*, 16, 175–85.

Staley, M. and Bonneaud, C. (2015). Immune responses of wild birds to emerging infectious diseases. *Parasite Immunology*, 37, 242–54.

Stearns, S.C. (1999). *Evolution in Health & Disease*. Oxford University Press, Oxford.

Szczepanek, S.M., Frasca, S., Schumacher, V.L., et al. (2010). Identification of lipoprotein MslA as a neoteric virulence factor of *Mycoplasma gallisepticum*. *Infection and Immunity*, 78, 3475–83.

Tardy, L., Giraudeau, M., Hill, G.E., McGraw, K.J., and Bonneaud, C. (2019). Contrasting evolution of virulence and replication rate in an emerging bacterial pathogen. *Proceedings of the National Academy of Sciences of the United States of America*, 116, 16927–32.

Taubenberger, J.K. and Morens, D.M. (2017). H5Nx panzootic bird flu—influenza's newest worldwide evolutionary tour. *Emerging Infectious Diseases*, 23, 340–2.

Tenaillon, O., Barrick, J.E., Ribeck, N., et al. (2016). Tempo and mode of genome evolution in a 50,000-generation experiment. *Nature*, 536, 165.

Thompson, J.N. (1994). *The Coevolutionary Process*. University of Chicago Press, Chicago.

Thompson, J.N. (2005). *The Geographic Mosaic of Coevolution*. University of Chicago Press, Chicago.

Van Riper, C., Van Riper, S.G., Goff, M.L., and Laird, M. (1986). The epizootiology and ecological significance of malaria in Hawaiian land birds. *Ecological Monographs*, 56, 327–44.

VanDalen, K.K., Hall, J.S., Clark, L., McLean, R.G., and Smeraski, C. (2013). West Nile virus infection in American robins: new insights on dose response. *PLoS ONE*, 8, e68537.

Velova, H., Gutowska-Ding, M.W., Burt, D.W., and Vinkler, M. (2018). Toll-like receptor evolution in birds: gene duplication, pseudogenization, and diversifying selection. *Molecular Biology and Evolution*, 35, 2170–84.

Venkatesh, D., Poen, M.J., Bestebroer, T.M., et al. (2018). Avian influenza viruses in wild birds: virus evolution in a multihost ecosystem. *Journal of Virology*, 92, e00433–18.

Vogl, G., Plaickner, A., Szathmary, S., Stipkovits, L., Rosengarten, R., and Szostak, M.P. (2008). *Mycoplasma gallisepticum* invades chicken erythrocytes during infection. *Infection and Immunity*, 76, 71–7.

Westerdahl, H., Waldenstrom, J., Hansson, B., Hasselquist, D., von Schantz, T., and Bensch, S. (2005). Associations between malaria and MHC genes in a migratory songbird. *Proceedings of the Royal Society B: Biological Sciences*, 272, 1511–18.

Westerdahl, H., Stjernman, M., Raberg, L., Lannefors, M., and Nilsson, J.A. (2013). MHC-I affects infection intensity but not infection status with a frequent avian malaria parasite in blue tits. *PLoS ONE*, 8, e72647.

Williams, P.D., Dobson, A.P., Dhondt, K.V., Hawley, D.M., and Dhondt, A.A. (2014). Evidence of trade-offs shaping virulence evolution in an emerging wildlife pathogen. *Journal of Evolutionary Biology*, 27, 1271–8.

Woodworth, B.L., Atkinson, C.T., LaPointe, D.A., et al. (2005). Host population persistence in the face of introduced vector-borne diseases: Hawaii amakihi and avian malaria. *Proceedings of the National Academy of Sciences of the United States of America*, 102, 1531–6.

Woolhouse, M.E.J., Webster, J.P., Domingo, E., Charlesworth, B., and Levin, B.R. (2002). Biological and biomedical implications of the co-evolution of pathogens and their hosts. *Nature Genetics*, 32, 569–77.

Woolhouse, M.E.J., Haydon, D.T., and Antia, R. (2005). Emerging pathogens: the epidemiology and evolution of species jumps. *Trends in Ecology & Evolution*, 20, 238–44.

Yaremych, S.A., Warner, R.E., Mankin, P.C., Brawn, J.D., Raim, A., and Novak, R. (2004). West Nile virus and high death rate in American crows. *Emerging Infectious Diseases*, 10, 709–11.

Yoder, H.S.J. (1991). *Mycoplasma gallisepticum* infection. In B.W. Cal-nek, H.J. Barnes, C.W. Beard, W.M. Reid, and H.W. Yoder, eds. *Diseases of Poultry*, 9th ed., pp. 198–212. Iowa State University Press, Ames.

Zehender, G., Ebranati, E., Bernini, F., et al. (2011). Phylogeography and epidemiological history of West Nile virus genotype 1a in Europe and the Mediterranean basin. *Infection Genetics and Evolution*, 11, 646–53.

Zylberberg, M., Klasing, K.C., and Hahn, T.P. (2013). House finches (*Carpodacus mexicanus*) balance investment in behavioural and immunological defences against pathogens. *Biology Letters*, 9, 4.

CHAPTER 6

Fitness Effects of Parasite Infection in Birds

Jenny C. Dunn, Dana M. Hawley, Kathryn P. Huyvaert, and Jennifer C. Owen

6.1 Introduction

By definition, pathogens and parasites (hereafter referred to collectively as 'parasites') harm their hosts by using the host's resources for their own survival and reproduction. However, the degree to which that harm causes notable fitness costs for an individual host can be highly variable. While some parasites cause only minimal harm to their hosts, others can have profound fitness consequences for the host organism by limiting reproductive success and survival (Figure 6.1). In Darwinian terms, a host's fitness is determined by the ability to pass its genes on to subsequent generations, and thus both survival and reproductive success are necessary components of individual fitness (e.g., Stearns 2000). While some effects of parasites on host fitness are drastic and immediate in nature (i.e., acute mortality), a growing body of evidence suggests that even parasites that cause chronic and subclinical infections can have notable fitness consequences for hosts by reducing the probability of long-term survival and reproductive success. Overall, understanding the individual-level fitness consequences of parasite infection is critical because these effects ultimately limit a population's growth potential (see Chapter 7 for more on populations) and determine the strength of selection that parasites place on host populations (see Chapter 5 for more on evolution).

In this chapter, we examine the impacts of parasites and pathogens on diverse fitness-relevant traits in birds, including survival and components of reproductive success (Figure 6.1). Because these fitness effects can vary with characteristics of the parasite, host, and environmental context in which the host–parasite interaction occurs, we highlight some of the characteristics thought to be important for mediating fitness effects of infection on avian hosts. We first discuss the effects of parasites on host survival, including ways in which these effects are measured and the influence of both parasite and host characteristics on survival impacts. We then discuss the diverse effects of parasites on reproductive parameters in birds and note areas for future consideration. We end with a brief discussion of the context-dependence of fitness effects in birds, which is just beginning to be explored. The potential influence of environmental factors such as climate and nutritional stress on fitness impacts of parasites underscores the ecological complexity of the host–parasite relationship and thus the heterogeneity in the degree to which parasites are important players in both the evolution and population dynamics of bird populations more broadly.

6.2 Effects of parasites on host survival

Parasites have long been recognized as important sources of mortality in birds, particularly in host populations that are naïve to a given parasite (e.g., Hawaiian honeycreepers and blood parasites that cause avian malaria) (van Riper et al. 1986) or during periods when birds are densely concentrated on breeding grounds. For example, avian cholera (caused by *Pasteurella multocida*) outbreaks have been linked to mortality rates of 43% and 95%,

Jenny C. Dunn, Dana M. Hawley, Kathryn P. Huyvaert, and Jennifer C. Owen, *Fitness Effects of Parasite Infection in Birds* In: *Infectious Disease Ecology of Wild Birds*. Edited by: Jennifer C. Owen, Dana M. Hawley, and Kathryn P. Huyvaert, Oxford University Press. © Oxford University Press 2021.
DOI: 10.1093/oso/9780198746249.003.0006

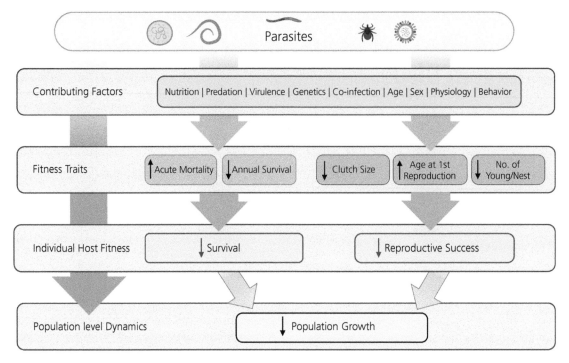

Figure 6.1 Conceptual diagram of effects of parasites and pathogens on host fitness. Parasites influence fitness traits (orange) that determine survival and reproductive success (green) at the level of the individual host, and these effects can scale up to affect population dynamics (yellow; further explored in Chapter 7). Effect of parasites on fitness traits is frequently mediated by one or more contributing factors (blue), though those listed here are not exhaustive. Effects of several of these factors are covered in more detail in Chapters 3–5. Parasitism can reduce host survival on long or short time scales by reducing annual survival or by causing acute mortality, respectively. Parasites can alter reproductive success by influencing several components of a bird's reproductive output including age at which reproduction begins (when infections occur early in life) or clutch size or brood size produced during a given reproductive bout. Parasites may also have effects on parentage and other diverse aspects of reproductive success (see text for details). (Created with Biorender.com.)

respectively, of breeding female common eiders (*Somateria mollissima*) from dense colonies in the Canadian Arctic and Denmark (Christensen et al. 1997; Iverson et al. 2016; Table 6.1). Although outbreak events such as those caused by avian cholera are examples of cases where parasites directly kill their hosts via mechanisms such as tissue damage, parasites can also reduce host survival through several indirect mechanisms. For example, parasites may cause indirect mortality by making hosts more vulnerable to predation (Adelman et al. 2017; Isomursu et al. 2008) or by depleting host nutritional resources (Smith et al. 2005). Regardless of

the mechanism, effects of infection on host survival can often be significant.

Outbreak events provide acute and dramatic illustrations of the extent to which parasites can affect host survival. However, mortality due to parasite infection, whether direct or indirect, is often challenging to detect for several reasons. First, avian carcasses are notoriously difficult to find, particularly for small birds (Wobeser 2013). Second, because parasite-mediated mortality is often indirect, effects of parasites on host survival often occur in combination with other ecological factors such as predation (e.g., Hudson et al. 1992a) and nutritional

Table 6.1 Examples of field studies examining differences in survival between uninfected and infected (or diseased) groups of free-living wild birds varying by age class (Ad=Adult, Juv=Juvenile), sex (F, M, or ND=not differentiated), and study approach (EXP=experimental; OBS=observational). Survival percentages are annual except in Faustino et al. 2004 and "Survival difference" is calculated as % Uninfected - % Infected surviving, such that positive values represent higher survival of uninfected relative to infected birds.

Pathogen/Disease	Host Species	Age Class	Study Duration (Years)	Approach	Uninfected [% surviving] (n)[a]	Infected [% surviving] (n)[a]	Survival difference [%] (n)	Source
Bacteria								
Pasteurella multocida/Avian cholera	Common eider (Somateria mollissima)	F	4	OBS			14 (741)	Tjørnløv et al. 2013
		Juv	6	OBS			90 (943)	Descamps et al. 2011
		F	2	OBS	86 (325)	77 (26)	9 (351)	Legagneux et al. 2014
	Common guillemot (Uria aalge)	ND	5	OBS			14.9 (111)[b]	Österblom et al. 2004
	Snow goose (Anser caerulescens)	ND	3[c]	EXP	88.5 (292)	84.2 (290)	4.3 (582)	Samuel et al. 1999
		ND	2[d]	EXP	77.9 (114)	70.3 (114)	7.6 (228)	Samuel et al. 1999
Mycoplasma gallisepticum/ Mycoplasmal conjunctivitis	House finch (Haemorhous mexicanus)	ND	0.25	OBS			10.5 (256)	Faustino et al. 2004
		ND	0.25	OBS			–23.8 (256)	Faustino et al. 2004
		ND	0.25	OBS			23.1 (725)	Faustino et al. 2004
		ND	0.25	OBS			36.4 (1187)	Faustino et al. 2004
Borrellia burgdorferi/ Lyme disease	Black-legged kittiwake (Rissa tridactyla)	Ad	1	EXP	75 (12)	80 (11)	–5 (23)	Chambert et al. 2012
Viruses								
Avian pox virus	Galápagos mockingbird (Mimus parvulus)	Juv	1	OBS	92 (18)	0 (14)	72 (32)	Vargas 1987
	Waved albatross (Phoebastria irrorata)	Juv	0.25	OBS	98.5 (199)	57 (14)	41.5 (213)	Tompkins et al. 2017
	Great tit (Parus major)	ND	2.5	OBS	28.5 (1004)	6 (82)	22.5 (1806)	Lachish et al. 2012a
	Red-legged partridge (Alectoris rufa)	Juv	0.75	OBS	42 (41)	31 (29)	11 (70)	Buenestado et al. 2004
	Serin (Serinus serinus)	ND	0.5	OBS	87 (1408)	46 (62)	41 (1470)	Senar and Conroy 2004
	Laysan albatross (Phoebastria immutabilis)	Juv	14	OBS	77.8 (349)	64.8 (128)	4 – 13 (477)	VanderWerf and Young 2016

(Continued)

Table 6.1 Continued

Pathogen/Disease	Host Species	Age Class	Study Duration (Years)	Approach	Uninfected [% surviving] (n)[a]	Infected [% surviving] (n)[a]	Survival difference [%] (n)	Source
Low-pathogenic avian influenza virus	Bewick's swan (Cygnus columbianus)	ND	0.25	OBS	63 (8)	75 (4)	−12 (12)	van Gils et al. 2007
	Ruddy turnstone (Arenaria interpres)	ND	1	OBS	46 (1070)	45 (163)	1 (1233)	Maxted et al. 2012
	Greater white-fronted goose (Anser albifrons)	ND	4	OBS	89.3 (1722)	93.7 (67)	−4.4 (1789)	Kleijn et al. 2010
West Nile virus	Greater sage grouse (Centrocercus urophasianus)	ND Ad F	0.25 0.17	OBS OBS	96 (61)	86 (40)	10 (533) 56 (44)	Naugle et al. 2005 Walker et al. 2004
	Northern cardinal (Cardinalis cardinalis)	ND	2	OBS	50.1 (159)	41.1 (24)	34.7 (183)	Ward et al. 2010
Protozoa								
Haemoproteus sp.	Blue tit (Parus caeruleus)	F	3	EXP	42 (47)	22 (48)	20 (95)	Martínez-de la Puente et al. 2010
		M	3	EXP	27 (47)	39 (45)	−12 [e] (92)	Martínez-de la Puente et al. 2010
	Crested tit (Lophophanes cristatus)	ND	0.83	OBS	97 (71)	13 (15)	84 (86)	Krama et al. 2015
	Seychelles warbler (Acrocephalus sechellensis)	Juv	9	OBS	86 (41)	71 (9)	15 (50)	van Oers et al. 2010
Haemoproteus beckeri	Mountain white-crowned sparrow (Zonotrichia leucophrys)	ND	10	OBS	28.6 (273)	50 (34)	−35 (307)	Zylberberg et al. 2015
Haemoproteus columbae	Feral pigeon (Columba livia)	ND	0.5	OBS	66 (46)	39 (18)	27 (64)	Sol et al. 2003
Haemoproteus GRW1	Great reed warbler (Acrocephalus arundinaceus)	ND	17	OBS	49 (373)	50 (69)	−1 (449)	Bensch et al. 2007
Haemoproteus SW1	Sedge warbler (Acrocephalus schoenobaenus)	M	9	OBS	7.3 (274)	9.9 (161)	−2.6 (435)	Bielański et al. 2017

Pathogen	Host	Sex		Study				Reference
Haemoproteus SW3	Sedge warbler	M	9	OBS	13 (297)	6.5 (138)	6.5 (435)	Bielański et al. 2017
Leucocytozoon marchouxi	Pink pigeon (*Nesoenas mayeri*)	ND	0.25	OBS	98.1 (268)	93.3 (60)	4.8 (328)	Bunbury et al. 2007
			1	OBS	88 (268)	79.4 (60)	8.6 (328)	
Plasmodium sp.	Blue tit	ND	9	OBS			17 (3424)	Lachish et al. 2011
Plasmodium GRW2	Great reed warbler	ND	17	OBS	49 (416)	55 (26)	−6 (442)	Bensch et al. 2007
Plasmodium GRW4	Great reed warbler	ND	17	OBS	49 (373)	50 (69)	−1 (442)	Bensch et al. 2007
Plasmodium or *Haemoproteus* sp.	Willow tit (*Poecile montanus*)	ND	0.83	OBS	96 (56)	23 (26)	73 (82)	Krama et al. 2015
	House martin (*Delichon urbicum*)	ND	3	OBS	32 (184)	15 (33)	17 (217)	Magallanes et al. 2017
		ND	6	OBS	47 (32)	23 (62)	24 (94)	Marzal et al. 2008
		ND	6	OBS	18.6 (70)	20.7 (232)	−2.1 (302)	Piersma and van der Velde 2012
Cryptosporidium sp./Respiratory cryptosporidiosis	Red grouse (*Lagopus lagopus scotica*)	F	1.5	OBS	70 (51)	44 (68)	26 (119)	Baines et al. 2018
		M	1.5	OBS	70 (16)	22 (43)	48 (43)	
Trichomonas gallinae	Pink pigeon	ND	1.7	OBS			13 – 23 (233)	Bunbury et al. 2008
Helminths								
Microfilaria	Purple martin (*Progne subis*)	ND	7	OBS	32 (149)	9 (23)	23 (172)	Davidar and Morton 2006
Trichostrongylus tenuis	Red grouse	ND	10	EXP	28.9 (173)	18.4 (136)	10.5 (306)	Hudson et al. 1992b

a Where infected and uninfected sample sizes are not provided, data are from studies comparing outbreak to non-outbreak years or sites or are from longitudinal studies.

b A marked increase in predation was also observed.

c Survival estimates are from the Wrangel Island, Russia study population.

d Survival estimates are from the Banks Island, Canada study population

e Medication did not reduce infection intensity in males.

stress (e.g., Chapman et al. 2006), making these effects challenging to elucidate. Third, while some parasite-mediated mortality is acute (as occurs with avian cholera, for example), reductions in survival due to parasite infection often occur over longer time scales and can only be tracked over the long term (e.g., Pigeault et al. 2018).

Given the difficulty in measuring the effects of parasite infection on survival in free-living birds, some studies infer potential fitness effects by documenting that parasite-infected individuals have lower body mass or poorer condition than uninfected conspecifics (e.g., Garvin et al. 2006). While these studies suggest potential negative effects of parasites on host survival, studies that frequently monitor marked individuals are critical for robustly quantifying the effects of parasites on individual host survival. Here, we first consider what is needed to determine survival differences associated with parasite infection. We then describe general evidence for the effects of parasites on avian survival, considering how characteristics of parasites, hosts, and other ecological mechanisms explain some of the heterogeneity in survival in the face of infection.

6.2.1 What is needed to show parasite-mediated differences in survival?

Estimating survival rates of free-living birds in the presence or absence of parasites is a challenging exercise. While counts of carcasses during an outbreak can give some estimate of the proportion of a breeding colony or population affected (and thus the degree of mortality from a given parasite), quantifying the effects of parasites on individual survival probabilities requires longitudinal data from marked birds (e.g., recaptures or encounters via standardized trapping and observations of marked birds or direct radiotracking of individuals over time). For long-lived birds and/or infection with parasites that are relatively chronic in nature, these longitudinal studies need to be conducted over a long time period to capture meaningful differences in survival. Finally, a key challenge for all studies attempting to quantify survival effects of parasitism is the ability to robustly assign infection status of a given individual using the tests available (see Chapter 2 and Chapter 7, Box 7.1 for more discussion of this issue). Imperfect detection of

disease or infection state, which is inherent to some degree in almost all host–parasite systems, will limit the accuracy of any estimated differences in survival between individuals for which parasites were or were not detected.

There are several ways that longitudinal studies of marked birds have been used to quantify survival differences due to parasite infection. Most simply, because many birds show high rates of annual breeding site fidelity, differences in presence at breeding sites for birds banded in a prior year can be used as one proxy for differences in annual survival in the interim. As such, many studies estimate whether infection state from the prior year is associated with differences in return rates to a breeding area or, for resident birds, the continued presence of marked individuals on the breeding grounds in the following season (Zylberberg et al. 2015). However, the use of return rates to estimate survival is confounded by potential differences in the ability of researchers to encounter infected versus uninfected individuals that are alive in a given area (see Box 7.1 for more on detection issues). For example, individuals with a given parasite infection may be less likely to maintain a territory even if present (e.g., Mougeot et al. 2005; Whiteman and Parker 2004) and thus may be less likely to be detected as still alive. Further, even if the likelihood of encounter of a given individual does not vary with infection state, differences in return rate can occur through emigration of birds to other breeding areas, which may, in itself, be influenced by parasite status. For example, in Indian yellow-nosed albatross (*Thalassarche carteri*) breeding on Amsterdam Island, individuals whose nesting attempts failed (potentially due to avian cholera) had higher rates of emigration to other breeding areas relative to individuals who fledged young (Rolland et al. 2009). Overall, robustly estimating differences in survival in light of parasitism based on annual return rates to breeding areas alone can be problematic.

Mark-recapture approaches are considered the gold standard for estimating survival probabilities of free-living animals, as they account for potential differences in our ability to encounter individuals based on a variety of factors, including their infection state (McClintock et al. 2010; see also Chapter 7). However, even traditional mark-recapture approaches often cannot distinguish mortality from permanent emigration from a study area, leading to estimates of 'apparent survival' rather than 'true survival.' Nonetheless, mark-recapture approaches generally allow for robust comparisons of survival between individuals of a particular disease or infection state (Table 6.1), and in some systems, these survival estimates have been shown to be close to those for experimentally inoculated birds in captivity (e.g., Ward et al. 2010). For example, using a mark-recapture approach, Ward et al. (2010) estimated West Nile virus (WNV)-related mortality in northern cardinals (*Cardinalis cardinalis*) as 34.6%, which did not differ significantly from an experimental study on cardinals that documented a mortality rate of 25% (Komar et al. 2005). Because mortality from experimental infection in captivity is likely an underestimate of what would occur in the wild, mark-recapture studies of free-living individuals typically yield the most robust predictions of parasite-mediated effects on host survival.

Longitudinal methods for quantifying survival probabilities in free-living birds are sometimes used in conjunction with experimental methods that directly manipulate infection state, such as vaccination or medication. Although these experimental manipulations are rarely perfect (e.g., vaccination rarely confers 100% protection and medication rarely completely clears infection), these approaches help to control for the fact that some mortality due to infection occurs in individuals that are already at high risk for mortality due to confounding factors (e.g., low body condition) which may also put that individual at risk for infection. Thus, experimental manipulations of infection status can be powerful approaches for isolating the effects of parasite infection *per se* on survival. For example, Samuel et al. (1999) radio-tagged and tracked lesser snow geese (*Anser caerulescens*) in two different populations affected by avian cholera. A subset of the geese in each population were vaccinated to protect them from *P. multocida*, while others were untreated controls. Overall, when controlling for other sources of mortality (i.e., hunting, overwinter weather) the authors found that avian cholera contributed to about half of the annual mortality in this species in most winters. Medication approaches are often used to estimate the survival impacts of parasite infections that are relatively chronic and readily treatable, though these treatments rarely com-

pletely eliminate the parasite. For example, Hudson et al. (1992b) gave a subset of red grouse (*Lagopus lagopus scotica*) an antihelminthic medication over a 3-year period to elucidate the effects of the common intestinal helminth *Trichostrongylus tenuis* on survival. While medication did not completely clear worm infections, medicated birds had significantly lower infection burdens and increased rates of winter survival (Hudson et al. 1992b).

6.2.2 What evidence is there that parasites influence individual survival in birds?

Despite the methodological challenges in quantifying survival differences associated with parasite infection, effects on survival have been documented for a diverse set of parasite infections in birds (Table 6.1). Here, we consider some of the many factors that can influence the heterogeneity in effects on host survival (Figure 6.1), including characteristics of the parasites themselves (microparasites vs. macroparasites; virulence), co-infection, host species differences, and individual host demographic characteristics (e.g., age, sex).

6.2.2.1 Parasite characteristics that impact effects on host survival

The time scale of survival effects caused by parasites that are classically defined as microparasites (i.e., bacteria, viruses, protozoa) versus macroparasites (i.e., helminths), based on parasite life histories, are expected to differ (see Chapter 2). Although the two categories of parasites are arguably somewhat arbitrary, microparasites generally cause acute, self-limiting infections that result in relatively immediate mortality or immune clearance for a given host. Macroparasites, on the other hand, often cause chronic infections that historically have been considered to not result in severe fitness impacts for hosts. Certain protozoal infections (e.g., many avian hemoparasites) do not fit cleanly into either category, as infections with these parasites often have both an acute/immediate and a chronic stage of infection. Nonetheless, these broad categories provide a useful distinction for considering whether and how survival effects differ for distinct groups of parasites (Table 6.1).

Some of the strongest evidence for reductions in survival due to parasite infection comes from studies of classic microparasite infections that recently emerged in a host that was previously naïve to infection (see Section 6.2.2.2 for further discussion of host susceptibility). For example, the bacterium *Mycoplasma gallisepticum* that emerged in the mid-1990s as a novel pathogen of North American populations of house finches (*Haemorhous mexicanus*) is associated with significant reductions of up to 36.5% in apparent survival of infected individuals, although effects on survival vary seasonally (Faustino et al. 2004; Table 6.1). Similarly, the recent emergence of avian poxvirus in the well-studied great tit (*Parus major*) population in Oxford, UK, allowed a detailed study of the epidemiology and fitness impacts of the parasite (Lachish et al. 2012a; Lachish et al. 2012b). This outbreak led to a 26.5% per month reduction in survival of infected juvenile birds and a 13.1% reduction in infected adults. Whilst disease prevalence within the population was relatively low, the reductions in survival of infected birds led to the important prediction that, if prevalence in the population persisted above 5%, then the parasite would reduce the host population growth rate (Lachish et al. 2012a; Lachish et al. 2012b; see also Chapter 7).

Nonetheless, not all microparasites are associated with notable survival impacts in birds (Table 6.1). For example, low pathogenic avian influenza viruses (LPAIV) are often associated with only negligible survival effects in many affected hosts (Kuiken 2013). In one study of white-fronted geese (*Anser albifrons albifrons*), the likelihood of resighting LPAIV-infected individuals was even slightly higher than, though not statistically different from, the likelihood of resighting LPAIV-negative geese (Kleijn et al. 2010). This lack of detectable effects on survival may be because avian influenza viruses have evolved extensively in wild birds, which act as an important reservoir for transmission. However, even microparasites that are relatively new to avian populations can cause negligible survival impacts in some cases. For example, despite high infection rates of avian poxvirus (a parasite likely introduced to Hawai'i in the early 1800s) in endemic Hawaiian birds, there was no evidence of survival impacts of pox over a 12-year study (Samuel et al. 2018).

Classic macroparasite infections, such as helminth infections, are generally considered to cause smaller reductions in host survival than their micro-parasite counterparts. Nonetheless, these parasite infections can still cause notable survival reductions in birds. For example, as discussed in Section 6.2.1, medication studies revealed that the helminth *T. tenuis* has notable effects on winter survival in red grouse. Further, second-year purple martins (*Progne subis*) infected with an unidentified filarial nematode had lower return rates over a seven-year study (Davidar and Morton 2006), suggesting that this parasite may reduce host survival.

Protozoan infections, which are often considered to fall between micro- and macroparasites in their characteristics, are not commonly associated with obvious mortality events in birds except in cases where the host, parasite, or vector is not native (Grilo et al. 2016; Sijbranda et al. 2017). An exception includes some pathogenic lineages of *Trichomonas gallinae* which are linked to mortality events in wild birds (Marx et al. 2017). However, even non-fatal protozoan infections can cause severe fitness impacts that have been underappreciated until relatively recently. One common set of protozoan parasites in birds are haemosporidian parasites (blood-borne protozoa comprising *Plasmodium*, *Haemoproteus*, and *Leucocytozoon* spp.), which are known to infect hundreds of bird species (Valkiunas 2004). Haemosporidian infections generally have both acute and chronic stages: in the acute stage, birds may exhibit overt clinical signs including lethargy and fever (Atkinson et al. 2008). The fitness effects of the acute stage of infection are challenging to study because the stage is short-lived. However, evidence from a long-term study of Eurasian blue tits (*Cyanistes caeruleus*; Figure 6.2A) infected by *Plasmodium relictum* suggest that the acute stage can be associated with significant mortality, even in systems where haemosporidian parasites are endemic, as survival rates for all birds were lower in areas of high parasite prevalence (Lachish et al. 2011; Figure 6.2B). Acute mortality from haemosporidian parasites also presents a challenge for detection of fitness effects associated with chronic infection, because this stage typically removes the most susceptible individuals from the longitudinal dataset (Lachish et al. 2011). Nonetheless, effects of chronic infection with haemosporidian parasites on long-term survival have been detected in several studies (e.g., Krama et al. 2015; Magallanes et al. 2017), although other studies have failed to find any effect (Piersma and van der Velde 2012). In some avian systems, chronic *Plasmodium* infection has even been documented to reduce host lifespan through degradation of telomeres (Asghar et al. 2015; Bensch et al. 2007), the non-coding DNA repeats that 'cap' the ends of chromosomes and shorten with each cellular division; this shortening can be driven by systemic 'stress' (Asghar et al. 2016) including that caused by parasitism.

The degree to which a parasite impacts host survival varies not only among types of parasites but also within the same parasite species. Several avian parasites show strain variation in virulence or the degree of infection-mediated mortality they cause in hosts of similar genetic 'background' (see Chapter 2). For example, both WNV in house sparrows (*Passer domesticus*; Duggal et al. 2014) and *M. gallisepticum* in house finches (*Haemorhous mexicanus*; Hawley et al. 2013) have evolved to become increasingly virulent over time following their emergence (see Chapter 5 for more discussion of host–parasite coevolution). Even chronic parasite infections such as those caused by *Plasmodium* spp. appear to show interlineage variation in their impacts on host survival (e.g., Lachish et al. 2011). These patterns emphasize the need to understand

not only what parasite type is affecting a given host under study but also the particular lineage, to fully understand the factors contributing to infection-mediated mortality.

Finally, parasite infections rarely occur in isolation in wild animals, and growing evidence suggests that concurrent infections of multiple parasites within a given individual host can interact with each other in ways that either suppress or augment parasite-mediated mortality. For instance, in great tits, over two-thirds of birds sampled in populations in Switzerland were co-infected with multiple types of haemosporidian parasites, and co-infections were associated with significantly lower survival than were single infections (Pigeault et al. 2018). Similarly, although survival *per se* was not measured, house finches with chronic *Plasmodium* infections showed more severe clinical disease from *M. gallisepticum* following experimental infection, suggesting that co-infections with *Plasmodium* may result in higher *M. gallisepticum*-mediated mortality in the wild (Dhondt et al. 2017). Overall, further study is needed to determine to what degree avian parasites interact in ways that alter host survival in the face of co-infection. Future work on the way that infections impact avian fitness should consider

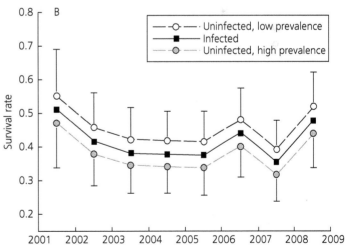

Figure 6.2 Acute blood parasite infections can cause significant mortality; however, due to transient high parasitemia their impacts are difficult to measure, especially in areas where the parasite is endemic. In a long-term study of Eurasian blue tits (*Cyanistes caeruleus*), annual survival of uninfected birds in areas with high prevalence of Plasmodium malarial parasites, (gray circles), was lower than the survival rates of infected birds (black squares), potentially because infected birds represent "survivors" of acute infections that became chronic. Thus, acute infection likely carries fitness costs that can only be measured indirectly. In areas where parasite prevalence is low (open circles), infected birds have lower annual survival than uninfected birds, indicating fitness costs of chronic infection as well. Photo credit iStock.comian600f. Figure from Lachish et al. 2011.

the suite of parasites that concurrently affect a given host and the ways in which they potentially interact to alter infection-mediated mortality.

6.2.2.2 Host characteristics that impact parasite effects on survival

Multiple characteristics of hosts will also impact the extent to which parasites reduce their survival. Interspecific variation in susceptibility to parasites commonly results in differential mortality across species. In particular, host populations that are naïve to an introduced or otherwise emerging parasite are often most susceptible to that parasite's survival impacts. For example, many honeycreeper species in Hawaii are highly susceptible to *P. relictum*, the causative agent of avian malaria, unlike non-native species exposed to the parasites in their native range (van Riper et al. 1986). Further, captive populations of species such as penguins that have not coevolved with *P. relictum* or other avian blood parasites in the wild can show very high mortality from infection (e.g., Bueno et al. 2010). In the Galápagos, exposure to avian poxvirus can result in significant reductions in survival in both juvenile endemic Galápagos mockingbirds (*Mimus parvulus*; Vargas 1987) and juveniles of the critically endangered waved albatross (*Phoebastria irrorata*; Tompkins et al. 2017). Finally, respiratory cryptosporidiosis, an emerging infectious disease in red grouse caused by the protozoan parasite *Cryptosporidium baileyi*, has strong negative effects on annual survival, although this study compared birds with and without clinical signs, rather than infected vs. uninfected birds (Baines et al. 2018).

Even when all species in an area are naïve to a given parasite, there can be significant inherent interspecific heterogeneity in survival rates in response to infection. For example, WNV was a novel parasite for all avian species in North America when it first arrived in the late 1990s. Nonetheless, North American bird species showed substantial heterogeneity in mortality in response to infection. While many species survive infection, some species exhibit high mortality, including corvids (crows and jays; Caffrey et al. 2003; Yaremych et al. 2004), raptors (Saito et al. 2007), pelicans (Sovada et al. 2008), and grouse (Clark et al. 2006; Nemeth et al. 2017). The reasons for this apparent taxon-related

variation in susceptibility are unknown but represent an important area for future work. There are also numerous ecological reasons why some species are particularly affected by disease outbreaks, including high host population density (e.g., Koenig et al. 2010). The ease of transmission in dense breeding colonies means that colonially breeding birds and birds that rely on supplemental food resources, for example, can be especially vulnerable to parasite-mediated mortality. For instance, mortality events from *T. gallinae*, a protozoan parasite thought to be spread largely in species that congregrate at bird feeders, have been associated with declines of up to 47% of European greenfinch (*Chloris chloris*) populations in Finland (Lehikoinen et al. 2013), as well as breeding chaffinch (*Fringilla coelebs*), and wintering goldfinch (*Carduelis carduelis*) populations (Chavatte et al. 2019; Lehikoinen et al. 2013).

Within species, certain demographic groups in a population can be more or less susceptible to an invading parasite and thus can show variation in the ability to survive infection. Sex and age differences in survival have been documented for several species. For example, Martínez-de la Puente et al. (2010) gave blue tits a medication that reduced the intensity of *Haemoproteus* blood parasites and found that treatment was associated with higher survival until the next breeding season of females but not males. In terms of age, young birds are often more likely to die from parasite infection than conspecific adults due to their relatively undeveloped immune systems (see Chapter 3). For example, Buggy Creek virus was associated with high nestling mortality in house sparrows (O'Brien and Brown 2012) and post-fledging survival of American crows (*Corvus brachyrhynchos*) was negatively associated with *Plasmodium* infection (Townsend et al. 2018). Double-crested cormorant (*Phalacrocorax auritus*) young of the year (< 16 weeks old) are highly susceptible to Newcastle disease virus, with mortality rates that range from 32 to 64%, while adults experience little to no mortality (Kuiken 1998). Further, as noted previously, the mortality rate of juvenile great tits infected with avian poxvirus (26.5%) is double that of infected adult birds (13.1%) (Lachish et al. 2012a). On the other hand, females that have been exposed to a given parasite can pass parasite-specific antibodies on to their offspring, potentially protecting the chick up to several weeks post-hatching (Garnier et al. 2012;

Nemeth and Bowen 2007; also see Chapter 2). Further study is needed to determine to what extent and for how long maternal antibodies provide a survival benefit against parasitism in wild birds.

6.3 Effect of parasites on reproductive success

The effects of parasites on host survival are not the only way in which parasites can impact individual fitness. Individuals that are actively infected with a parasite often have compromised reproductive success, which will impact their Darwinian fitness and, ultimately, the population's growth potential. Variations in numerous components of breeding success shape the overall reproductive success of an individual bird (Figure 6.1), from mate acquisition to the production and survival of young (Clutton-Brock 1988). Here, we focus on the effects of parasite infection on a suite of breeding parameters that are well studied in free-living birds and contribute to measures of annual reproductive success (Figure 6.1), including pairing success and parentage, clutch size, hatching success or brood size, and fledging success.

For each type of reproductive parameter, we discuss the documented effects of parasite infection as well as give examples, where available, of cases where parasites had no apparent effect on reproductive parameters. We also briefly discuss cases of parasite infection having apparent positive effects on reproductive parameters in birds. We consider how these patterns can arise from several possible mechanisms, both causative and correlational. Note that although we focus here on components of annual fecundity that are well studied in birds (e.g., clutch size), parasite infection may also influence reproductive traits such as age at first reproduction (e.g., Lanciani 1975) that are more challenging to study in birds but can contribute greatly to an individual's lifetime reproductive success.

6.3.1 Breeding phenology, pairing success, and parentage

Parasite infection can cause diverse effects on breeding phenology, pairing success, and parentage that are potentially relevant for reproductive

success and, ultimately, fitness (Table 6.2). For example, infection by blood parasites is correlated with later return dates in migrant birds (Møller et al. 2004; Rätti et al. 1993) and later laying dates in both migratory and resident species (Allander and Bennett 1995; Votýpka et al. 2003). Given that early breeding augments reproductive success for many bird species (e.g., Verhulst and Nilsson 2008), parasite-mediated delays to timing of breeding are likely to have negative consequences for the reproductive parameters discussed below (clutch size, hatching success, etc.) and thus for overall reproductive success. Numerous studies have also documented reduced pairing success or parentage in infected birds relative to uninfected conspecifics. For example, infection with avian poxvirus in male small ground finches (*Geospiza fuliginosa*) reduces their pairing success (from 77% paired in healthy birds to 17% paired in those with pox lesions; Kleindorfer and Dudaniec 2006) and infection by trypanosomes reduces paternity in aquatic warblers (*Acrocephalus paludicola*; Dyrcz et al. 2005). In some cases, infection status can alter the appearance of sexually selected traits in male birds, thereby reducing their mating success (see Chapter 4).

6.3.2 Clutch size

One of the most obvious ways that parasites impact reproductive success in birds is by reducing clutch size—the number of eggs that a female will lay during a given breeding bout. The association between infection and clutch size has been examined for several avian host–parasite systems, with most studies investigating the effects of protozoan (blood parasite) infections in particular (Table 6.2). For example, female blue tits naturally infected with blood parasites (most commonly *H. majoris*) laid significantly smaller clutches than uninfected conspecifics (Merilä and Andersson 1999). It can be challenging in wild systems to determine whether associations between clutch size and infection are causative, but a number of studies have treated infected birds to clear or reduce infection, allowing examination of causal links between parasitism and clutch size. For example, house martins (*Delichon urbicum*) treated with the antimalarial drug primaquine had clutch sizes that were, on average, 18% larger than those of control individuals (Marzal et al. 2005; Figure 6.3).

Table 6.2 Examples of studies examining differences in reproductive metrics between uninfected and uninfected groups of free-living, wild birds studied using an experimental (EXP) or observational (OBS) approach. Studies comparing seropositive with seronegative birds were not included.

Metric	Pathogen/Disease	Host Species	Location	Group	Approach	Uninfected mean (n)	Infected mean (n)	Source
Parentage/lifetime reproductive metrics								
Number of sired nestlings	Trypanosoma sp.	Aquatic warbler (Acrocephalus paludicola)	Poland	Male	OBS	3.60 (15)	2.06 (16)	Dyrcz et al. 2005
Lifetime eggs laid	Haemoproteus beckeri	Mountain white-crowned sparrow (Zonotrichia leucophrys)	USA	Female	OBS	5.6 (67)	11.6 (16)	Zylberberg et al. 2015
Lifetime eggs hatched	Haemoproteus beckeri	Mountain white-crowned sparrow	USA	Female	OBS	4.7 (67)	8.3 (16)	Zylberberg et al. 2015
Lifetime chicks fledged	Haemoproteus beckeri	Mountain white-crowned sparrow	USA	Female	OBS	3.2 (58)	7.3 (15)	Zylberberg et al. 2015
Clutch size/number of eggs laid								
Clutch size	Haemoproteus sp.	Kestrel (Falco tinnunculus)	Finland	Male	OBS	5.7 (71)	5.2 (23)	Korpimäki et al. 1995
	Haemoproteus prognei	House martin (Delichon urbica)	Spain		EXP	4.44 (16)	3.64 (14)	Marzal et al. 2005
	Haemoproteus or Plasmodium	House martin	Spain		OBS	4.23 (32)	4.09 (62)	Marzal et al. 2008
	Haemosporidians	Great tit (Parus major)	Switzerland		OBS	8.30 (47)	8.17 (267)	Pigeault et al. 2018
	Plasmodium sp.	'Amakihi (Chlorodrepanis virens)	Hawai'i	Female / Male	OBS / OBS	2.68 (44) / 2.64 (45)	2.54 (12) / 2.69 (13)	Kilpatrick et al. 2006 / Kilpatrick et al. 2006
		Barn swallow (Hirundo rustica)	Italy	Yearling female	OBS	6.06 (86)	4.81 (16)	Romano et al., 2019
		Red-billed gull (Chroicocephalus novaehollandiae)	New Zealand	Female / Male	OBS / OBS	1.90 (71) / 1.93 (59)	1.90 (10) / 1.89 (9)	Cloutier et al. 2011 / Cloutier et al. 2011
	Respiratory cryptosporidiosis	Red grouse (Lagopus lagopus scotica)	UK	2014 / 2015	OBS / OBS	9.20 (14) / 8.20 (25)	8.70 (6) / 7.40 (23)	Baines et al. 2018 / Baines et al. 2018
	Trichostrongylus tenuis	Red grouse	UK	1982 / 1983	EXP / EXP	8.25 (8) / 8.00 (13)	7.93 (14) / 5.28 (11)	Hudson 1986 / Hudson 1986
Hatching success/brood size metrics								
Hatching success	Plasmodium sp.	'Amakihi	Hawai'i	Female / Male	OBS	0.80 (39) / 0.80 (33)	0.74 (6) / 0.82 (12)	Kilpatrick et al. 2006 / Kilpatrick et al. 2006
	Haemoproteus prognei	House martin	Spain		EXP	0.75 (16)	0.58 (14)	Marzal et al. 2005

	Respiratory cryptosporidiosis	Red grouse	UK		OBS	0.85 (39)	0.77 (30)	Baines et al. 2018
	T. tenuis	Red grouse	UK	1982 / 1983	EXP / EXP	0.97 (8) / 0.75 (13)	0.77 (14) / 0.38 (11)	Hudson 1986 / Hudson 1986
Brood size	Trypanosoma sp.	Great tit	Sweden	Female	OBS	0.93 (44)	0.67 (2)	Dufva 1996
	Haemosporidia	Blue tit (Cyanistes caeruleus)	Spain		EXP	7.36 (33)	8.03 (33)	Merino et al. 2000
		Great tit	Switzerland		OBS	7.14 (47)	7.29 (267)	Pigeault et al. 2018
	Plasmodium sp.	Red-billed gull	New Zealand	Female / Male	OBS / OBS	1.55 (71) / 1.68 (59)	1.90 (10) / 1.44 (9)	Cloutier et al. 2011 / Cloutier et al. 2011
	Respiratory cryptosporidiosis	Red grouse	UK		OBS	3.70 (39)	2.10 (36)	Baines et al. 2018
	T. tenuis	Red grouse	UK	Female	EXP	3.60 (12)	1.70 (12)	Redpath et al. 2006
Fledging success/ independent young								
Proportion brood fledged (males)	Anisakis	European shag (Phalacrocorax aristotelis)	UK	Females	EXP	0.91 (34)	0.81 (83)	Reed et al. 2008
Proportion brood fledged	Haemoproteus prognei	House martin	Spain		EXP	0.85 (16)	0.82 (14)	Marzal et al. 2005
	Haemosporidia	Blue tit	Sweden / Spain		OBS / EXP	0.41 (30) / 0.98 (33)	0.56 (22) / 0.92 (33)	Merilä & Andersson 1999 / Merino et al. 2000
	Respiratory cryptosporidiosis	Red grouse	UK		OBS	0.62 (32)	0.37 (51)	Baines et al. 2018
Number of young fledged	Avian pox virus	Great tit	UK	2010 / 2011	OBS / OBS	6.87 (233) / 7.08 (166)	6.00 (20) / 6.83 (6)	Lachish et al. 2012a / Lachish et al. 2012a
	Haemoproteus prognei	Purple martin (Progne subis)	USA	Female	OBS	3.81 (48)	4.13 (32)	Davidar & Morton 1993
	Haemosporidia	Great tit	Switzerland		OBS	4.10 (47)	5.38 (267)	Pigeault et al. 2018
		House martin	Spain		OBS	3.60 (32)	3.26 (62)	Marzal et al. 2008
	Plasmodium sp.	Red-billed gull	New Zealand	Female / Male	OBS / OBS	0.70 (71) / 0.88 (59)	1.00 (10) / 0.89 (9)	Cloutier et al. 2011 / Cloutier et al. 2011

(Continued)

Table 6.2 Continued

Metric	Pathogen/Disease	Host Species	Location	Group	Approach	Uninfected mean (n)	Infected mean (n)	Source
		'Amakihi	Hawai'i, USA	Female	OBS	1.30 (49)	1.00 (14)	Kilpatrick et al. 2006
				Male	OBS	1.07 (50)	1.75 (16)	Kilpatrick et al. 2006
Fledglings per successful nest	Plasmodium sp.	'Amakihi	Hawai'i, USA	Female	OBS	2.32 (33)	2.50 (6)	Kilpatrick et al. 2006
				Male	OBS	2.34 (28)	2.31 (13)	Kilpatrick et al. 2006
Proportion pairs fledging young	Avian pox virus	Great tit	UK	2010	OBS	0.87 (233)	0.75 (20)	Lachish et al. 2012a
				2011	OBS	0.93 (166)	0.83 (6)	Lachish et al. 2012a
Proportion pairs producing independent young	Avian pox virus	Great tit	UK	2010	OBS	0.54 (233)	0.20 (20)	Lachish et al. 2012a
				2011	OBS	0.25 (166)	0.00 (6)	Lachish et al. 2012a
Number of independent young	Avian pox virus	Great tit	UK	2010	OBS	0.94 (233)	0.30 (20)	Lachish et al. 2012a
				2011	OBS	0.58 (166)	0.00 (6)	Lachish et al. 2012a
Number of recruits	Plasmodium GRW1	Great reed warbler (Acrocephalus arundinaceus)	Sweden		OBS	0.34 [a]	0.57 [a]	Bensch et al. 2007
	Plasmodium GRW2	Great reed warbler	Sweden		OBS	0.36 [a]	0.47 [a]	Bensch et al. 2007
	Plasmodium GRW4	Great reed warbler	Sweden		OBS	0.36 [a]	0.38 [a]	Bensch et al. 2007

[a] Longitudinal study (17 years); infected and uninfected sample sizes are not provided.

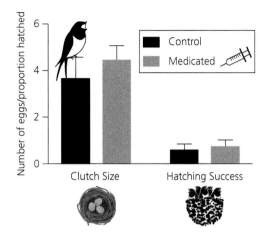

Figure 6.3 Experimental manipulations such as medication treatments that reduce or eliminate chronic parasites are commonly used in free-living birds to reveal fitness effects of parasites. Here, medication of house martins (*Delichon urbica*) with the drug primaquine (n = 16; gray bars) to reduce levels of the haemosporidian parasite *Haemoproteus prognei* led to statistically significantly higher clutch sizes (number of eggs per nest) and hatching success (proportion of eggs hatched) relative to control, unmedicated martins (n = 14; black bars), which harbored significantly higher parasite loads. Means and standard deviations are shown. (Data from Marzal et al. 2005.)

Similarly, red grouse females treated with an anti-helminthic drug that reduced burdens of the macro-parasite *T. tenuis* had significantly larger clutches than control conspecifics in two out of three years of study (Hudson 1986).

While infection with many types of parasite appears to result in reduced clutch sizes, clutch size is not always altered by infection. For instance, in Hawai'i 'amakihi (*Chlorodrepanis virens*), natural infection of either the male or female with *P. relictum* was not associated with differences in clutch size (Kilpatrick et al. 2006). Further, in red-billed gulls (*Chroicocephalus novaehollandiae scopulinus*), *Plasmodium* infection of males or females was not associated with variation in clutch size, even though one lineage of *Plasmodium* caused declines in body condition for both sexes (Cloutier et al. 2011). Overall, the effects of parasite infection on clutch size appear to be variable, and many confounding factors likely contribute to clutch size variation in birds, making it challenging to determine the effects of infection *per se*. Future work using a mixture of experimental approaches or, in systems where

infection status cannot be manipulated, longitudinal approaches where covariates can be measured consistently will help to tease apart what factors predict the extent to which a given parasite infection influences clutch size in birds.

6.3.3 Hatching success and/or brood size

The effect of a female's infection status on hatching success (defined as the proportion of eggs laid that produce a chick) or brood size (defined as the number of young that successfully hatch in a given clutch) is not as well studied as clutch size. Nevertheless, several studies show significant effects of infection on hatching success or brood size (Table 6.2). Dufva found that trypanosome infection of female great tits was associated with lower hatching success (Dufva 1996). Similarly, red grouse infected with *T. tenuis* (Redpath et al. 2006) and with *Cryptosporidium* spp. had significantly smaller brood sizes (Baines et al. 2018) than uninfected individuals. In studies of blood parasites, infected birds tended to have larger broods than uninfected birds, which may be consistent with enhanced investment in reproduction by infected birds (see Section 6.3.6). On the other hand, as two out of three studies were observational, anticipated causality may be reversed, meaning that a higher reproductive effort can lead to higher parasitemia and thus an increased likelihood of detecting hemoparasite infection (Knowles et al. 2009). Further experimental work is needed to deepen our understanding of the underlying causality in the infection–brood size relationships described to date.

6.3.4 Fledging success

Fledging success, which is defined as either the proportion of the brood that successfully leaves the nest or the number of fledglings, can be notably impacted by infection in some systems. For example, treatment of *Plasmodium* infection in female blue tits positively influenced multiple breeding parameters, including fledging success (Knowles et al. 2010). Further, the mean proportion and numbers of great tits fledging young and producing independent young were all lower in birds with pox lesions (Lachish et al. 2012a). Despite these notable negative associations between parasite infection and fledging success in some systems, several studies to date (Table 6.2) show little evidence that infection status of the female affects fledging success. For example, female blue tits infected with blood parasites (primarily *H. majoris*) had equivalent fledging success to that of uninfected females, though young of infected females fledged in significantly poorer condition (Merilä and Andersson 1999). This suggests that simple metrics such as the percentage of young that leave the nest may not fully capture the downstream fitness effects of parasite infection with respect to reproductive success. Further, studies of effects of parasites on fledging success are largely limited to *Plasmodium* and *Haemoproteus* parasites, for which heterogeneity in reproductive impacts between infections with different strains or lineages of the same parasite is often evident (Bensch et al. 2007; Ortego et al. 2008).

Negative impacts of parasite infection on fledging success in birds are likely to vary with the degree of infection and co-infection, but these are less well studied. For example, the intensity of *Haemoproteus* infection in male yellowhammers (*Emberiza citrinella*) was negatively correlated with the number of fledglings produced (Sundberg 1995), suggesting that metrics that take variation in parasite load into account (rather than just 'infected' or 'not') may be important in detecting parasite impacts on reproduction in some systems. Further, just as with survival impacts, co-infection of diverse parasites can alter reproduction-related fitness outcomes of infection. For example, in the endangered Mauritian pink pigeon (*Nesoenas mayeri*), birds co-infected by *T. gallinae* and *Leucocytozoon marchouxii* failed to fledge any young (Bunbury et al. 2008).

6.3.5 Parental provisioning rates, parental care, and transgenerational effects

Detected effects of infection status on hatching and/or fledging success can result from several potential mechanisms, including insufficient incubation or brooding by infected parents or reduced parental provisioning rate to young (e.g., Marzal et al. 2005). For example, medication to reduce blood parasite load was associated with higher parental provisioning rates to nestling blue tits (Knowles et al. 2010; Tomás et al. 2007), and *Anisakis* nematode infection in European shags (*Phalacrocorax aristotelis*) reduces the survival of male offspring through reduced maternal investment (Reed et al. 2008).

The effects of infection on parental provisioning rates and resulting effects on nestling growth and condition may extend well into adulthood for nestlings that successfully survive and fledge. If young individuals with infected parents experience poor conditions in the nest that affect, for example, body condition or growth (e.g., Calero-Riestra and García 2016; Palacios et al. 2012), then life history theory suggests that these individuals may have reduced reproductive success or survival as an adult (Metcalfe and Monaghan 2001; Taborsky 2006). This idea has gained increasing empirical support over the years (Criscuolo et al. 2008; Fisher et al. 2006; Lee et al. 2012; Spencer et al. 2010) and suggests that fitness impacts of disease have the potential to act across generations.

6.3.6 Apparent positive effects of parasite infection on fecundity

Overall, most of the detected associations between parental infection status and reproductive parameters to date have been negative in nature (Table 6.2). However, significant and positive associations between breeding parameters and parasite infection do exist. In several systems, recovered or infected birds unexpectedly appear to have higher fecundity compared to those that had never been infected (Pigeault et al. 2018; Zylberberg et al. 2015). For example, Zylberberg et al. (2015) conducted an 11-year study of mountain white-crowned sparrows (*Zonotrichia leucophrys*) and found that females infected with *Haemoproteus beckeri* at some point in their lives had higher reproductive success than uninfected females and had higher return rates the next breeding season. Fledging success and female *Haemoproteus* infection were positively associated in purple martins (Davidar and Morton 1993), which the authors attribute to chronic infection serving as an indicator of immune systems that are sufficiently strong and thus have not been 'weeded out' from the population.

Whilst the mechanisms behind these apparent positive associations remain unclear, these are generally correlational studies where the influence of another factor (e.g., behavior influencing exposure where more 'bold' birds are more likely to be infected [Dunn et al. 2011] and may also have a higher breeding success [Dingemanse et al. 2004]) cannot be ruled out. Nonetheless, there are a few potential mechanisms that can produce causal positive relationships between parasite status and reproductive investment. First, due to the classic life history trade-off between current and future reproduction, hosts that are actively infected with parasites are hypothesized to show what is called 'terminal investment' (see Chapters 4 and 5). In these cases, birds that are actively infected and thus unlikely to live long enough to invest in future reproduction are expected to increase their investment in current reproduction, as has been documented in blue tits, although parasite-mediated selection might also explain such findings (Podmokła et al. 2014). On the other hand, positive correlations between reproductive success and parasite infection can also arise from several non-adaptive mechanisms, including increased reproductive investment leading to higher loads of parasites and thus higher parasite detection (Knowles et al. 2009). Overall, the majority of correlational and experimental evidence points to parasite infection having a negative impact on a wide range of ecological and demographic parameters in birds (e.g., Asghar et al. 2011; Knowles et al. 2010; Martínez-de la Puente et al. 2010; Merino et al. 2000; Tomás et al. 2007; van Oers et al. 2010).

6.4 Context-dependence of fitness effects

The degree of any fitness effects of infection can vary with the environmental context in which the host–parasite interaction occurs (see Chapter 2). Multiple stressors, including parasites, can act synergistically within individuals to determine resulting fitness impacts (Clinchy et al. 2004; Sih et al. 2004; Zanette et al. 2003). Equally, a pre-existing environmental stressor can increase the susceptibility of birds to infection (Echaubard et al. 2010; Friend et al. 2001). For example, clinical signs of beak and feather disease tend to be evident in Cape parrots (*Poicephalus robustus*) only

during extreme climatic events (Downs et al. 2015).

Anthropogenic food provisioning, which is very commonly directed at birds, may also alter the fitness impacts of parasitism (see Chapter 9), potentially allowing for a larger proportion of infected individuals to survive acute infection by providing additional resources for them (Fischer and Miller 2015). However, the role of nutritional stress in interacting with parasite infection to cause mortality remains largely unknown in wild birds and is an important area for future work. There is some evidence that nutritional stress can impact the extent to which parasites reduce reproductive success. For example, *Leucocytozoon ziemanni* infection in Tengmalm's owls (*Aegolius funereus*) was associated with a reduced clutch size only in years where food abundance was lower than usual (Korpimäki et al. 1993). Future work should examine how environmental contexts like abiotic factors and nutritional factors alter the impacts of parasites on both survival and reproductive success in birds.

Long-term studies are an extremely valuable resource in assessing the context-dependence of the impacts of infection on host fitness, but currently these are limited to very few populations. For example, avian blood parasite infections are highly ecologically complex, both temporally and spatially, and long-term studies are invaluable for elucidating causes and consequences of infection (e.g., Knowles et al. 2011; Lachish et al. 2011). Similarly, the recent emergence of avian poxvirus in the intensively studied population of great tits in Wytham Woods, Oxford, UK, allowed the rapid assessment of individual-level impacts of infection and the way in which demographic factors influence these impacts (Lachish et al. 2012a; Lachish et al. 2012b). Such datasets, where available, ultimately allow the examination of the role of abiotic factors in driving heterogeneity in the impacts of parasite infection on hosts.

6.5 Summary

In conclusion, parasites can have diverse and significant effects on key components of individual fitness—survival and reproductive success—which then have important consequences for population-scale processes (see Chapter 7). However, we still have much more to learn about the ways in which parasites and pathogens influence avian host fitness. Studies of wild populations are fraught with sources of potential error in measurement (e.g., uncertainty in detecting infection status or hosts; see Box 7.1) and multiple sources of uncertainty can propagate within a study (Lachish and Murray 2018). For example, many studies do not identify parasite lineages—or even genera—which, given the potential for interstrain variation in host impacts (Ortego et al. 2008), may mask effects on individual fitness. Further, the role of co-infection and the ecological context in mediating the effects of parasites on avian hosts is just beginning to be studied. Despite these inherent sources of variation and ecological complexity, studies to date suggest that parasite infections are important sources of variation in avian fitness, and the effects of parasites on fitness operate through a diverse set of mechanisms. Furthermore, effects of avian parasites on host fitness are not limited to microparasites with acute, immediate effects, but also appear pervasive in more chronic infections such as those with helminths and blood-borne protozoa. Overall, the effects of parasites on avian fitness and their context-dependence are critical to understand, given the implications of fitness effects for both avian evolutionary responses and conservation in the face of old and new parasites.

Literature cited

Adelman, J.S., Mayer, C., and Hawley, D.M. (2017). Infection reduces anti-predator behaviors in house finches. *Journal of Avian Biology*, 48, 519–28.

Allander, K. and Bennett, G.F. (1995). Retardation of breeding onset in great tits (*Parus major*) by blood parasites. *Functional Ecology*, 9, 677.

Asghar, M., Hasselquist, D., and Bensch, S. (2011). Are chronic avian haemosporidian infections costly in wild birds? *Journal of Avian Biology*, 42, 530–7.

Asghar, M., Hasselquist, D., Hansson, B., Zehtindjiev, P., Westerdahl, H., and Bensch, S. (2015). Hidden costs of infection: chronic malaria accelerates telomere degradation and senescence in wild birds. *Science*, 347, 436–8.

Asghar, M., Palinauskas, V., Zaghdoudi-Allan, N., et al. (2016). Parallel telomere shortening in multiple body tissues owing to malaria infection. *Proceedings of the Royal Society B: Biological Sciences*, 283, 20161184.

Atkinson, C.T., Thomas, N.J., and Hunter, D.B. (2008). *Parasitic Diseases of Wild Birds*. Wiley-Blackwell, Ames, IA.

Baines, D., Allinson, H., Duff, J.P., Fuller, H., Newborn, D., and Richardson, M. (2018). Lethal and sub-lethal impacts of respiratory cryptosporidiosis on red grouse, a wild gamebird of economic importance. *Ibis*, 160, 882–91.

Bensch, S., Waldenström, J., Jonzén, N., et al. (2007). Temporal dynamics and diversity of avian malaria parasites in a single host species. *Journal of Animal Ecology*, 76, 112–22.

Bielański, W., Biedrzycka, A., Zając, T., Ćmiel, A., and Solarz, W. (2017). Age-related parasite load and longevity patterns in the sedge warbler *Acrocephalus schoenobaenus*. *Journal of Avian Biology*, 48, 997–1004.

Buenestado, F., Gortázar, C., Millán, J., Höfle, U., and Villafuerte, R. (2004). Descriptive study of an avian pox outbreak in wild red-legged partridges (Alectoris rufa) in Spain. *Epidemiology and Infection*, 132, 369–74.

Bueno, M.G., Lopez, R.P.G., de Menezes, R.M.T., et al. (2010). Identification of *Plasmodium relictum* causing mortality in penguins (*Spheniscus magellanicus*) from São Paulo Zoo, Brazil. *Veterinary Parasitology*, 173, 123–7.

Bunbury, N., Barton, E., Jones, C., Greenwood, A., Tyler, K., and Bell, D. (2007). Avian blood parasites in an endangered columbid: Leucocytozoon marchouxi in the Mauritian pink pigeon *Columba mayeri*. *Parasitology*, 134, 797–804.

Bunbury, N., Jones, C.G., Greenwood, A.G., and Bell, D.J. (2008). Epidemiology and conservation implications of *Trichomonas gallinae* infection in the endangered Mauritian pink pigeon. *Biological Conservation*, 141, 153–61.

Caffrey, C., Weston, T.J., and Smith, S.C.R. (2003). High mortality among marked crows subsequent to arrive of West Nile virus. *Wildlife Society Bulletin*, 31, 870–2.

Calero-Riestra, M. and García, J.T. (2016). Sex-dependent differences in avian malaria prevalence and consequences of infections on nestling growth and adult condition in the Tawny pipit, *Anthus campestris*. *Malaria Journal*, 15, 178.

Chambert, T., Staszewski, V., Lobato, E., et al. (2012). Exposure of black-legged kittiwakes to Lyme disease spirochetes: dynamics of the immune status of adult hosts and effects on their survival: exposure and effects of Borrelia on kittiwakes. *Journal of Animal Ecology*, 81, 986–95.

Chapman, C.A., Wasserman, M.D., Gillespie, T.R., et al. (2006). Do food availability, parasitism, and stress have synergistic effects on red colobus populations living in forest fragments? *American Journal of Physical Anthropology*, 131, 525–34.

Chavatte, J.-M., Giraud, P., Esperet, D., Place, G., Cavalier, F., and Landau, I. (2019). An outbreak of trichomonosis in European greenfinches *Chloris chloris* and European goldfinches *Carduelis carduelis* wintering in northern France. *Parasite*, 26, 21.

Christensen, T.K., Bregnballe, T., Andersen, T.H., and Dietz, H.H. (1997). Outbreak of pasteurellosis among wintering and breeding common eiders *Somateria mollissima* in Denmark. *Wildlife Biology*, 3, 125–8.

Clark, L., Hall, J., McLean, R., et al. (2006). Susceptibility of greater sage-grouse to experimental infection with West Nile virus. *Journal of Wildlife Diseases*, 42, 14–22.

Clinchy, M., Zanette, L., Boonstra, R., Wingfield, J.C., and Smith, J.N.M. (2004). Balancing food and predator pressure induces chronic stress in songbirds. *Proceedings of the Royal Society B: Biological Sciences*, 271, 2473–9.

Cloutier, A., Mills, J., Yarrall, J., and Baker, A. (2011). *Plasmodium* infections of red-billed gulls (*Larus scopulinus*) show associations with host condition but not reproductive performance. *Journal of the Royal Society of New Zealand*, 41, 261–77.

Clutton-Brock, T.H. (1988). Introduction. In T.H. Clutton-Brock, ed. *Reproductive Success: Studies of Individual Variation in Contrasting Breeding Systems*, pp. 1–6. University of Chicago Press, Chicago.

Criscuolo, F., Monaghan, P., Nasir, L., and Metcalfe, N.B. (2008). Early nutrition and phenotypic development: 'catch-up' growth leads to elevated metabolic rate in adulthood. *Proceedings of the Royal Society B: Biological Sciences*, 275, 1565–70.

Davidar, P. and Morton, E.S. (1993). Living with parasites: prevalence of a blood parasite and its effect on survivorship in the purple martin. *The Auk*, 110, 109–16.

Davidar, P. and Morton, E.S. (2006). Are multiple infections more severe for purple martins (*Progne subis*) than single infections? *The Auk*, 123, 141–7.

Descamps, S., Forbes, M.R., Gilchrist, H.G., Love, O.P., and Bêty, J. (2011). Avian cholera, post-hatching survival and selection on hatch characteristics in a long-lived bird, the common eider *Somateria mollisima. Journal of Avian Biology*, 42, 39–48.

Dhondt, A.A., Dhondt, K.V., and Nazeri, S. (2017). Apparent effect of chronic *Plasmodium* infections on disease severity caused by experimental infections with *Mycoplasma gallisepticum* in house finches. *International Journal for Parasitology: Parasites and Wildlife*, 6, 49–53.

Dingemanse, N.J., Both, C., Drent, P.J., and Tinbergen, J.M. (2004). Fitness consequences of avian personalities in a fluctuating environment. *Proceedings of the Royal Society B: Biological Sciences*, 271, 847–52.

Downs, C.T., Brown, M., Hart, L., and Symes, C.T. (2015). Review of documented beak and feather disease virus cases in wild Cape parrots in South Africa during the last 20 years. *Journal of Ornithology*, 156, 867–75.

Dufva, R. (1996). Blood parasites, health, reproductive success, and egg volume in female great tits *Parus major. Journal of Avian Biology*, 27, 83–7.

Duggal, N.K., Bosco-Lauth, A., Bowen, R.A., et al. (2014). Evidence for co-evolution of West Nile virus and house sparrows in North America. *PLoS Neglected Tropical Diseases*, 8, e3262.

Dunn, J.C., Cole, E.F., and Quinn, J.L. (2011). Personality and parasites: sex-dependent associations between avian malaria infection and multiple behavioural traits. *Behavioral Ecology and Sociobiology*, 65, 1459–71.

Dyrcz, A., Wink, M., Kruszewicz, A., and Leisler, B. (2005). Male reproductive success is correlated with blood parasite levels and body condition in the promiscuous aquatic warbler (*Acrocephalus paludicola*). *The Auk*, 122, 558–65.

Echaubard, P., Little, K., Pauli, B., and Lesbarrères, D. (2010). Context-dependent effects of ranaviral infection on northern leopard frog life history traits. *PLoS ONE*, 5, e13723.

Faustino, C.R., Jennelle, C.S., Connolly, V., et al. (2004). *Mycoplasma gallisepticum* infection dynamics in a house finch population: seasonal variation in survival, encounter and transmission rate. *Journal of Animal Ecology*, 73, 651–69.

Fischer, J.D. and Miller, J.R. (2015). Direct and indirect effects of anthropogenic bird food on population dynamics of a songbird. *Acta Oecologica*, 69, 46–51.

Fisher, M.O., Nager, R.G., and Monaghan, P. (2006). Compensatory growth impairs adult cognitive performance. *PLoS Biology*, 4, e251.

Friend, M., McLean, R.G., and Dein, F.J. (2001). Disease emergence in birds: challenges for the twenty-first century. *The Auk*, 118, 290–303.

Garnier, R., Ramos, R., Staszewski, V., et al. (2012). Maternal antibody persistence: a neglected life-history trait with implications from albatross conservation to comparative immunology. *Proceedings of the Royal Society B: Biological Sciences*, 279, 2033–41.

Garvin, M.C., Szell, C.C., and Moore, F.R. (2006). Blood parasites of nearctic–neotropical migrant passerine birds during spring trans-gulf migration: impact on host body condition. *Journal of Parasitology*, 92, 990–6.

Grilo, M.L., Vanstreels, R.E.T., Wallace, R., et al. (2016). Malaria in penguins—current perceptions. *Avian Pathology*, 45, 393–407.

Hawley, D.M., Osnas, E.E., Dobson, A.P., Hochachka, W.M., Ley, D.H., and Dhondt, A.A. (2013). Parallel patterns of increased virulence in a recently emerged wildlife pathogen. *PLoS Biology*, 11, e1001570.

Hudson, P.J. (1986). The effect of a parasitic nematode on the breeding production of red grouse. *Journal of Animal Ecology*, 55, 85–92.

Hudson, P.J., Dobson, A.P., and Newborn, D. (1992a). Do parasites make prey vulnerable to predation? Red grouse and parasites. *Journal of Animal Ecology*, 61, 681–92.

Hudson, P.J., Newborn, D., and Dobson, A.P. (1992b). Regulation and stability of a free-living host–parasite system: *Trichostrongylus tenuis* in red grouse. I. Monitoring and parasite reduction experiments. *Journal of Animal Ecology*, 61, 477–86.

Isomursu, M., Rätti, O., Helle, P., and Hollmén, T. (2008). Parasitized grouse are more vulnerable to predation as revealed by a dog-assisted hunting study. *Annales Zoologici Fennici*, 45, 496–502.

Iverson, S.A., Forbes, M.R., Simard, M., Soos, C., and Gilchrist, H.G. (2016). Avian cholera emergence in Arctic-nesting northern common eiders: using community-based, participatory surveillance to delineate disease outbreak patterns and predict transmission risk. *Ecology and Society*, 21, 12.

Kilpatrick, A.M., LaPointe, D.A., Atkinson, C.T., et al. (2006). Effects of chronic avian malaria (*Plasmodium relictum*) infection on reproductive success of Hawai'i 'amakihi (*Hemignathus virens*). *The Auk*, 123, 764–74.

Kleijn, D., Munster, V., Ebbinge, B., et al. (2010). Dynamics and ecological consequences of avian influenza virus infection in greater white-fronted geese in their winter staging areas. *Proceedings of the Royal Society B: Biological Sciences*, 277, 2041–8.

Kleindorfer, S. and Dudaniec, R.Y. (2006). Increasing prevalence of avian poxvirus in Darwin's finches and its effect on male pairing success. *Journal of Avian Biology*, 37, 69–76.

Knowles, S.C.L., Nakagawa, S., and Sheldon, B.C. (2009). Elevated reproductive effort increases blood parasitaemia and decreases immune function in birds: a meta-regression approach. *Functional Ecology*, 23, 405–15.

Knowles, S.C.L., Palinauskas, V., and Sheldon, B.C. (2010). Chronic malaria infections increase family inequalities and reduce parental fitness: experimental evidence from a wild bird population. *Journal of Evolutionary Biology*, 23, 557–69.

Knowles, S.C.L., Wood, M.J., Alves, R., Wilkin, T.A., Bensch, S., and Sheldon, B.C. (2011). Molecular epidemiology of malaria prevalence and parasitaemia in a wild bird population: molecular epidemiology of avian malaria. *Molecular Ecology*, 20, 1062–76.

Koenig, W.D., Hochachka, W.M., Zuckerberg, B., and Dickinson, J.L. (2010). Ecological determinants of American crow mortality due to West Nile virus during its North American sweep. *Oecologia*, 163, 903–9.

Komar, N., Panella, N.A., Langevin, S.A., et al. (2005). Avian hosts for West Nile virus in St. Tammany Parish, Lousiana, 2002. *American Journal of Tropical Medicine and Hygiene*, 73, 1031–7.

Korpimäki, E., Hakkarainen, H., and Bennett, G.F. (1993). Blood parasites and reproductive success of Tengmalm's owls: detrimental effects on females but not on males? *Functional Ecology*, 7, 420.

Korpimäki, E., Tolonen, P., and Bennett, G. (1995). Blood parasites, sexual selection and reproductive success of European kestrels. *Ecoscience*, 2, 335–43.

Krama, T., Krams, R., Cīrule, D., Moore, F.R., Rantala, M.J., and Krams, I.A. (2015). Intensity of haemosporidian infection of parids positively correlates with proximity to water bodies, but negatively with host survival. *Journal of Ornithology*, 156, 1075–84.

Kuiken, T. (1998). *Newcastle Disease and Other Causes of Mortality in Double-Crested Cormorants* (Phalacrocorax auritus). University of Saskatchewan, Saskatoon.

Kuiken, T. (2013). Is low pathogenic avian influenza virus virulent for wild waterbirds? *Proceedings of the Royal Society B: Biological Sciences*, 280, 2013099.

Lachish, S. and Murray, K.A. (2018). The certainty of uncertainty: potential sources of bias and imprecision in disease ecology studies. *Frontiers in Veterinary Science*, 5, 90.

Lachish, S., Knowles, S.C.L., Alves, R., Wood, M.J., and Sheldon, B.C. (2011). Fitness effects of endemic malaria infections in a wild bird population: the importance of ecological structure. *Journal of Animal Ecology*, 80, 1196–206.

Lachish, S., Bonsall, M.B., Lawson, B., Cunningham, A.A., and Sheldon, B.C. (2012a). Individual and population-level impacts of an emerging poxvirus disease in a wild population of great tits. *PLoS ONE*, 7, e48545.

Lachish, S., Lawson, B., Cunningham, A.A., and Sheldon, B.C. (2012b). Epidemiology of the emergent disease paridae pox in an intensively studied wild bird population. *PLoS ONE*, 7, e38316.

Lanciani, C. (1975). Parasite-induced alterations in host reproduction and survival. *Ecology*, 56, 689–95.

Lee, W.-S., Monaghan, P., and Metcalfe, N.B. (2012). The pattern of early growth trajectories affects adult breeding performance. *Ecology*, 93, 902–12.

Legagneux, P., Berzins, L.L., Forbes, M., et al. (2014). No selection on immunological markers in response to a highly virulent pathogen in an Arctic breeding bird. *Evolutionary Applications*, 7, 765–73.

Lehikoinen, A., Lehikoinen, E., Valkama, J., Väisänen, R.A., and Isomursu, M. (2013). Impacts of trichomonosis epidemics on greenfinch *Chloris chloris* and chaffinch *Fringilla coelebs* populations in Finland. *Ibis*, 155, 357–66.

Magallanes, S., García-Longoria, L., López-Calderón, C., et al. (2017). Uropygial gland volume and malaria infection are related to survival in migratory house martins. *Journal of Avian Biology*, 48, 1355–9.

Martínez-de la Puente, J., Merino, S., Tomás, G., et al. (2010). The blood parasite *Haemoproteus* reduces survival in a wild bird: a medication experiment. *Biology Letters*, 6, 663–5.

Marx, M., Reiner, G., Willems, H., et al. (2017). High prevalence of *Trichomonas gallinae* in wild columbids across western and southern Europe. *Parasites and Vectors*, 10, 1–11.

Marzal, A., Bensch, S., Reviriego, M., Balbontin, J., and De Lope, F. (2008). Effects of malaria double infection in birds: one plus one is not two. *Journal of Evolutionary Biology,* 21, 979–87.

Marzal, A., De Lope, F., Navarro, C., and Møller, A.P. (2005). Malarial parasites decrease reproductive success: an experimental study in a passerine bird. *Oecologia,* 142, 541–5.

Maxted, A.M., Porter, R.R., Luttrell, M.P., et al. (2012). Annual survival of ruddy turnstones is not affected by natural infection with low pathogenicity avian influenza viruses. *Avian Diseases,* 56, 567–73.

McClintock, B.T., Nichols, J.D., Bailey, L.L., MacKenzie, D.I., Kendall, W., and Franklin, A.B. (2010). Seeking a second opinion: uncertainty in disease ecology. *Ecology Letters,* 13, 659–74.

Merilä, J. and Andersson, M. (1999). Reproductive effort and success are related to haematozoan infections in blue tits. *Écoscience,* 6, 421–8.

Merino, S., Moreno, J., José Sanz, J., and Arriero, E. (2000). Are avian blood parasites pathogenic in the wild? A medication experiment in blue tits (*Parus caeruleus*). *Proceedings of the Royal Society B: Biological Sciences,* 267, 2507–10.

Metcalfe, N.B. and Monaghan, P. (2001). Compensation for a bad start: grow now, pay later? *Trends in Ecology and Evolution,* 16, 254–60.

Møller, A.P., de Lope, F., and Saino, N. (2004). Parasitism, immunity, and arrival date in a migratory bird, the barn swallow. *Ecology,* 85, 206–19.

Mougeot, F., Evans, S.A., and Redpath, S.M. (2005). Interactions between population processes in a cyclic species: parasites reduce autumn territorial behaviour of male red grouse. *Oecologia,* 144, 289–98.

Naugle, D.E., Aldridge, C.L., Walker, B.L., et al. (2005). West Nile virus and sage-grouse: what more have we learned? *Wildlife Society Bulletin,* 33, 616–23.

Nemeth, N.M. and Bowen, R.A. (2007). Dynamics of passive immunity to West Nile virus in domestic chickens (*Gallus gallus domesticus*). *American Journal of Tropical Medicine and Hygiene,* 76, 310–17.

Nemeth, N.M., Bosco-Lauth, A.M., Williams, L.M., Bowen, R.A., and Brown, J.D. (2017). West Nile virus infection in ruffed grouse (*Bonasa umbellus*): experimental infection and protective effects of vaccination. *Veterinary Pathology,* 54, 901–11.

O'Brien, V.A. and Brown, C.R. (2012). Arbovirus infection is a major determinant of fitness in house sparrows (*Passer domesticus*) that invade cliff swallow (*Petrochelidon pyrrhonota*) colonies. *The Auk,* 129, 707–15.

Ortego, J., Cordero, P.J., Aparicio, J.M., and Calabuig, G. (2008). Consequences of chronic infections with three different avian malaria lineages on reproductive performance of lesser kestrels (*Falco naumanni*). *Journal of Ornithology,* 149, 337–43.

Osterblom, H., Van Der Jeugd, H.P., and Olsson, O. (2004). Adult survival and avian cholera in common guillemots *Uria aalge* in the Baltic Sea. *Ibis,* 146, 531–4.

Palacios, M.J., Valera, F., and Barbosa, A. (2012). Experimental assessment of the effects of gastrointestinal parasites on offspring quality in chinstrap penguins (*Pygoscelis antarctica*). *Parasitology,* 139, 819–24.

Piersma, T. and van der Velde, M. (2012). Dutch house martins *Delichon urbicum* gain blood parasite infections over their lifetime, but do not seem to suffer. *Journal of Ornithology,* 153, 907–12.

Pigeault, R., Cozzarolo, C.S., Choquet, R., et al. (2018). Haemosporidian infection and co-infection affect host survival and reproduction in wild populations of great tits. *International Journal for Parasitology,* 48, 1079–87.

Podmokła, E., Dubiec, A., Drobniak, S.M., Arct, A., Gustafsson, L., and Cichoń, M. (2014). Avian malaria is associated with increased reproductive investment in the blue tit. *Journal of Avian Biology,* 45, 219–24.

Rätti, O., Dufva, R., and Alatalo, R.V. (1993). Blood parasites and male fitness in the pied flycatcher. *Oecologia,* 96, 410–14.

Redpath, S.M., Mougeot, F., Leckie, F.M., Elston, D.A., and Hudson, P.J. (2006). Testing the role of parasites in driving the cyclic population dynamics of a gamebird: parasites and population dynamics. *Ecology Letters,* 9, 410–18.

Reed, T.E., Daunt, F., Hall, M.E., Phillips, R.A., Wanless, S., and Cunningham, E.J.A. (2008). Parasite treatment affects maternal investment in sons. *Science,* 321, 1681–2.

Rolland, V., Barbraud, C., and Weimerskirch, H. (2009). Assessing the impact of fisheries, climate and disease on the dynamics of the Indian yellow-nosed albatross. *Biological Conservation,* 142, 1084–95.

Romano, A., Nodari, R., Bandi, C., et al. (2019). Haemosporidian parasites depress breeding success and plumage coloration in female barn swallows *Hirundo rustica*. *Journal of Avian Biology,* 50, 1–14.

Saito, E.K., Sileo, L., Green, D.E., et al. (2007). Raptor mortality due to West Nile virus in the United States, 2002. *Journal of Wildlife Diseases,* 43, 206–13.

Samuel, M.D., Takekawa, J.Y., Baranyuk, V.V., and Orthmeyer, D.L. (1999). Effects of avian cholera on survival of lesser snow geese *Anser caerulescens*: an experimental approach. *Bird Study,* 46, S239–47.

Samuel, M.D., Woodworth, B.L., Atkinson, C.T., Hart, P.J., and LaPointe, D.A. (2018). The epidemiology of avian pox and interaction with avian malaria in Hawaiian forest birds. *Ecological Monographs,* 88, 621–37.

Senar, J.C. and Conroy, M.J. (2004). Multi-state analysis of the impacts of avian pox on a population of serins (*Serinus serinus*): the importance of estimating recapture rates. *Animal Biodiversity and Conservation,* 27, 133–46.

Sih, A., Bell, A.M., and Kerby, J.L. (2004). Two stressors are far deadlier than one. *Trends in Ecology and Evolution*, 19, 274–6.

Sijbranda, D., Hunter, S., Howe, L., Lenting, B., Argilla, L., and Gartrell, B. (2017). Cases of mortality in little penguins (*Eudyptula minor*) in New Zealand associated with avian malaria. *New Zealand Veterinary Journal*, 65, 332–7.

Smith, V.H., Jones, T.P., and Smith, M.S. (2005). Host nutrition and infectious disease: an ecological view. *Frontiers in Ecology and the Environment*, 3, 268–74.

Sol, D., Jovani, R., and Torres, J. (2003). Parasite mediated mortality and host immune response explain age-related differences in blood parasitism in birds. *Oecologia*, 135, 542–7.

Sovada, M.A., Pietz, P.J., Converse, K.A., et al. (2008). Impact of West Nile virus and other mortality factors on American white pelicans at breeding colonies in the northern plains of North America. *Biological Conservation*, 141, 1021–31.

Spencer, K.A., Heidinger, B.J., D'Alba, L.B., Evans, N.P., and Monaghan, P. (2010). Then versus now: effect of developmental and current environmental conditions on incubation effort in birds. *Behavioral Ecology*, 21, 999–1004.

Stearns, S.C. (2000). Life history evolution: successes, limitations, and prospects. *Naturwissenschaften*, 87, 476–86.

Sundberg, J. (1995). Parasites, plumage coloration and reproductive success in the yellowhammer, *Emberiza citrinella*. *Oikos*, 74, 331.

Taborsky, B. (2006). The influence of juvenile and adult environments on life-history trajectories. *Proceedings of the Royal Society B: Biological Sciences*, 273, 741–50.

Tjørnløv, R.S., Humaidan, J., and Frederiksen, M. (2013). Impacts of avian cholera on survival of common eiders *Somateria mollissima* in a Danish colony. *Bird Study*, 60, 321–6.

Tomás, G., Merino, S., Moreno, J., Morales, J., and Martínez-De La Puente, J. (2007). Impact of blood parasites on immunoglobulin level and parental effort: a medication field experiment on a wild passerine. *Functional Ecology*, 21, 125–33.

Tompkins, E.M., Anderson, D.J., Pabilonia, K.L., and Huyvaert, K.P. (2017). Avian pox discovered in the critically endangered waved albatross (*Phoebastria irrorata*) from the Galápagos Islands, Ecuador. *Journal of Wildlife Diseases*, 53, 891.

Townsend, A.K., Wheeler, S.S., Freund, D., Sehgal, R.N.M., and Boyce, W.M. (2018). Links between blood parasites, blood chemistry, and the survival of nestling American crows. *Ecology and Evolution*, 8, 8779–90.

Valkiunas, G. (2004). *Avian Malaria Parasites and Other Haemosporidia*. CRC Press, Boca Raton, FL.

van Gils, J.A., Munster, V.J., Radersma, R., Liefhebber, D., Fouchier, R.A., and Klaassen, M. (2007). Hampered foraging and migratory performance in swans infected with low-pathogenic avian influenza a virus. *PLoS ONE*, 2, e184.

van Oers, K., Richardson, D.S., Sæther, S.A., and Komdeur, J. (2010). Reduced blood parasite prevalence with age in the Seychelles warbler: selective mortality or suppression of infection? *Journal of Ornithology*, 151, 69–77.

van Riper III, C., van Riper, S.G., Goff, M.L., and Laird, M. (1986). The epizootiology and ecological significance of malaria in Hawaiian land birds. *Ecological Monographs*, 56, 327–44.

VanderWerf, E.A. and Young, L.C. (2016). Juvenile survival, recruitment, population size, and effects of avian pox virus in Laysan albatross (*Phoebastria immutabilis*) on Oahu, Hawaii, USA. *The Condor*, 118, 804–14.

Vargas, H. (1987). Frequency and effect of pox-like lesions in Galapagos mockingbirds. *Journal of Field Ornithology*, 58, 101–2.

Verhulst, S. and Nilsson, J.-Å. (2008). The timing of birds' breeding seasons: a review of experiments that manipulated timing of breeding. *Philosophical Transactions of the Royal Society B: Biological Sciences*, 363, 399–410.

Votýpka, J., Šimek, J., and Tryjanowski, P. (2003). Blood parasites, reproduction and sexual selection in the red-backed shrike (*Lanius collurio*). *Annales Zoologici Fennici*, 40, 431–9.

Walker, B.L., Naugle, D.E., Doherty, K.E., and Cornish, T.E. (2004). Outbreak of West Nile virus in greater sage grouse and guidelines for monitoring, handling, and submitting dead birds. *Wildlife Society Bulletin*, 32, 1000–6.

Ward, M.P., Beveroth, T.A., Lampman, R., Raim, A., Enstrom, D., and Novak, R. (2010). Field-based estimates of avian mortality from West Nile virus infection. *Vector-borne and Zoonotic Diseases*, 10, 909–13.

Whiteman, N.K. and Parker, P.G. (2004). Body condition and parasite load predict territory ownership in the Galapagos hawk. *The Condor*, 106, 915–21.

Wobeser, G.A. (2013). *Essentials of Disease in Wild Animals*. John Wiley & Sons, Inc., Ames, IA.

Yaremych, S.A., Warner, R.E., Mankin, P.C., Brawn, J.D., Raim, A., and Novak, R. (2004). West Nile virus and high death rate in American crows. *Emerging Infectious Diseases*, 10, 709–11.

Zanette, L., Smith, J.N.M., Oort, H.v., and Clinchy, M. (2003). Synergistic effects of food and predators on annual reproductive success in song sparrows. *Proceedings of the Royal Society B: Biological Sciences*, 270, 799–803.

Zylberberg, M., Derryberry, E.P., Breuner, C.W., Macdougall-Shackleton, E.A., Cornelius, J.M., and Hahn, T.P. (2015). *Haemoproteus* infected birds have increased lifetime reproductive success. *Parasitology*, 142, 1033–43.

CHAPTER 7

Wild Bird Populations in the Face of Disease

Kathryn P. Huyvaert

7.1 Introduction

Parasites and pathogens typically have detectable negative fitness impacts on their individual hosts (see Chapter 6), but the role of parasites in driving avian population dynamics is less straightforward. In fact, whether and under what conditions parasites influence host population dynamics have been long-standing questions in the field of infectious disease ecology more broadly (Tompkins et al. 2002). Understanding the role of parasites in host population dynamics requires estimating statistical parameters such as infection prevalence and host abundance at population scales. Mathematical approaches such as process-based models are also often used to simulate population-level dynamics of host and parasite interactions over time.

In this chapter, I will first describe some key tools used in disease ecology to estimate important parameters for elucidating the effects of parasites and pathogens (used interchangeably in this chapter) in populations of wild birds, with a focus on accounting for imperfect detection of individual animals or their disease or infection status. I will then briefly summarize some of the mathematical approaches, including SIR models, that are commonly used to simulate and predict the population dynamics of host–parasite interactions. With a general idea of these tools in hand and through a series of case studies, the chapter will end by considering whether and under what conditions parasites affect the overall growth of populations, whether parasites have a tendency to regulate populations of wild birds, and some examples of parasite-induced local extinctions.

7.2 Tools for assessing effects of parasites on avian populations

In an ecological context, a **population** is a group of individuals of the same species in the same place at the same time (e.g., Begon et al. 1996), allowing for the potential to interbreed and produce viable offspring. While the borders of the population, both in space and time, are often delimited by the researcher, the definition is useful for identifying tractable units for investigating and managing disease in the wild (Williams et al. 2002). For example, a population could be defined as the adult, breeding waved albatrosses (*Phoebastria irrorata*) on the island of Española in mid-May or all of the tree swallows (*Tachycineta bicolor*) in a high mountain valley during summer. This more traditional, ecological definition of a population reminds us that there is uncertainty in the wild—uncertainty about which individuals are detected and included in the population—because of processes such as births, deaths, movement, and, of course, infectious disease.

Many of the quantitative and modeling tools used to examine the effects of parasites in wild bird populations are built from approaches developed within population and community ecology, so they share elements of a lexicon that is used in this chapter and elsewhere in this volume. When working on wild bird populations, the first important limitation that we encounter is the idea that it is usually impossible to collect data from all of the individuals in the population. Because the goal is to have an *estimate* of a parameter of interest—mean probability

Kathryn P. Huyvaert, *Wild Bird Populations in the Face of Disease* In: *Infectious Disease Ecology of Wild Birds.* Edited by: Jennifer C. Owen, Dana M. Hawley, and Kathryn P. Huyvaert, Oxford University Press. © Oxford University Press 2021.
DOI: 10.1093/oso/9780198746249.003.0007

of annual survival, mean fledging success, prevalence, etc.—data are collected from a set of individuals that serve as a representative **sample** of the population. A number of probability-based designs exist for sampling populations (too many to adequately cover here; see, for example, Thompson 2012), but all rely on the idea that the **sample units**—the individuals or plots that make up the representative sample—are randomly selected and replicated, and sources of variation in the sampling design are 'controlled' to reduce bias and enhance precision of the parameter estimates (Williams et al. 2002).

Sampling issues to be considered include variation in the distribution of sampling units over space and time (e.g., many birds move within and across seasons and within and across habitats), variability arising from the actual process of taking the random sample (i.e., sampling variability), and detectability (Williams et al. 2002). Detectability is the idea that, depending on the system, all sample units are detected, a portion are detected with constant probability, or the variation in detectability is not uniform across all of the units; sources of and methods to deal with imperfect detection of indi-

vidual hosts and other detection issues related to avian disease ecology are addressed in Box 7.1.

7.2.1 Describing infection status in avian populations

The tools and methods of the field of **epidemiology** or 'what is upon the people' apply equally well to birds and other non-human animals. Here, the field is termed **epizootiology**, defined as the study of the patterns associated with disease—causes, occurrence, distribution in time and space, and others—in non-human animal populations. Regardless of the focal species, a key parameter of interest to disease ecologists is **prevalence**, or the proportion of hosts in a sample (from a population) that is infected or exhibits disease (see also Chapter 2).

7.2.1.1 Prevalence

To estimate prevalence, researchers collect biological specimens (e.g., blood samples, cloacal swabs) or observations of clinical signs of infection (e.g., degree of swelling, lesions; Figure 7.1) from each of the individuals in a sample of the population of interest, determine the infection or disease status

---

Box 7.1 Imperfect detection in avian disease ecology

Uncertainty in estimates of the impact of parasites and/or associated disease on components of avian host fitness (see Chapter 6) and their subsequent impacts on avian populations can emerge from **imperfect detection**, the idea that the probability of detecting individual hosts or their infection or disease state is < 1.0. While imperfect detection poses challenges at multiple levels in both the field and the laboratory (McClintock et al. 2010), the focus here is on detection of individual hosts that comprise populations, as well as uncertainty in establishing an individual host's infection or disease state (see Chapter 2).

Imperfect detection of hosts—Individual hosts may be detected imperfectly simply because infected or diseased individuals are less available for counting or are more difficult to catch than uninfected birds. For example, house finches (*Haemorhous mexicanus*) exhibiting clinical signs of mycoplasmal conjunctivitis are less likely to be encountered at feeders (Faustino et al. 2004), probably because they are lethargic and move less frequently within and between sites.

Lower detection of infected or diseased hosts can result in estimates of prevalence that are biased low (Lachish and Murray 2019). On the other hand, overestimates of prevalence might result if infected or diseased individuals have higher probabilities of capture (detection), as was the case for serins (*Serinus serinus*) with avian pox lesions, who may depend more on bird feeders and are more likely to be detected than birds without pox lesions (Senar and Conroy 2004).

Imperfect detection of infection or disease state—Issues with the diagnostic assays or other methods used to establish a bird's infection or disease state can also lead to biases in estimates of prevalence. For infection state, bias can result in cases where assays are developed for domestic species and applied to wild counterparts who are often not closely related and may show distinct immune responses (Pedersen and Babayan 2011), potentially leading to lower test sensitivity, concomitant false negatives (classifying a bird as not infected when it is), and underestimates of prevalence (Lachish and Murray 2019). Further, application of tests with low specificity can lead to overestimates of prevalence because tests reflect a positive detection in the absence of infection. While classification of disease state (e.g., pox lesions, conjunctivitis) is often less prone to detection issues than infection state (e.g., serological data; see Chapter 2), classification of disease can also be imperfect, particularly if clinical signs are mild or are similar to those caused by other agents, or observations of these signs are made from afar (e.g., using binoculars) or by observers with varying degrees of experience (Boxed Figure 7.1).

Accounting for imperfect detection—When the probability of detecting a host, parasite, or clinical signs of disease is imperfect, parameters like infection prevalence or parasite load that rely on estimates of the abundance of those hosts or parasites will be underestimated (Cooch et al. 2012), and this bias can propagate to our understanding of the impact of parasites on bird populations and associated conservation or management decisions. Information derived from the detection histories (see Section 7.2.1.3) developed from repeated sampling of individual hosts or repeated tests of a tissue sample can be used to directly estimate detection probabilities and to incorporate detection into critical parameter estimates (McClintock et al. 2010) such as host abundance, disease state-specific survival probabilities, and prevalence. Many of the key analytical approaches for addressing imperfect detection of hosts or parasites are detailed in the text (see Section 7.2.1).

Boxed Figure 7.1 A house finch (*Haemorhous mexicanus*) with detectable eye swelling, indicative of the disease mycoplasmal conjunctivitis, perches on a feeder. Birds like this are often less likely to be detected by researchers than non-diseased hosts. Imperfect detection of diseased hosts can alter estimates of prevalence and of the impacts of parasites on avian populations. (Photo credit: Marie Read Wildlife Photography.)

of each individual in the sample based on an assay or observation (see Table 2.1 for techniques to establish infection status; see Box 7.1 for detection issues), and then calculate the proportion of the total sample that was considered 'positive' (Equation 7.1):

$$\text{Prevalence} = \frac{\text{Number of positive hosts}}{\text{Number of hosts examind}} \quad (7.1)$$

This form of prevalence is sometimes called **sample prevalence** or **naïve prevalence** to recognize that the true size of the population from which the sample of individuals was taken is often unknown. For example, Thinh et al. (2012) placed mist nets in the forests of two national parks in northern Vietnam and captured a total of 193 different forest birds in one of the parks in 2008. The authors collected a small blood sample from each individual and tested aliquots of blood from each for the presence of avian influenza A virus matrix gene by real-time reverse transcription polymerase chain reaction (rRT-PCR); 14 samples were positive, translating to a sample prevalence of 7.25% for infection with avian influenza viruses (Thinh et al. 2012). The overall size of the 'population' of forest dwellers is unknown in this example, so the estimate of prevalence is only for the sample.

Similarly, terms such as **point prevalence** and **period prevalence** are used to indicate that the estimate of prevalence was calculated at a given point in time or between two points of time, respectively (Gordis 2013). These modifiers highlight several additional areas of uncertainty that are especially important when considering parasitism in most populations of free-ranging birds; handling this uncertainty will be explored later in this chapter.

The focus of many surveillance programs is estimating the proportion of birds that are positive for

Figure 7.1 Estimates of disease prevalence of 'avian poxvirus' are typically based on observations of clinical signs of infection that indicate disease, like pox lesions on the leg of a Hawai'i 'amakihi (left) and on the bill of a juvenile waved albatross (right). (Photo credits: C. Gesmundo/ USGS [left]; E. Tompkins [right].)

antibodies specific to a particular pathogen; the term **seropositive** is often used because the antibodies are detected in the serum sample collected from the bird. Thus, estimates of **seroprevalence** provide a snapshot of the proportion of the sampled individuals that had been infected with the pathogen at some point in the past but do not provide more precise information about when or where the birds in the sampled population were infected (see Chapter 2). Nevertheless, **serosurveillance** programs often serve as the first indication of a potential incursion of a parasite into a host population, which can trigger more intensive surveillance and disease management programs (Wobeser 2007). Further, serological sampling may be more likely to reflect the prevalence of infection over a broader temporal window because active infections caused by some pathogens are only detectable for several days, such as eastern equine encephalitis virus (EEEV) (Komar et al. 1999) and West Nile virus (WNV) (Komar et al. 2003). Yet, it is important to have some understanding of the persistence of pathogen-specific antibodies to interpret the significance of estimates of seroprevalence (see Chapter 2 for more on serology).

7.2.1.2 Estimating host population size

In the forest bird example described in Section 7.2.1.1, the denominator was the size of the sample of birds that had been captured in the national park in 2008. In that case, the sample size was relatively large (n = 193)—suggesting that it is likely a representa-

tive sample of the unknown population—such that we may be tempted to assume that the sample prevalence is a good reflection of the 'true' prevalence of avian influenza viruses in the population of forest birds in the national park in Vietnam. Ideally, researchers conduct a sample size calculation in advance of the study to determine, for example, the number of individual forest birds that are needed to estimate infection prevalence within some 'distance' of the true population mean (see Williams et al. 2002 for more on sampling design). Nevertheless, there is value in having a more precise estimate of the population size of the host from which the sample came so that we have a better idea of the real impact of the parasites on the population.

Several approaches exist for estimating bird population size and variation around the estimate (i.e., variance) while accounting for imperfect detection. Counts of birds that are seen or heard within a circular plot of a determined diameter (avian point counts; Ralph et al. 1995) or while walking along a transect have been used to monitor bird abundance, density, or species richness. In an avian disease context, such point count data may be useful in documenting broad trends in changes in host population size, as was the case for WNV for a suite of North American bird species (e.g., LaDeau et al. 2007). Index approaches correspond to actual abundance when we assume that detection of individuals is perfect or at least constant from site to site and over time (see Nichols et al. 2000; Box 7.1).

Distance sampling methods (Buckland et al. 2015) have found wide use in estimating bird abundance when the probability of detection is < 1.0 (see Box 7.1). In the field, distances to individual detected birds are recorded and the frequency distribution of these distances comprises the 'detection function' which is then used to estimate the probability of detection. The counts at a point or for a line (c) and their corresponding detection probabilities (p) can then be applied in a general equation (Equation 7.2) to estimate population size ($\hat{N}$) for the portion of the total area sampled (p_a) or the entire area for which inferences are being made ($p_a = 1$).

$$\hat{N} = \frac{c}{p * p_a} \qquad (7.2)$$

When distance data are not available, another approach for estimating detection probability and population size from counts involves two observers: a primary observer points out to a secondary observer all of the birds that the primary detects. The secondary observer records all of the primary observer's detections in addition to any that the primary observer did not detect. Data collected using this double-observer approach (Nichols et al. 2000) can then be used to estimate detection probabilities and population size for the site; this approach also permits assessment of variation in detection based on the observer or bird species, as might be the case when some species are more difficult to detect because they do not vocalize readily, are cryptic, or because of a species-specific or individual behavioral response to factors such as infection status (see Chapter 4; Box 7.1).

7.2.1.3 Mark-recapture methods

Capturing and placing bands (rings) on birds has a deep and rich history in ornithology (Jackson 2008) and mark-recapture methods to estimate avian population size have found extensive use, in large part because of the emergence of accessible software (e.g., Program MARK; White and Burnham 1999) and growing long-term mark-recapture datasets in birds. For example, datasets exist for some species (e.g., black brant, *Brant bernicla nigricans*; Sedinger et al. 2019) that now extend over 30 years, across more than one researcher's tenure, and simi-

larly long-term datasets that include host infection status are emerging (e.g., species of tits in Wytham Woods, Oxford, UK; Perrins and Gosler 2011).

Mark-recapture methods for estimating population size (and other demographic parameters like survival probabilities; see Chapter 6 and Box 7.1) rely on an initial capture and placement of marks (i.e., bird bands) on individuals followed by subsequent recaptures or resighting of those banded individuals. The record of times when an individual bird is caught, recaptured, or resighted (recorded as a 1) versus not recaptured or resighted (recorded as a 0) constitutes the bird's **detection** or **encounter history**. For example, a detection history of 1001 indicates that the individual bird was captured on the first of four occasions, was not captured on the second or third occasions, and was captured on the final occasion. Populations that do not lose (through deaths or emigration) or gain (through births or immigration) additional members from one sampling session to the next meet the assumption of demographic closure. The foundation of closed capture models for estimating population size is the two-sample Lincoln–Petersen estimator (Williams et al. 2002) requiring counts of individuals only caught (detected) in the first occasion (n_1), those only caught in the second occasion (n_2), and those marked individuals caught during the second sampling occasion (m_2); these counts are then used to estimate population size ($\hat{N}$) as:

$$\hat{N} = n_1 * n_2 / m_2. \qquad (7.3)$$

Using the Lincoln–Petersen estimator, we can also obtain a detection (or capture) probability for the first session ($\hat{p}_1$) from the ratio of the count statistic (n_1) to the estimate ($\hat{N}$). However, because the estimator assumes that all animals have the same probability of being captured within the sampling session—and we expect that infected or diseased individuals have detection probabilities that are different than their uninfected counterparts (Box 7.1)—avian disease researchers turn to more sophisticated closed and open capture-recapture models that incorporate individual heterogeneity in detection. Closed capture-recapture models extend the Lincoln–Petersen estimator to allow for variation in detection probability due to time, behavior, and individual heterogeneity (Otis et al. 1978).

Open capture-recapture models acknowledge that populations can change through the loss or gain of individuals between capture periods. Individual covariates representing a host's infection or disease state can be included in standard Cormack–Jolly–Seber (CJS) models to estimate the effects of infection/disease status or infection intensity on apparent survival (e.g., Dadam et al. 2019), while information on the number of individuals that are caught but not marked is used in Jolly–Seber models to estimate population size in addition to survival and detection probabilities (Pollock et al. 1990; Williams et al. 2002).

Importantly, information about an individual's infection or disease state can sometimes be unknown or missing, resulting in the need to censor the data (e.g., drop cases with missing data) when using a CJS model. Censoring leads to loss of valuable information about other factors, such as survival probability, that are being investigated. Multistate mark-recapture (MSMR) models incorporating hidden Markov chains allow for data from birds of unknown infection or disease state to be included. This inclusion can lead to more precise estimates of infection or disease state-specific survival probabilities as well as estimates about the probability of entering or leaving a state (e.g., Conn and Cooch 2009). Handling data from birds of unknown state is a facet of these models that will become important when tracking states of infectiousness using SIR models and extensions (see Section 7.2.3.1). Understanding how survival probabilities differ for infected/diseased versus subclinical/uninfected birds is addressed in detail in Chapter 6.

In terms of mark-recapture approaches to estimate host population size, models incorporating data collected using a robust design (repeated sampling occasions within a period of interest; e.g., Kendall 2009) allow for estimation of population size within the period of interest while accounting for imperfect detection of individual hosts. Recent advances in these 'multi-event' mark-recapture models allow for imperfect detection of known hosts during a particular sampling visit, uncertainty about the infection or disease state a host is in, and misclassification of the actual infection or disease state (Benhaeim et al. 2018). This is an exciting advance-

ment as it gets us a step closer to handling the many sources of uncertainty present in avian disease studies carried out in the wild (Box 7.1).

7.2.1.4 Occupancy methods to estimate prevalence

Similar to the mark-recapture encounter histories that we can build for individual birds, a detection history can be constructed for repeated visits to a 'site' where the presence or absence—detection or non-detection—of a species of interest is recorded at each visit to the site. In an avian infectious disease setting, the 'site' (or 'sample unit' *sensu* Mosher et al. 2019) could be a subpopulation, colony, or other grouping of hosts, an individual host, or a relevant tissue sample collected from an individual host (MacKenzie et al. 2018; McClintock et al. 2010). The 'species of interest' is the parasite/pathogen infecting an individual or group of individuals, or a proxy that indicates the state of disease or infection from a relevant tissue sample (e.g., antigen, antibody, nucleic acid; see Chapter 2).

Analytically, maximum likelihood methods are then used to estimate the probabilities of detection (p) of the species of interest (in this case, the parasite) given that it is present at the site/sample unit, and occupancy (ψ), where this probability is an estimate of the occurrence of the parasite, disease, or infection at the level of the sample unit. When the sample unit is an individual host and the detection/non-detection data indicate detection or non-detection of the disease or infection state of the host, occupancy can be interpreted as infection or disease prevalence (e.g., Kendall 2009; MacKenzie et al. 2018; McClintock et al. 2010).

In avian systems, single-season occupancy models have been used to examine avian blood parasite prevalence in taxa such as northwestern crows (*Corvus caurinus*; Van Hemert et al. 2019) and waterfowl where prevalence can vary across a season and with host attributes such as species, age, sex, body condition, and parasite co-infection status (Meixell et al. 2016). Beyond precise estimates of infection prevalence, however, the primary advantage to using occupancy approaches comes in accounting for imperfect detection (see Box 7.1) of the host's infection or disease state, which we anticipate could vary not just by host attributes but also by aspects of the parasite's life history and the assays used to detect infection or disease (see Chapter 2). Avian blood parasites provide an interesting example. Explicitly accounting for **parasite load**, defined as the number of parasites enumerated in a tissue sample, when using quantitative PCR (qPCR) to detect *Plasmodium* infection in blue tits (*Cyanistes caeruleus*) improved estimates of both test sensitivity and prevalence (Lachish et al. 2012). Similarly, Ishtiaq et al. (2017) showed that different blood parasite assays produce varying estimates of prevalence and that lower and more variable estimates occurred when the number of parasites is low, a likely scenario when studying parasitism in wild bird populations.

Beyond detection of the parasite itself, issues of imperfect detection have been illuminated—and estimates of prevalence improved—through application of multistate occupancy approaches to *Toxoplasma gondii* serology data in arctic geese (Elmore et al. 2014). In this system, estimates of seroprevalence were ~10% higher when accounting for imperfect detection compared to naïve estimates that assumed that the test was perfect, a strong assumption given the limited set of diagnostic assays developed specifically for wild systems (Pedersen and Babayan 2011). The Elmore et al. (2014) study also highlighted differences in detection of antibodies between two different serological tests, an insight that could be important when allocating resources to disease surveillance or management programs.

7.2.2 Parasite population size

The finding that the probability of detection of blood parasite infection varied with parasite load (Lachish et al. 2012) reminds us that the parasite is a key member of the epidemiological tetrad (see Chapter 2). Here, I focus briefly on metrics of *parasite* population size given their critical importance in evaluating the impacts of infection or disease on host populations.

7.2.2.1 Parasite load

Where prevalence describes the number of hosts that are infected, **parasite load** (sometimes called parasite burden or infection intensity) is the

number of parasites that an infected individual host has (Equation 7.4) or the density of the parasite population on a single host (Bush et al. 1997).

$$\text{Parasite load} = \frac{\text{Number of parasites}}{\text{Infected host}} \quad (7.4)$$

An infected host (the denominator) is included if it has at least one parasite. Infection intensity is often quantified for macroparasites, including helminths and ectoparasites (see Chapter 2), where the number of parasites per individual host may be more important than simply whether or not a host is infected (Shaw and Dobson 1995).

Estimating infection intensity can involve detection challenges akin to those faced when designating a host as infected or not. For example, while estimating intensity is straightforward for readily visible ectoparasites, many avian ectoparasites are difficult to detect because of their size, adaptations for avoiding removal by a bird (e.g., feather lice on pigeons and doves; Bush et al. 2006), or location in sites (e.g., ears) not accessible to a grooming bird or easily found by a researcher. Dust-ruffling the feathers with a pyrethroid insecticide improves detection probability by making the clinging ectoparasite loosen its grip and drop onto a prepared surface for easier enumeration (Walther and Clayton 1997; see also Chapter 2). The number of ectoparasites is expected to decline over three repeated bouts of dust-ruffling, and typically bouts are continued until the return of additional ectoparasites is < 5% of the cumulative count for the first three bouts for that bird (Clayton et al. 1992). Total counts tallied with this method are useful as estimates of the population size of 'permanent' ectoparasites like adult feather lice (Walther and Clayton 1997).

Survey methods for ectoparasites that are less readily detected than adult feather lice (e.g., larval ticks), for endoparasites such as helminths, or for protozoa such as *Cryptosporidium*, *Trichomonas*, or *Toxoplasma* all pose similar issues with imperfect detection of parasites, which can confound both evaluation of whether or not a host is infected and quantification of the parasites. The planning phase of work on infection intensity, including consideration of the parasite taxa that are of interest and that could be encountered, is important for shaping data collection (Doster and Goater 1997). In addition to estimates of infection prevalence, multiscale occupancy approaches can identify which tissues to sample for the parasite, as in *T. gondii* in geese (Elmore et al. 2016)—information that can be helpful in shaping field and sampling efforts, particularly when data on detectability of particular parasites is scarce.

7.2.2.2 Parasite aggregation

The patterns of distribution of parasites among hosts—that is, which individuals have many parasites and which have few or none—provide a useful framework for predicting the degree of parasite transmission, understanding the potential for host population regulation (see Section 7.3.3), and formulating effective disease management actions. These topics rely on the idea that the majority of parasites are found in (or on) a minority of affected hosts (e.g., Shaw and Dobson 1995); that is, many parasites such as macroparasites tend to have an uneven or aggregated distribution across hosts.

Parasite aggregation can be quantified as the ratio of the variance (s^2) to the mean number of parasites per host (Equation 7.5) in the sample.

$$\frac{s^2}{\text{mean number of parasites}} \quad (7.5)$$

If the number of parasites were randomly distributed among hosts, the variance and the mean would be equal, and the ratio would be 1 because the intensity data come from a random distribution. When the variance is more than the mean, the ratio will be > 1, indicating aggregation, and a ratio of 0 indicates a uniform distribution of parasites across hosts where everyone is infected with the same number (see overview in Wilson et al. 2002).

Considerable work has gone into identifying the index that best expresses the degree of deviation from a random distribution, but, as McVinish and Lester (2019) argue, the lack of a common definition of aggregation among studies translates to a set of different metrics measuring different things and likely generating different conclusions about the ecology of the system. In birds, internal macroparasites (often in the intestinal tract) from the groups Digenea, Cestoda, and Nematoda in great cormorants (*Phalacrocorax carbo*) were all considered 'aggregated' based on the fact that regression coefficients (slopes of the regression lines) from Taylor's Power

Law exceeded 1 (Kanarek and Zalesny 2014). Using discrepancy indices (D) to describe the discrepancy between the observed distribution and a hypothetical, uniform distribution of parasites among hosts (Poulin 1993), Schei et al. (2005) showed that most parasites infecting willow grouse (*Lagopus lagopus*) were considered aggregated.

Interestingly, while aggregation has traditionally been studied in the context of macroparasites, such as helminths and ectoparasites, work on avian viruses suggests that aggregation of infection load can also occur for microparasites (see Chapter 2 for discussion of macro- vs. microparasites). A meta-analysis including 24 experimental infection studies of zoonotic, viral pathogens in 17 bird species found that approximately 20% of individuals in an experiment account for ~80% of the viral load of all individuals for a particular experiment (Jankowski et al. 2013), a pattern consistent with the degree of aggregation shown in many macroparasites (Shaw and Dobson 1995). Similarly, Woolhouse and colleagues (1997) showed that for several sexually transmitted and vector-borne pathogens in humans, 20% of the hosts accounted for ~80% of the transmission events (the 20/80 'rule' or the Pareto principle), suggesting that there is substantial heterogeneity in contributions to parasite transmission that may relate to microparasite aggregation within hosts (discussed here) or variation in contact rates (see Chapter 4). Overall, these findings have important ramifications for systems like birds, where managing disease spread by tracing contacts might be considerably more difficult than enacting management measures (e.g., treating or culling) based instead on evidence of 'supershedding' (i.e., birds with high parasite loads), which is considered one key component of the likelihood of a host contributing disproportionately to transmission in a population (i.e., 'superspreading;' VanderWaal and Ezenwa 2016).

7.2.3 Epidemiological modeling

Prevalence and aggregation estimates provide straightforward metrics of the impact of parasites on a host population by providing snapshots of the proportion of the population (or the sample) that are in the infected or diseased 'state' and the proportion potentially contributing to transmission dynamics. Studies incorporating the infected or diseased state as an individual covariate then allow comparisons with animals that are not in that state (i.e., those in all other states). Often, though, we want to know how parasites move through and affect the entire population; thus, we are interested in the dynamics of all of the states of host infection—susceptible, infectious, recovered—and we turn to a variety of mathematical models. Many of these models allow simulation of theoretical host and parasite population dynamics over time, furthering our understanding of the conditions under which parasites influence host population dynamics (Tompkins et al. 2002).

7.2.3.1 Compartmental models

In simple SIR models, the states of infectiousness that were established using the techniques described briefly in Chapter 2 are used to build 'compartmental' models. One way to conceptualize this classic method for modeling the dynamics of microparasites in a host population (Anderson and May 1991; Kermack and McKendrick 1927) is as a set of boxes (or compartments) representing the host infection states (most simply: susceptible [denoted S], infected [I] (or 'infectious', see below), and recovered [R]) and arrows representing parameters expressing the transitions from box to box (Box 7.2). Notably, these simple SIR models (as opposed to the SEIR models described in Chapter 2) assume that all individuals in the 'infected' state are also 'infectious,' an assumption that will be violated in many systems (see Chapter 2). Nevertheless, no distinction between infected and infectious is made here to illustrate the simplest SIR deterministic model and relevant equations.

Deterministic SIR models are used to describe what happens 'on average' in the population for a hypothetical avian host–parasite system. The model is expressed with a system of differential equations to represent changes in the number of individual hosts in a particular state (i.e., state variables) in continuous time. In the simplest case, we assume that the host population size (N) is constant and the sum of the three state variables (S + I + R) equals N; because the quantity of individuals in the compartments is being tracked and not each and every individual, we assume that mixing, contacts, and the likelihood of transmission given a contact are homogeneous (Keeling and Rohani 2008; Vynnycky and White 2010).

Imagine a directly transmitted parasite in birds—akin to, for example, *Mycoplasma gallisepticum* in a house finch (*Haemorhous mexicanus*). Individuals within the population are assigned to the susceptible state when they are naïve to infection and have no immunity or prior experience with the pathogen, an infected/infectious state once successful transmission occurs, or a recovered/immune state when the host is no longer infectious (this does not occur for every host–pathogen system, as some individuals will be infectious for life [models denoted SI] or immediately become susceptible again [models denoted SIS]; see Box 7.2). The structure of this classic SIR compartmental model can be described with this set of differential equations (following Keeling and Rohani 2008):

$$\frac{dS}{dt} = -\beta SI \tag{7.6a}$$

$$\frac{dI}{dt} = \beta S - \gamma I \tag{7.6b}$$

$$\frac{dR}{dt} = \gamma I \tag{7.6c}$$

where β denotes the transmission term and is the product ($\beta = k \times b$) of the contact rate, k, between susceptible and infected/infectious individuals and the probability of transmission upon contact (b) (Keeling and Rohani 2008). The parameters in the transmission term are shaped by the biology of the pathogen, the host, and their interactions. The recovery or removal rate is denoted with the parameter γ in this model.

To depict the transition or movement of individuals from one compartment to the next, we can think about subtracting individuals from one compartment and adding them to the next compartment. Thus, the susceptible compartment changes when infected individuals are 'subtracted' from it because they transition from the susceptible to the infected or exposed (see Chapter 2) state at the rate β. Similarly, the change in the infectious compartment will include 'additions' of those hosts that become infected, or transition from S to I, and 'subtraction' of individuals that transition from infected to recovered at rate γ. Finally, the recovered compartment changes through the transition of individuals from the infected compartment to the recovered compartment through the addition of those in I to the R compartment at the rate of recovery (γ). Recall that this is the system of differential equations for a simple SIR model under the assumptions that host population size is constant and every infected individual is infectious, but, as noted above, how a population is compartmentalized varies with the different host–microparasite systems, their infection and immune outcomes, and the overall infection timeline (see Chapters 2 and 3; Box 7.2).

When host population size in an SIR model is allowed to change with births and deaths of hosts (whether natural or parasite-mediated), these models can be used to understand when parasites are likely to alter or regulate the population sizes of their hosts (Figure 7.2), a topic considered in Section 7.3. Finally, the SIR models considered above, while traditionally used for modeling microparasites, can be adapted to model vector-borne and macroparasite infections, incorporating the population dynamics of vectors and parasites with external life stages.

In the simple SIR model described above, a system of differential equations (Equations 7.6a–c) express the transitions as rates of change of a compartment over continuous time; differential equations tend to be preferred because the rates are easier to calculate, issues associated with the choice of time step are avoided, and more realistic predictions about the dynamics of the host–parasite system emerge from the models (Vynnycky and White 2010). A

Box 7.2 Compartmental model structures and example pathogens affecting birds

S = Susceptible; E = Exposed or Pre-infectious; I = Infectious; R = Recovered/Immune; C = Carrier/Latent.

Description & Examples in Birds	Model Structure
SEI, SI: These structures reflect pathogens in which the host remains infectious until death – either natural or pathogen-induced. For SEI models, the host is not initially infectious and thus passes through an exposed ('E') state. **Examples**: Herpesvirus (duck plague)	
SEIR, SIR: These structures apply to parasites in which the host develops immunity to the pathogen that persists for the remainder of the host's life. The host may or may not go through the pre-infectious/ exposed ('E') state. **Examples**: avian influenza virus (AIV), West Nile virus (WNV), Eastern equine encephalitis virus (EEEV), St. Louis encephalitis virus (SLEV), tick-borne encephalitis virus (TBEV), avian poxvirus, *Salmonella* spp.	
SIS: This model structure applies when hosts do not develop any immunity and immediately re-enter the susceptible state after clearing the pathogen. **Examples**: None documented in birds, though often the degree to which recovered individuals are protected against re-infection is not known.	
SEIRS, SIRS: This model structure depicts infections for which the host acquires immunity but eventually returns to the susceptible state for that same pathogen due to waning immunity. **Examples**: *Pastuerella multocida, Mycoplasma gallisepticum*	
SIR – C: This model structure applies when some individuals that recover have chronic or latent infections in their tissues. A reactivation of a latent infection can make a host infectious again. **Examples**: *Borrelia burgdorferi, Plasmodium relictum*	

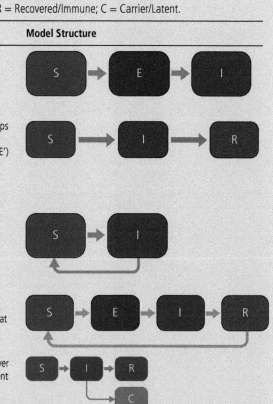

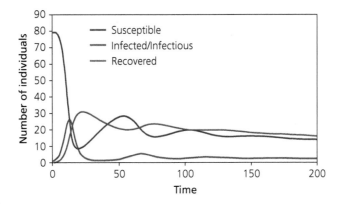

Figure 7.2 SIR models that incorporate the demographic processes of births and deaths (both natural and due to infection) can be used to track host population changes over time in response to parasitism. In this example, total host population size (N) starts at 80; 79 individuals are susceptible (S; blue) and one individual is infected (I; red). As parasite transmission occurs, individuals in S transition to I, leading to declines in S and a classic 'epidemic curve' for I (red; see Chapter 2). Infected individuals are lost rapidly due to death or transition to the recovered state (R). Initial decline in the susceptible population is restored via births over time, which, combined with parasite-induced deaths, can lead to a leveling out of the host population observable in the S curve. These patterns demonstrate the population regulatory effects that pathogens can have on avian hosts over time.

system of difference equations can be used to express the risk of individuals transitioning to the next compartment over discrete, user-defined time steps (e.g., day, week, month) when that is of interest.

A key determinant of the change in the susceptible compartment is the transmission term (β), but its derivation and usage are marred by debate and varying terminology (Keeling and Rohani 2008). **Force of infection**, often denoted as λ, is a useful concept for understanding the transmission term. Force of infection measures the per capita probability of becoming infected (Anderson and May 1991) and depends on the numbers of infectious individuals (I) that are already in the system, the average number of contacts that susceptible hosts have with those infectious individuals per unit time (k), and the probability that one of these contacts leads to an individual becoming infected (b) (Keeling and Rohani 2008). In other words, the risk or force of infection at time t (λ_t) is related to the number of infectious individuals in the population at that point in time (Vynnycky and White 2010; Equation 7.7), shown as

$$\lambda_t = \beta I_t \qquad (7.7)$$

All else being equal, the rate at which an individual enters and the duration they remain in the infectious state are the most important determinants

of the potential of a microparasite-caused disease becoming an epidemic. This is because the rate of entry to the infected/infectious state (λS or βIS) and duration a host is infectious ($1/\gamma$, where γ is the rate of recovery) determine the **R-naught (R_0)** or basic reproduction number of the pathogen, which is defined as 'the average number of secondary infections produced when one infected individual is introduced into a host population where everyone is susceptible' (Anderson and May 1991; see Chapter 2), or

$$R_0 = \frac{\beta N}{\gamma} = \frac{kbN}{\gamma} \qquad (7.8)$$

where N is the host population size. This is just one of a number of formulations of R_0 that have been used to describe the conditions under which an epidemic is established and grows within a population of hosts (e.g., Keeling and Rohani 2008; Vynnycky and White 2010). Regardless of the specific formation of R_0, parasites are expected to spread within a population when $R_0 > 1$, or when the mean number of new infections caused by a single infected individual is more than one. Again, this will depend on how frequently hosts encounter each other, whether those contacts are effective in transmitting the pathogen, and how long an individual remains in the infectious state.

As with other wildlife host–pathogen systems (e.g., McCallum et al. 2009),very few estimates of R_0 exist for modeling avian disease dynamics, in part because it would require monitoring populations before, during, and after the onset of epidemics to obtain the parameters needed. Nevertheless, mathematical models for WNV produced R_0 values ranging from 1.78 using parameters for resident bird species to 2.99 for migratory host species to 3.46 when models incorporated the presence of both host types (Bergsman et al. 2016). The net or effective reproduction number (R_n) provides an estimate of R_0 in light of the fact that some susceptible animals will transition to infectious and then become immune; this process changes the proportion of the population that is susceptible, effectively reducing the transmissions that occur, and can be written as:

$$R_n = R_0 \times S / N \qquad (7.9)$$

where S/N is the proportion of susceptible individuals in the population (Vynnycky and White 2010). This becomes important in the management of avian infectious disease because epidemic conditions will wane or be controlled when $R_n < 1$, which can be achieved through culling, vaccination, or other methods to reduce transmission (e.g., Wobeser 2007). For example, the effective or real-time reproduction number (sometimes denoted R_t) for an avian cholera epidemic in common eiders (*Somateria mollissima*) was 2.5 in 2006 but fell to values closer to 1.0 as the epidemic progressed over several years and herd immunity was induced (Iverson et al. 2016; see also Chapter 2).

In estimating values of R_0 for various avian microparasites, a final important aspect to consider is whether the force of infection increases in tandem with increases in host population size and concomitant increases in contacts, characterizing **density-dependent transmission** (Anderson and May 1978), or whether the force of infection depends on the proportion of hosts or vectors that are infected, characterizing **frequency-dependent transmission**. For many avian disease systems, transmission is assumed to be largely density-dependent such that transmission dynamics track variation in host density, as occurs in house finches and conjunctivitis (Altizer et al. 2004). Pathogens with density-dependent transmission are theoretically predicted

to go extinct before their hosts, as the force of infection becomes too low to sustain epidemics at low host densities. As a consequence, density-dependent pathogens are generally considered less likely to result in host population declines that are sufficiently severe to cause extinction (e.g., de Castro and Bolker 2005). In contrast, persistence of parasites in the population and, importantly, parasite-mediated declines in host population size can continue, even at low host densities, under frequency-dependence. Importantly, this can be the case even in systems with some degree of density-dependent transmission (Ryder et al. 2007). Vector-borne pathogens of birds like WNV, EEEV, and the avian blood parasites are transmitted mainly in a frequency-dependent manner.

A challenge with a simple SIR model as described above is that it treats individuals as homogenous within compartments, while the reality of most, if not all, wildlife disease systems is that significant heterogeneity in both pathogen shedding (see Section 7.2.2.2) and contact rates exists (e.g., White et al. 2017); this heterogeneity might complicate matters, and better understanding it—and accounting for it—enhances the insights we can make about how parasites affect individual hosts and the populations from which they come.

7.2.3.2 Agent- and individual-based models

Agent-based models (ABMs) and individual-based models (IBMs) comprise a class of simulation models that allows the investigator to incorporate another layer of ecological realism to disease models by accommodating individual heterogeneity and chance variation (e.g., Keeling and Rohani 2000). ABMs are more general and reflect the behavior of autonomous individuals or groups of individuals, while IBMs solely reflect individuals (Grimm et al. 2006). In a host–parasite context, ABMs/IBMs are stochastic models that track the progression of infection for every individual in the host population, either singly or as groups. Under this scenario, the risk of infection is now written as

$$\lambda_t = 1 - (1 - p)^{I_t} \qquad (7.10)$$

to indicate that risk (at a particular time step, t) is no longer proportional to the number of infected individuals in the population but, instead, is random, is applied to each individual, and is applied at each time step instead of as a rate across continuous time (i.e., the Reed–Frost equation; Vynnycky and White 2010). In birds, ABMs have been applied extensively to understand the dynamics of WNV (see Nasrinpour et al. 2019 for review), given the impact of infection on bird and human populations. For example, Bouden et al. (2008) built an ABM to describe the dynamics of WNV in the aggregate 'agents,' populations of the American crow (*Corvus brachyrhynchos*) and vector mosquitoes, under different climate and mosquito larvicide application regimes—information that could be useful in disease control decision-making processes. In a non-traditional application of these models, Tonelli and Dearborn (2019) used an IBM to investigate the influence of individual-level factors of migrating songbirds (e.g., migration timing) on the magnitude of tick dispersal across the North American continent. They found that two songbird species (ovenbird, *Seiurus aurocapillus*; wood thrush, *Hylocichla mustelina*) may be responsible for the movement of upwards of four million deer ticks (*Ixodes scapularis*) per year, an insight that has important human health implications in terms of the distribution of the tick and *Borrelia burgdorferi*, the Lyme disease-causing bacteria carried by these ticks (see also Chapter 12).

7.2.3.3 Social network analysis and network models

So far, we have considered models whereby parasites move through host 'states,' to gain an idea of how portions of the population move from one compartment to the next at an 'average' rate, and ABM/IBMs that allow for individuals or 'agents' to move through the infection process with an individual-specific risk of transmission, recovery, or immunity. While IBMs/ABMs allow for additional realism in models, a useful method for capturing the substantial variation in contact rate among individual birds (see Chapter 4) is network approaches that allow us to capture behavioral variation in contact frequency and duration among hosts (White et al. 2017).

Contact networks used in network modeling are typically generated or inspired by social network analyses of contact data, whereby behavioral interactions between individuals (e.g., aggressive interactions, allogrooming) are observed and used to generate contact networks. Contact networks rely on the properties of graphs and use points to represent individual units or 'nodes' (individual animals, farms, sites, etc.) that can be connected by lines that represent 'edges' or links (e.g., Croft et al. 2008), where an edge could indicate a behavioral or other interaction between the two nodes that leads to pathogen transmission (White et al. 2017). Once contact networks are generated, pathogen spread across these networks can then be simulated via network models. In wildlife, network models have been used to address questions ranging from whether heterogeneity in contacts and superspreading are present in wild animal populations to whether some populations of animals are more vulnerable to epidemics than others (see review by White et al. 2017). These are questions that are especially germane to birds, given substantial heterogeneity among individuals in contact behaviors (see Chapter 4) and given that the group of birds facing the most dire conservation outlook (albatrosses and petrels; Croxall et al. 2012) are also social, colonial breeders, making them especially susceptible to outbreaks with severe pathogens such as *Pasteurella multocida*, the causative agent of avian cholera (e.g., Jaeger et al. 2018). Nevertheless, among the 39 studies presented by White et al. (2017), the majority (32) of studies were on mammals and the rest were on lizard or fish populations. A number of the studies in this review included behavioral observations or capture-mark-recapture data, sources of data that are straightforward to obtain in bird study systems with existing long-term, banded populations; therefore, substantial opportunity exists to apply network modeling techniques to pressing disease questions in wild bird populations.

7.2.4 Additional tools and resources

The goal in these sections was to provide an overview of the many quantitative techniques and modeling approaches—tools—used to estimate key population-level parameters and to deepen understanding about the effects of parasites on populations of avian hosts. The peer-reviewed scientific literature provides excellent references to specific applications of techniques, some of which are cited above.

For in-depth guidance on sampling design and estimating population size and other demographic parameters in wild birds, interested readers can refer to relevant chapters in the comprehensive volumes *Analysis and Management of Animal Populations* (Williams et al. 2002), *Modeling Demographic Processes in Marked Populations* (Thomson et al. 2009), and *Population Ecology in Practice* (Murray and Sandercock 2019). In addition, approach-specific works (e.g., distance sampling; Buckland et al. 2015) and web-based resources and user networks provide guidance on the collection (e.g., EURING; https://euring.org/) and analysis (http://phidot.org/) of data from marked or unmarked birds. Readers can turn to *Occupancy Estimation and Modeling: Inferring Patterns and Dynamics of Species Occurrence* (MacKenzie et al. 2018) to explore the occupancy approaches in more depth; Chapters 2 and 6 are especially germane to applications of occupancy approaches to questions about prevalence, test efficacy, and disease dynamics.

For broad treatments of infectious disease modeling, readers can find more in the excellent volumes *Modeling Infectious Disease in Animals and Humans* (Keeling and Rohani 2008), *An Introduction to Infectious Disease Modelling* (Vynnycky and White 2010), and *Mathematical Tools for Understanding Infectious Disease Dynamics* (Diekmann et al. 2013). These volumes all provide in-depth guidance on compartmental models and their derivatives, as well as some treatment of IBMs and network models. For specific treatment of ABMs and IBMs, interested readers can turn to Railsback and Grimm's (2019) *Agent-Based and Individual-Based Modeling: A Practical Introduction (2nd edition)* which includes guidance on the NetLogo software. While not focused solely on modeling disease, *Exploring Animal Social Networks* (Croft et al. 2008) provides an introductory treatment of contact networks for wild animal populations.

7.3 Effects of parasites and pathogens on avian host populations

Parasites and pathogens, by definition, have negative impacts on components of individual fitness in birds, including declines in metrics of reproduction like clutch size and lower survival of infected birds compared to their uninfected counterparts (see

Chapter 6). However, the way these individual-level effects scale up to influence host population dynamics remains a subject of debate. The aim of this chapter, in addition to describing some important tools for analyzing the effects of parasites in birds, is to discuss the evidence—or lack thereof—for population-level impacts of parasites in wild avian systems. As with individual fitness, a key prediction is that parasitism will be linked to declines in host population size, concomitant regulation of bird populations, and the potential for parasite-mediated extinctions of avian populations and even species. At the same time, compensatory mortality—the idea that reductions in density via mortality from predation, harvest, or parasites, for example, are compensated by reduced effects on the population by another density-dependent factor (e.g., Wobeser 2006)—can complicate interpretation of data collected to evaluate these key predictions. Further, aspects of parasites such as their transmission ecology (the relative importance of frequency- versus density-dependent transmission) and virulence (whether parasites affect survival or reproduction and to what degree) can complicate predictions of the way in which parasites will influence host population dynamics. In short, the population ecology of avian hosts in the face of infectious disease is complex (see Figure 2.1).

7.3.1 Parasites and declines in avian host population size

Observations of large numbers of carcasses from catastrophic, large-scale mortality events, like outbreaks of avian cholera in shorebirds and waterfowl (e.g., Botzler 1999), implicate parasites and pathogens as a potentially major factor contributing to declines in avian populations. However, direct connections between parasite-caused mortality events and subsequent declines in bird population sizes are rare, in part because affected host populations are not always marked before the arrival of the parasite, such that changes in abundance rely on counts or indices that cannot account for variation in detection (Box 7.1) or other sources of heterogeneity affecting bird abundance when the count is conducted. These declines are more likely with emerging or reemerging pathogens in novel host populations (LaDeau et al. 2007), a fact that may also slow the detection of the pathogen or the affected host population (Box 7.1). Despite these limitations, immediate or short-term changes in abundance, as demonstrated by the case studies on trichomoniasis and WNV in the following sections, support the notion that pathogens can result in notable declines in bird populations.

7.3.1.1 Case study: *Trichomonas gallinae*

One of the oldest known wildlife diseases is trichomoniasis, a disease of the respiratory and digestive tracts in birds caused by the protozoal parasite *Trichomonas gallinae* (Forrester and Foster 2008). Trichomoniasis is widespread geographically and its host distribution appears to be expanding (see below). Species reported to have trichomoniasis include the columbiforms, strigiforms, and falconiforms (Forrester and Foster 2008), but recent emergence of 'finch trichomoniasis' provides an apt example to evaluate the evidence for declines in host population size.

An outbreak of finch trichomoniasis affected populations of chaffinches (*Fringilla coelebs*) and greenfinches (*Chloris chloris*) in 2006 (Figure 7.3). Chaffinch populations appeared to recover, but population numbers of greenfinches plummeted from 4.3 to 2.8 million birds in the span of about half a decade (Lawson et al. 2012). Subsequent tracking using data from the community-based Garden Bird Feeding Survey showed continued population-level effects for greenfinches with the rate of population decline persisting at more than 7% per year but not for chaffinches, despite the initial impact on both species (Lawson et al. 2012). The strengths in this study lie in its geographic scale at which declines were shown (the British Isles) and the concomitant large amount of data from two independent national efforts. However, weaknesses emerge in the relatively short timescale of the study and reliance on indices of bird abundance (e.g., counts of birds visiting feeders) to establish changes in population size. Nevertheless, evidence for similar declines of up to 47% in European greenfinches in Finland (Lehikoinen et al. 2013) supports the idea that newly emerging infectious diseases can contribute to striking, rapid changes in population sizes of hosts.

7.3.1.2 Case study: West Nile virus

West Nile virus (WNV) in North American bird populations presents a similar opportunity to evaluate the link between observed emergence of an infectious disease in wild birds and impacts on host population size. WNV was responsible for a striking outbreak—and mortality—affecting humans, two species of crow, and a Chilean flamingo (in the Bronx Zoo) in New York in 1999 (Lanciotti et al. 1999). Subsequent pathogen-induced morbidity and mortality was recorded in over 300 wild bird species (Marra et al. 2004) across North America and into South America by 2006 (Kilpatrick et al. 2007) and WNV persists as a serious health threat to humans in some parts of the world (e.g., California; Snyder et al. 2020).

Since the initial avian outbreak in New York City in 1999, substantial research effort has gone into better understanding many facets of WNV in wild birds, from transmission dynamics to impacts on populations. Early work by LaDeau et al. (2007) showed significant declines in populations of six different bird species—but especially populations of American crow—using data from the community-based Breeding Bird Survey and estimates of viral burden based on human infection data. A more recent review drawing together studies evaluating the impact of WNV on wild bird populations (Kilpatrick and Wheeler 2019) suggests that lasting, widespread population impacts, though, are limited to only a few species (e.g., American crow; yellow-billed magpie, *Pica nuttalli*) while also highlighting some important gaps in how effects of pathogens on bird populations are shown. Kilpatrick and Wheeler (2019) suggested that inferences can be improved by combining field counts of birds with experimental demonstration of mortality to strengthen the connection between the pathogenic cause and population effect. Further, spatiotemporal indices of transmission (e.g., infected mosquito density), rather than counts pre- and post-pathogen arrival, should be used to establish parasite-mediated changes in population size to avoid issues with type II errors from multiple comparisons, in addition to overlooking other ecological factors that might contribute to changes in avian abundance (Kilpatrick and Wheeler 2019).

Figure 7.3 Parasites affect populations of a diversity of bird species including European greenfinches (*Chloris chloris*; top left) common to bird feeders in gardens throughout Europe, house sparrows (*Passer domesticus*; top right), considered ubiquitous worldwide and tightly associated with humans, red grouse (*Lagopus lagopus scotica*; bottom right), a game species in Britain, and the critically endangered mangrove finch (*Cactospiza heliobates*; bottom left), endemic to the Galápagos archipelago, Ecuador and whose population numbers less than 100. (Photo credits, clockwise from top left: Estormiz [reproduced under Creative Commons Attribution 1.0 International (CC0 1.0) license]; Adamo [CC BY 2.0 de]; MPF—own work [CC BY-SA 3.0]; and M. Dvorak, Charles Darwin Foundation.)

7.3.2 Parasites and avian host population growth

While numerous, immediate negative effects of parasites on individual birds and populations have been highlighted throughout this volume, population ecologists, managers, and conservation biologists are keenly interested in understanding whether these immediate effects scale up to affect host population growth over the long term. Changes in the size of a population over time can be expressed using a simple equation (sometimes called the BIDE model) where change in population size at time $t + 1(N_{t+1})$ is a function of the population size in the time period before (N_t) plus any additions from births (B) or immigration (I) and any losses from deaths (D) or emigration (Williams et al. 2002). This formulation can be rewritten with per capita rates and then simplified as

$$N_{t+1} = \lambda_t N_t \qquad (7.11)$$

where λ_t represents the **finite rate of population growth**. Accordingly, when $\lambda_t > 1$, the population is increasing (growing), when $\lambda_t < 1$, the population is decreasing, and the population size remains constant when $\lambda_t = 1$ (e.g., Williams et al. 2002). From this, it makes sense that changes in vital rates—the components of the BIDE model—due to parasites (see Chapter 6) could contribute to similar changes (declines) in host population growth.

Similar to the concept of density-dependent transmission (see Section 7.2.3.1), host population growth is density-dependent when the growth of the population is influenced by its size; the form of the density-dependence will depend on factors contributing to population gains or losses. A population whose growth rate is related in a monotonic way with its size, for example, will have faster growth when a population suffers losses due to hunter harvest or parasitism, and this faster replacement of individuals compensates for the losses (i.e., compensatory mortality; Williams et al. 2002; Wobeser 2006). In the absence of density-dependence, the population will grow exponentially until some factor 'limits' or dampens the population's growth. Intraspecific factors like competition for food, nest space, and even mating opportunities or interspecific factors such as predation and parasitism are thought to limit population growth in birds (Newton 2007).

To study factors affecting population growth, we can build models of population growth using difference equations for discrete time steps or using differential equations to build continuous time versions. However, if there is age- or stage-structure in vital rates that are relevant to estimates of population growth rate, population projection matrix models (Caswell 2006) can be applied to provide a picture of the trajectory of the population under varying scenarios. Oli et al. (2006) present a modeling framework where projection matrices correspond to 'life-cycle graphs' for the discrete-time case of SIR-type compartment models. Using this framework, investigators can estimate the growth rate of the pathogen (i.e., basic reproduction number, R_0) or host population growth rate and ask about the sensitivity of these values to changes in model parameters describing the infection process and host population vital rates (Oli et al. 2006).

If the population of interest has a well-developed mark-recapture dataset, Pradel models (Pradel 1996) can be used to estimate population size by relying on the idea that survival probabilities (Φ) have a reverse time equivalent, the seniority probability (γ; note that this is a distinct parameter from that used to denote the recovery rate in Equations 7.6b, 7.6c, and 7.8), and that the population rate of change at time i (λ_i) is the ratio of the survival and seniority probabilities at times i and i+1, respectively, as in Equation 18.42 in Williams et al. (2002):

$$\hat{\lambda}_i = \hat{\Phi}_i / \hat{\gamma}_{i+1} \qquad (7.12)$$

A recent application of Pradel models to a long-term black-throated blue warbler (*Setophaga caerulescens*) dataset revealed that population growth was affected by climate metrics in interesting and complex ways, including that population growth was lower following El Niño years but that higher temperatures and late-season food were related to population increases (Townsend et al. 2016). Application of analogous metrics of infection or disease occurrence to these sorts of long-term mark-recapture datasets might also reveal similar complexity in the effects of parasites on avian host population growth.

7.3.2.1 Case study—avian blood parasites

The avian Haemosporida or avian blood parasites represent an important group for examining the effects of parasites on avian host population growth because they infect a wide diversity of birds, include well over 200 species in the genera *Haemoproteus*, *Plasmodium*, and *Leucocytozoon* (Valkiunas 2004), and have been relatively well studied over the long term. Indeed, while avian malaria is not thought to cause major die-offs in host populations that have coevolved with the pathogen, the introductions of a competent mosquito vector (*Culex quinquefasciatus*) and the avian malarial parasite *Plasmodium relictum* to Hawaii led to important negative effects of disease on populations of immunologically naïve honeycreepers (Van Riper et al. 1986; Warner 1968; see also Chapters 5, 6, 8, and 10).

The population-level effects of avian blood parasites such as *P. relictum* can be complex because infection involves both an acute phase of rapid parasite proliferation, where fitness effects such as mortality often occur, followed by a longer-term chronic phase in hosts that survive acute infection. The mortality of endemic Hawaiian honeycreepers (Van Riper et al. 1986; Warner 1968) juxtaposed with the lack of evidence for negative effects on components of fitness for chronically infected Hawai'i 'amakihi (*Hemignathus virens*) (Kilpatrick et al. 2006; see also Chapter 6) suggests that understanding the population-level effects of avian malaria requires study of both acute and chronic effects. In the years since the initial detection of avian malarial parasite-mediated issues in Hawaiian birds, substantial advances have been made in analytical approaches needed to unravel the differences between short- and long-term effects. Samuel et al. (2015) combined multistate mark-recapture models with age-prevalence approaches in a Bayesian framework to again examine infection and mortality rates along with population-level impacts for the Hawai'i 'amakihi, 'apapane (*Himatione sanguinea*), and 'i'iwi (*Vestiaria coccinea*). These authors confirmed that avian malaria continues to affect populations of Hawaiian honeycreepers: prevalence was highest at low elevations, infection rates were highest for 'i'iwi and 'apapane, and all populations of all three species suffered reductions in population size (Samuel et al. 2015), though reductions were smallest for Hawai'i 'amakihi (e.g., 5% for adult birds at high elevations). Finally, McClure and colleagues (2020) used information on variation in survival, fecundity, force of infection, and fatality ratio to project per capita population growth rates (λ) for Hawai'i 'amakihi. Transmission of *P. relictum* was connected to reductions of 7–14% in λ, but, interestingly, λ never dropped below 1.0, even when the probability of infection was 1.0 (McClure et al. 2020). Taken together, this work suggests that avian malaria has important impacts on populations of native Hawaiian forest birds that required decades to elucidate and that, over time, these effects have been reduced for the Hawai'i 'amakihi, potentially due to host evolution to minimize the impacts of infection (see Chapter 5).

The population declines—and concomitant effects on λ—caused by avian malaria in Hawaii were likely exacerbated by the lack of any coevolutionary history between endemic Hawaiian birds and *P. relictum*. Nonetheless, avian malaria may result in notable effects on populations even when the host–parasite relationship is unlikely to be novel. The observation that house sparrow (*Passer domesticus*; Figure 7.3) numbers in Europe appeared to be in decline motivated work evaluating the effects of parasites on sparrow survival and abundance (Dadam et al. 2019). These authors tracked house sparrow populations at 11 sites from 2006 to 2009 using CJS models to estimate impacts on individual sparrow survival, counts to establish changes in population size (as a measure of λ), and samples of blood and feces to establish infection status with various parasites including *Plasmodium*. While trends in the counts of chirping males declined over

the course of the study at seven of the study sites, changes in population size were not related to prevalence of *Plasmodium*. Interestingly and importantly, however, population growth was negatively associated with the intensity of infection with *Plasmodium* for juveniles and juveniles and adults combined but not for adults alone (Dadam et al. 2019). Coupled with the finding that overwinter survival probability was also related to infection intensity (higher intensity, lower survival), it may be that the impact of parasites on host population growth is mediated through declines in recruitment, highlighting, again, the complexity of disease ecological systems.

In sum, studies on avian malaria in wild birds show that effects on population growth (λ) are not simply a function of mortality in adults due to disease but that effects on other vital rates and other age classes are important. Further, these studies highlight the value of long-term work in establishing effects of parasites on host population growth, a factor that is also important in showing whether parasitism effectively causes cycling or even regulates avian host populations.

7.3.3 Host population regulation and fluctuation linked to parasites

From the work on native honeycreepers in Hawaii and house sparrows in the UK, we see that disease can affect rates of population growth, but whether any dampening on growth imposed by parasites scales up to population regulation is unclear. As mentioned above, under ideal conditions a host population will grow exponentially unless it is limited by (usually density-dependent) factors such as a finite food supply, adverse weather conditions, or parasites. The classic conceptualization of **population regulation** relies on density-dependent factors, such as parasites, that slow per capita growth as the density increases (Newton 1998). When growth slows because of density-dependent effects, the population drops towards a threshold population size (i.e., carrying capacity), which eases the limits on population growth such that growth begins to increase again (e.g., Williams et al. 2002). Apparent population cycles like those described for voles and lemmings have often been attributed to the density-dependent effects of a predator on the prey population (see Oli 2019 for review), but parasites can also contribute to these cyclic dynamics by impacts on reproduction or survival that cause the population to cycle through increases and decreases in population size.

7.3.3.1 Case study—*Mycoplasma gallisepticum*

One of the only clear examples of potential parasite-mediated population regulation comes from a common backyard songbird in North America, the house finch. As has been described in more detail in Chapter 5, mycoplasmal conjunctivitis is an important disease of house finches caused by the bacterium *Mycoplasma gallisepticum*. Tracking of both house finch abundance and observed disease at feeders by community scientists in eastern North America enabled study of the pathogen-mediated changes in house finch abundance over this broad geographic extent (Hochachka and Dhondt 2000). This study demonstrated pathogen-mediated declines in bird abundance that were density-dependent. While abundance declined by up to 60% in the region of highest pre-disease density, no decline or even a slight increase in density was found in the region of lowest pre-disease density, such that populations converged to similar densities following pathogen emergence (Figure 3 in Hochachka and Dhondt 2000). This suggested that *M. gallisepticum* was acting in a density-dependent way to regulate house finch population sizes. Hochachka and Dhondt (2006) examined whether similar patterns occurred at the local scale. In all regions examined in the eastern US, sizes of groups counted at individual bird feeding sites declined rapidly and remained low throughout the remaining period for which data were available, which was almost a decade following *M. gallisepticum* emergence (for regions hit earliest by the pathogen). Further, declines in group sizes were strongly associated with local density of house finches, with the biggest declines at sites that started with the largest group sizes. This work suggests that house finch population sizes are regulated by *M. gallisepticum* and that this regulation can continue for a relatively long period following initial pathogen emergence.

7.3.3.2 Case study—*Trichostrongylus tenuis*

The red grouse (*Lagopus lagopus scotica*) is a subspecies of the willow grouse (*Lagopus lagopus*) distributed throughout Britain (Figure 7.3). Like many other grouse species around the world, the red grouse is a popular game animal, such that bag numbers reflecting deep population cycling in the 1970s raised alarms for managers and generated critical questions about the cause of the declines and rebounds. Central to explanations for the grouse cycles was the nematode, *Trichostrongylus tenuis*, a gastrointestinal parasite typically infecting the ceca and small intestine of galliforms and anseriforms, leading to morbidity, especially when infections are of high intensity, as occurs in the red grouse (Tompkins 2008).

To evaluate the hypothesis that *T. tenuis* was responsible for the population fluctuations in red grouse, Hudson et al. (1998) treated populations with antihelminthics which reduced parasite burden and appeared to be effective in reducing the extent of the fluctuations for several years. This and earlier work (Hudson 1986) showed a negative association between infection intensity and ratios of young to adult grouse, a metric of breeding success, 'so that population crashes are associated with high parasite intensities' (Hudson et al. 1998). While this manipulative experiment was effective in revealing a connection between a parasite and host population dynamics, two important criticisms have been levied about the results that highlight, again, that the ecology of infectious diseases in bird populations is complex. The first critique is that the assessment of the impact of the anthelmintic on grouse numbers was based on numbers of birds that were shot, but zero birds were shot in some years, which will bias estimates of differences in the effects of interest (Lambin et al. 1999, Redpath et al. 2006). Second, cycles persisted in treated areas, which may mean that some other factor(s) is involved (Lambin et al. 1999; Redpath et al. 2006). The evidence accumulated in this focal study of *T. tenuis* in red grouse certainly points to population-level effects of a parasite on a host population but also illustrates that these population-level effects can be complex.

7.3.4 Local extinctions of host populations

The most striking effects of disease on bird populations may be disease-mediated population declines that are so severe that they lead to local—or even species-level—extinctions. Indeed, given that, by definition, parasites 'harm' hosts, it makes sense that parasitism is implicated in the extinction of host species. Disease-linked extinctions are difficult to document, however, because the simplest models predict that parasites should go extinct before their hosts (de Castro and Bolker 2005). Stepping beyond this prediction, parasite-mediated extinction is most likely to occur when transmission of the parasite is frequency-dependent and thus can still occur at low host densities, when the parasite is a host generalist and thus has other options for maintenance (i.e., reservoirs) other than the host species that it is rapidly depleting, when the host population density drops, and when the host population is already threatened by other stressors before the onset of disease, including a small population size (de Castro and Bolker 2005).

The clearest case of likely parasite-caused extinction in wild birds is the loss of many iconic Hawaiian honeycreeper species due to avian malaria (Van Riper et al. 1986; Warner 1968), which is both a vector-borne (frequency-dependent transmission) and multihost parasite. In this case, native, immunologically naïve host species that were isolated to an island archipelago (i.e., relatively small pre-disease population sizes) were faced with the introduction of novel vectors, novel parasites, and potentially competing non-native hosts. Thus, a suite of factors including low host population size, frequency-dependent transmission, and the presence of reservoirs likely combined to facilitate parasite-mediated extinctions in honeycreepers. While no extinctions of bird species have yet been documented in another archipelago, the Galápagos Islands of Ecuador, larvae of the introduced and multihost parasitic fly *Philornis downsi* cause severe impacts on host fitness in many species of Darwin's finch (Geospizinae; see overview in Kleindorfer and Dudaniec 2016). Significant concern exists over the persistence of the rarest of these, the mangrove finch (*Cactospiza heliobates*; Figure 7.3), whose total population numbers < 100 individuals (Cunninghame et al. 2017), limited to two sites on a single island, and which is also threatened by predation by the introduced black rat (*Rattus rattus*).

7.4 Lessons learned

The studies from which we have learned the most about the effects of parasites on avian host populations have a few 'common denominators.' The first of these comes from our collective fascination with birds as a research subject. Birds are relatively easy to capture and band safely and economically and, as we have seen from the sections on mark-recapture approaches, these aspects allow for the collection of copious data over the long term and with which we can learn a great deal about vital rates, population size, and population dynamics. Large amounts of population data in birds are also collected by community scientists who enthusiastically observe birds using standardized protocols along set transects or in their own backyards (see also Chapter 13). The strength of these community science datasets is that the population data have typically been collected for many years prior to an outbreak, as in the case of mycoplasmal conjunctivitis and house finches, allowing the study of phenomena like population regulation that are challenging to study in the absence of pre-outbreak data. Further, our deepest understanding of population ecology in avian host–parasite systems has come from the use of multiple avenues of inquiry. For example, intensive field observational studies coupled with detailed experimental infections shed important light on the likely population consequences of avian malaria in Hawaiian forest birds, as did pairing SIR model predictions and field experiments to test hypotheses about parasite-driven population cycling for red grouse and *T. tenuis*.

Many advances in our understanding of the effects of parasites on bird populations have emerged by making use of the array of quantitative techniques building from the mark-recapture framework, including occupancy approaches. The critical lesson in applying these techniques is that detection—of antibodies, of individual birds, etc.—is imperfect and that we learn the most by acknowledging that uncertainty and accounting for it directly. Finally, the lessons we have learned by examining population effects of disease in birds reveals the best (and worst) about ecology—that systems can be complex, confusing, and often complicated—but when we embrace that complexity rather than ignore it, we learn the most.

Literature cited

Altizer, S., Hochachka, W.M., and Dhondt, A.A. (2004). Seasonal dynamics of mycoplasmal conjunctivitis in eastern North American house finches. *Journal of Animal Ecology*, 73, 309–22.

Anderson, R.M. and May, R.M. (1978). Regulation and stability of host–parasite population interactions: I. Regulatory processes. *Journal of Animal Ecology*, 47, 219–47.

Anderson, R.M. and May, R.M. (1991). *Infectious Disease of Humans*. Oxford University Press, Oxford.

Begon, M., Harper, J.L., and Townsend, C.R. (1996). *Ecology: Individuals, Populations, and Communities*, 3rd ed. Blackwell Publishing, Oxford.

Benhaiem, S., Marescot, L., Hofer, H., et al. (2018). Robustness of eco-epidemiological capture-recapture parameter estimates to variation in infection state uncertainty. *Frontiers in Veterinary Science*, 5, 197.

Bergsman, L.D., Hyman, J.M., and Manore, C.A. (2016). A mathematical model for the spread of West Nile virus in migratory and resident birds. *Mathematical Biosciences and Engineering*, 13, 401–24.

Botzler, R.G. (1999). Epizootiology of avian cholera in wildfowl. *Journal of Wildlife Diseases*, 27, 367–95.

Bouden, M., Moulin, B., and Gosselin, P. (2008). The geo-simulation of West Nile virus propagation: a multi-agent and climate sensitive tool for risk management in public health. *International Journal of Health Geographics*, 7, 35.

Buckland, S.T., Rexstad, E.A., Marques, T.A., and Oedekoven, C.S. (2015). *Distance Sampling: Methods and Applications*. Springer, New York.

Bush, A.O., Lafferty, K.D., Lotz, J.M., and Shostak, A.W. (1997). Parasitology meets ecology on its own terms: Margolis et al. revisited. *Journal of Parasitology*, 83, 575–83.

Bush, S.E., Sohn, E., and Clayton, D.H. (2006). Ecomorphology of parasite attachment: experiments with feather lice. *Journal of Parasitology*, 92, 25–31.

Caswell, H. (2006). *Matrix Population Models: Construction, Analysis, and Interpretation*, 2nd ed. Sinauer Associates, Sunderland, MA.

Clayton, D.H., Gregory, R.D., and Price, R.D. (1992). Comparative ecology of neotropical bird lice (Insecta: Phthiraptera). *Journal of Animal Ecology*, 61, 781–95.

Conn, P.B. and Cooch, E.G. (2009). Multistate capture-recapture analysis under imperfect state observation: an application to disease models. *Journal of Applied Ecology*, 46, 486–92.

Cooch, E.G, Conn, P.B., Ellner, S.P., Dobson, A.P., and Pollock, K.H. (2012). Disease dynamics in wild populations: modeling and estimation: a review. *Journal of Ornithology*, 152(S2), 485–509.

Croft, D.P., James, R., and Krause, J. (2008). *Exploring Animal Social Networks*. Princeton University Press, Princeton, NJ.

Croxall, J.P., Butchart, S.H.M., Lascelles, B., et al. (2012). Seabird conservation status, threats and priority actions: a global assessment. *Bird Conservation International*, 22, 1–34.

Cunninghame, F., Fessl, B., Sevilla, C., Young, G., and La Greco, N. (2017). Long-term conservation management to save the Critically Endangered mangrove finch (*Camarhynchus heliobates*). In *Galapagos Report 2015–2016*, pp. 161–8. GNPD, GCREG, CDF, and GC, Puerto Ayora, Galápagos, Ecuador.

Dadam, D., Robinson, R.A., Clements, A., et al. (2019). Avian malaria-mediated population decline of a widespread iconic bird species. *Royal Society Open Science*, 6, 182197.

de Castro, F. and Bolker, B. (2005). Mechanisms of disease-induced extinction. *Ecology Letters*, 8, 117–26.

Diekmann, O., Heesterbeek, H., and Britton, T. (2013). *Mathematical Tools for Understanding Infectious Disease Dynamics*. Princeton University Press, Princeton, NJ.

Doster, G.L. and Goater, C.P. (1997). Collection and quantification of avian helminths and protozoa. In D.H. Clayton and J. Moore, eds. *Host–Parasite Evolution*, pp. 396–418. Oxford University Press, Oxford.

Elmore, S.A., Huyvaert, K.P., Bailey, L.L., et al. (2014). *Toxoplasma gondii* exposure in arctic-nesting geese: a multi-state occupancy framework and comparison of serological assays. *International Journal for Parasitology: Parasites and Wildlife*, 3, 147–53.

Elmore, S.A., Huyvaert, K.P., Bailey, L.L., et al. (2016). Multi-scale occupancy approach to estimate *Toxoplasma gondii* prevalence and detection probability in tissues: an application and guide for field sampling. *International Journal for Parasitology*, 46, 563–70.

Faustino, C.R., Jennelle, C.S., Connolly, V., et al. (2004). *Mycoplasma gallisepticum* infection dynamics in a house finch population: seasonal variation in survival, encounter and transmission rate. *Journal of Animal Ecology*, 73, 651–69.

Forrester, D.J. and Foster, G.W. (2008). Trichomonosis. In C.T. Atkinson, N.J. Thomas, and D.B. Hunter, eds. *Parasitic Disease of Wild Birds*, pp. 120–53. Wiley-Blackwell, Ames, IA.

Gordis, L. (2013). *Epidemiology*, 5th ed. Elsevier, Philadelphia.

Grimm, V., Berger, U., Bastiansen, F., et al. (2006). A standard protocol for describing individual-based and agent-based models. *Ecological Modelling*, 198, 115–26.

Hochachka, W.M. and Dhondt, A.A. (2000). Density-dependent decline of host abundance resulting from a new infectious disease. *Proceedings of the National Academy of Sciences of the United States of America*, 97, 5303–6.

Hochachka, W.M., and Dhondt, A.A. (2006). House Finch population- and group-level responses to a bacterial disease. In R.K. Barraclough, ed. *Current topics in avian disease research: understanding endemic and invasive diseases. Ornithological Monographs*, 60, 30–43.

Holmstad, P.R., Hudson, P.J., and Skorping, A. (2005). The influence of a parasite community on the dynamics of a host population: a longitudinal study on willow ptarmigan and their parasites. *Oikos*, 111, 377–91.

Hudson, P.J. (1986). The effect of a parasitic nematode on the breeding production of red grouse. *Journal of Animal Ecology*, 55, 85–92.

Hudson, P.J., Dobson, A.P., and Newborn, D. (1998). Prevention of population cycles by parasite removal. *Science*, 282, 2256–8.

Ishtiaq, F., Rao, M., Huang, X., and Bensch, S. (2017). Estimating prevalence of avian haemosporidians in natural populations: a comparative study on screening protocols. *Parasites & Vectors*, 10, 127.

Iverson, S.A., Gilchrist, H.G., Soos, C., Buttler, I.I., Harms, N.J., and Forbes, M.R. (2016). Injecting epidemiology into population viability analysis: avian cholera transmission dynamics at an arctic seabird colony. *Journal of Animal Ecology*. 85, 1481–90.

Jackson, J.L. (2008). The early history of bird banding in North America. In J.A. Jackson, W.E. Davis Jr, and J. Tautin, eds. *Bird Banding in North America—The First 100 Years*, pp. 1–30. Memoirs of the Nuttall Ornithological Club, no. 15, Cambridge, MA.

Jaeger, A., Lebarbenchon, C., Bourret, V., et al. (2018) Avian cholera outbreaks threaten seabird species on Amsterdam Island. *PLoS ONE*, 13, e0197291.

Jankowski, M.D., Williams, C.J., Fair, J.M., and Owen, J.C. (2013). Bird shed RNA-viruses according to the Pareto principle. *PLoS ONE*, 8, e72611.

Kanarek, G. and Zaleśny, G. (2014). Extrinsic- and intrinsic-dependent variation in component communities and patterns of aggregations in helminth parasites of great cormorant (*Phalacrocorax carbo*) from N.E. Poland. *Parasitology Research*, 113, 837–50.

Keeling, M.J. and Rohani, P. (2000). Individual-based perspectives on R_0. *Journal of Theoretical Biology*, 203, 51–61.

Keeling, M.J. and Rohani, P. (2008). *Modeling Infectious Diseases in Humans and Animals*. Princeton University Press, Princeton, NJ.

Kendall, W.L. (2009). One size does not fit all: adapting mark-recapture and occupancy models for state uncertainty.

In D.L. Thomson, E.G. Cooch, and M.C. Conroy, eds. *Modeling Demographic Processes in Marked Populations*, pp. 765–80. Springer Science+Business Media, New York.

Kermack, W.O. and McKendrick, A.G. (1927). A contribution to the mathematical theory of epidemics. *Proceedings of the Royal Society A*, 115, 700–21.

Kilpatrick, A.M., LaPointe, D.A., Atkinson, C.T., et al. (2006). Effects of chronic avian malaria (*Plasmodium relictum*) infection on reproductive success of Hawai'i amakihi (*Hemignathus virens*). *The Auk*, 123, 764–74.

Kilpatrick, A.M., LaDeau, and Marra, P.P. (2007). Ecology of West Nile virus transmission and its impact on birds in the western hemisphere. *The Auk*, 124, 1121–36.

Kilpatrick, A.M. and Wheeler, S.S. (2019). Impact of West Nile Virus on bird populations: Limited lasting effects, evidence for recovery, and gaps in our understanding of impacts on ecosystems. *Journal of Medical Entomology*, 56, 1491–97.

Kleindorfer, S. and Dudaniec, R.Y. (2016). Host–parasite ecology, behavior, and genetics: a review of the introduced fly parasite *Philornis downsi* and its Darwin's finch hosts. *BMC Zoology* 1, 1.

Komar, N., Dohm, D.J., Turell, M.J., and Spielman, A. (1999). Eastern equine encephalitis virus in birds: relative competence of European starlings (*Sturnus vulgaris*). *American Journal of Tropical Medicine and Hygiene*, 60, 387–91.

Komar, N., Langevin, S., Hinten, S., et al. (2003). Experimental infection of North American birds with the New York 1999 strain of West Nile virus. *Emerging Infectious Diseases*, 9, 311–22.

Lachish, S. and Murray, K. (2019). The certainty of uncertainty: potential sources of bias and imprecision in disease ecology studies. *Frontiers in Veterinary Science*, 5, 1–14.

Lachish, S., Gopalaswamy, A.M., Knowles, S.C.L., Sheldon, B.C. (2012). Site-occupancy modelling as a novel framework for assessing test sensitivity and estimating wildlife disease prevalence from imperfect diagnostic tests. *Methods in Ecology and Evolution*, 3, 339–48.

LaDeau, S.L., Kilpatrick, A.M., and Marra, P.P. (2007). West Nile virus emergence and large-scale declines of North American bird populations. *Nature*, 447, 710–14.

Lambin, X., Krebs, C.J., Moss, R., Stenseth, N.C., and Yoccoz, N.G. (1999). Population cycles and parasitism. *Science*, 286, 2425.

Lanciotti, R.S., Roehrig, J.T., Deubel, V. et al. (1999). Origin of the West Nile virus responsible for an outbreak of encephalitis in the northeastern United States. *Science*, 286, 2333–7.

Lawson, B., Robinson, R.A., Colvile, K.M., et al. (2012). The emergence and spread of finch trichomonosis in the British Isles. *Philosophical Transactions of the Royal Society B: Biological Sciences*, 367, 2852–63.

Lehikoinen, A., Lehikoinen, E., Valkama, J., Väisänen, R.A., and Isomurus, M. (2013). Impacts of trichomonosis epidemics on greenfinch *Chloris chloris* and chaffinch *Fringilla coelebs* populations in Finland. *Ibis*, 155, 357–66.

MacKenzie, D.I., Nichols, J.D., Royle, J.A., Pollock, K.H., Bailey, L.L., and Hines, J.E. (2018). *Occupancy Estimation and Modeling: Inferring Patterns and Dynamics of Species Occurrence*. Elsevier, Burlington, MA.

Marra, P.P., Griffing, S., Caffrey, C., et al. (2004). West Nile virus and wildlife. *BioScience*, 54, 393–402.

McCallum, H., Jones, M., Hawkins, C., et al. (2009). Transmission dynamics of Tasmanian devil facial tumor disease may lead to disease-induced extinction. *Ecology*, 90, 3379–92.

McClintock, B. T., Nichols, J.D., Bailey, L.L., MacKenzie, D.I., Kendall, W., and Franklin, A.B. (2010). Seeking a second opinion: uncertainty in disease ecology. *Ecology Letters*, 13, 659–74.

McClure, K.M., Fleischer, R.C., and Kilpatrick, A.M. (2020). The role of native and introduced birds in transmission of avian malaria in Hawaii. *Ecology*, 101, e03038.

McVinish, R. and Lester, R.J.G. (2019). Measuring aggregation in parasite populations. *Journal of the Royal Society Interface*, 17, 20190886.

Mosher B.A., Brand, A.B., Wiewel, A.N., et al. (2019). Estimating occurrence, prevalence, and detection of amphibian pathogens: insights from occupancy models. *Journal of Wildlife Diseases*, 5, 563–75.

Murray, D.L. and Sandercock, B.K. (2019). *Population Ecology in Practice*. Wiley-Blackwell, Hoboken, NJ.

Nasrinpour, H.R., Friesen, M.R., and McLeod, R.D. (2019). Agent based modelling and West Nile virus: a survey. *Journal of Medical and Biological Engineering*, 39, 178–83.

Newton, I. (1998). *Population Limitation in Birds*. Academic Press, San Diego, CA.

Newton, I. (2007). Population limitation in birds: the last 100 years. *British Birds*, 100, 518–39.

Nichols, J.D., Hines, J.E., Sauer, J.R., Fallon, F.W., Fallon, J.E., and Heglund, P.J. (2000). A double-observer approach for estimating detection probability and abundance from point counts. *The Auk*, 117, 393–408.

Oli, M.K., Venkataraman, M., Klein, P.A., Wendland, L.D., and Brown, M.B. (2006). Population dynamics of infectious disease: a discrete time model. *Ecological Modelling*, 198, 183–94.

Oli, M.K. (2019). Population cycles in voles and lemmings: state of the science and future directions. *Mammal Review*, 49, 226–39.

Otis, D.L., Burnham, K.P., White, G.C., and Anderson, D.R. (1978). Statistical inference from capture data on closed animal populations. *Wildlife Monographs*, 62, 1–135.

Pedersen, A.B. and Babayan, S.A. (2011). Wild immunology. *Molecular Ecology* 20, 872–80.

Perrins, C.M. and Gosler, A.G. (2011). Birds. In P. Savill, C.M. Perrins, K. Kirby, and N. Fisher, eds. *Wytham Woods: Oxford's Ecological Laboratory*, pp. 145–71. Oxford University Press, Oxford.

Pollock, K.H., Nichols, J.D., Brownie, C., and Hines, J.E. (1990). Statistical inference for capture-recapture experiments. *Wildlife Monographs*, 107, 3–97.

Poulin, R. (1993). The disparity between observed and uniform distributions: a new look at parasite aggregation. *International Journal for Parasitology*, 7, 937–44.

Pradel, R. (1996). Utilization of capture-mark-recapture for the study of recruitment and population growth rate. *Biometrics*, 52, 703–9.

Railsback, S.F. and Grimm, V. (2019). *Agent-Based and Individual-Based Modeling: A Practical Introduction*, 2nd ed. Princeton University Press, Princeton, NJ.

Ralph, C.J., Sauer, J.R., and Droege, S. (1995). Monitoring bird populations by point counts. In US Department of Agriculture Forest Service, ed. *General Technical Report PSW-GTR-149*. Pacific Southwest Research Station, Albany, CA.

Redpath, S.M., Mougeot, F., Leckie, F.M., Elston, D.A., and Hudson, P.J. (2006). Testing the role of parasites in driving the cyclic population dynamics of a gamebird. *Ecology Letters*, 9, 410–18.

Ryder, J.J., Miller, M.R., White, A., Knell, R.J., and M. Boots. (2007). Host–parasite population dynamics under combined frequency- and density-dependent transmission. *Oikos*, 116, 2017–26.

Samuel, M.D., Woodworth, B.L., Atkinson, C.T., Hart, P.J., and LaPointe, D.A. (2015). Avian malaria in Hawaiian forest birds: infection and population impacts across species and elevations. *Ecosphere*, 6, 104.

Schei, E., Homstad, P.R., and Skorping, A. (2005) Seasonal infection patterns in willow grouse (*Lagopus lagopus L.*) do not support the presence of parasite-induced winter losses. *Ornis Fennica*, 82, 137–46.

Sedinger, J.S., Riecke, T.V., Leach, A.G., and Ward, D.H. (2019). The black brant population is declining based on mark recapture. *Journal of Wildlife Management*, 83, 627–37.

Senar JC. (2004). Multi-state analysis of the impacts of avian pox on a population of serins (*Serinus serinus*): the importance of estimating recapture rates. *Animal Biodiversity and Conservation*, 27, 133–46.

Shaw, D.J. and Dobson, A.P. (1995). Patterns of macroparasite abundance and aggregation in wildlife populations: a quantitative review. *Parasitology*, 111, S111–33.

Snyder, R.E., Sondermeyer Cooksey, G., Kramer, V., Jain, S., and Vugia, D.J. (2020). West Nile virus-associated hospitalizations, California, 2004–2017. *Clinical Infectious Diseases*, Jun 11, ciaa749.

Thinh, T.V., Gilbert, M., Bunpapong, N., et al. (2012). Avian influenza viruses in wild land birds in northern Vietnam. *Journal of Wildlife Diseases*, 48, 195–200.

Thompson, S.K. (2012). *Sampling*, 3rd ed. Wiley, Hoboken, NJ.

Thomson, D.L., Cooch, E.G., and Conroy, M.J. (eds) (2009). *Modeling Demographic Processes in Marked Populations*. Springer, New York.

Tompkins, D.M. (2008). *Trichostrongylus*. In C.T. Atkinson, N.J. Thomas, and D.B. Hunter, eds. *Parasitic Disease of Wild Birds*, pp. 316–25. Wiley-Blackwell, Ames, IA.

Tompkins, D.M., Dobson, A.P., Arneberg, P. et al. (2002). Parasites and host population dynamics. In P.J. Hudson, A. Rizzoli, B.T. Grenfell, H. Heesterbeek, and A.P. Dobson, eds. *The Ecology of Wildlife Diseases*, pp. 6–44. Oxford University Press, Oxford.

Tonelli, B.A. and Dearborn, D.C. (2019). An individual-based model for the dispersal of *Ixodes scapularis* by ovenbirds and wood thrushes during fall migration. *Ticks and Tick-borne Diseases*, 10, 1096–1104.

Townsend, A.K., Cooch, E.G., Sillett, T.S., Rodenhouse, N.L., Holmes, R.T., Webster, M.S. (2016). The interacting effects of food, spring temperature, and global climate cycles on population dynamics of a migratory songbird. *Global Change Biology*, 22, 544–55.

Valkiunas, G. (2004). *Avian Malaria Parasites and Other Haemosporidia*. CRC Press, Boca Raton, FL.

van Riper, C., van Riper, S.G., Goff, M.L., and Laird, M. (1986). The epizootiology and ecological significance of malaria in Hawaiian land birds. *Ecological Monographs*, 56, 327–44.

VanderWaal, K.L. and Ezenwa, V.O. (2016). Heterogeneity in pathogen transmission: mechanisms and methodology. *Functional Ecology*, 30, 1606–22.

Vynnycky, E. and White, R.G. (2010). *An Introduction to Infectious Disease Modelling*. Oxford University Press, Oxford.

Walther, B.A. and Clayton, D.H. (1997). Dust-ruffling: a simple method for quantifying ectoparasite loads of live birds. *Journal of Field Ornithology*, 68, 509–18.

Warner, R.E. (1968). The role of introduced diseases in the extinction of the endemic Hawaiian avifauna. *The Condor*, 70, 101–20.

White, G.C. and Burnham, K.P. (1999). Program MARK: survival estimation from populations of marked animals. *Bird Study*, 46, S120–39.

White, L.A., Forester, J.D., and Craft, M.E. (2017). Using contact networks to explore mechanisms of parasite transmission in wildlife. *Biological Reviews*, 92, 389–409.

Williams, B.K., Nichols, J.D., and Conroy, M.J. (2002). *Analysis and Management of Animal Populations*. Academic Press, San Diego, CA.

Wilson, K., Bjornstad, O.N., Dobson, A.P., et al. (2002). Heterogeneities in macroparasite infections: patterns and processes. In P.J. Hudson, A. Rizzoli, B.T. Grenfell, H. Heesterbeek, and A.P. Dobson, eds. *The Ecology of Wildlife Diseases*, pp. 6–44. Oxford University Press, Oxford.

Wobeser, G.A. (2006). *Essentials of Disease in Wild Animals*. Blackwell Publishing, Ames, IA.

Wobeser, G.A. (2007). *Disease in Wild Animals: Investigation and Management*, 2nd ed. Springer, Heidelberg.

Woolhouse, M.E.J., Dye, C., Etard, J.-F., et al. (1997). Heterogeneities in the transmission of infectious agents: implications for the design of control programs. *Proceedings of the National Academy of Sciences of the United States of America*, 94, 338–42.

CHAPTER 8

Community-Level Interactions and Disease Dynamics

Karen D. McCoy

8.1 Introduction

Understanding host and parasite ecology is crucial for inferring mechanisms of parasite transmission and for predicting its short- and long-term consequences for host population dynamics (see Chapter 7). This basic information becomes more challenging to obtain, and the outcome more difficult to predict, in complex communities where hosts and non-hosts interact directly and indirectly in different ways. Current shifts in biodiversity due to global change, and the associated modifications to biotic interactions within communities, may result in profound changes in the probability of disease emergence, its dynamics over time, and its population- and community-level impacts at different spatial and temporal scales. Apprehending and predicting the impacts of these changes has become increasingly important for human health, for animal production, and for conservation. Birds represent an integral part of almost all natural communities. Due to this ubiquity and their intrinsic biological characteristics, this taxonomic group plays a major role in the ecology, evolution, and epidemiology of parasitic species at the community level.

In this chapter, I examine the role of birds as hosts to parasites in natural communities and how changes in these communities can alter disease dynamics and emergence. I start by considering the relative importance of birds and parasites in natural communities, highlighting some basic notions in community ecology. I then consider how general changes in diversity can alter disease dynamics: first, how the addition or removal of parasite diversity can alter overall community structure and function; then, how changes in host diversity can affect parasite and pathogen transmission, with a specific consideration of dilution and amplification effects. Finally, I briefly consider the importance of combining data from host and parasite communities to understand metacommunity dynamics. The different elements in each section are illustrated with brief, concrete examples from the literature on avian species, when possible. Near the end of the chapter, I provide a more detailed example from my own work that examines the structure of ticks and Lyme disease bacteria in marine bird communities. The chapter finishes with a brief discussion of diverse human impacts on avian communities and the feedback effects they can have on public health and disease risk. Throughout the chapter, I refer to parasitic organisms as '**parasites**' which includes both **microparasites** and **macroparasites**. The term '**pathogen**' is reserved for microorganisms that cause disease: the debilitating manifestations of infection.

8.2 Community structure and function

8.2.1 Community ecology and the place of birds

An **ecological community** comprises all individuals of all species that potentially interact within a single patch or local area of habitat (Lawton 1999; Leibold et al. 2004; Wiens 1992). These interactions can take different forms: antagonistic (predation or parasitism), competitive, or mutualistic—but all can ultimately affect individual fitness and local species persistence. **Community ecology** is the

Karen D. McCoy, *Community-Level Interactions and Disease Dynamics* In: *Infectious Disease Ecology of Wild Birds*. Edited by: Jennifer C. Owen, Dana M. Hawley, and Kathryn P. Huyvaert, Oxford University Press. © Oxford University Press 2021.
DOI: 10.1093/oso/9780198746249.003.0008

study of these interactions and attempts to understand how they lead to the patterns of diversity we observe at different spatial and temporal scales. It considers aspects such as variation in species richness, evenness of abundance, and productivity, as well as food web structure and community assembly rules. Not only do birds make up a significant part of most natural communities, but also studies of avian communities have been of primary importance in the development of general theories in community ecology (Wiens 1992).

Community ecology is inherently complex due to the range of species present at a given place and time and the changing conditions from one place to another; it has been thoughtfully referred to as an ecological 'mess' (Lawton 1999). To deal with this complexity, communities are often subdivided into 'guilds' or assemblages of different sizes and specificities, where species are grouped by similarities in the way they exploit resources (e.g., predators), by the types of resources they use (e.g., nectar-feeding species or granivores), by habitat units (e.g., intertidal), or by taxonomy (e.g., Class Aves). While this results in more manageable units for study, important interactions can be missed and calls for caution have been made when trying to understand community dynamics based solely on restricted criteria, particularly taxonomic (Wiens 1983). Vellend (2010) argued that patterns of species diversity, abundance, and composition within a community can be understood via four key organizing processes: selection, drift, speciation, and dispersal. The impacts on community assembly and function of processes like drift (i.e., stochastic changes in species population sizes), speciation (i.e., addition of novel species to a community), and dispersal (i.e., species connections from outside the community) are relatively straightforward to understand. The last, more enigmatic process, selection, encompasses all the varied and complex interactions within and among species within a community that will ultimately lead to fitness differences among individuals and species.

Two major questions in community ecology have been the focus of debate over the last 50 years. The first is, *How do communities assemble?* This is an essential question to answer if we are to understand how quickly communities may reassemble after perturbation or how readily novel species can invade (Pearson et al. 2018). Community assembly rules actually originated as a biogeographic approach to explain island bird assemblages (Diamond 1975). They were then later adapted as a general framework for understanding the structure of all natural communities (Weiher and Keddy 2001). Traditional community assembly theory was niche-based, postulating that the composition and relative abundance of species within a community is determined by a series of hierarchical biotic and abiotic filters related to the functional traits of each species (Weiher and Keddy 2001). In contrast, Hubbell's (2001) neutral theory assumed that all individuals at a particular trophic level in a community are essentially equivalent regarding their chance of survival and reproduction and thus accumulate locally through random processes—drift, speciation, and immigration—from the global species pool (Rosindell et al. 2011). More recent syntheses have highlighted the complementarity of the two theories for explaining assembly patterns at different spatial scales (Chave 2004; Pearson et al. 2018; Rosindell et al. 2011; Seabloom et al. 2015). For example, species arrive by random dispersal from the general species pool and then only establish if they pass local environmental filters.

The second question asks, *What characterizes community stability?* (e.g., Lawton 1999; Loreau and de Mazancourt 2013; Mougi and Kondoh 2012; Wiens, 1983). Communities are considered to have a dynamic stability that leads assemblages towards an equilibrium composition (Wiens 1992). Maintaining stability is then thought to depend on both the distribution and strength of interactions (May 2006) and on the level of biodiversity, with higher diversity increasing community stability (Loreau and de Mazancourt 2013). This latter relationship occurs via several complementary mechanisms: asynchrony in species' phenologies, interspecific differences in temporal responses to perturbations, reductions in the strength of competition, and the occurrence of multiple types of biotic interactions within and across trophic levels (Loreau and de Mazancourt 2013; Mougi and Kondoh 2012). Dispersal across communities is also a stabilizing force, replacing lost individuals and species and maintaining community function (Mouquet and

Loreau 2003). Birds are particularly important in this sense, both via their own dispersal and by transporting other species during movement. **Keystone species** are thought to occur in most systems as these are species whose presence is crucial for maintaining organization and community diversity (Mills et al. 1993). These species can act via bottom-up effects, where changes in resources alter competition intensity and thus community structure, or top-down effects, where predation impacts prey population sizes, for example. The role of keystone species can also be more subtle, modifying habitats to create new niches or linking trophic systems by maintaining key resources. Birds may participate in many of these functions. Although few avian species have been identified as keystones *per se*, their role in community function as a whole is substantial (Şekercioğlu 2006). For example, Daily et al. (1993) studied the keystone role of the red-naped sapsucker (*Sphyrapicus nuchalis*) within subalpine Rocky Mountain communities and found that these birds are an integral part of a keystone complex, playing a dual role in which they both create food resources for numerous other sap-feeding species and produce breeding habitats for secondary cavity nesters.

Community theory has traditionally focused on a single spatial scale, assuming that local communities were closed and isolated and that populations interacted directly to alter each other's birth and death rates (Leibold et al. 2004). However, some species interactions are indirect and occur across a network of local communities such that they can affect large-scale colonization and/or extinction probabilities (e.g., source-sink effects). For example, individual lakes in a region might be considered to have fairly independent and isolated local fish communities, but some of their avian predators may have demographic rates that are regulated over larger spatial scales, involving sets of lakes; the community dynamics of fish within a single lake may thus depend on the presence and abundance of a piscivorous bird species, but the reverse may not be true (Leibold et al. 2004). Such observations led scientists to develop the concept of the '**metacommunity**,' that is, a set of local communities linked by the dispersal of potentially interacting species (Wilson 1992). Birds, with their capacity to move quickly over long distances, are essential components of these metacommunities, connecting local communities across different spatial scales (Figure 8.1).

8.2.2 Parasites in communities

Although often overlooked, parasites are ubiquitous and active members of all natural communities, playing a major role in shaping community structure and function at different spatial scales (Freeland 1983; Minchella and Scott 1991; Poulin 1999; Price et al. 1986). By definition, a **parasite** has a negative impact on the host individual. However, it is the relative virulence and transmissibility of a parasite that will define its overall impact on other members of the community (e.g., McMahon et al. 2016).

Parasites can have direct, debilitating effects on a host's fitness (see Chapter 6), resulting in its extinction from a community or changing its phenotype and thereby modifying the nature of local interactions (Poulin 1999; Thomas et al. 2000). Raptors infected with avian poxvirus may, for instance, exhibit reduced activity and/or food intake due to the weakening effects of infection, releasing predation pressure on the prey community (Morishita et al. 1997). Infected prey may, in contrast, become more readily available to predators, improving predator fitness. Pacific killifish (*Fundulus parvipinnis*) infected by the trematode *Euhaplorchis californiensis* are, for example, are more susceptible to predation by piscivorous birds than uninfected fish (Lafferty and Morris 1996). However, in this case, the fitness advantage afforded by reduced foraging costs could be offset by the costs of parasitism because increasing the susceptibility of parasitized fish will result in higher infection loads in the avian predator, the parasite's definitive host. Parasites can also alter the dispersal propensity of their hosts, either reducing it due to poor body condition (Sánchez et al. 2018) or increasing it to escape parasitism. Indeed, parasites that lower the reproductive success of their hosts can provoke host dispersal to new locations for subsequent reproduction, drastically changing immigration and emigration rates within the metacommunity; this type of dynamic has been documented in several bird species (Boulinier et al. 2001; Brown and Brown 1996; Heeb et al. 1999). For

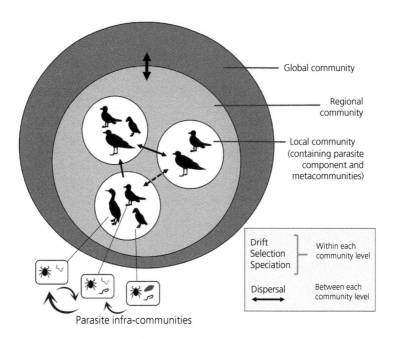

Parasite infra-communities

Figure 8.1 Structure of natural communities and general processes that characterize different levels of biological organization (after Vellend 2010). Communities can form at different levels of biological organization, starting with the community of parasites found within a single host individual (infra-community) and extending to regional and global communities (metacommunities) of different interacting species. At each level, the processes of ecological drift, selection, and speciation will occur. Dispersal links communities across scales, defining their distinctness, and can vary in intensity and direction. The global species pool and nature of dispersal will determine which biological units arrive at local scales, where intrinsic biotic and abiotic processes will then act to determine final community composition (Pearson et al. 2018).

example, high infestation levels of nest ectoparasites have been suggested to cause colony abandonment in several colonially breeding bird species (Duffy 1983; Feare 1976; King et al. 1977). Brown and Brown (1992) used monitoring data to show that fledgling cliff swallows (*Petrochel idon pyrrhonota*) from nests with heavy infestations of hematophagous nest parasites including fleas (*Ceratophyllus celsus*) and swallow bugs (*Oeciacus vicarius*) had a lower probability of recruiting to the natal colony than fledglings from uninfested nests. Similarly, in black-legged kittiwakes (*Rissa tridactyla*), reproductive success and breeding philopatry were found to be significantly lower when nests were heavily infested by the seabird tick *Ixodes uriae* (Boulinier et al. 2001).

Interspecific differences in infection and virulence can also mediate outcomes of interspecific competition among species sharing similar ecological niches within a community. Early experimental studies on competing species of flour beetles

(*Tribolium* spp.) demonstrated, for example, how the presence of a shared pathogen (*Adelina tribolii*) could invert the results of resource competition, leading to the extinction of the competitive dominant species (Park 1948). **Apparent competition** can also occur in communities, where indirect interactions between host species that do not directly compete arise because of a shared parasite (Holt and Bonsall 2017). In such cases, an overall increase in parasite abundance can cause shifts in community structure and function because of asymmetrical effects of the parasite on different host species (Bonsall and Hassell 1997). Tompkins et al. (2000) found, for example, that declines in grey partridge (*Perdix perdix*) populations in the UK were associated with parasite-mediated apparent competition with the ring-necked pheasant (*Phasianus colchicus*). Pheasants carry heavy loads of the intestinal nematode *Heterakis gallinarum* whose eggs are released into the environment. Partridge that become infected with these parasites while feeding the next season suffer high negative fitness consequences (Tompkins et al. 2000). The observed negative correlation in the abundance of the two avian species in this case was not due to direct competition for resources but rather to the shared enemy that had stronger virulence in partridge compared to pheasants.

Pathogens can also promote species coexistence by regulating the relative abundance of dominant species. This has been experimentally demonstrated in plant communities where parasite removal can result in reduced community diversity, both in terms of evenness and species richness (e.g., Grewell 2008). While community-level impacts of infection of an avian species remain undocumented, birds and the entire community can be directly impacted by the cascading effects of an infected keystone species (Getz 2009; Selakovic et al. 2014). For example, the Spanish Imperial eagle (*Aquila adalberti*), an apex predator, has suffered severe fecundity declines (–45.5%) since 2011 due to the emergence of a novel variant of the rabbit hemorrhagic disease virus (RHDV) in southern Europe that drastically reduced the population size of European rabbits (*Oryctolagus cuniculus*), a keystone species of Iberian-Mediterranean ecosystems. Changes in the demographics of this avian predator are now, in turn, expected to alter the top-down regulation of other members of this trophic community (Monterroso et al. 2016).

As is evident throughout this book, birds host a rich diversity of parasites that can be more or less shared with different community members depending on their host range (strict specialists to broad generalists) and life cycle (from direct one-host to complex three- to four-host transmission cycles). Given this diversity and the ubiquity of avian species in natural communities, birds can play a major role in determining the presence, persistence, and impact of parasites and pathogens on overall community structure and function (Figure 8.1).

8.3 Community diversity and parasite/pathogen circulation

8.3.1 Interspecific host heterogeneity

Some hosts are better than others. Indeed, we know that only a few individuals within a population

typically give rise to the majority of secondary infections; these individuals are sometimes referred to as **superspreaders** (Lloyd-Smith et al. 2005; Woolhouse et al. 1997; see also Chapters 3, 4, and 7). The same notion can be applied at the community level, where different species are more or less competent reservoirs or transmitters of a given parasite. Some species may be completely resistant to infection, whereas others are highly susceptible and infectious. A parasite may be highly virulent for one host species, potentially eliminating it from the community before transmission can take place, whereas another host may be highly tolerant to infection, even at high infection intensities, maintaining the infection locally (i.e., **reservoir hosts**). This interspecific variation will greatly modify local disease dynamics in the community.

In the presence of a highly competent host species for the parasite, amplification should occur. The overall prevalence of infection will increase either because the parasite is better able to exploit this host species and produce more secondary infections (i.e., amplifying hosts) or because this host species contributes to the overall pool of competent individuals such that the threshold density of hosts for parasite persistence is reached (i.e., the critical community size; Bartlett 1960; Begon 2008). In contrast, a low-competent host species (i.e., one that cannot maintain an infection when exposed to a parasite) can reduce overall parasite prevalence by reducing the density of susceptible hosts below the persistence threshold and/or by absorbing transmission events (i.e., **dilution hosts**). This latter effect has been termed the 'dilution effect' and was primarily conceived to understand the epidemiology of vector-borne diseases, such as Lyme disease (Ostfeld and Keesing 2000a; Ostfeld and Keesing 2000b). The dilution effect hypothesis (DEH) has received much attention over the last decade and is still considered controversial with respect to its generality (Box 8.1).

The circulation of West Nile virus (WNV) provides an excellent example of the importance of interspecific host heterogeneity within avian communities for determining disease risk. WNV is a mosquito-borne pathogen for which birds are the primary reservoir hosts. Humans, or other incidental host groups, become infected via **spillover** when

Box 8.1 The dilution effect hypothesis

The dilution effect hypothesis (DEH) was initially conceived to explain the link between biodiversity and human disease risk of vector-borne zoonoses, particularly Lyme disease (Ostfeld and Keesing 2000a; Ostfeld and Keesing 2000b), but then was extended more generally across disease systems (Johnson and Thieltges 2010; Keesing et al. 2006). Under this model, high biodiversity is good because it results in lower disease prevalence. The idea is logical: if there is variation in pathogen reservoir competence among species within a community, competent hosts will be diluted among non-competent hosts when biodiversity is high, reducing transmission success of the pathogen and thus the overall density of infected individuals (Ostfeld and Keesing 2000a; Ostfeld and Keesing 2000b).

However, to work, this hypothesis relies on several key assumptions: (1) variation in competence for a pathogen occurs among species within the community, (2) the most competent species are present in low-diversity communities, and (3) transmission is only horizontal. In the specific case of vector-borne disease, vectors are also assumed to be generalists in terms of the host species they exploit, with no vertical transmission of the pathogen.

Some empirical evidence for this model exists (e.g., Allan et al. 2003; Civitello et al. 2015; Ezenwa et al. 2006; Haas et al. 2011; Johnson et al. 2013b; LoGiudice et al. 2003), but criticisms of its universality are numerous (Randolph and Dobson 2012; Salkeld et al. 2013). For example, it remains difficult to distinguish between true dilution and a simple reduction in reservoir density when additional species are added to a community (Begon 2008); the end result is the same but the mechanism is different (Keesing et al. 2006). Likewise, with vector-borne pathogens, the DEH assumes that vector abundance remains the same with increases in host biodiversity (i.e., there is no vector amplification) (Randolph and Dobson 2012). If vector abundance increases, the number of infected vectors in the community may remain the same, even if infection prevalence *per se* is lower. Finally, several studies have now shown that host species are not used indiscriminately by vectors and that some species within communities contribute disproportionately to vector blood-meals (e.g., Kilpatrick et al. 2006; Montgomery et al. 2011; see also Section 8.4). If strong host preferences occur within a community, the applicability of the DEH is reduced because, regardless of local species diversity, vectors will use the preferred host (McCoy et al. 2013). A recent meta-analysis suggests that the DEH may be strongly scale-dependent, working well at local spatial scales but becoming weaker over larger areas (Halliday and Rohr 2019).

they come in contact with an infected vector, different species of *Culex* mosquitoes. Field studies that have examined infection rates in relation to avian diversity have found a negative correlation between the species richness of non-passerine birds in the avian community and the prevalence and density of infected *Culex* mosquitoes (Ezenwa et al. 2006). This correlation did not occur for passerine species, suggesting that non-passerines, poor hosts for the virus, act as alternative blood-meal sources for mosquitoes, reducing contact rates with highly competent passerine hosts and thus densities of infected vectors (DIV). The reduced DIV then translates to lower global disease risk for humans; areas with a high diversity of non-passerine species had lower human WNV disease incidence (Ezenwa et al. 2006). Other studies have identified specific passerine species whose presence significantly amplifies DIV due to both high competence for the virus and a high rate of exploitation by mosquitoes (Kilpatrick

et al. 2006; Simpson et al. 2012). For example, American robins (*Turdus migratorius*) have been shown to feed a disproportionate number of mosquitoes compared to other avian species within a given community (Apperson et al. 2004; Montgomery et al. 2011) and to increase DIV (Kilpatrick et al. 2006; Hamer et al. 2009; Figure 8.2). In regions where robins are less common, other passerine species are more likely to be used as blood-meal sources, such as the northern mockingbird (*Mimus polyglottos*) in the southeastern US (Apperson et al. 2004) and the house finch (*Haemorhous mexicanus*) in the western US (Molaei et al. 2010; Montgomery et al. 2011).

When attempting to predict changes in parasite prevalence or disease emergence within communities, additional complexity arises from heterogeneities in the environment, among individuals within a host and/or vector species, and from the level of host specificity and heterogeneity in the parasites

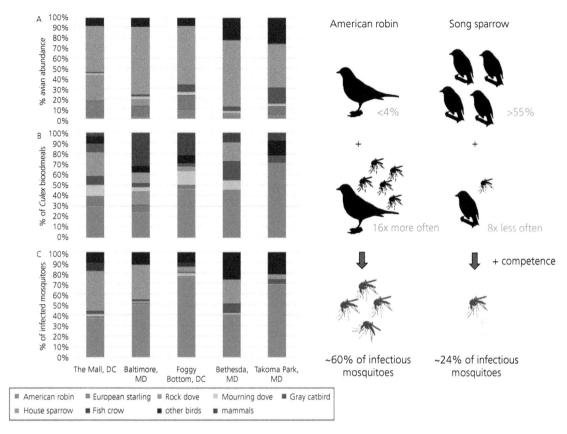

Figure 8.2 Relationship between avian community composition, mosquito feeding preference, and amplification fraction of WNV infections. Among the five studied sites (x-axis), American robins made up less than 4% of the avian community (A) but accounted for over 40% of mosquito blood-meals (B) such that mosquitoes fed on robins 16× more often than expected than if they had fed randomly. Taken together with species-specific competence estimated from the lab, viremic robins were estimated to produce approximately 60% of infected mosquitoes, compared to only about 24% in dominant song sparrows (C), if initial infection rates were similar among avian species within the community. (From Kilpatrick et al. 2006.)

themselves (Jones et al. 2018). Unravelling the intricacies of these systems thus demands careful comparative analyses across a wide range of systems and spatial-temporal scales (Tompkins et al. 2011) and has become a pressing challenge, given ongoing rapid changes in natural communities.

8.3.2 Changing biodiversity

We are living in a period of rapid and unprecedented changes to biodiversity (Tittensor et al. 2014). What predictions can we make for the circulation of parasitic organisms if community diversity increases or decreases? What kinds of impacts can we expect from invasive species in terms of disease

emergence? What role might non-host species have within communities? What are the mechanisms underlying the link between biodiversity and disease prevalence? These diverse questions have been the focus of some research effort, but much is left to do (Gibb et al. 2020; Halsey 2019).

Due to the complex nature of interactions within natural communities and the high diversity of parasitic species and life cycles (see Section 8.2.2), the relationship between host diversity and disease risk can vary in both intensity and direction. At a global scale, we expect that increased host biodiversity will increase parasite diversity and exposure (Jones et al. 2008). However, at more local scales, the effect of adding biodiversity can be more variable (Salkeld et al. 2013) and will depend on the nature of ecological interactions within the community (Roche et al. 2012). If a highly competent host for either an existing pathogen or vector is added to a community, transmission risk may increase (Ogden and Tsao 2009; Roche et al. 2013) and disease amplification occurs. Although intuitive, examples of this phenomenon in the literature remain surprisingly scarce. The closest example in birds comes from cliff swallow colonies in the American Midwest invaded by the European house sparrow (*Passer domesticus*). In the presence of house sparrows, both infestation levels by swallow bugs and the prevalence of Buggy Creek virus (Togaviridae: Alphavirus), a pathogen transmitted by the bugs, increases (O'Brien et al. 2011). This increase is seemingly due to changes in prevalence and infection intensity in house sparrows as these novel hosts are more susceptible to both the vector and the pathogen than native cliff swallows (Fassbinder-Orth et al. 2013). More detailed studies are now required to assess the impact of the invading host on infection levels in other native hosts within these communities.

In the case that a weakly competent species is added to a community, rates of transmission may be reduced via different 'dilution' mechanisms (see Box 8.1). For example, amphibian host diversity has been shown to inhibit transmission of the trematode *Ribeiroia ondatrae*, a parasite that uses aquatic snails as first intermediate hosts, fish or amphibians as second immediate hosts, and mammals or birds as final hosts (Johnson et al. 2013a, 2013b). In species-rich communities, realized transmission can be reduced by almost 80% due to the addition of less-competent amphibian species. Few examples of a dilution effect have been demonstrated with respect to avian diversity within communities, with the exception of WNV. As outlined above, the presence of incompetent non-passerine species reduces the density of infectious mosquitoes within an area (Ezenwa et al. 2006). Swaddle and Calos (2008) analysed data on avian species richness and human incidence of WNV disease from a series of paired counties in the eastern US and demonstrated that higher avian diversity was clearly associated with a lower incidence of human cases. However, these authors did not identify the precise mechanism at play. Examples of the impact of adding weakly

competent or incompetent hosts to a community are still lacking for other avian disease/parasite systems.

Even species that do not participate directly in transmission may alter infection rates within communities. Non-host species can, for example, prey directly on free-living parasite stages or can destroy them during their normal activities. For instance, free-swimming cercariae are readily consumed by non-host predators such as shrimp, crabs, barnacles, and oysters (Welsh et al. 2017), affecting trematode transmission to final avian hosts (e.g., bivalve-consuming species such as gulls [*Larus* spp.] or common eiders [*Somateria mollissima*]). Likewise, snails can kill encysted metacercariae while grazing on aquatic vegetation, removing them from the transmission cycle (Vielma et al. 2019). Other species may alter habitat conditions for competent host species and thus may increase or decrease parasite prevalence; for example, introduced aquatic plants can alter habitat suitability for foraging shorebirds, final hosts for many parasites with complex life cycles (Crooks 2002).

With increasing global movements of flora and fauna, the number of invasive species in natural communities is also increasing. Not only can these novel species have major impacts on community structure and function through direct competitive and trophic interactions as outlined above, but also they frequently arrive in the invaded zone with their cortege of parasites and pathogens from the native range (Dunn and Hatcher 2015). These parasites may then spill over to native community members, directly affecting their survival and reproduction and/or altering community-level interactions. Indeed, by modifying host–host interactions (including competition and predation), these parasites may be key factors determining the success of an invasion. Martin-Albarracin et al. (2015) carried out a global scale meta-analysis on the ecological impacts of invasions by non-native bird species and found that disease transmission to native fauna had one of the highest impacts on native community structure. Surprisingly, disease transmission has been relatively less studied than other effects of invasive species.

The introduction of non-native birds to the Hawaiian islands demonstrates well how invasive species, in interaction with parasites, can impact native fauna. During the last century, over 50% of the native avifauna has been lost from this archipelago due to a variety of anthropogenic causes (Atkinson et al. 2005). Exotic birds started to be introduced to the Hawaiian islands in the 1800s, and the bird-associated mosquito *Culex quinquefasciatus* arrived shortly thereafter (Warner 1968). *Plasmodium relictum*, a blood parasite transmitted by mosquitoes that causes avian malaria, likely arrived at the same time but was not considered a factor in declines until the late 1920s when a sufficient number of introduced avian species were established to maintain the pathogen locally (Van Riper et al. 1986). This parasite is particularly virulent for endemic bird species, causing acute mortality, but is relatively well tolerated by many of the introduced species (Van Riper et al. 1986). Avian malaria has been identified as a major cause of mortality in low- to mid-altitude regions where mosquitoes are present (Atkinson et al. 2000). Current climate-related changes are now enabling the expansion of the mosquito vector to higher altitudes and thus the spread of the parasite to zones previously lacking it (Atkinson et al. 2005), threatening the last refuges for certain native species, notably Hawaiian honeycreepers, an emblematic clade of endemic passerines.

8.3.3 Parasite communities within avian communities

Within any local community, there exists a series of hierarchical parasite communities that can impact other community component species through diverse effects. These parasite communities start at the scale of the host individual—the parasite infra-community—affecting individual host fitness and go all the way up to the level of the global metacommunity, composed of distinct parasite communities linked through dispersal (Figure 8.1). These different nested communities and their impacts are outlined in the following sections.

8.3.3.1 Parasite infra-community

The individual is increasingly recognized as being its own ecosystem, hosting a range of parasitic, commensal, and mutualistic microorganisms. These internal communities, or infra-communities, which

include all infecting organisms (or co-infections), can have an array of important effects on the host via the nature of their interactions—facilitative, neutral, or antagonistic. Indeed, this internal community has the same basic traits as the broader community in which it is nested, including resources (from the host), consumers (different types of parasites), competitors (different parasites using the same resource), and predators (host immune responses) (Pedersen and Fenton 2007). The diversity and composition of the infra-community will affect how a host responds to other biotic and abiotic factors in the environment, whether infection (invasion) by other parasitic organisms is successful, and, ultimately, this will determine individual host fitness (e.g., Seabloom et al. 2015; Telfer et al. 2010). The key difference with the broader free-living community is that, in infra-communities, resources are directly linked to predation because the host supplies both ends of the trophic chain (resources and predation). By considering the within-host environment as a community, the mechanisms defining interactions can be elucidated, variation among individuals can be better explained, and *a priori* hypotheses on the response of individuals to different factors can be predicted (Graham 2008). Likewise, as individual hosts represent well-defined and self-contained habitats, they can be considered as suitable replicate units for testing general theories in community ecology (Poulin 1999; Rynkiewicz et al. 2015). Although several studies have examined interactions among ectoparasites infecting individual avian hosts (Choe and Kim 1988; Clayton et al. 2015; Matthews et al. 2018; Stefan et al. 2015), for logistical reasons relatively few have focused on internal parasites (but see Biard et al. 2015; Leung and Koprivnikar 2016; Stock and Holmes 1988). Likewise, the link between diversity at the infra-community level and broader community dynamics still requires specific consideration.

8.3.3.2 Parasite component community

Moving up a level of biological organization, the parasite component community includes all infra-communities, that is, all parasite species found within a subset of host species or within a particular abiotic microhabitat (Bush et al. 1997). When researchers consider parasite communities, studies are often focused at this scale, examining prevalence of diverse parasites and pathogens at a given place and time for a focal host species/group. For example, a component community would include all parasites infesting/infecting black-legged kittiwakes within a specific seabird colony (see Section 8.4.1).

As outlined above, the infra-community of each individual host can dictate the host's relative quality and life history traits. In contrast, it is the component community that will define host exposure potential to a given parasite. The evenness of the parasite component community among hosts will depend on the exposure rate, the nature of parasite–parasite interactions (i.e., whether current infections facilitate or constrain infection by other parasites; Telfer et al. 2010), and the degree of host susceptibility to infection. For example, similarity in helminth communities of ring-billed gulls (*Larus delawarensis*) shows strong age structure, with high similarity among chicks within populations that declines with age (Locke et al. 2012). In addition to host-associated factors such as habitat use and diet (Gutiérrez et al. 2017; Leung and Koprivnikar 2016), host exposure rates will depend on the parasite life cycle (Poulin 1997). For example, parasites with environmental stages or active intermediate life stages should have more even distributions within the host population than parasites that rely on direct contact of territorial species. The evenness of the parasite community will condition its stability and the efficacy of control measures. Indeed, component communities of helminths in gulls were found to be highly conserved among years, after accounting for seasonal changes (Levy 1997). As for all communities, the removal of a single parasite species can destabilize relationships within a community (Ives and Cardinale 2004), potentially increasing the virulence and dissemination of existing parasites or facilitating invasion by novel parasites from outside the component community (Johnson et al. 2013a).

8.3.3.3 Parasite metacommunity (or supra-community)

A pressing question in disease ecology concerns how host and parasite assemblages interact to determine transmission and disease severity within ecological communities (Rigaud et al. 2010). To address these aspects, we need to increase the scale of biological organization to the metacommunity level or parasite supra-community (Bush et al. 1997). The parasite metacommunity includes all parasites from all hosts (or microhabitats) at a defined spatial scale, such as all parasite species (including microparasites, helminths, and ectoparasites) from different passerine birds within a farm. A complete understanding of transmission dynamics often demands detailed information on both host and parasite communities. However, to be tractable, studies at this level of biological organization need to specify the spatial scale and the limits of the host community being considered. For example, a parasite metacommunity could include all parasites infecting different avian species present within a single seabird colony (see Section 8.4.1). This could also be extended to include all potential intermediate hosts and free-living stages within the same geographic region. At this level of biological organization, it is important to keep in mind that parasite richness is not directly equivalent to disease risk, because parasites vary in virulence and relative abundance, hosts vary in competence, and the presence of one parasite may alter the propensity for subsequent infections (Telfer et al. 2010). Thus, key aspects to consider at the parasite metacommunity scale are the degree to which parasites are shared among local host types, such as the degree of host specialization, the potential for host shifts, and the trophic links that occur within the host community.

8.3.4 Modified communities

Urbanization can be defined as concentrated human presence in residential and industrial settings and its associated effects (Chace and Walsh 2006; Marzluff 2001). For the purpose of ecological studies, urban centers have been quantified as containing more than 2,500 people (Dumouchel 1975 as cited by Chace and Walsh 2006). Urbanization of the landscape has had and is continuing to have a dramatic effect on the floral and faunal communities that previously occupied these spaces (see Chapter 9). It is generally associated with a reduction in overall species richness and diversity

(Beissinger and Osborne 1982; Marzluff 2001) due to direct habitat loss or fragmentation of intact habitat such that sufficient breeding habitat or food resources are no longer available for some species. For example, in a review of studies on bird responses to human settlement, Marzluff (2001) showed that, although bird density increased in response to urbanization, species richness and evenness decreased, largely due to the loss of cavity and interior-nesting species and raptors. Reduced diversity can also be associated with the increased presence of a few species preadapted to exploit human-associated resources; these are species that outcompete and/or dominate urbanized areas. Several factors determine which species can coexist and flourish with humans: for example, the presence and patch size of remnant (native) habitat, the presence of exotics—competitors, predators, or parasites—the structure and composition of the vegetation, supplementary feeding by humans, or the presence of pollutants (reviewed in Chace and Walsh 2006).

In urbanized communities, several avian species are known to thrive, particularly omnivores, granivores, and artificial cavity-nesting species like house sparrows (*Passer domesticus*), European starlings (*Sturnus vulgaris*), Canada geese (*Branta canadensis*), or yellow-legged gulls (*Larus michahellis*). These birds can reach very high artificial densities, increasing the transmission potential for parasites and maintaining them within the community. If virulence is sufficiently low in these common host species, the hosted parasites can spill over into other less well-adapted host species, potentially impacting their population sizes and viability and thus altering community structure. Urbanized communities also are typically more homogeneous geographically (Chace and Walsh 2006), meaning that parasites present in one locality may rapidly disseminate and establish in new areas with bird dispersal and migration. In addition to the risk posed to more vulnerable species in the community, the presence of certain parasites may be of concern for human and domestic animal health. For example, high densities of birds at backyard feeders can lead to increased transmission of diverse pathogens (Brobey et al. 2017), including *Mycoplasma gallisepticum*, a bacterium responsible for severe conjunctivitis in some avian species (Adelman et al. 2015; Dhondt et al.

2007; see also Chapters 3, 4, 5, and 7). This bacterium emerged in songbirds in the 1990s after a host switch from domestic poultry (Fischer et al. 1997). It causes substantial mortality in its primary wild host, the house finch, but is also carried by other birds with variable clinical symptoms (Farmer et al. 2005). Results of winter disease surveys from backyard feeders in northeastern North America found that higher disease prevalence in house finches correlated with the increased presence of alternative host species, suggesting that community composition in urban zones plays a direct role in disease risk for finches (States et al. 2009) and that this enlarged community may be a source of infection for poultry farms (Sawicka et al. 2020).

Wild birds have also been directly implicated in the dissemination of agents such as influenza viruses, *Salmonella*, *Campylobacter*, and WNV to domesticated animals (e.g., Brobey et al. 2017; Craven et al. 2000; Ezenwa et al. 2006; Lycett et al. 2016; see also Chapter 11). Artificial feeding stations used in animal production may act in the same way as backyard feeders, attracting a high diversity of wild birds and increasing pathogen transmission, both within the wild bird community and at the wild–domestic interface (Caron et al. 2010). Bird droppings left in urbanized areas can also contain pathogenic bacteria or viruses that may be of public health concern (Feare et al. 1999). Indeed, reviews of available literature have implicated birds in the global epidemiology of numerous pathogenic bacterial species for humans and have documented cases in which certain species maintain and disseminate antibiotic-resistant strains (e.g., Benskin et al. 2009; Tsiodras et al. 2008; Vittecoq et al. 2017; Zurfluh et al. 2019; see Chapter 11). Finally, landscape modifications that change the vegetation structure and composition can cascade down to the local community. For example, floristic changes due to modifications in agricultural practices in northern England resulted in an increase in local tick populations and an associated increase in the transmission of the tick-borne virus that causes looping ill disease in wild birds. The presence of this pathogen contributed to declines of a dominant avian species in the landscape and one of particular human interest, the red grouse (*Lagopus lagopus scoticus*), and thus to a shift in the structure of the local avian community (Dobson and Hudson 1986).

8.4 An example of Lyme borreliosis in seabird communities

Many of the aspects related to community-level interactions and disease ecology mentioned in this chapter and elsewhere in this book can be illustrated by the example of Lyme disease bacteria that circulate naturally in polar seabird communities.

8.4.1 The biological system

8.4.1.1 Seabird communities

Seabirds are the most numerous, diverse, and widespread of marine megafauna (Hoekstra et al. 2010). Although they include an eclectic group from an evolutionary perspective, all species are adapted to life within the marine environment, typically feeding at sea. This group includes gulls, terns, skimmers, skuas, auks, and selected phalaropes (Charadriiformes), tropicbirds (Phaethontiformes), penguins (Sphenisciformes), tubenoses like albatrosses and shearwaters (Procellariiformes), cormorants, frigatebirds, boobies, gannets (Suliformes), and pelicans (Pelecaniformes). Species distributions vary considerably, with some species found around the globe (e.g., great cormorant, *Phalacrocorax carbo*) and others limited to a single island archipelago (e.g., Balearic shearwater, *Puffinus mauretanicus*). Seabirds are also among the most threatened groups of birds (Croxall et al. 2012) and many populations are in decline (Paleczny et al. 2015), making this group particularly important to consider when examining marine ecosystem structure and the potential implications of disease in seabird communities (Provencher et al. 2019).

The greatest majority of seabirds are colonial breeders, meaning that they aggregate in large numbers (hundreds to hundreds of thousands of pairs) for several months per year in order to reproduce. The location of these colonies tends to be stable over long periods and birds often return to the same colony (and sometimes to the exact same nest site) year after year to breed (i.e., breeding site fidelity; Furness and Monaghan 1987). The high density of individuals within colonies and their predictable seasonal occurrence provide excellent conditions for the maintenance and transmission of parasites and pathogens (McCoy et al. 2016). In addition to colonial breeding, seabirds are long-lived; once reaching maturity at 3–6 years old, these birds can breed for decades (Furness and Monaghan 1987). Given this longevity, chronic infections of non-lethal parasites may be maintained and transmitted over long periods (e.g., *Borrelia* spp. bacteria; Gylfe et al. 2000). Similarly, with the regular reuse of nesting sites, temporary ectoparasites such as fleas, ticks, and flies can build up large local populations. Indeed, infestation levels can increase over time until bird reproductive success becomes so low that birds abandon the colony (e.g., Danchin 1992; Duffy 1983) and prospect for new breeding locations (Boulinier et al. 2016).

From a community ecology perspective, marine birds are interesting biological models because both breeding colonies and foraging hotspots at sea are frequently heterospecific (Figure 8.3), with different species breeding or feeding in complete sympatry. Indeed, seabirds use a variety of terrestrial habitats, with some species nesting in trees (if available), others on the ground (with or without nests) or on cliff ledges, and still others using underground burrows or rock crevices. Seabirds also frequently interact at sea while foraging in response to the aggregated, but temporally variable, position of their marine prey (Fauchald et al. 2000). Although each species can occupy specific habitat niches and/or specialize on particular prey items, intra- and interspecific competition for breeding sites and food can be fierce (Ballance et al. 1997; Kokko et al. 2004; Oro et al. 2009). The presence of multiple interacting species can result in the interspecific transmission of parasites and pathogens that can then favor host shifts, alter the outcome of interspecific interactions (e.g., competition, predation), and ultimately modify the circulation of parasitic agents at different spatial scales (e.g., Dabert et al. 2015; Dietrich et al. 2014a; Gómez-Díaz et al. 2007; Gómez-Díaz et al. 2010; Stefan et al. 2018). These modifications can then trickle down to alter the dynamics and composition of species at different levels of community organization, from the parasite infra-community to the global metacommunity (Figure 8.1). Finally, in response to their particular life history traits, seabirds host an extremely wide array of parasites and pathogens—temporary and

Figure 8.3 Community-level interactions among seabird communities and their ectoparasites. (A) Heterospecific colony of seabirds on Hornoya, an isolated island in northern Norway. Five different species breed in sympatry at this site and are all exploited by the hard tick *Ixodes uriae*. (B) Two different seabird species found on the small island of Ile de Lys (Scattered Islands) in the western Indian Ocean: sooty terns (*Onychoprion fuscatus*) and lesser noddies (*Anous tenuirostris*). In colonies of this region, seabirds are frequently exploited by both hard (*Amblyomma loculosum*, left inset) and soft (*Ornithodoros capensis*, right inset) ticks. (Photo credits: (A) T. Boulinier; inset: K.D. McCoy; (B) K.D. McCoy; left inset: K.D. McCoy; right inset: P. Landmann.)

permanent ectoparasites, macroparasites obtained via trophic transmission, and diverse viral, bacterial, and protozoan pathogens transmitted both directly and indirectly by different vector organisms. A broad list of these infecting organisms can be found in Khan et al. (2019).

8.4.1.2 Ticks and the circulation of Lyme borreliosis-causing agents

Among seabird ectoparasites, ticks are among the most common and have the greatest impact on seabird body condition and reproductive success (Boulinier and Danchin 1996; Duffy 1983; Feare 1976). These ticks can include both hard (Ixodidae) and soft (Argasidae) tick species (Figure 8.3). Hard ticks are found in colonies around the globe, from continental Antarctica to the high Arctic, whereas soft ticks are limited to tropical and temperate regions (Dietrich et al. 2011). The greatest part of the tick life cycle takes place in the substrate of the host colony, with host contact being limited to the time required for the blood-meal. In the case of hard ticks, infestation occurs during seabird reproduction when birds remain in the same general location for prolonged periods of time. A single, long blood-meal of several days is taken in each of the three active life stages of these ticks (larva, nymph, and adult female). In the case of soft ticks, blood-meals are short (minutes to a few hours) and thus can be

taken any time a host is available locally but generally happen at night when the host is immobile (Vial 2009). To compensate for the limited size of the blood-meal, soft ticks feed several times in each life stage and nymphs will undergo several molts before developing into the adult life stage. Both groups of ticks are known to carry diverse microorganisms, including human pathogens (e.g., Dietrich et al. 2011; Dietrich et al. 2014b; Wilkinson et al. 2014), but the full diversity and pathogenicity of these infectious agents for the birds are largely unknown.

In the polar regions of the world, seabird colonies are frequently infested by hard ticks of the genus *Ixodes*, such as *I. kergulensis*, *I. kohlsi*, *I. eudyptidis*, *I. philipi*, *I. auritulus*, and *I. uriae*. Among these species, *I. uriae* is the most widely distributed, found in polar colonies of both hemispheres and on a range of over 50 different seabird species (Dietrich et al. 2011). Different infectious agents have been described circulating via this tick species (e.g., Great Island virus: Nunn et al. 2006; *Rickettsia*, *Rickettsiella*, *Coxiella*, and *Spiroplasma* bacteria: Duron et al. 2016; Duron et al. 2017), but among the most prevalent and widespread are bacteria of the *Borrelia burgdorferi* sensu lato (Bbsl) complex responsible for human Lyme disease (Dietrich et al. 2008; Duneau et al. 2008; Gómez-Díaz et al. 2011; Munro et al. 2019; Olsen et al. 1993).

Lyme disease is the most common vector-borne zoonosis in temperate regions of the northern hemisphere (Centers for Disease Control and Prevention 2019; Sykes and Makiello 2017). The Bbsl complex currently includes over 20 genospecies, seven of which are thought to be pathogenic for humans (Margos et al. 2011; Stanek and Reiter 2011). The obligate enzootic life cycle of these bacteria involves *Ixodes* ticks and a variety of vertebrate hosts, including mammals, reptiles, and birds (Gray et al. 2002). Despite a wide host/vector spectrum, some degree of ecological specialization exists among the different species of the complex, particularly towards the vertebrate host (Kurtenbach et al. 2006). Although rarely considered in the global epidemiology of the disease, at least four genospecies have been reported circulating in seabirds via the tick *I. uriae* (Dietrich et al. 2008; Duneau et al. 2008), the most prevalent by far being *B. garinii*, a pathogenic species generally associated with forest birds (Hanincova et al. 2003). Indeed, this genospecies represents > 80% of tick infections in the marine system (Duneau et al. 2008; Gylfe et al. 1999; Munro et al. 2019). Serological studies on diverse seabird species show that exposure rates can be high within colonies (> 90%) but can vary strongly among seabird species (Lobato et al. 2011; Staszewski et al. 2008). Direct genetic typing of bacteria in sampled ticks confirms high prevalence in the vector populations (up to 40% on average in some colonies; Gómez-Díaz et al. 2010). Although difficult to test directly in natural populations, studies combining exposure rates and capture-recapture modeling have suggested that infection has little impact on adult survival (Chambert et al. 2012). However, as yet, potential effects of infection on juvenile seabirds have not been assessed. Understanding the epidemiology of Lyme disease in seabird communities, and more globally the potential role of seabirds as disease reservoirs for terrestrial systems, requires a clear comprehension of the infection dynamics within and among colonies in both the different seabird host species and their respective tick populations.

8.4.2 Population structure of *Ixodes uriae* within and among seabird communities

As outlined in Section 8.4.1, *I. uriae* is a widespread generalist tick of marine birds, found in polar seabird communities of both hemispheres and on a wide range of seabird species from across different avian families (Dietrich et al. 2011). As seabird colonies are often heterospecific in nature (Figure 8.3), one would expect that sympatrically breeding species should share the same tick population. Likewise, as this tick only exploits the seabird host during the breeding season, when bird movement is largely limited to the nest site and the foraging area at sea, we would expect the probability of tick dispersal among colonies to be low. Population genetic studies carried out on this tick species at different spatial and temporal scales have falsified these general predictions and can help us better understand the circulation of infectious agents in this system.

Initial field observations of tick infestation dynamics on sympatrically breeding black-legged kittiwakes and Atlantic puffins (*Fratercula arctica*) suggested that two distinct tick populations, one associated with each seabird species, occurred within colonies (McCoy et al. 1999). Later population genetic studies employing highly polymorphic microsatellite markers demonstrated that this was indeed the case (McCoy et al. 2001); *I. uriae* structures locally into distinct host races recurrently across the globe (Dietrich et al. 2012; McCoy et al. 2005; Figure 8.4). Experimental studies have since

Figure 8.4 *(Opposite)* The tick *Ixodes uriae* is a regular member of seabird communities in polar regions of both hemispheres. These communities commonly include multiple seabird species. (A) Population genetic studies have shown that ticks have evolved distinct local populations associated with each type of seabird in the local community: tick host races. Here, ticks were sampled from six seabird species in nine heterospecific colonies of the North Pacific and were typed at a series of microsatellite markers. (B) At broader spatial and temporal scales, divergence levels between tick host races varied across the different regions of its distribution (orange dots indicate colony locations) but not in relation to the time since tick colonization. On the figure, colonization order is shown based on 37 tick populations from 18 colony locations around the globe: *I. uriae* first evolved in the southern hemisphere, where it expanded along with its seabird hosts approximately 22 million years ago (mya) (1). Ticks then invaded the northern hemisphere independently via the two oceans, first in the Pacific (2) and later in the Atlantic (3). Although populations in the North Atlantic were among the last colonized, host-associated divergence among sympatric tick populations is strongest, as shown by average estimates of F_{ST} (± SE) between host-associated tick populations within colonies. (Figures modified from (A) Dietrich et al. 2012 and (B) Dietrich et al. 2014a.)

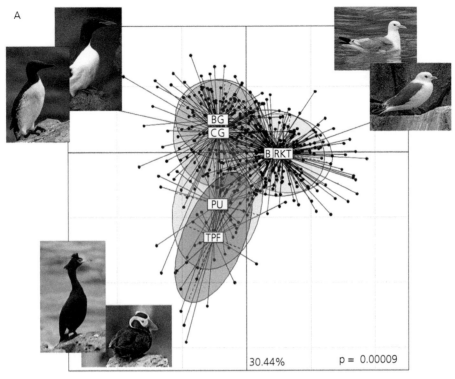

Between-group analysis

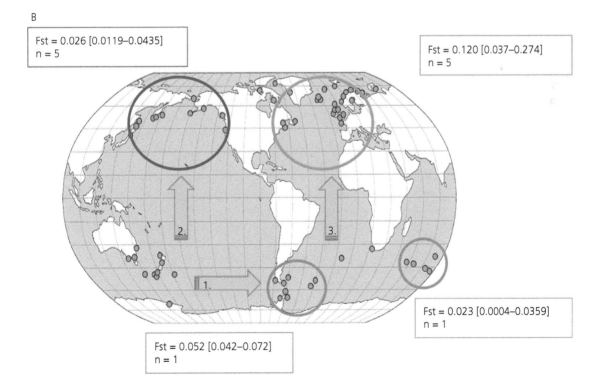

demonstrated that this divergence is not a simple consequence of genetic isolation but that different races are adapted to exploit their particular host type (Dietrich et al. 2014c) and have different morphologies (Dietrich et al. 2013). The formation of these host races is temporally dynamic; distinct races are not present in all colonies and can vary in their degree of divergence in different isolated regions. Interestingly, initial divergence into host races does not seem to lead to full speciation in this system (Dietrich et al. 2014a); more ancient areas of the tick distribution do not show stronger patterns of host-associated divergence than more recently colonized zones (Figure 8.4). The reason for this is currently unknown but may be linked to the relative long-term stability of host populations and/or genetic constraints on adaptation.

In addition to host-associated genetic structure, *I. uriae* populations show variable patterns of geographic structure, patterns that depend on the host species being exploited. For example, ticks exploiting Atlantic puffins in the North Atlantic start to show signatures of population structure when colonies are separated by more than 700 km; that is, it is only when breeding colonies are greater than 700 km apart that tick gene flow is low enough for genetic divergence by drift to occur. For ticks exploiting black-legged kittiwakes or common murres (*Uria aalge*), population genetic structure occurs at a smaller spatial scale, when colonies are approximately 200 km apart (McCoy et al. 2003a). Given the limited movement of birds during the breeding season, when ticks actively feed on birds, tick dispersal likely occurs with juveniles or failed breeders that prospect for new breeding locations (Boulinier et al. 2016). Tracking data from kittiwakes match these predictions: successfully breeding kittiwakes only move between the nest site and foraging areas, whereas failed breeders tend to visit different colonies within an approximate 200-km radius around the initial breeding site (Ponchon et al. 2014). From these results, we expect that the composition and relative stability of parasite component communities will differ across spatial scales according to the seabird host species; that is, parasite communities of puffins should be more homogeneous at large spatial scales compared to those exploiting kittiwakes or murres.

8.4.3 Understanding spatial patterns of Lyme borreliosis within and among avian communities

In the marine cycle of Lyme borreliosis, we know that vector populations are at least partially isolated among different component species of the seabird community and that tick dispersal across colonies varies depending on the seabird species exploited. When a parasite is also a vector, host use conditions not only the parasite's own population dynamics and evolutionary trajectory but also that of the associated microparasite. We now examine patterns of Bbsl structure in marine communities at different spatial scales in light of our knowledge of tick population structure and other heterogeneities that may be present in these communities.

8.4.3.1 Within and among seabird colonies

As outlined in Section 8.3.1, heterogeneity within the host community can mean that different species contribute more or less to the maintenance and transmission of an infectious agent. In the case of Bbsl spirochaetes circulating in marine birds, this seems to be the case, although placing the mechanism in the correct biological compartment is not yet obvious.

Studies that have examined both prevalence of Bbsl in vector populations and exposure rates of kittiwakes have shown that pathogen presence varies at both within and among colony scales (Dietrich et al. 2008; Staszewski et al. 2008). Exposure rates in kittiwakes have been shown to depend on local levels of tick infestation (Gasparini et al. 2001), which may vary due to heterogeneity in individual host susceptibility (Boulinier et al. 1997), the quality and/or topology of the breeding area (McCoy et al. 2003b), and colony history (Danchin 1992). The overall dynamic of a particular breeding area will also alter infection risk. Seabirds such as kittiwakes use public information on local breeding success to select future breeding locations (Boulinier et al. 2008). When success is high, new individuals, with their infracommunity of parasites, will arrive more readily than in patches where breeding success is low.

These patterns become even stronger at the level of the seabird community. As outlined in Section 8.4.2, the tick vector in the marine system has diverged locally into distinct seabird-specific host races. This means that Bbsl harbored by one tick host race will not necessarily be shared with another tick host race, even if these races occur sympatrically (Figure 8.5). Thus, we could expect distinct infection dynamics among different members of the seabird community. Ticks of different sympatric races examined for Bbsl infection have shown that this is indeed the case. Data from three colonies in Iceland and one colony in Norway indicated that prevalence and infection intensities of Bbsl vary significantly among tick host races at the colony scale, with ticks from Atlantic puffins showing higher prevalence and greater infection intensities compared to kittiwake and murre races (Gómez-Díaz et al. 2010). This suggests that puffins and/or their specific ticks are more competent hosts for Bbsl than the other seabird species/tick races in the community. Genetic analyses of the most prevalent bacterial species of the complex, *B. garinii*, further demonstrated that bacterial strains were locally isolated among seabird species/tick host races; at the colony level, Bbsl strains do not circulate freely among the different component species of the community (Figure 8.5).

This structure does not seem to be due to adaptation of the infectious agent to different seabird species, as no pattern of divergence could be found in genes thought to be related to host use (OSPs genes), and host-specific patterns of strain divergence at the colony level were not repeatable at larger spatial scales; that is, a strain infecting a puffin tick in one colony could be found in a murre tick in a different isolated location (Gómez-Díaz et al. 2011). Patterns of pathogen structure at the colony level therefore seem to be due to isolation by vector host use rather than adaptation *per se*. Because of this, increasing or decreasing seabird diversity within a given seabird community would not be expected to alter the infection dynamics of Bbsl; that is, no dilution effect would be expected in this system. The question now is the degree to which the patterns found for ticks and Bbsl bacteria holds for other parasites of the metacommunity. Evidence from another tick-borne pathogen, Great Island virus (Orbivirus), supports results from Bbsl that parasite component communities may not be readily shared among sympatric seabirds. Indeed, viral prevalence in the one colony was found to be high in both murres and their ticks but low in kittiwake ticks and absent in kittiwakes themselves (Nunn et al. 2006). Too few studies currently exist for other types of parasites to examine this question more generally.

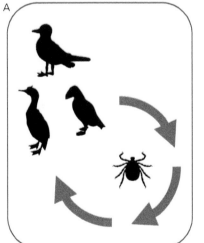
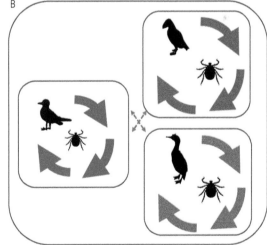

Figure 8.5 Circulation of vector-borne pathogens within host communities depends on vector host use. (A) All available species in the community are used by the vector (i.e., the vector is a true host generalist). Pathogen transmission will depend on the relative abundance of each host species in the community and their competence for the pathogen. (B) Each vertebrate host species is exploited by a distinct vector population resulting in the presence of locally independent or semi-independent disease cycles. Pathogen transmission in this case will depend on the degree of vector specificity (dotted arrows indicate potential contacts). (After McCoy 2008 and McCoy et al. 2013.)

8.4.3.2 Linking marine and terrestrial enzootic cycles

Results from several studies have suggested that Bbsl bacteria are shared between marine and terrestrial systems (Bunikis et al. 1996; Gómez-Díaz et al. 2011; Munro et al. 2019). This link leads to several important questions about the epidemiology of Lyme disease and the evolutionary trajectory of these infectious agents. (1) How important are seabirds as reservoirs for Lyme disease spirochaetes? (2) At what frequency do seabirds disperse these infectious agents over large spatial scales? (3) And where are the contact points between terrestrial and marine ecosystems?

The importance of the first two questions depends on the answer to the third. Indeed, from the patterns described above we know that Bbsl circulates in a natural enzootic transmission cycle in marine bird communities and that seabirds could be excellent reservoirs (and potential generators) of pathogen diversity. We also know that seabirds can travel rapidly over substantial distances, taking Bbsl, and the rest of their parasite infra-community, with them. Dispersal propensity in colonial seabirds depends on both individual reproductive success and overall colony dynamics, with failed individuals leaving unsuccessful colonies to find better locations for future reproduction. This dynamic means that any extrinsic factor affecting seabird reproductive success, including anthropogenic effects such as reductions in food resources due to overfishing or colony disturbance due to ecotourism, may increase pathogen dispersal to alternative colony locations. However, to understand the importance of marine Bbsl for human health, there needs to be a link between ecosystems. When present in seabird colonies, humans can be bitten by *I. uriae* ticks, and cases of Bbsl transmission have occurred under these circumstances (pers. obs.). The increasing presence of humans in seabird colonies with the rise in ecotourism therefore poses a substantiated risk for Bbsl exposure. However, as humans are accidental and dead-end hosts for Bbsl, infective strains received from seabirds cannot be transmitted to the terrestrial cycle.

The bridge between systems may thus lie in the presence of other alternative host species within colonies. For example, other animals, such as rodents or passerine birds can sometimes be present in seabird colonies and may serve as accidental hosts to *I. uriae*. In such cases, ticks may transmit seabird-derived Bbsl strains to these animals which may then circulate outside the colony location. In more temperate zones, it is possible to have additional tick species within seabird colonies, ticks that can feed successfully on a wider range of host species than *I. uriae*, including seabirds, mammals, and passerine birds. These links still require specific testing but suggest that transmission dynamics within the marine cycle of Bbsl should be considered more explicitly for understanding the global epidemiology of Lyme disease. More generally, the detailed work on Bbsl circulation within seabird communities highlights the importance of accounting for different spatial scales for understanding pathogen transmission dynamics and parasite community composition. It has also identified the key role of parasite specialization (here, host race formation in ticks) in determining local exposure rates and infra-community composition within host communities.

8.5 Summary and future directions

We are currently experiencing a period of rapid global change that will have strong impacts on natural communities and their dynamics. Organisms are shifting their geographic distributions and phenologies in relation to climate change (Gorman et al. 2010; Jaenson et al. 2012; Matthysen et al. 2011), species are going extinct due to changes in food resources (Grémillet et al. 2019), and natural habitats are becoming more and more fragmented with expanding human populations (Chace and Walsh 2006). These different effects may alter overall community structure and function, including avian diversity and that of their associated parasites and infectious agents.

Parasites are fundamental components of all natural communities. This diversity is structured at different biological scales and can define individual host fitness and lead to increases or decreases in overall community diversity. However, in order to determine the functional role of a parasite at the level of the free-living community, it is necessary to first understand how it impacts the parasite infra-community, because it is this community that will alter individual host fitness. Studies on within-host interactions in co-infecting parasites are becoming more frequent (e.g., Graham 2008; Telfer et al. 2010), but work at the level of the infra-community is still missing. The possibility that a parasite species simultaneously affects structure at all levels of a community, from the parasite community up to the community of free-living animals (Poulin 1999), has not yet been evaluated for avian communities but may be interesting to explore more explicitly in zones of recent parasite invasion (e.g., the spread of *P. relictum* to high-altitude zones of the Hawaiian islands).

Throughout this chapter, examples have high-lighted the direct relationship between avian community structure and parasite diversity. Adding host biodiversity to communities can shift parasite community structure and stability—in some cases amplifying parasite presence in the community, in other cases reducing transmission and lowering overall infection prevalence. The direction this takes is thought to be associated with the spatial scale of the community being considered (Halliday and Rohr 2019), with higher biodiversity at local scales often being associated with reduced disease risk and amplification occurring at broader spatial scales. More rigorous tests of this hypothesis are now required, as well as a consideration of the temporal stability of such changes for both parasite and host community dynamics.

Much of our understanding of the links between avian community structure and function and parasite and pathogen diversity have come from field studies of natural populations, but data are still limited. For example, the generality of the dilution effect model requires data demonstrating that shifts in diversity and community composition directly affect vector infection rates. Surveillance data from communities undergoing rapid changes in community structure, such as island populations following the removal of an invasive species (Le Corre et al. 2015) or communities found in forested areas with planned logging activities (e.g., Schmiegelow et al. 1997), may be particularly useful in this respect. Similarly, the spatial scale at which parasites might alter host community functioning will depend on relative dispersal rates and distances. Such information is now becoming more readily available for avian species with recent technical advances in animal tracking (Kays et al. 2015) and rapid genetic typing of infectious agents (e.g., Michelet et al. 2014).

Finally, although the notion of spatial variation is often explicit in analyses of community structure in relation to pathogen transmission and disease risk, temporal variation in species presence and activity within the community are less frequently considered but are essential to estimate contact rates, particularly in seasonal species (Caron et al. 2010). Thus, to better predict risk, surveillance data on the activity of different species within a community over time are required. When combined with data on pathogen competence, this spatial and temporal data can then be combined to estimate parasite spillover within natural communities or to human and domestic animals.

More generally, one thing is evident: if biodiversity buffers ecosystems against environmental variation, its loss, be it avian or parasitic, can destabilize natural communities, potentially altering transmission cycles and leading to disease emergence. Although the mechanisms underlying such changes are diverse and difficult to pinpoint (e.g., dilution effects, threshold density effects, parasite–parasite interactions, and other idiosyncratic effects), maintaining high diversity within natural communities should be a priority to help retain their stability and may reduce disease risk for humans and their domestic animals under the One Health paradigm (Zinsstag et al. 2012). This aim will be challenging to meet in the near future.

Literature cited

Adelman, J.S., Moyers, S.C., Farine, D.R., and Hawley, D.M. (2015). Feeder use predicts both acquisition and transmission of a contagious pathogen in a North American songbird. *Proceedings of the Royal Society B: Biological Sciences*, 282, 20151429.

Allan, B.F., Keesing, F., and Ostfeld, R.S. (2003). Effect of forest fragmentation on Lyme disease risk. *Conservation Biology*, 17, 267–72.

Apperson, C. S., Hassan, H. K., Harrison, B. A., et al. (2004). Host feeding patterns of established and potential mosquito vectors of West Nile virus in the eastern United States. *Vector-Borne and Zoonotic Diseases*, 4, 71–82.

Atkinson, C.T., Dusek, R.J., Woods, K.L., and Iko, W.M. (2000). Pathogenicity of avian malaria in experimentally-infected Hawaii amakihi. *Journal of Wildlife Diseases*. 36, 197–201.

Atkinson, C.T., Lease, J.K., Dusek, R.J., and Samuel, M.D. (2005). Prevalence of pox-like lesions and malaria in forest bird communities on leeward Mauna Loa Volcano, Hawaii. *Condor* 107, 537–46.

Ballance, L.T., Pitman, R.L., and Reilly, S.B. (1997). Seabird community structure along a productivity gradient: importance of competition and energetic constraint. *Ecology*, 78, 1502–18.

Bartlett, M. (1960). The critical community size for measles in the United States. *Journal of the Royal Statistical Society: Series A* (General), 123, 37–44.

Begon, M. (2008). Effects of host diversity on disease dynamics. In R. Ostfeld, F. Keesing, V. Eviner, eds. *Infectious Disease Ecology: Effects of Ecosystems on Disease and of Disease on Ecosystems*, pp. 12–29. Princeton University Press, Princeton, NJ.

Beissinger, S.R. and Osborne, D.R. (1982). Effects of urbanization on avian community organization. *The Condor*, 84, 75–83.

Benskin, C.M.H., Wilson, K., Jones, K., and Hartley, I.R. (2009). Bacterial pathogens in wild birds: a review of the frequency and effects of infection. *Biological Reviews*, 84, 349–73.

Biard, C., Monceau, K., Motreuil, S., and Moreau, J. (2015). Interpreting immunological indices: the importance of taking parasite community into account. An example in blackbirds *Turdus merula*. *Methods in Ecology and Evolution*, 6, 960–72.

Bonsall, M. and Hassell, M. (1997). Apparent competition structures ecological assemblages. *Nature*, 388, 371.

Boulinier, T. and Danchin, E. (1996). Population trends in kittiwake *Rissa tridactyla* colonies in relation to tick infestation. *Ibis*, 138, 326–34.

Boulinier, T., Sorci, G., Monnat, J.-Y., and Danchin, E. (1997). Parent–offspring regression suggests heritable susceptibility to ectoparasites in a natural population of kittiwake *Rissa tridactyla*. *Journal of Evolutionary Biology*, 10, 77–85.

Boulinier, T., McCoy, K.D., and Sorci, G. (2001). Dispersal and parasitism. In J. Clobert, E. Danchin, A. Dhondt, and J.D. Nichols, eds. *Dispersal*, pp. 169–79. Oxford University Press, Oxford.

Boulinier, T., McCoy, K.D., Yoccoz, N.G., Gasparini, J., and Tveraa, T. (2008). Public information affects breeding dispersal in a colonial bird: kittiwakes cue on neighbours. *Biology Letters*, 4, 538–40.

Boulinier, T., Kada, S., Ponchon, A., et al. (2016). Migration, prospecting, dispersal? What host movement matters for infectious agent circulation? *Integrative and Comparative Biology*, 56, 330–42.

Brobey, B., Kucknoor, A., and Armacost, J. (2017). Prevalence of *Trichomonas, Salmonella,* and *Listeria* in wild birds from southeast Texas. *Avian Diseases*, 61, 347–52.

Brown, C.R. and Brown, M.B. (1992). Ectoparasitism as a cause of natal dispersal in cliff swallows. *Ecology*, 73, 1718–23.

Brown, C.R. and Brown, M.B. (1996). *Coloniality in the Cliff Swallow*. University of Chicago Press, Chicago.

Bunikis, J., Olsen, B., Fingerle, V., Bonnedahl, J., Wilske, B., and Bergstrom, S. (1996). Molecular polymorphism of the Lyme disease agent *Borrelia garinii* in northern Europe is influenced by a novel enzootic Borrelia focus in the North Atlantic. *Journal of Clinical Microbiology*, 34, 364–8.

Bush, A.O., Lafferty, K.D., Lotz, J.M., and Shostak, A.W. (1997). Parasitology meets ecology on its own terms: Margolis et al. revisited. *Journal of Parasitology*, 83, 575–83.

Caron, A., Garine-Wichatitsky, D., Gaidet, N., Chiweshe, N., and Cumming, G.S. (2010). Estimating dynamic risk factors for pathogen transmission using community-level bird census data at the wildlife/domestic interface. *Ecology and Society*, 15, 25.

Centers for Disease Control and Prevention (2019). Data and Surveillance. Lyme Disease. https://www.cdc.gov/lyme/stats/. Accessed 22 November 2019.

Chace, J.F. and Walsh, J.J. (2006). Urban effects on native avifauna: a review. *Landscape and Urban Planning*, 74, 46–69.

Chambert, T., Staszewski, V., Lobato, E., et al. (2012). Exposure of black-legged kittiwakes to Lyme disease spirochetes: dynamics of the immune status of adult hosts and effects on their survival. *Journal of Animal Ecology*, 81, 986–95.

Chave, J. (2004). Neutral theory and community ecology. *Ecology Letters*, 7, 241–53.

Choe, J.C. and Kim, K.C. (1988). Microhabitat preference and coexistence of ectoparasitic arthropods on Alaskan seabirds. *Canadian Journal of Zoology*, 66, 987–97.

Civitello, D.J., Cohen, J., Fatima, H., et al. (2015). Biodiversity inhibits parasites: broad evidence for the dilution effect. *Proceedings of the National Academy of Sciences of the United States of America*, 112, 8667–71.

Clayton, D.H., Bush, S.E., and Johnson, K.P. (2015). Coevolution of life on hosts: integrating ecology and history. University of Chicago Press, Chicago.

Craven, S., Stern, N., Line, E., Bailey, J., Cox, N., and Fedorka-Cray, P. (2000). Determination of the incidence of *Salmonella* spp., *Campylobacter jejuni*, and *Clostridium perfringens* in wild birds near broiler chicken houses by sampling intestinal droppings. *Avian Diseases*, 44, 715–20.

Crooks, J.A. (2002). Characterizing ecosystem-level consequences of biological invasions: the role of ecosystem engineers. *Oikos*, 97, 153–66.

Croxall, J.P., Butchart, S.H.M., Lascelles, B., et al. (2012). Seabird conservation status, threats and priority actions: a global assessment. *Bird Conservation International*, 22, 1–34.

Dabert, M., Coulson, S.J., Gwiazdowicz, D.J., et al. (2015). Differences in speciation progress in feather mites (Analgoidea) inhabiting the same host: the case of *Zachvatkinia* and *Alloptes* living on arctic and long-tailed skuas. *Experimental and Applied Acarology*, 65, 163–79.

Daily, G.C., Ehrlich, P.R., and Haddad, N.M. (1993). Double keystone bird in a keystone species complex. *Proceedings of the National Academy of Sciences of the United States of America*, 90, 592–4.

Danchin, E. (1992). The incidence of the tick parasite *Ixodes uriae* in kittiwake *Rissa tridactyla* colonies in relation to the age of the colony and the mechanism of infecting new colonies. *Ibis*, 134, 134–41.

Dhondt, A.A., Dhondt, K.V., Hawley, D.M., and Jennelle, C.S. (2007). Experimental evidence for transmission of *Mycoplasma gallisepticum* in house finches by fomites. *Avian Pathology*, 36, 205–8.

Diamond, J.M. (1975). Assembly of species communities. In ML Cody and JM Diamond, eds. *Ecology and Evolution of Communities*, pp. 342–444. Harvard University Press, Cambridge, MA.

Dietrich, M., Gómez-Díaz, E., Boulinier, T., and McCoy, K.D. (2008). Local distribution and genetic structure of tick-borne pathogens: an example involving the marine cycle of Lyme disease. In M. Bertrand, S. Kreiter, M.S. Tixier, et al., eds. *Integrative Acarology. 6th European Congress of Acarology*, EURAAC, pp. 33–41.

Dietrich, M., Gómez-Díaz, E., and McCoy, K.D. (2011). Worldwide distribution and diversity of seabird ticks: implications for the ecology and epidemiology of tick-borne pathogens. *Vector-Borne and Zoonotic Diseases*, 11, 453–70.

Dietrich, M., Kempf, F., Gómez-Díaz, E., et al. (2012). Inter-oceanic variation in patterns of host-associated divergence in a seabird ectoparasite. *Journal of Biogeography*, 39, 545–55.

Dietrich, M., Beati, L., Elguero, E., Boulinier, T., and McCoy, K.D. (2013). Body size and shape evolution in host races of the tick *Ixodes uriae*. *Biological Journal of the Linnean Society*, 108, 323–34.

Dietrich, M., Kempf, F., Boulinier, T., and McCoy, K.D. (2014a). Tracing the colonization and diversification of the worldwide seabird ectoparasite *Ixodes uriae*. *Molecular Ecology*, 23, 3292–305.

Dietrich, M., Lebarbenchon, C., Jaeger, A., et al. (2014b). *Rickettsia* spp. in seabird ticks from western Indian Ocean islands. *Emerging Infectious Diseases*, 20, 838–42.

Dietrich, M., Lobato, E., Boulinier, T., and McCoy, K.D. (2014c). An experimental test of host specialization in a ubiquitous polar ectoparasite: a role for adaptation? *Journal of Animal Ecology*, 83, 576–87.

Dobson, A.P. and Hudson, P.J. (1986). Parasites, disease and the structure of ecological communities. *Trends in Ecology & Evolution*, 1, 11–15.

Duffy, D.C. (1983). The ecology of tick parasitism on densely nesting Peruvian seabirds. *Ecology*, 64, 110–19.

Duneau, D., Boulinier, T., Gómez-Díaz, E., et al. (2008). Prevalence and diversity of Lyme borreliosis bacteria in marine birds. *Infection, Genetics and Evolution*, 8, 352–9.

Dunn, A.M. and Hatcher, M.J. (2015). Parasites and biological invasions: parallels, interactions, and control. *Trends in Parasitology*, 31, 189–99.

Duron, O., Cremaschi, J., and McCoy, K.D. (2016). The high diversity and global distribution of the intracellular bacterium *Rickettsiella* in the polar seabird tick *Ixodes uriae*. *Microbial Ecology*, 71, 761–70.

Duron, O., Binetruy, F., Noel, V., et al. (2017). Evolutionary changes in symbiont community structure in ticks. *Molecular Ecology*, 26, 2905–21.

Ezenwa, V.O., Godsey, M.S., King, R.J., and Guptill, S.C. (2006). Avian diversity and West Nile virus: testing associations between biodiversity and infectious disease risk. *Proceedings of the Royal Society B: Biological Sciences*, 273, 109–17.

Farmer, K., Hill, G., and Roberts, S. (2005). Susceptibility of wild songbirds to the house finch strain of *Mycoplasma gallisepticum*. *Journal of Wildlife Diseases*, 41, 317–25.

Fassbinder-Orth, C.A., Barak, V.A., and Brown, C.R. (2013). Immune responses of a native and an invasive bird to Buggy Creek virus (Togaviridae: Alphavirus) and its arthropod vector, the swallow bug (*Oeciacus vicarius*). *PLoS ONE* 8, 2.

Fauchald, P., Erikstad, K.E., and Skarsfjord, H. (2000). Scale-dependent predator–prey interactions: the hierarchical spatial distribution of seabirds and prey. *Ecology*, 81, 773–83.

Feare, C.J. (1976). Desertion and abnormal development in a colony of sooty terns *Sterna fuscata* infested by virus-infected ticks. *Ibis*, 118, 112–15.

Feare, C.J., Sanders, M., Blasco, R., and Bishop, J. (1999). Canada goose (*Branta canadensis*) droppings as a potential source of pathogenic bacteria. *Journal of the Royal Society for the Promotion of Health*, 119, 146–55.

Fischer, J.R., Stallknecht, D.E., Luttrell, P., Dhondt, A.A., and Converse, K.A. (1997). Mycoplasmal conjunctivitis in wild songbirds: the spread of a new contagious disease in a mobile host population. *Emerging Infectious Diseases*, 3, 69.

Freeland, W.J. (1983). Parasites and the coexistence of animal host species. *American Naturalist*, 121, 223–36.

Furness, R.W., Monaghan, P. (1987). *Seabird Ecology*. Blackie and Son, Glasgow.

Gasparini, J., McCoy, K.D., Haussy, C., Tveraa, T., and Boulinier, T. (2001). Induced maternal response to the Lyme disease spirochaete *Borrelia burgdorferi sensu lato* in a colonial seabird, the kittiwake *Rissa tridactyla*. *Proceedings of the Royal Society B: Biological Sciences*, 268, 647–50.

Getz, W.M. (2009). Disease and the dynamics of food webs. *PLoS Biology*, 7, e1000209.

Gibb, R., Redding, D.W., Chin, K.Q., et al. (2020). Zoonotic host diversity increases in human-dominated ecosystems. *Nature*, 2020, 1–5.

Gómez-Díaz, E., Gonzalez-Solis, J., Peinado, M.A., and Page, R.D.M. (2007). Lack of host-dependent genetic structure in ectoparasites of *Calonectris* shearwaters. *Molecular Ecology*, 16, 5204–15.

Gómez-Díaz, E., Doherty Jr., P.F., Duneau, D., and McCoy, K.D. (2010). Cryptic vector divergence masks vector-specific patterns of infection: an example from the marine cycle of Lyme borreliosis. *Evolutionary Applications*, 3, 391–401.

Gómez-Díaz, E., Boulinier, T., Sertour, N., Cornet, M., Ferquel, E., and McCoy, K.D. (2011). Genetic structure of marine *Borrelia garinii* and population admixture with the terrestrial cycle of Lyme borreliosis. *Environmental Microbiology*, 13, 2453–67.

Gorman, K.B., Erdmann, E.S., Pickering, B.C., et al. (2010). A new high-latitude record for the macaroni penguin (*Eudyptes chrysolophus*) at Avian Island, Antarctica. *Polar Biology*, 33, 1155–8.

Graham, A.L. (2008). Ecological rules governing helminth–microparasite coinfection. *Proceedings of the National Academy of Sciences of the United States of America*, 105, 566–70.

Gray, J., Kahl, O., Lane, R.S., and Stanek, G. (2002). *Lyme Borreliosis: Biology, Epidemiology, and Control*. CABI Publishing, New York.

Grémillet, D., Ponchon, A., Paleczny, M., Palomares, M.-L.D., Karpouzi, V., and Pauly, D. (2019). Persisting worldwide seabird–fishery competition despite seabird community decline. *Current Biology*, 28, 4009–4013.e2.

Grewell, B.J. (2008). Parasite facilitates plant species coexistence in a coastal wetland. *Ecology*, 89, 1481–8.

Gutiérrez, J.S., Rakhimberdiev, E., Piersma, T., and Thieltges, D.W. (2017). Migration and parasitism: habitat use, not migration distance, influences helminth species richness in Charadriiform birds. *Journal of Biogeography*, 44, 1137–47.

Gylfe, Å., Olsen, B., Strasevicius, D., et al. (1999). Isolation of Lyme disease Borrelia from puffins (*Fratercula arctica*) and seabird ticks (*Ixodes uriae*) on the Faeroe Islands. *Journal of Clinical Microbiology*, 37, 890–6.

Gylfe, Å., Bergström, S., Lundström, J., and Olsen, B. (2000). Reactivation of Borrelia infection in birds. *Nature*, 403, 724–5.

Haas, S.E., Hooten, M.B., Rizzo, D.M., and Meentemeyer, R.K. (2011). Forest species diversity reduces disease risk in a generalist plant pathogen invasion. *Ecology Letters*, 14, 1108–16.

Halliday, F.W. and Rohr, J.R. (2019). Measuring the shape of the biodiversity–disease relationship across systems reveals new findings and key gaps. *Nature Communications*, 10, 1–10.

Halsey, S. (2019). Defuse the dilution effect debate. *Nature Ecology & Evolution*, 3, 145.

Hamer, G. L., Kitron, U. D., Goldberg, T. L., et al. (2009). Host selection by *Culex pipiens* mosquitoes and West Nile virus amplification. *American Journal of Tropical Medicine and Hygiene*, 80(2), 268–78.

Hanincova, K., Taragelova, V., Koci, J., et al. (2003). Association of *Borrelia garinii* and *B. valaisiana* with songbirds in Slovakia. *Applied and Environmental Microbiology*, 69, 2825–30.

Heeb, P., Werner, I., Mateman, A.C., et al. (1999). Ectoparasite infestation and sex-biased local recruit;ment of hosts. *Nature*, 400, 63–5.

Hoekstra, J.M., Molnar, J.L., Jennings, M., Revenga, C., and Spaulding, M. (2010). *The Atlas of Global Conservation*. University of California Press, Berkeley, CA.

Holt, R.D. and Bonsall, M.B. (2017). Apparent competition. *Annual Review of Ecology, Evolution, and Systematics*, 48, 447–71.

Hubbell, S.P. (2001). *The Unified Neutral Theory of Biodiversity and Biogeography*. Princeton University Press, Princeton, NJ.

Ives, A.R. and Cardinale, B.J. (2004). Food-web interactions govern the resistance of communities after nonrandom extinctions. *Nature*, 429, 174.

Jaenson, T.G.T., Jaenson, D.G.E., Eisen, L., Petersson, E., and Lindgren, E. (2012). Changes in the geographical distribution and abundance of the tick *Ixodes ricinus* during the past 30 years in Sweden. *Parasites & Vectors*, 5, 8.

Johnson, P.T.J. and Thieltges, D.W. (2010). Diversity, decoys and the dilution effect: how ecological communities affect disease risk. *Journal of Experimental Biology*, 213, 961–70.

Johnson, P.T., Preston, D.L., Hoverman, J.T., and LaFonte, B.E. (2013a). Host and parasite diversity jointly control disease risk in complex communities. *Proceedings of the National Academy of Sciences of the United States of America*, 110, 16916–21.

Johnson, P.T.J., Preston, D.L., Hoverman, J.T., and Richgels, K.L.D. (2013b). Biodiversity decreases disease through predictable changes in host community competence. *Nature*, 494, 230–3.

Jones, K.E., Patel, N.G., Levy, M.A., et al. (2008). Global trends in emerging infectious diseases. *Nature*, 451, 990–3.

Jones, S.M., Cumming, G.S., and Peters, J.L. (2018). Host community heterogeneity and the expression of host specificity in avian haemosporidia in the Western Cape, South Africa. *Parasitology*, 145, 1876–83.

Kays, R., Crofoot, M.C., Jetz, W., and Wikelski, M. (2015). Terrestrial animal tracking as an eye on life and planet. *Science* 348, aaa2478.

Keesing, F., Holt, R.D., and Ostfeld, R.S. (2006). Effects of species diversity on disease risk. *Ecology Letters*, 9, 485–98.

Khan, J.S., Provencher, J.F., Forbes, M.R., Mallory, M.L., Lebarbenchon, C., and McCoy, K.D. (2019). Parasites of seabirds: a survey of effects and ecological implications. *Advances in Marine Biology*, 82, 1.

Kilpatrick, A.M., Daszak, P., Jones, M.J., Marra, P.P., and Kramer, L.D. (2006). Host heterogeneity dominates West Nile virus transmission. *Proceedings of the Royal Society B: Biological Sciences*, 273, 2327–33.

King, K.A., Keith, J.O., Mitchell, C.A., and Keirans, J.E. (1977). Ticks as a factor in nest desertion of California brown pelicans. *The Condor*, 79, 507–9.

Kokko, H., Harris, M.P., and Wanless, S. (2004). Competition for breeding sites and site-dependent population regulation in a highly colonial seabird, the common guillemot *Uria aalge*. *Journal of Animal* Ecology, 73, 367–76.

Kurtenbach, K., DeMichelis, S., Etti, S., et al. (2002). Host association of *Borrelia burgdorferi sensu lato*—the key role of host complement. *Trends in Microbiology*, 10, 74–9.

Kurtenbach, K., Hanincova, K., Tsao, J.I., Margos, G., Fish, D., and Ogden, N.H. (2006). Fundamental processes in the evolutionary ecology of Lyme borreliosis. *Nature Reviews Microbiology*, 4, 660–9.

Lafferty, K.D. and Morris, A.K. (1996). Altered behavior of parasitized killifish increases susceptibility to predation by bird final hosts. *Ecology*, 77, 1390–7.

Lawton, J.H. (1999). Are there general laws in ecology? *Oikos*, 84, 177–92.

Le Corre, M., Danckwerts, D.K., Ringler, D., et al. (2015). Seabird recovery and vegetation dynamics after Norway rat eradication at Tromelin Island, western Indian Ocean. *Biological Conservation*, 185, 85–94.

Leibold, M.A., Holyoak, M., Mouquet, N., et al. (2004). The metacommunity concept: a framework for multiscale community ecology. *Ecology Letters*, 7, 601–13.

Leung, T.L. and Koprivnikar, J. (2016). Nematode parasite diversity in birds: the role of host ecology, life history and migration. *Journal of Animal Ecology*, 85, 1471–80.

Levy, M.S. (1997). *Helminth communities of ring billed gulls* (Larus delawarensis) *collected along the St. Lawrence River and Estuary*. MSc dissertation, Concordia University, Montreal.

Lloyd-Smith, J.O., Schreiber, S.J., Kopp, P.E., and Getz, W.M. (2005). Superspreading and the effect of individual variation on disease emergence. *Nature*, 438, 355–9.

Lobato, E., Pearce, J., Staszewski, V., et al. (2011). Seabirds and the circulation of Lyme borreliosis bacteria in the North Pacific. *Vector-Borne and Zoonotic Diseases*, 11, 1521–7.

Locke, S.A., Levy, M.S., Marcogliese, D.J., Ackerman, S., and McLaughlin, J.D. (2012). The decay of parasite community similarity in ring-billed gulls *Larus delawarensis* and other hosts. *Ecography*, 35, 530–8.

LoGiudice, K., Ostfeld, R.S., Schmidt, K.A., and Keesing, F. (2003). The ecology of infectious disease: effects of host diversity and community composition on Lyme disease risk. *Proceedings of the National Academy of Sciences of the United States of America*, 100, 567–71.

Loreau, M. and de Mazancourt, C. (2013). Biodiversity and ecosystem stability: a synthesis of underlying mechanisms. *Ecology Letters*, 16, 106–15.

Lycett, S.J., Bodewes, R., Pohlmann, A., et al. (2016). Role for migratory wild birds in the global spread of avian influenza H5N8. *Science*, 354, 213–17.

Margos, G., Vollmer, S.A., Ogden, N.H., and Fish, D. (2011). Population genetics, taxonomy, phylogeny and evolution of *Borrelia burgdorferi sensu lato*. *Infection, Genetics, Evolution*, 11, 1545–63.

Martin-Albarracin, V.L., Amico, G.C., Simberloff, D., and Nuñez, M.A. (2015). Impact of non-native birds on native ecosystems: a global analysis. *PLoS ONE*, 10, e0143070.

Marzluff, J.M., 2001. Worldwide urbanization and its effects on birds. In J.M. Marzluff, R. Bowman, and R. Donnelly, eds. *Avian Ecology and Conservation in an Urbanizing World*, pp. 19–47. Springer, New York.

Matthews, A.E., Larkin, J.L., Raybuck, D.W., Slevin, M.C., Stoleson, S.H., and Boves, T.J. (2018). Feather mite abundance varies but symbiotic nature of mite–host relationship does not differ between two ecologically dissimilar warblers. *Ecology and Evolution*, 8, 1227–38.

Matthysen, E., Adriaensen, F., and Dhondt, A.A. (2011). Multiple responses to increasing spring temperatures in the breeding cycle of blue and great tits (*Cyanistes caeruleus, Parus major*). *Global Change Biology*, 17, 1–16.

May, R.M. (2006). Network structure and the biology of populations. *Trends in Ecology & Evolution*, 21, 394–9.

McCoy, K.D. (2008). The population genetic structure of vectors and our understanding of disease epidemiology. *Parasite*, 15, 444–8.

McCoy, K.D., Boulinier, T., Chardine, J.W., Danchin, E., and Michalakis, Y. (1999). Dispersal and distribution of the tick *Ixodes uriae* within and among seabird host populations: the need for a population genetic approach. *Journal of Parasitology*, 85, 196–202.

McCoy, K.D., Boulinier, T., Tirard, C., and Michalakis, Y. (2001). Host specificity of a generalist parasite: genetic evidence of sympatric host races in the seabird tick *Ixodes uriae*. *Journal of Evolutionary Biology*, 14, 395–405.

McCoy, K.D., Boulinier, T., Tirard, C., and Michalakis, Y. (2003a). Host-dependent genetic structure of parasite populations: differential dispersal of seabird tick host races. *Evolution*, 57, 288–96.

McCoy, K.D., Tirard, C., and Michalakis, Y. (2003b). Spatial genetic structure of the ectoparasite *Ixodes uriae* within breeding cliffs of its colonial seabird host. *Heredity*, 91, 422–9.

McCoy, K.D., Chapuis, E., Tirard, C., et al. (2005). Recurrent evolution of host-specialized races in a globally distributed parasite. *Proceedings of the Royal Society B: Biological Sciences*, 272, 2389–95.

McCoy, K.D., Leger, E., and Dietrich, M. (2013). Host specialization in ticks and transmission of tick-borne diseases: a review. *Frontiers in Cellular and Infection Microbiology*, 3, 12.

McCoy, K.D., Dietrich, M., Jaeger, A., et al. (2016). The role of seabirds of the Iles Eparses as reservoirs and disseminators of parasites and pathogens. *Acta Oecologica*, 72, 98–109.

McMahon, D.P., Natsopoulou, M.E., Doublet, V., et al. (2016). Elevated virulence of an emerging viral genotype as a driver of honeybee loss. *Proceedings of the Royal Society B: Biological Sciences*, 283, 20160811.

Michelet, L., Delannoy, S., Devillers, E., et al. (2014). High-throughput screening of tick-borne pathogens in Europe. *Frontiers in Cellular and Infection Microbiology*, 4, 13.

Mills, L.S., Soulé, M.E., and Doak, D.F. (1993). The keystone-species concept in ecology and conservation. *BioScience*, 43, 219–24.

Minchella, D.J. and Scott, M.E. (1991). Parasitism: a cryptic determinant of animal community structure. *Trends in Ecology and Evolution*, 6, 250–4.

Molaei, G., Cummings, R.F., Su, T., et al. (2010). Vector–host interactions governing epidemiology of West Nile virus in southern California. *American Journal of Tropical Medicine and Hygiene*, 83, 1269–82.

Monterroso, P., Garrote, G., Serronha, A., et al. (2016). Disease-mediated bottom-up regulation: an emergent virus affects a keystone prey, and alters the dynamics of trophic webs. *Scientific Reports*, 6, 36072.

Montgomery, M.J., Thiemann, T., Macedo, P., Brown, D.A., and Scott, T.W. (2011). Blood-feeding patterns of the *Culex pipiens* complex in Sacramento and Yolo Counties, California. *Journal of Medical Entomology*, 48, 398–404.

Morishita, T.Y., Aye, P.P., and Brooks, D.L. (1997). A survey of diseases of raptorial birds. *Journal of Avian Medicine and Surgery*, 11, 77–92.

Mougi, A. and Kondoh, M. (2012). Diversity of interaction types and ecological community stability. *Science*, 337, 349–51.

Mouquet, N. and Loreau, M. (2003). Community patterns in source-sink metacommunities. *The American Naturalist*, 162, 544–57.

Munro, H.J., Ogden, N.H., Mechai, S., et al. (2019). Genetic diversity of *Borrelia garinii* from *Ixodes uriae* collected in seabird colonies of the northwestern Atlantic Ocean. *Ticks and Tick-borne Diseases*, 10, 101255.

Nunn, M.A., Barton, T.R., Wanless, S., Hails, R.S., Harris, M.P., and Nuttall, P.A. (2006). Tick-borne Great Island virus: (I) Identification of seabird host and evidence for co-feeding and viraemic transmission. *Parasitology*, 132, 233–40.

O'Brien, V.A., Moore, A.T., Young, G.R., Komar, N., Reisen, W.K., and Brown, C.R. (2011). An enzootic vector-borne virus is amplified at epizootic levels by an invasive avian host. *Proceedings of the Royal Society B: Biological Sciences*, 278, 239–46.

Ogden, N.H. and Tsao, J.I. (2009). Biodiversity and Lyme disease: dilution or amplification? *Epidemics*, 1, 196–206.

Olsen, B., Jaenson, T.G.T., Noppa, L., Bunikis, J., and Bergström, S. (1993). A Lyme borreliosis cycle in seabirds and *Ixodes uriae* ticks. *Nature*, 362, 340–2.

Oro, D., Pérez-Rodríguez, A., Martínez-Vilalta, A., Bertolero, A., Vidal, F., and Genovart, M. (2009). Interference competition in a threatened seabird community: a paradox for a successful conservation. *Biological Conservation*, 142, 1830–5.

Ostfeld, R. and Keesing, F. (2000a). The function of biodiversity in the ecology of vector-borne zoonotic diseases. *Canadian Journal of Zoology*, 78, 2061–78.

Ostfeld, R.S. and Keesing, F. (2000b). Biodiversity and disease risk: the case of Lyme disease. *Conservation Biology*, 14, 722–8.

Paleczny, M., Hammill, E., Karpouzi, V., and Pauly, D. (2015). Population trend of the world's monitored seabirds, 1950–2010. *PLoS ONE*, 10, e0129342.

Park, T. (1948). Experimental studies of interspecies competition. I. Competition between populations of the flour beetles, *Tribolium confusum* Duval and *Tribolium castaneum* Herbst. *Ecological Monographs*, 18, 265–308.

Pearson, D.E., Ortega, Y.K., Eren, Ö., and Hierro, J.L. (2018). Community assembly theory as a framework for biological invasions. *Trends in Ecology & Evolution*, 33, 313–25.

Pedersen, A.B. and Fenton, A. (2007). Emphasizing the ecology in parasite community ecology. *Trends in Ecology & Evolution*, 22, 133–9.

Ponchon, A., Gremillet, D., Christensen-Dalsgaard, S., et al. (2014). When things go wrong: intra-season dynamics of breeding failure in a seabird. *Ecosphere*, 5, 1–19.

Poulin, R. (1997). Species richness of parasite assemblages: evolution and patterns. *Annual Reviews in Ecology and Systematics*, 28, 341–58.

Poulin, R. (1999). The functional importance of parasites in animal communities: many roles at many levels? *International Journal for Parasitology*, 29, 903–14.

Price, P.W., Westoby, M., Rice, B., et al. (1986). Parasite mediation in ecological interactions. *Annual Reviews in Ecology and Systematics*, 17, 487–505.

Provencher, J.F., Borrelle, S., Sherley, R.B., et al. (2019). Seabirds. In C. Sheppard, ed. *World Seas: An Environmental Evaluation, Vol III: Ecological Issues and Environmental Impacts*, pp. 133–62. Academic Press, London.

Randolph, S.E. and Dobson, A.D.M. (2012). Pangloss revisited: a critique of the dilution effect and the biodiversity-buffers-disease paradigm. *Parasitology*, 139, 847–63.

Rigaud, T., Perrot-Minnot, M.J., and Brown, M.J.F. (2010). Parasite and host assemblages: embracing the reality will improve our knowledge of parasite transmission and virulence. *Proceedings of the Royal Society B: Biological Sciences*, 277, 3693–702.

Roche, B., Dobson, A.P., Guegan, J.F., and Rohani, P. (2012). Linking community and disease ecology: the impact of biodiversity on pathogen transmission. *Philosophical Transactions of the Royal Society B: Biological Sciences*, 367, 2807–13.

Roche, B., Rohani, P., Dobson, A.P., and Guegan, J.F. (2013). The impact of community organization on vector-borne pathogens. *The American Naturalist*, 181, 1–11.

Rosindell, J., Hubbell, S.P., and Etienne, R.S. (2011). The unified neutral theory of biodiversity and biogeography at age ten. *Trends in Ecology & Evolution*, 26, 340–8.

Rynkiewicz, E.C., Pedersen, A.B., and Fenton, A. (2015). An ecosystem approach to understanding and managing within-host parasite community dynamics. *Trends in Parasitology*, 31, 212–21.

Salkeld, D.J., Padgett, K.A., and Jones, J.H. (2013). A meta-analysis suggesting that the relationship between biodiversity and risk of zoonotic pathogen transmission is idiosyncratic. *Ecology Letters*, 16, 679–86.

Sánchez, C.A., Becker, D.J., Teitelbaum, C.S., et al. (2018). On the relationship between body condition and parasite infection in wildlife: a review and meta-analysis. *Ecology Letters*, 21, 1869–84.

Sawicka, A., Durkalec, M., Tomczyk, G., and Kursa, O. (2020). Occurrence of *Mycoplasma gallisepticum* in wild birds: a systematic review and meta-analysis. *PLoS ONE*, 15, e0231545.

Schmiegelow, F.K., Machtans, C.S., and Hannon, S.J. (1997). Are boreal birds resilient to forest fragmentation? An experimental study of short-term community responses. *Ecology*, 78, 1914–32.

Seabloom, E.W., Borer, E.T., Gross, K., et al. (2015). The community ecology of pathogens: coinfection, coexistence and community composition. *Ecology Letters*, 18, 401–15.

Şekercioğlu, C.H. (2006). Increasing awareness of avian ecological function. *Trends in Ecology & Evolution*, 21, 464–71.

Selakovic, S., de Ruiter, P.C., and Heesterbeek, H. (2014). Infectious disease agents mediate interaction in food webs and ecosystems. *Proceedings of the Royal Society B: Biological Sciences*, 281, 20132709.

Simpson, J.E., Hurtado, P.J., Medlock, J., et al. (2012). Vector host-feeding preferences drive transmission of multi-host pathogens: West Nile virus as a model system. *Proceedings of the Royal Society B: Biological Sciences*, 279, 925–33.

Stanek, G. and Reiter, M. (2011). The expanding Lyme Borrelia complex—clinical significance of genomic species? *Clinical Microbiology & Infection*, 17, 487–93.

Staszewski, V., McCoy, K.D., and Boulinier, T. (2008). Variable exposure and immunological response to Lyme disease Borrelia among North Atlantic seabird species. *Proceedings of the Royal Society B: Biological Sciences*, 275, 2101–9.

States, S.L., Hochachka, W.M., and Dhondt, A.A. (2009). Spatial variation in an avian host community: implications for disease dynamics. *EcoHealth*, 6, 540–5.

Stefan, L.M., Gómez-Díaz, E., Elguero, E., Proctor, H.C., McCoy, K.D., and González-Solís, J. (2015). Niche partitioning of feather mites within a seabird host, *Calonectris borealis*. *PLoS ONE*, 10, e0144728.

Stefan, L.M., Gómez-Díaz, E., Mironov, S.V., González-Solís, J., and McCoy, K.D. (2018). 'More than meets the eye': cryptic diversity and contrasting patterns of host-specificity in feather mites inhabiting seabirds. *Frontiers in Ecology and Evolution*, 6, 97.

Stock, T. and Holmes, J.C. (1988). Functional relationships and microhabitat distributions of enteric helminths of grebes (Podicipedidae): the evidence for interactive communities. *Journal of Parasitology*, 74, 214–27.

Swaddle, J.P. and Calos, S.E. (2008). Increased avian diversity is associated with lower incidence of human West Nile infection: observation of the dilution effect. *PLoS ONE*, 3, e2488.

Sykes, R.A. and Makiello, P. (2017). An estimate of Lyme borreliosis incidence in Western Europe. *Journal of Public Health*, 39, 74–81.

Telfer, S., Lambin, X., Birtles, R., et al. (2010). Species interactions in a parasite community drive infection risk in a wildlife population. *Science*, 330, 243–6.

Thomas, F., Guégan, J.F., Michalakis, Y., and Renaud, F. (2000). Parasites and host life-history traits: implications for community ecology and species co-existence. *International Journal for Parasitology*, 30, 669–74.

Tittensor, D.P., Walpole, M., Hill, S.L., et al. (2014). A mid-term analysis of progress toward international biodiversity targets. *Science*, 346, 241–4.

Tompkins, D.M., Draycott, R.A.H., and Hudson, P.J. (2000). Field evidence for apparent competition mediated via shared parasites of two gamebird species. *Ecology Letters*, 3, 10–14.

Tompkins, D.M., Dunn, A.M., Smith, M.J., and Telfer, S. (2011). Wildlife diseases: from individuals to ecosystems. *Journal of Animal Ecology*, 80, 19–38.

Tsiodras, S., Kelesidis, T., Kelesidis, I., Bauchinger, U., and Falagas, M.E. (2008). Human infections associated with wild birds. *Journal of Infection*, 56, 83–98.

Van Riper III, C., Van Riper, S.G., Goff, M.L., and Laird, M. (1986). The epizootiology and ecological significance of malaria in Hawaiian land birds. *Ecological Monographs*, 56, 327–44.

Vellend, M. (2010). Conceptual synthesis in community ecology. *Quarterly Review of Biology*, 85, 183–206.

Vial, L. (2009). Biological and ecological characteristics of soft ticks (Ixodida: Argasidae) and their impact for predicting tick and associated disease distributions. *Parasite*, 16, 191–202.

Vielma, S., Lagrue, C., Poulin, R., and Slebach, C. (2019). Non-host organisms impact transmission at two different life stages in a marine parasite. *Parasitology Research*, 118, 111–17.

Vittecoq, M., Laurens, C., Brazier, L., et al. (2017). VIM-1 carbapenemase-producing *Escherichia coli* in gulls from southern France. *Ecology and Evolution*, 7, 1224–32.

Warner, R.E. (1968). The role of introduced diseases in the extinction of the endemic Hawaiian avifauna. *The Condor*, 70, 101–20.

Weiher, E. and Keddy, P. (2001). *Ecological Assembly Rules: Perspectives, Advances, Retreats*. Cambridge University Press, Cambridge.

Welsh, J.E., Liddell, C., Van Der Meer, J., and Thieltges, D.W. (2017). Parasites as prey: the effect of cercarial density and alternative prey on consumption of cercariae by four non-host species. *Parasitology*, 144, 1775–82.

Wiens, J.A. (1983). Avian community ecology: an iconoclastic view. In A.H. Brush and G.A. Clark, eds. *Perspectives in Ornithology: Essays Presented for the Centennial of the American Ornitholgists' Union*, pp. 355–403. Cambridge University Press, Cambridge.

Wiens, J.A. (1992). *The Ecology of Bird Communities*. Cambridge University Press, Cambridge.

Wilkinson, D.A., Dietrich, M., Lebarbenchon, C., et al. (2014). Massive infection of seabird ticks with bacterial species related to *Coxiella burnetii*. *Applied and Environmental Microbiology*, 80, 3327–33.

Wilson, D.S. (1992). Complex interactions in metacommunities, with implications for biodiversity and higher levels of selection. *Ecology*, 73, 1984–2000.

Woolhouse, M.E., Dye, C., Etard, J.-F., et al. (1997). Heterogeneities in the transmission of infectious agents: implications for the design of control programs. *Proceedings of the National Academy of Sciences of the United States of America*, 94, 338–42.

Zinsstag, J., Mackenzie, J.S., Jeggo, M., Heymann, D.L., Patz, J.A., and Daszak, P. (2012). Mainstreaming one health. *EcoHealth*, 9, 107–10.

Zurfluh, K., Albini, S., Mattmann, P., et al. (2019). Antimicrobial resistant and extended-spectrum β-lactamase producing *Escherichia coli* in common wild bird species in Switzerland. *MicrobiologyOpen*, 8, e845.

CHAPTER 9

Land Use Change and Avian Disease Dynamics

Maureen H. Murray and Sonia M. Hernandez

9.1 Introduction

Birds live on a human-dominated planet. Over 75% of the Earth's ice-free land has been disturbed by human land uses (Ellis and Ramankutty 2008), impacting ecosystems around the globe. Disturbances associated with human activities such as land clearing, fragmentation, and pollution influence the relationships between birds, their pathogens, and the environment they share. Such shifts in disease dynamics can arise through the impacts of land use change on aspects of hosts, vectors, and/or pathogens, including vector and host abundance, behavior, and physiology, and through pathogen persistence in the environment. To address this complexity, we first describe the major causes of anthropogenic land use change that can impact birds and their pathogens across diverse ecosystems. We next discuss key changes associated with land use, such as habitat loss or fragmentation, pollution, and anthropogenic resources, that are relevant to avian disease ecology. We then synthesize documented associations between avian health and urbanization, the fastest growing type of land use change on Earth (Antrop and Van Eetvelde 2000). We close this chapter with a consideration of land use changes in a One Health context and future directions for advancing avian disease ecology in rapidly changing landscapes.

9.1.1 Major drivers of land use change

Global declines of many avian species can be attributed to habitat loss and degradation, primarily from urbanization, deforestation, and agriculture (Rosenberg et al. 2019). Urban and agricultural areas are typically associated with deforestation to create cropland or built environments, although logging and other extractive industries contribute to deforestation globally. Agriculture is the most widespread land use on Earth with approximately 51 million km² of land—50% of Earth's habitable land area—used for cropland or pasture (Ritchie and Roser 2020). To accommodate global demand for crops, land clearing for agriculture typically involves clearing forest or other native vegetation through logging, burning, or draining wetlands, managing soil through tilling and fertilization, and maintaining plant monocultures and/or production animals. Based on its extent and disturbance, agricultural intensification is considered the largest driver of biodiversity loss globally (Dudley and Alexander 2017) and avian species that inhabit agricultural areas are declining faster than birds associated with any other biome (Murphy 2003).

Although urban areas only account for a small proportion of Earth's habitable land area, they are the fastest growing habitat type on Earth due to a rapidly urbanizing human population (Venter et al. 2016). With urban development comes profound changes in abiotic and biotic conditions that can impact the ecology of avian hosts, vectors, and pathogens (reviewed in Bradley and Altizer 2007). Increased impervious surfaces associated with roads and buildings absorb solar radiation and lead to higher temperatures in urban centers than the surrounding countryside, a phenomenon known as the heat island effect (Oke 1973). Increasing

Maureen H. Murray and Sonia M. Hernandez, *Land Use Change and Avian Disease Dynamics* In: *Infectious Disease Ecology of Wild Birds.*
Edited by: Jennifer C. Owen, Dana M. Hawley, and Kathryn P. Huyvaert, Oxford University Press. © Oxford University Press 2021.
DOI: 10.1093/oso/9780198746249.003.0009

temperatures could impact diverse components of disease dynamics including the behavior or physiology of hosts and vectors and/or pathogen persistence in the environment. Urban areas are also more likely to contain non-native species than surrounding landscapes from either intentional or unintentional human introductions (Gaertner et al. 2017), leading to altered host and vector communities.

The global loss of natural wetlands has forced aquatic birds to become increasingly dependent on alternative and artificial habitats, but the use of such wetlands by waterbirds is considered opportunistic, and little consideration has been given for the potential implications to pathogen dynamics in these environments. Sewage treatment plants with waste ponds and wastewater treatment plants have been documented as significant waterbird habitat (Murray et al. 2014). However, these modified wetland environments can be heavily polluted (Hamilton 2007) and may alter pathogen dynamics by facilitating environmental persistence of pathogens, increased production of toxins, or antimicrobial resistance.

9.2 Key changes associated with land use and influences on disease dynamics

Anthropogenic land use, regardless of cause, is typically accompanied by a set of key changes that can impact disease dynamics for birds (Figure 9.1). Land conversion for the purposes of urbanization, cropland, or resource extraction typically causes habitat loss (e.g., removal of a forest stand), habitat degradation (i.e., a reduction in quality), and fragmentation (e.g., bisecting habitat patches with roads or other linear features). These processes are also typically accompanied by increases in human disturbance, which can generate pollution from industrial processes and anthropogenic resources provided intentionally (e.g., bird feeding) or unintentionally (e.g., landfills) to birds. These landscape changes can have several effects on avian hosts, vectors, and pathogens relevant to disease transmission. Altered habitat conditions can lead to changes in the abundance, community composition, or behavior of hosts and vectors, ultimately changing the likelihood of contact with other hosts or vectors, and thus pathogen transmission

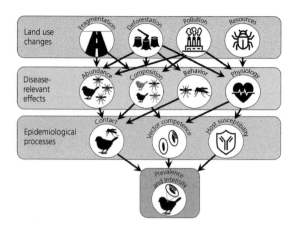

Figure 9.1 A conceptual framework using the example of avian malaria (caused by *Plasmodium* spp.) to illustrate how prevalence and intensity of a vector-borne pathogen (purple) can be influenced by some of the key changes associated with land use (green bar), their downstream effects on individuals and populations (teal bar), and associated consequences for several epidemiological processes (blue bar). Here we conceptualize how habitat degradation can alter abundance and community composition of avian hosts and mosquito vectors, biting behavior of mosquitoes, and physiology of hosts and vectors in ways that alter their susceptibility and competence, respectively. While we selectively show the arrows likely to be most important for avian malaria, changes to 'Resources' from human land use can also alter host and vector abundance, community composition, and behaviors in ways relevant for many other avian–pathogen systems. (Icons were created by Norbert Kucsera, Audrey Laymik, Iconographer, Bohdan Burmich, Symbolon, Alex Tai of Noun Project.)

(Figure 9.1). Habitat changes can also impact host and vector physiology, potentially increasing susceptibility to infection or vector competence. Finally, while not illustrated in our conceptual model of a vector-borne pathogen, land use changes can alter the ability of a parasite or pathogen to persist in the external environment. Because of the myriad potential effects on avian disease ecology, we will discuss each key change and their respective impacts on hosts, vectors, and pathogens in turn.

9.2.1 Habitat loss, degradation, and fragmentation

The primary mechanisms by which habitat loss, degradation, and fragmentation can alter disease dynamics and risk include changes in opportunities for contact among hosts or between hosts and vectors and changes in host and vector competence/susceptibility due to environmental stress. It can be challenging to interpret and predict the individual effects of habitat loss, degradation, and fragmentation on avian disease dynamics because fragmentation is nearly always accompanied by habitat loss (Fahrig 2003). For example, although fragmentation primarily alters opportunities for contact between host and vector species by introducing edges between habitat types, fragmentation is often associated with habitat degradation due to edge effects and the human disturbance that accompanies development. The term habitat loss is also species-specific; while fragmentation reduces the habitat available to interior species (e.g., northern spotted owls, *Strix occidentalis caurina*), it creates more habitat for edge-loving species (e.g., common blackbirds, *Turdus merula*). These factors complicate our understanding of the underlying mechanisms driving changes in avian disease dynamics with land use and emphasize that a synthetic view is necessary.

One of the primary ways that habitat loss and degradation alter disease dynamics is via changes to host and/or vector abundance and thus host and vector community composition. Bird abundance typically declines in degraded landscapes due to lack of habitat-associated resources and, perhaps to a lesser extent, habitat connectivity. For example, a loss of forest cover from 95% to 50% was associated with the extirpation of 33 forest bird species in the Colombian Andes (Palacio et al. 2019). A reduction in host density likely reduces opportunities for pathogens that have density-dependent transmission (see Chapter 7). However, extinction risk from habitat loss is higher for habitat specialists (Owens and Bennett 2000) and disturbed landscapes are often colonized by non-native species such as house sparrows (*Passer domesticus*) and rock doves (*Columba livia*). These changes in host community composition and novel opportunities for contact with non-native species can lead to enhanced exposure to pathogens. For example, urban raptors have a higher infection prevalence of *Trichomonas gallinae*, the causative agent of trichomoniasis, than their non-urban counterparts because urban raptors prey on higher proportions of Columbiformes that are significant reservoirs (Boal et al. 1998; see also Chapter 8). Thus, although habitat loss may reduce the likelihood of conspecific contact by reducing

host abundance (but see barriers to movement below), it can sometimes increase pressure from novel host contact.

Habitat loss and fragmentation can also create opportunities for contact between birds and insect vectors (Figure 9.1). These opportunities could arise because of shifts in host communities or microclimate conditions suitable for vectors. For example, the proportion of birds with attached ticks in Brazil's Atlantic forest was higher in smaller, degraded forest patches where bird species diversity was lower (Ogrzewalska et al. 2011). Degraded patches were dominated by relatively few generalist bird species, potentially promoting vector transmission; however, the environmental conditions in these patches may also be favorable for free-living tick life stages (Ogrzewalska et al. 2011). At a microhabitat scale, the removal of canopy cover typically associated with land clearing and fragmentation could impact the diversity and abundance of insect vectors through the availability of direct sunlight, humidity, edge habitat, soil structure, and standing water. For example, in Southeast Asia, survival of *Anopheles* mosquitoes was higher in deforested areas, where ambient temperatures were higher (Zhong et al. 2016). Relevant to these rapid changes in microhabitat for insect vectors, a growing body of work has examined changes in avian blood parasites in birds with deforestation (Sehgal 2010; Sehgal 2015). The prevalence of avian malaria can often be relatively higher in disturbed forests, likely due to an increase in the range of the mosquito vector. For example, deforestation in Alakaʻi Wilderness Preserve on Kauaʻi Island in Hawaiʻi may shift seasonal pulses in malaria prevalence for Hawaiian honeycreepers to a year-round risk of infection by increasing the altitudinal range of vector populations (Atkinson and LaPointe 2009). However, the relationship between deforestation and blood parasites can be complex, with some *Plasmodium* morphospecies increasing and others decreasing in disturbed forests (Chasar et al. 2009). Thus, the interactions between hosts and vectors in disturbed forests are complicated by species-specific differences in mosquito responses to forest disturbance in interior, edge, and open habitats.

The configuration of disturbed and conservation areas may have important consequences for wildlife persistence and pathogen transmission in areas with dynamic land use change. For example, although more integration between agricultural and natural areas (a 'land sharing' system) may lead to proportionately more fragmented habitat patches, model simulations demonstrate that such an approach may mitigate risk of avian malaria relative to large-scale nature reserves (a 'land sparing' approach) because small remnant forest patches are important in reducing malaria prevalence in surrounding non-forested habitats (Mendenhall et al. 2013). Dipteran vectors also respond to the configuration of forest management, with haemosporidian infections in Diptera insects increasing with more edge habitat, lower shrub density, less leaf litter, and lower humidity, all characteristics of harvested forest stands (van Hoesel et al. 2019). Overall, anticipating changes in vector-borne pathogens for birds with land use change will require species-specific knowledge of vector life history and integrating such knowledge with avian ecology, including habitat requirements.

Although fragmentation can create opportunities for contact between hosts and vectors, it can also introduce barriers to movement. Many species of birds are less likely to move through open areas or gaps in the canopy relative to forested habitat, which can reduce functional connectivity for birds or even persistence in fragmented landscapes. Reduced connectivity can lead to population isolation and inbreeding, which has been linked with reduced immune function in birds including song sparrows (*Melospiza melodia*; Reid et al. 2007). Small remnant habitat patches may promote host contact if subpopulations become more spatially clustered. Reduced movement capacity from fragmentation may decrease opportunities for contact between hosts and between hosts and vectors across the landscape; however, hosts may also have fewer opportunities to leave habitat patches with high host or vector densities.

In addition to these effects on contact rates, habitat loss and fragmentation can impact pathogen dynamics through shifts in host and vector physiology. Habitat degradation associated with land use change can increase susceptibility to infection in hosts through interactions between physiological stress responses and immune function

(see also Chapter 3), though acute stressors such as those stimulated by predators can also enhance immune responses in some cases (Martin 2009). Habitat loss can be associated with increased stress due to the higher levels of disturbance or nutritional stress in altered landscapes. For example, nestling Eurasian treecreepers (*Certhia familiaris*) sampled from small forest patches had higher heterophil:lymphocyte ratios (a marker of stress) relative to nestlings in large patches (Suorsa et al. 2004). Moreover, their ratios were positively associated with mortality and poor body condition (Suorsa et al. 2004). Similarly, willow tits (*Poecile montanus*) in younger forests had higher baseline stress levels and decreased stress responses than tits in old growth forest patches, indicating chronic stress (Cīrule et al. 2017). While these studies point to potential impacts of land use change on host physiology and thus susceptibility, the immune or stress endpoints examined in these studies may not always be predictive of avian susceptibility to infection (see Chapter 3). Thus, further work is needed to understand how land use impacts host physiology in ways directly relevant for susceptibility to pathogens. Finally, habitat clearing and any subsequent increases in temperature can impact vector physiology, including greater vectorial capacity (Lefèvre et al. 2013) and more aggressive biting behavior (Vittor et al. 2006).

9.2.2 Chemical pollutants (toxicants)

Virtually all types of anthropogenic land use change are accompanied by increases in local concentrations of various pollutants in the air, water, and soil. Toxicants are toxic substances introduced to the environment or produced in artificially high quantities by industrial activities such as manufacturing or transportation. The exact types of pollutants that serve as toxicants depend on the presence of industrial activity, control of unwanted plants or animals, and other anthropogenic activities. For instance, urban areas can have higher concentrations of lead in nearby soil due to vehicles on roads (Qiao et al. 2014), which may explain parallel increases in lead concentrations in some urban birds (e.g., kestrel, *Falco tinnunculus*; Garcia-Fernandez et al. 2005). In agricultural areas, large quantities of pesticides are applied to promote the growth of some species and inhibit others, and these products are associated with declines in aerial insectivorous birds (Spiller and Dettmers 2019). Although different pollutants enact their damaging effects on host physiology using different mechanisms, many interfere with metabolism and biochemical reactions, causing immunotoxic or immunosuppressive effects (Fernie et al. 2005; see also Chapter 3).

9.2.2.1 Heavy metals

Metals are deposited into the air, soil, and water through industrial activities such as mining, smelting, metallurgy, and vehicles on roads and the application of fertilizers, manures, and wastewater (He et al. 2005). Of these, heavy metals such as mercury (Hg), lead (Pb), cadmium (Cd), arsenic (As), chromium (Cr), and thallium (Tl) are harmful to wildlife and humans because they are toxic at low doses and cause organ damage, reduced reproductive success, and mortality in a wide variety of species (reviewed in Sánchez-Virosta et al. 2015). These elements are especially harmful because they are persistent in the environment, mobile, and able to bioaccumulate in tissues (Bauerová et al. 2017). Birds are often exposed to heavy metals through consuming contaminated food items such as fish and insects (Godwin et al. 2016). Likely due to this environmental and dietary exposure, the concentration of heavy metals is often higher in birds sampled in contaminated aquatic environments (Hollamby et al. 2004).

Increased exposure to heavy metals in contaminated landscapes can alter avian disease dynamics directly or indirectly. First, heavy metals can decrease host abundance, and thus contact rates between hosts and/or vectors, by causing direct mortality. Historically, the heavy metal most frequently associated with avian mortality has been lead and its distribution correlates to regions where hunting is legal and a common activity. Lead toxicity, through the ingestion of lead pellets used in shotgun shells, has (Pain 1996) and continues to kill thousands of waterfowl every year. Although mortalities have declined due to the 1985 US federal ban on the use of lead pellets for hunting waterfowl, lead pellets are still found in a wide range of waterfowl and predatory birds (Anderson et al.

2000). Lead toxicity is one of the main obstacles for the recovery of the California condor (*Gymnogyps californianus*; Kelly et al. 2014), which consume them by scavenging on remnants of game animals left by hunters. Likewise, birds of prey suffer significant mortality because they consume lead fragments contained in smaller prey, often left behind by hunters (e.g., prairie dogs [*Cynomys ludovicianus*] hunted as pest control; Stephens et al. 2005). Finally, population-level effects from lead toxicity have been documented for aquatic species that swallow fishing gear (e.g., loons; Haig et al. 2014). Although this has not been explicitly documented in birds, in some cases direct mortality from heavy metals may preferentially occur for hosts that are parasitized, which may have the effect of reducing disease transmission by removing infectious hosts from a population.

Sublethal effects to birds from heavy metals can result in the form of an increase in host susceptibility to pathogens. In experimental settings, birds fed metal-contaminated feed exhibited suppressed antibody production, leukocyte proliferation, and humoral and cell-mediated immune responses (Whitney and Cristol 2017). In free-living birds, individuals with greater metal contamination in their feathers can have higher stress hormone levels (e.g., cadmium and lead in common blackbirds; Meillère et al. 2016) and higher heterophil:lymphocyte ratios (e.g., great tits [*Parus major*]; Bauerová et al. 2017), indicating chronic stress. This relationship can be complicated and scale-dependent. Tree swallows (*Tachycineta bicolor*) feeding at mercury-contaminated sites exhibited reduced immune responses, but this relationship was not correlated with individual mercury loads (Hawley et al. 2009). Similarly, great tits sampled closer to a smelter exhibited lower humoral immune responses (Snoeijs et al. 2004). While exposure to lead, cadmium, and arsenic reduces host immune responses, Vermeujlen et al. (2015) found that there was no population-level impact. Still other studies find no effect of heavy metal exposure on immunocompetence (Baos et al. 2006) and some find that metal exposure stimulates immune activity (Eeva et al. 2005). This variation may be at least partially explained by the timing of exposure as heavy metal exposure is more damaging to immature birds than adults.

The consequences of these pollutant-driven physiological shifts for host susceptibility to infection are not well studied. In zebra finches (*Taeniopygia guttata*), experimental exposure to methylmercury did not significantly alter coccidian parasite load or clearance rate, although for a short period during the course of infection, methylmercury-exposed birds had slightly higher coccidian counts (Ebers Smith et al. 2018). In France, house sparrows in highly urban habitats had higher lead concentrations in their feathers and were more likely to be infected with a malarial parasite, *Plasmodium relictum*, although in that study they did not find a clear relationship between lead concentrations, infection risk, and body condition (Bichet et al. 2013). Given the variability in avian host responses to metal exposure and the levels of contamination in disturbed areas, this is a rich area for future study.

9.2.2.2 Persistent organic pollutants

Persistent organic pollutants (POPs) are organic compounds typically used as pesticides (e.g., organochlorine pesticides, OCPs), industrial chemicals (e.g., polychlorinated biphenyls, PCBs), and pharmaceuticals that are resistant to degradation in the environment (Jones and de Voogt 1999). One of the most infamous POPs is dichlorodiphenyltrichloroethane (DDT), the negative effects of which are well known for birds, most notably decreased reproductive success through eggshell thinning (Bitman et al. 1970). Other POP pesticides, including heptachlor and diazinon, have been implicated in large bird 'die-off' events, likely through acute toxicity (Mitra et al. 2011). POPs are also able to bioaccumulate in fatty animal tissues, rendering them more harmful to carnivorous birds such as urban Cooper's hawks (*Accipiter cooperii*; Brogan et al. 2017).

In addition to reduced reproductive success and mortality, which directly reduce host abundance, chronic exposure to POPs can alter disease dynamics by altering host immunity and thus susceptibility to parasites or pathogens. Organochlorine compounds such as PCBs have demonstrated immunotoxic effects in several wildlife species. For example, glaucous gulls (*Larus hyperboreus*) can exhibit impaired humoral and cell-mediated immune responses (Bustnes et al. 2004) and higher parasite loads (Sagerup et al. 2000) following PCB exposure. Because these compounds bioaccumulate in tissues, it is expected that these health effects will be more severe in predatory species at higher trophic levels.

While the application of pesticides in urban and agricultural lands will likely increase disease risks for birds by increasing susceptibility to infection, pesticides may also decrease the likelihood of encountering pathogen vectors. The impacts of pesticides on insect communities can be profound, reducing biomass by half in some contexts (Rioux et al. 2013). A decrease in vector abundance or diversity may explain why blood parasite prevalence is lower in urban birds (Santiago-Alarcon et al. 2020) and blood parasite richness is lower in disturbed (i.e., agricultural and urban) landscapes (Hernández-Lara et al. 2020). Studies that integrate vector abundance, host abundance, and host susceptibility in altered landscapes will be critical in disentangling these complex and sometimes opposing effects of anthropogenic land use change.

9.2.2.3 Other chemical pollutants

Second-generation anticoagulant rodenticides (SGARS) are potent rodenticides with long half-lives that increasingly appear to present a significant risk to wildlife. The US Environmental Protection Agency (EPA) determined that SGARs present a significant risk to wildlife (Erickson and Urban 2004), leading to the ban of most SGARs for consumer use, although use by pest managers and agricultural managers is still allowed. Thus toxicity from SGARs is associated with areas where pest control is needed and heavily managed. Primary mortality from consumption of prey affected by rodenticides has been recognized for some time. However, through improvement in diagnostic techniques and the recognition of sublethal lesions, secondary rodenticide toxicity is now seen as a significant health problem for birds of prey, especially species whose diet contains a large proportion of pest rodents (Murray 2017). In at least one study, rodenticide exposure has been associated with infection risk. In Spain, liver concentrations of rodenticides clorophacinone and flocoumafen were positively correlated with parasite richness (i.e., number of parasite species) in great bustards (*Otis tarda*; Lemus et al. 2011). Further study is needed to assess the impacts of rodenticide exposure for avian immune function and susceptibility to infection.

The impact of plastics on birds from consumption and entanglement cannot be ignored. Although population-level effects have not been clearly elucidated, ongoing research may bring to light the potential for plastics to be one of the most significant causes of direct and indirect mortality for all organisms, including birds. The consumption of plastic by birds has been documented since the early 1960s; however, only recently has the frequency of mortality events, particularly affecting seabirds, come to light (Wilcox et al. 2015). In addition to large particles that cause inflammation, malabsorption, and blockage of the gastrointestinal tract, microplastics have been found in seawater and in fish and various other items consumed by birds (Savoca et al. 2016). Further, microplastics can bind to other pollutants, such as heavy metals and PCBs, and thus their consumption can have an additive effect by leading to a higher consumption of pollutants (Jinhui et al. 2019). Lastly, compounds leaching from plastics themselves (e.g., OCPs and PCBs) have been known to cause a variety of health effects for birds (Colabuono et al. 2010; Teuten et al. 2009). Due to poor waste management systems, plastic products can end up in regions far from their production and use (Eriksen et al. 2014). The impact of plastics on pathogen dynamics is an area ripe for future study in birds and other wildlife.

9.2.3 Light and noise pollution

Although not a pollutant in the traditional sense, light pollution is gaining attention for its pervasive effects on ecosystem function (Navara and Nelson 2007). Light pollution, otherwise known as artificial light at night, occurs when light levels at night exceed 10% of natural illumination from lighting roads, homes, commercial buildings, and vehicles (Le Tallec 2019). Light pollution is now considered to be the fastest growing type of pollution on Earth, increasing on average by 6% of its global extent per year (Hölker et al. 2010).

Wildlife exposed to light pollution undergo a variety of behavioral and physiological shifts in the timing of activity, reproduction, and migration

(Ouyang et al. 2017). Although relatively much less is known about the impacts of light pollution on disease ecology, it is plausible that such conditions could impact avian susceptibility to infection. Artificial light at night has been associated with higher baseline corticosterone in great tits (Ouyang et al. 2015), which may decrease host immune function if such stressors are long term. Few studies to date have linked artificial light at night with infection risk in birds. Kernbach et al. (2019) experimentally exposed house sparrows to artificial light at night to test its effects on West Nile virus (WNV) infections. Sparrows exposed to artificial light for most of the night were on average infectious for two days longer than control birds, which the authors estimated could increase the risk of a WNV outbreak by an estimated 41% (Kernbach et al. 2019). In the same study, birds exposed to artificial light at night did not appear to have altered stress responses; however, immune responses appeared to be dysregulated based on changes in gene expression. In a related study, also on house sparrows, artificial light at night suppressed melatonin levels and increased WNV-induced mortality (Kernbach et al. 2020). In addition to physiological changes in hosts, artificial light at night may change the timing or frequency of interactions between hosts and vectors (Kernbach et al. 2018). For example, the crepuscular periods when both mosquitoes and sparrows are active may be extended in illuminated areas.

Noise pollution, or the elevation of background noise by anthropogenic activities such as vehicular traffic, also has diverse demonstrated effects on birds. In terms of effects potentially relevant to disease dynamics, studies have shown that birds exposed to traffic noise have elevated stress levels (Injaian et al. 2018). Few studies have examined the effect of traffic noise on immune function and what has been done shows no clear relationship (Crino et al. 2013). Nevertheless, if traffic noise promotes chronic stress in birds, it may ultimately increase avian susceptibility to infection over time.

9.2.4 Anthropogenic resources

A natural consequence of land transformation, including urbanization, is the creation and concentration of resources, such as water and food, that attract birds. These resource sites for wild birds can be either intentional (e.g., bird feeders, zoos, parks where birds are fed) or unintentional (e.g., water features, landfills, wastewater treatment sites, constructed wetlands). Bird feeding is a multi-billion dollar industry in the US, where approximately 60 million households purchase bird feed and related items annually (US Department of the Interior et al. 2018). Common to these sites is the concentration of large numbers of birds of various species comingling in small areas, which can have specific implications for contact rates within and between host species, nutrition, and physical condition (Becker et al. 2015; Murray et al. 2016). Some types of anthropogenic resources also concentrate pathogen loads on inanimate surfaces ('fomites') that hosts contact and/or facilitate longer pathogen survival in the external environment, thus increasing the rate at which hosts contact pathogens in the environment (see also Chapter 4).

9.2.4.1 Supplemental feeding

Bird feeders are known to facilitate pathogen transmission for what are commonly known as 'backyard birds' (Passeriformes, Columbiformes, Piciformes, etc.). Evidence exists for bird feeders facilitating the transmission of the pathogens that cause at least five notorious 'feeder-associated' diseases: mycoplasmosis, aspergillosis, trichomoniasis, salmonellosis, and avian pox (Adelman et al. 2015); however, others such as avian mange and other types of infestations with macroparasites are likely to also have associations with bird feeders. For these diseases, the spatiotemporal overlap between feeding activities and outbreaks and the specific characteristics of the etiologic agent (e.g., the environmental persistence of *Salmonella*) have lent evidence to this association (Altizer et al. 2004; Lawson et al. 2018; Tizard 2004). For example, avian salmonellosis outbreaks tend to occur in the winter, when bird feeding activities peak, and the persistence of *Salmonella* on avian feces on feeders, particularly those that allow for defecation on food items, has been well demonstrated (Lawson et al. 2014). Unfortunately, pathogen transmission at bird feeders can also have public health significance (see also Chapter 12), as *Salmonella* strains circulating among passerines during outbreaks in both the UK and

US match genotypes also known to cause human salmonellosis

Feeders can impact pathogen dynamics in birds in myriad ways. In addition to behaviorally concentrating hosts, which augments contact rates within and among bird species, feeders modulate one of the major drivers of population regulation—food. Thus, supplemental resources from feeders can result in a higher abundance of avian hosts (Plummer et al. 2019), potentially increasing the magnitude of infectious disease outbreaks. Bird feeders and water baths, another form of supplemental resource for wildlife, also serve as environmental fomites for the persistence and spread of many types of pathogens, including fecal–oral (e.g., *Salmonella*), those spread by salivary or crop secretions (e.g., *Trichomonas gallinae*; Purple et al. 2015), and those spread from conjunctival secretions (i.e., *Mycoplasma gallisepticum*; Dhondt et al. 2007; Moyers et al. 2018).

Feeders may also indirectly influence disease dynamics by altering avian community composition. A recent study in the UK found that bird feeders were associated with increasing bird diversity and community evenness over time, which the authors attribute to the diversification of bird food types offered in backyards (Plummer et al. 2019). In New Zealand, Galbraith et al. (2015) found that supplemental feeding heavily impacted avian assemblages, with an increase in introduced species and a negative effect on a native insectivore. Changes in community structure due to bird feeders may have important consequences for pathogen transmission, particularly for multihost pathogens, because some species are more competent reservoirs than others. For example, in North America, pine siskins (*Spinus pinus*) have been suspected to be important in the epidemiology of avian salmonellosis because they form large flocks and are very aggressive at feeders (Daoust and Prescott 2007; Hernandez et al. 2012). The ability of feeders to amplify or dilute pathogen transmission by altering avian community composition is an important area for future work.

Although feeders are generally predicted to augment disease outbreaks through the above mechanisms, supplemental feeding can also potentially dampen disease outbreaks by augmenting host immunocompetence through improved nutritional condition (Becker et al. 2018). Although nutritional status has long been associated with immunocompetence in birds (Klasing 1998; see also Chapter 3), the relationships among backyard feeders, nutritional status, and infection prevalence are likely complex and species-specific (Galbraith et al. 2017). For example, Wilcoxen et al. (2015) manipulated bird feeder presence at forested sites and found that birds at feeder-supplemented sites had higher body condition but also higher prevalence of clinical signs of parasite and pathogen infection. Thus, while feeders may augment nutritional status for some species, the increased contact rates associated with feeders may overwhelm any nutritional benefits. Additionally, commercial bird seed diets are likely deficient in important micronutrients and overdependence might negatively impact immunity, although this has not been specifically tested.

The role of individual heterogeneity in the use of bird feeders and the consequences of this variation for disease transmission are areas ripe for further study. Adelman et al. (2015) found that house finches (*Haemorhous mexicanus*) that spend the longest amount of time on feeders per day are more likely to both acquire and transmit *M. gallisepticum*. In fact, behavioral plasticity is a characteristic of birds that successfully live near humans, such as those that take advantage of backyard feeders. Yet innovative foraging strategies are associated with higher contact with pathogens (Audet et al. 2016). Future work should examine how heterogeneity within and among avian species in feeder use impacts pathogen dynamics via diverse mechanisms. Although there is growing evidence that backyard bird feeders augment pathogen risk for a suite of bird species, the social impact it has on people and their connection with nature may be significant enough to discourage restrictive management (Cox and Gaston 2015).

9.2.4.2 Unintentional anthropogenic resources

The role of unintentional anthropogenic resources, such as water features, landfills, wastewater treatment sites, animal feedlots, and constructed wetlands, in driving avian disease dynamics is relatively less studied than the role of backyard bird feeding. Nonetheless, unintentional resources likely have very similar effects to those of backyard bird feeders by augmenting contact between individuals within and among species. Here, we briefly highlight two characteristics that are unique to some forms of unintentional anthropogenic resources. First, sites like animal feedlots can bring wild birds into close proximity with agricultural animals, increasing the risk of transmission of pathogens across the agricultural–wild bird interface (see also Chapter 11). For example, sites like dairy farms are preferred foraging sites for American crows (*Corvus brachyrhynchos*), and foraging on highly concentrated resources likely puts them at risk of acquiring *Campylobacter jejuni* via fecal–oral transmission (Taff et al. 2016). Notably, American crows carry strains of *C. jejuni* that are similar to human and livestock strains (Weis et al. 2014), suggesting the potential for transmission across the wild bird–human–agricultural interface (see also Chapter 12).

Unintentional anthropogenic resources such as wastewater treatment sites also have the potential to augment pathogen persistence and virulence. For example, *Clostridium botulinum* toxicity is common in wastewater treatment plants because key environmental factors for toxin production (high temperatures, low oxygen tension, and decay of organic material) are inherent in these systems. Similarly, in the southeast US, mortality of wading birds, particularly of juveniles, from high loads of the intestinal nematode *Eustrongylides ignotus* is known to occur in sites with high nutrient inputs and is considered a risk for birds using wastewater ponds (Coyner et al. 2002). Antimicrobial resistance development in wastewater treatment ponds and the role that birds play in their carriage and dissemination is also a recently recognized concern (Martín-Maldonado et al. 2020). Although unlikely to cause outright mortality, much like *C. jejuni* transmission, it provides a One Health framework with which to examine how avian and public health are rooted in the health of the environment (see Section 9.4).

9.3 Overall consequences of land use change for avian parasitism: urbanization as a case study

Identifying generalizable relationships between land use change and avian disease dynamics is important for predicting the impacts of future development for avian conservation, community ecology, and zoonotic disease risk. As may be evident from the previous sections, characterizing such an overall relationship is difficult due to the myriad effects of land use change that simultaneously affect contact rates between birds, contact between birds and vectors, and infection competence and susceptibility. Previous reviews have examined the impacts of land use change on wildlife disease prevalence across many taxa (Brearley et al. 2013; Gottdenker et al. 2014) and for birds in specific habitat types (Martin and Boruta 2014; Sehgal 2010; Sehgal 2015). Such efforts have found marked variation in the impacts of land use change on wildlife health depending on the type of land use change and mechanism of pathogen transmission. For example, Brearley et al. (2013) reported that just over half (53%) of wildlife studies detected an increase in infection prevalence due to human-associated land use change, especially in urban landscapes, while Chung et al. (2018) found that proportionately more bacterial pathogens had been reported in birds sampled in undisturbed landscapes relative to agricultural and urbanizing areas. Such variation is expected given the many mechanisms that may alter disease dynamics following changes in land use (Figure 9.1). To reconcile contrasting results between studies, mechanistic approaches that combine changes in communities (e.g., community composition) and individuals (e.g., susceptibility) with relevant abiotic factors (e.g., temperature gradients) could help identify the most salient changes that promote infection. Meta-analyses could also help identify more generalizable patterns to overcome site-specific effects.

There has been a recent surge in interest in the effects of urbanization, in particular on avian disease ecology, due to the rapid increase in urban areas around the globe. Much like the broader studies of land use changes noted above, studies of urbanization and avian health have found positive, negative, or no detectable relationships between urbanization and parasite prevalence, potentially due to variation in urban landscapes and host, vector, and pathogen life histories. Parasite prevalence in some systems is higher for urban birds. For example, urban house finches had a higher prevalence of poxvirus infections and more severe coccidial infections relative to conspecifics in less urban habitats (Giraudeau et al. 2014). In other systems, urban birds fare better than their rural counterparts (e.g., Eurasian blackbirds and *Haemoproteus* spp.; Geue and Partecke 2008). In several studies, the relationships between urbanization and disease risk are complex. For example, urban blackbirds (*Turdus merula*) in Poland were less likely to carry ticks, but ticks in urban areas were more likely to be infected with *Borrelia* spp. (Gryczyńska 2018). For perhaps the quintessential urban bird, the pigeon or rock dove (*Columba livia*), the prevalence of haemosporidian parasites (*Haemoproteus* spp. and *Plasmodium* spp.) increased with urbanization for paler but not darker morphotypes (Jacquin et al. 2013). Because urbanization can affect a variety of biotic and abiotic conditions that impact host ecology, host physiology, and vector ecology, synthetic approaches to urban avian disease ecology are helpful in disentangling this complexity.

Given the diverse sources of variation underlying these host–pathogen relationships, a meta-analysis was conducted to identify any generalizable relationships between urbanization and wildlife health (Murray et al. 2019). This analysis included 106 studies found through a systematic search that compared urban and non-urban populations of the same species in terms of four measures of health: stress biomarkers, body condition, toxicant concentrations, and parasite prevalence or intensity (see Murray et al. 2019 for details on metrics). Of these, 46 studies focused on birds and included 159 different health comparisons (Figure 9.2). We analyzed this subset focused on birds using mixed effects models that included the study's standardized effect size (r) as the response variable (with positive values representing a positive health effect), health metric as a fixed effect, and study as a random effect. Urban birds had significantly higher toxicant concentrations than their rural counterparts, as reflected by negative effect sizes (Figure 9.2A), but there were no statistically significant differences in the other examined traits, including body condition, stress biomarkers, or parasitism (either prevalence or intensity) (for details on methods and these results see Murray et al. 2019).

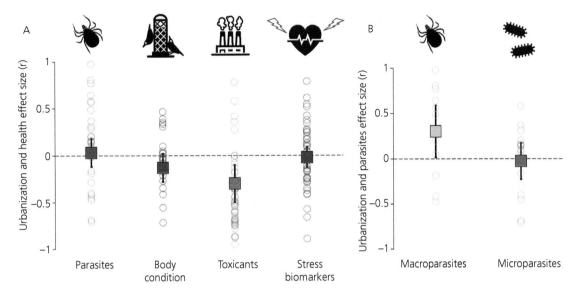

Figure 9.2 Living in urban areas might affect several metrics of avian health in positive (i.e., higher r values) or negative (i.e., lower r values) ways. (A) Relationships between urbanization and four metrics of health from 46 studies comparing the health of urban and non-urban bird populations (see Murray et al. 2019 for full results and explicit definitions of health metrics used for each category). Squares represent predicted means and 95% confidence intervals from mixed-effects models. (B) The 26 studies that quantified parasite prevalence in urban and non-urban populations are divided into macroparasites (e.g., arthropods, helminths) and microparasites (e.g., bacteria, viruses; see Chapter 2 for more on micro- and macroparasites). For both graphs, a positive value of r indicates a positive health effect (i.e., lower parasite prevalence, higher body condition, lower toxicant load, lower stress biomarkers) associated with urbanization. (Icons were created by Stewart Lamb Crowar, Maxim Kulikov, Iconographer, and Symbolon from Noun Project.)

To determine whether effects of urbanization emerge only for certain parasite types, we examined the subset of avian studies that quantified differences in parasite prevalence (n = 26) to determine whether urban versus rural birds differ in the prevalence of microparasites (bacteria: n = 1 studies, viruses: n = 3, protozoa: n = 14) versus macroparasites (helminths: n = 3, arthropods: n = 9). Within this subset, urban birds were more likely to have a lower prevalence of macroparasite infections relative to rural birds (i.e., a positive health effect and positive r value), but we detected no statistically significant differences in microparasite prevalence between the two groups (Figure 9.2B, n = 26; macroparasites: $\beta = 0.31 \pm 0.14$, p = 0.05; microparasites: $\beta = -0.37 \pm 0.17$, p = 0.15). This is consistent with the hypothesis that urban birds may benefit from the anthropogenic resources associated with urbanization in some way—perhaps because they enable hosts more time to remove macroparasitic anthropods (i.e., ectoparasites) via grooming (Murray et al. 2018). Thus, though urban birds may

experience negative health effects via increased exposure to toxicants, these potential negative health consequences are not reflected in parasite prevalence, which instead was lower for some types of parasites for urban birds. The relationships between urbanization and avian disease are complex and require more understanding of the underlying mechanisms associated with urban land use change.

Part of the observed variation in responses to urbanization likely stems from shifts in abiotic conditions and their impacts on vector populations. For instance, the heat island effect (see Section 9.1.1) may increase the risk of vector-borne pathogens for birds via mosquito abundance and biting behavior (Vittor et al. 2006). Novel microhabitats in urban environments can also promote standing water and provide breeding habitat for *Aedes aegypti* and *Culex quinquefasciatus*. These changes in vector ecology may be driving observed increases in avian blood parasites, such as West Nile virus (e.g., Bradley et al. 2008), in urban areas. Other vectors can be negatively impacted by urbanization if they rely on particular vegetation types to find hosts, for example as occurs for *Ixodes* ticks in unmanaged grassy areas. Blackbirds sampled in urban sites in Germany were less likely to be infested with *Ixodes* ticks than in non-urban forests, potentially because urban habitats are less suitable for tick populations (Geue and Partecke 2008). Greater collaboration between ornithologists and entomologists will help identify such links between vector ecology and avian disease prevalence in urban areas.

9.4 Synthesis: need for multidisciplinary approaches

To understand the broader consequences of land use change for avian pathogen prevalence, it can be useful to apply the concept of One Health which encourages viewing health challenges through a multidisciplinary lens that emphasizes the commonalities among human, animal, and environmental health. In contexts in which land use change promotes higher pathogen prevalence in avian populations, there will likely be several impacts on other sectors, such as public, domestic, and environmental health. These effects are most notable for multi-host pathogens that, in addition to wild birds, can infect other wildlife, domestic animals, and/or humans (see also Chapters 11 and 12). With an increase in infected avian hosts, the prevalence of infected vectors (for vector-borne systems) will likely also increase, potentially leading to a higher infection prevalence in any non-avian hosts for a given system. For pathogens that are persistent in the environment (e.g., *Salmonella*), shifts in land use that promote infection in birds may lead to a higher environmental pathogen pool through fecal contamination, resulting in potential health risks for both domestic animals and humans. Additionally, birds can act as important and tractable means of monitoring other environmental sources of *Salmonella* spp. For example, a high diversity of *Salmonella* genotypes isolated from white ibis (*Eudocimus albus*) in urban areas indicates that they are reflecting environmental sources, which also infect people (Hernandez et al. 2016).

This increased shedding of pathogens in local environments can act in concert with increased opportunities for direct contact between wild birds

and domestic animals in urban and agricultural settings, reinforcing positive feedback cycles between land use change and pathogen dynamics that favor avian disease outbreaks. Furthermore, if high infection prevalence leads to high mortality rates for birds (e.g., from botulism), these processes may lead to disruptions in ecosystem services provided by wild birds such as seed dispersal, pollination, and consuming pests and pathogen vectors. All of these processes (vector infection rates, environmental contamination, domestic animal health, ecosystem services) impact public health through the transmission of zoonotic pathogens or environmental health. Thus, the health of birds in changing landscapes is intimately connected to the health of other wildlife, domestic animals, and humans and is fundamentally a One Health issue (see Chapter 13).

A growing body of work has documented the impacts of land use change for avian disease ecology in varied contexts, but the complexity and variability of study results highlight areas for future research. As previously mentioned, more studies are needed that link abiotic and biotic factors accompanying land use change with specific epidemiological processes such as host contact or physiology (Figure 9.1). If studies are timed appropriately, land use changes can provide 'natural' experiments and opportunities to measure changes in prevalence or other relevant factors following a before–after control–impact design. An increased focus on particular mechanisms (e.g., temperature gradients) could also help tease apart the effects of land use change from global climate change (see Chapter 10). Importantly, land use changes do not act in isolation from other anthropogenic effects, and interactions between land use and climate-driven changes in disease dynamics are thus an important avenue for future research.

Given the variability in avian health with urbanization, systematic multicity comparisons are needed to examine the role of urban development and study location as contributing factors (Magle et al. 2019). Such studies will also benefit from greater knowledge about biologically relevant thresholds for toxicants in avian tissues arising from air and water pollutants. The effects of food provisioning are multifaceted, with potential benefits (e.g., lower nutritional stress) and harmful effects (e.g., aggregation) on avian health. Field studies and experimental approaches can help identify the relative importance of diet quality, contact rates, and movement on pathogen exposure for urban birds. Studies addressing the mechanisms for the issues listed above will help predict changes in avian disease dynamics to advance bird conservation and public health. Finally, it is important to consider that here we have primarily discussed the consequences of land use change for parasite prevalence or intensity in birds; however, relatively few studies have directly linked parasite load to wild bird health or fitness (Delgado-V. and French 2012; see also Chapter 3). Ultimately, linking host–pathogen dynamics with avian survival and reproduction can inform management strategies to mitigate bird declines in changing landscapes.

Acknowledgments

We thank Cecilia A. Sánchez, Daniel J. Becker, Kaylee A. Byers, Katherine E.L. Worsley-Tonks, and Meggan E. Craft for their roles in collecting the data associated with the urban bird meta-analysis and Figure 9.2.

Literature cited

Adelman, J.S., Moyers, S.C., Farine, D.R., and Hawley, D.M. (2015). Feeder use predicts both acquisition and transmission of a contagious pathogen in a North American songbird. *Proceedings of the Royal Society B: Biological Sciences*, 282, 20151429.

Afrane, Y.A., Little, T.J., Lawson, B.W., Githeko, A.K., and Yan, G. (2008). Deforestation and vectorial capacity of *Anopheles gambiae* Giles mosquitoes in malaria transmission, Kenya. *Emerging Infectious Diseases*, 14(10), 1533–8.

Altizer, S., Hochachka, W.M., and Dhondt, A.A. (2004). Seasonal dynamics of mycoplasmal conjunctivitis in eastern North American house finches. *Journal of Animal Ecology*, 73(2), 309–22.

Anderson, W. L., Havera, S. P., and Zercher, B. W. (2000). Ingestion of lead and nontoxic shotgun pellets by ducks in the Mississippi Flyway. *Journal of Wildlife Management*, 64(3), 848–57.

Antrop, M. and Van Eetvelde, V. (2000). Holistic aspects of suburban landscapes: visual image interpretation and landscape metrics. *Landscape and Urban Planning*, 50(1–3), 43–58.

Atkinson, C.T. and LaPointe, D.A. (2009). Introduced avian diseases, climate change, and the future of Hawaiian honeycreepers. *Journal of Avian Medicine and Surgery*, 23(1), 53–63.

Audet, J.-N., Ducatez, S., and Lefebvre, L. (2016). The town bird and the country bird: problem solving and immunocompetence vary with urbanization. *Behavioral Ecology*, 27(2), 637–44.

Baos, R., Jovani, R., Forero, M.G., et al. (2006). Relationships between T-cell-mediated immune response and Pb, Zn, Cu, Cd, and As concentrations in blood of nestling white storks (*Ciconia ciconia*) and black kites (*Milvus migrans*) from Doñana (southwestern Spain) after the Aznalcóllar toxic. *Environmental Toxicology and Chemistry*, 25(4), 1153–9.

Bauerová, P., Vinklerová, J., Hraníček, J., et al. (2017). Associations of urban environmental pollution with health-related physiological traits in a free-living bird species. *Science of the Total Environment*, 601–602, 1556–65.

Becker, D.J., Streiker, D.G., and Altizer, S. (2015). Linking anthropogenic resources to wildlife–pathogen dynamics: a review and meta-analysis. *Ecology Letters*, 18(5), 483–95.

Becker, D.J., Hall, R.J., Forbes, K.M., Plowright, R.K., and Altizer, S.M. (2018). Anthropogenic resource subsidies and host–parasite dynamics in wildlife. *Philosophical Transactions of the Royal Society B: Biological Sciences*, 373(1745).

Bichet, C., Scheifler, R., Cœurdassier, M., Julliard, R., Sorci, G., and Loiseau, C. (2013). Urbanization, trace metal pollution, and malaria prevalence in the house sparrow. *PLoS One*, 8(1), e53866.

Bitman, J., Cecil, H., and Fries, G. (1970). DDT-induced inhibition of avian shell gland carbonic anhydrase: a mechanism for thin eggshells. *Science*, 168(3931), 594–6.

Boal, C.W., Mannan, W., and Hudelson, K.S. (1998). Trichomoniasis in Cooper's hawks from Arizona. *Journal of Wildlife Diseases*, 34(3), 590–3.

Bradley, C.A. and Altizer, S. (2007). Urbanization and the ecology of wildlife diseases. *Trends in Ecology and Evolution*, 22(2), 95–102.

Bradley, C.A., Gibbs, S.E.J., and Altizer, S. (2008). Urban land use predicts West Nile virus exposure in songbirds. *Ecological Applications*, 18(5), 1083–92.

Brearley, G., Rhodes, J., Bradley, A., et al. (2013). Wildlife disease prevalence in human-modified landscapes. *Biological Reviews*, 88(2), 427–42.

Brogan, J.M., Green, D.J., Maisonneuve, F., and Elliott, J.E. (2017). An assessment of exposure and effects of persistent organic pollutants in an urban Cooper's hawk (*Accipiter cooperii*) population. *Ecotoxicology*, 26(1), 32–45.

Bustnes, J.O., Hanssen, S.A., Folstad, I., Erikstad, K.E., Hasselquist, D., and Skaare, J.U. (2004). Immune function and organochlorine pollutants in arctic breeding glaucous gulls. *Archives of Environmental Contamination and Toxicology*, 47(4), 530–41.

Chasar, A., Loiseau, C., Valkiunas, G., Iezhova, T., Smith, T.B., and Sehgal, R.N.M. (2009). Prevalence and diversity patterns of avian blood parasites in degraded African rainforest habitats. *Molecular Ecology*, 18(19), 4121–33.

Chung, D.M., Ferree, E., Simon, D.M., and Yeh, P.J. (2018). Patterns of bird–bacteria associations. *EcoHealth*, 15(3), 627–41.

Cīrule, D., Krama, T., Krams, R., et al. (2017). Habitat quality affects stress responses and survival in a bird wintering under extremely low ambient temperatures. *Die Naturwissenschaften*, 104(11–12), 99.

Colabuono, F.I., Taniguchi, S., and Montone, R.C. (2010). Polychlorinated biphenyls and organochlorine pesticides in plastics ingested by seabirds. *Marine Pollution Bulletin*, 60(4), 630–4.

Cox, D.T.C. and Gaston, K.J. (2015). Likeability of garden birds: Importance of species knowledge and richness in connecting people to nature. *PLoS ONE*, 10(11), 1–14.

Coyner, D.F., Spalding, M.G., and Forrester, D.J. (2002). Epizootiology of *Eustrongylides ignotus* in Florida: transmission and development of larvae in intermediate hosts. *Journal of Wildlife Diseases*, 38(3), 483–99.

Crino, O.L., Johnson, E.E., Blickley, J.L., Patricelli, G.L., and Breuner, C.W. (2013). Effects of experimentally elevated traffic noise on nestling white-crowned sparrow stress physiology, immune function and life history. *Journal of Experimental Biology*, 216(11), 2055–62.

Daoust, P. and Prescott, J. (2007). Salmonellosis. In N. Thomas, D. Hunter, and C.T. Atkinson, eds. *Infectious Diseases of Wild Birds*, pp. 270–88. Blackwell Publishing, Oxford.

Delgado-V., C.A. and French, K. (2012). Parasite–bird interactions in urban areas: current evidence and emerging questions. *Landscape and Urban Planning*, 105(1–2), 5–14.

Dhondt, A.A., Dhondt, K.V., Hawley, D.M., and Jennelle, C.S. (2007). Experimental evidence for transmission of *Mycoplasma gallisepticum* in house finches by fomites. *Avian Pathology*, 36(3), 205–8.

Dudley, N. and Alexander, S. (2017). Agriculture and biodiversity: a review. *Biodiversity*, 18, 45–9.

Ebers Smith, J.H., Cristol, D.A., and Swaddle, J.P. (2018). Experimental infection and clearance of coccidian parasites in mercury-exposed zebra finches. *Bulletin of Environmental Contamination and Toxicology*, 100, 89–94.

Eeva, T., Hasselquist, D., Langefors, Å., Tummeleht, L., Nikinmaa, M., and Ilmonen, P. (2005). Pollution related effects on immune function and stress in a free-living population of pied flycatcher *Ficedula hypoleuca*. *Journal of Avian Biology*, 36(5), 405–12.

Ellis, E.C. and Ramankutty, N. (2008). Putting people in the map: anthropogenic biomes of the world. *Frontiers in Ecology and the Environment*, 6(8), 439–47.

Erickson, W. and Urban, D. (2004). *Potential Risks of Nine Rodenticides to Birds and Nontarget Mammals: A Comparative Approach*. US EPA, Washington, DC.

Eriksen, M., Lebreton, L.C.M., Carson, H.S., et al. (2014). Plastic pollution in the world's oceans: more than 5 trillion plastic pieces weighing over 250,000 tons afloat at sea. *PLoS ONE*, 9(12), e111913.

Fahrig, L. (2003). Effects of habitat fragmentation on biodiversity. *Annual Review of Ecology, Evolution and Systematics*, 34, 487–515.

Fernie, K.J., Mayne, G., Shutt, J.L., et al. (2005). Evidence of immunomodulation in nestling American kestrels (*Falco sparverius*) exposed to environmentally relevant PBDEs. *Environmental Pollution*, 138(3), 485–93.

Gaertner, M., Wilson, J.R.U., Cadotte, M.W., MacIvor, J.S., Zenni, R.D., and Richardson, D.M. (2017). Non-native species in urban environments: patterns, processes, impacts and challenges. *Biological Invasions*, 19(12), 3461–9.

Galbraith, J.A., Beggs, J.R., Jones, D.N., and Stanley, M.C. (2015). Supplementary feeding restructures urban bird communities. *Proceedings of the National Academy of Sciences of the United States of America*, 112(20), E2648–57.

Galbraith, J. A., Stanley, M. C., Jones, D. N., and Beggs, J. R. (2017). Experimental feeding regime influences urban bird disease dynamics. *Journal of Avian Biology*, 48, 700–13.

Garcia-Fernandez, A., Romero, D., Martinez-Lopez, E., Navas, I., Pulido, M., and Maria-Mojica, P. (2005). Environmental lead exposure in the European kestrel (*Falco tinnunculus*) from southeastern Spain: the influence of leaded gasoline regulations. *Bulletin of Environmental Contamination and Toxicology*, 74, 314–19.

Geue, D. and Partecke, J. (2008). Reduced parasite infestation in urban Eurasian blackbirds (*Turdus merula*): a factor favoring urbanization? *Canadian Journal of Zoology*, 86, 1419–25.

Giraudeau, M., Mousel, M., Earl, S., and McGraw, K. (2014). Parasites in the city: degree of urbanization predicts poxvirus and coccidian infections in house finches (*Haemorhous mexicanus*). *PLoS ONE*, 9(2), e86747.

Godwin, C.M., Smits, J.E.G., and Barclay, R.M.R. (2016). Metals and metalloids in nestling tree swallows and their dietary items near oilsands mine operations in northern Alberta. *Science of the Total Environment*, 562, 714–23.

Gottdenker, N.L., Streicker, D.G., Faust, C.L., and Carroll, C.R. (2014). Anthropogenic land use change and infectious diseases: a review of the evidence. *EcoHealth*, 11, 619–32.

Gryczyńska, A. (2018). Urban and forest-living blackbirds *Turdus merula* as hosts of *Borreliella* spp. infected ticks. *Polish Journal of Ecology*, 66(3), 309–14.

Haig, S.M., D'Elia, J., Eagles-Smith, C., et al. (2014). The persistent problem of lead poisoning in birds from ammunition and fishing tackle. *The Condor*, 116(3), 408–28.

Hamilton, A.J. (2007). Potential microbial and chemical hazards to waterbirds at the Western Treatment Plant. *Ecological Management and Restoration*, 8(1), 38–41.

Hawley, D.M., Hallinger, K.K., and Cristol, D.A. (2009). Compromised immune competence in free-living tree swallows exposed to mercury. *Ecotoxicology*, 18, 499–503.

He, Z.L., Yang, X.E., and Stoffella, P.J. (2005). Trace elements in agroecosystems and impacts on the environment. *Journal of Trace Elements in Medicine and Biology*, 19(2–3), 125–40.

Hernandez, S.M., Keel, K., Sanchez, S., et al. (2012). Epidemiology of a *Salmonella enterica* subsp. *enterica* serovar *typhimurium* strain associated with a songbird outbreak. *Applied and Environmental Microbiology*, 78(20), 7290–8.

Hernandez, S.M., Welch, C.N., Peters, V.E., et al. (2016). Urbanized white ibises (*Eudocimus albus*) as carriers of *Salmonella enterica* of significance to public health and wildlife. *PLoS ONE*, 11(10), 1–22.

Hernández-Lara, C., Carbó-Ramírez, P., and Santiago-Alarcon, D. (2020). Effects of land use change (rural–urban) on the diversity and epizootiological parameters of avian *Haemosporida* in a widespread neotropical bird. *Acta Tropica*, 209, 105542.

Hölker, F., Moss, T., Griefahn, B., et al. (2010). The dark side of light: a transdisciplinary research agenda for light pollution policy. *Ecology and Society*, 15(4), 13–24.

Hollamby, S., Afema-azikuru, J., Sikarskie, J.G., et al. (2004). Mercury and persistent organic pollutant concentrations in African fish eagles, Marabou storks, and Nile tilapia in Uganda. *Journal of Wildlife Diseases*, 40(3), 501–14.

Injaian, A.S., Taff, C.C., Pearson, K.L., Gin, M.M.Y., Patricelli, G.L., and Vitousek, M.N. (2018). Effects of experimental chronic traffic noise exposure on adult and nestling corticosterone levels, and nestling body condition in a free-living bird. *Hormones and Behavior*, 106(July), 19–27.

Jacquin, L., Récapet, C., Prévot-Julliard, A.C., et al. (2013). A potential role for parasites in the maintenance of color polymorphism in urban birds. *Oecologia*, 173(3), 1089–99.

Jinhui, S., Sudong, X., Yan, N., Xia, P., Jiahao, Q., and Yongjian, X. (2019). Effects of microplastics and attached heavy metals on growth, immunity, and heavy metal accumulation in the yellow seahorse, *Hippocampus kuda* Bleeker. *Marine Pollution Bulletin*, 149(August), 110510.

Jones, K. and de Voogt, P. (1999). Persistent organic pollutants (POPs): state of the science. *Environmental Pollution*, 100, 209–21.

Kelly, T.R., Grantham, J., George, D., et al. (2014). Spatiotemporal patterns and risk factors for lead exposure in endangered California condors during 15 years of reintroduction. *Conservation Biology*, 28(6), 1721–30.

Kernbach, M.E., Hall, R.J., Burkett-Cadena, N.D., Unnasch, T.R., and Martin, L.B. (2018). Dim light at night: physiological effects and ecological consequences for infectious disease. *Integrative and Comparative Biology*, 58(5), 995–1007.

Kernbach, M.E., Newhouse, D.J., Miller, J.M., et al. (2019). Light pollution increases West Nile virus competence of a ubiquitous passerine reservoir species. *Proceedings of the Royal Society B: Biological Sciences*, 286(1907), 20191051.

Kernbach, M.E., Cassone, V.M., Unnasch, T.R., and Martin, L.B. (2020). Broad-spectrum light pollution suppresses melatonin and increases West Nile virus-induced mortality in house sparrows (*Passer domesticus*). *The Condor*, 122, 1–13.

Klasing, K.C. (1998). Nutritional modulation of resistance to infectious diseases. *Poultry Science*, 77(8), 1119–25.

Lawson, B., De Pinna, E., Horton, R.A., et al. (2014). Epidemiological evidence that garden birds are a source of human salmonellosis in England and Wales. *PLoS ONE*, 9(2), 1–10.

Lawson, B., Robinson, R.A., Toms, M.P., Risely, K., Macdonald, S., and Cunningham, A.A. (2018). Health hazards to wild birds and risk factors associated with anthropogenic food provisioning. *Philosophical Transactions of the Royal Society B: Biological Sciences*, 373(1745).

Le Tallec, T. (2019). *What is the ecological impact of light pollution?* Encyclopedia of the Environment, https://www.encyclopedie-environnement.org/en/life/what-is-the-ecological-impact-of-light-pollution/.

Lefèvre, T., Vantaux, A., Dabiré, K.R., Mouline, K., and Cohuet, A. (2013). Non-genetic determinants of mosquito competence for malaria parasites. *PLoS Pathogens*, 9(6).

Lemus, J.A., Bravo, C., García-Montijano, M., et al. (2011). Side effects of rodent control on non-target species: rodenticides increase parasite and pathogen burden in great bustards. *Science of the Total Environment*, 409(22), 4729–34.

Magle, S.B., Fidino, M., Lehrer, E.W., et al. (2019). Advancing urban wildlife research through a multi-city collaboration. *Frontiers in Ecology and the Environment*, 17(4), 232–9.

Martin, L.B. (2009). Stress and immunity in wild verte-brates: timing is everything. *General and Comparative Endocrinology*, 163(1–2), 70–6.

Martin, L.B. and Boruta, M. (2014). The impacts of urbanization on avian disease transmission and emergence. In D. Gil and H. Brumm, eds. *Avian Urban Ecology*, Issue I, pp. 116–28. Oxford University Press, Oxford.

Martín-Maldonado, B., Vega, S., Mencía-Gutiérrez, A., et al. (2020). Urban birds: an important source of antimicrobial resistant *Salmonella* strains in central Spain. *Comparative Immunology, Microbiology and Infectious Diseases*, 72, 101519.

Meillère, A., Brischoux, F., Bustamante, P., et al. (2016). Corticosterone levels in relation to trace element contamination along an urbanization gradient in the common blackbird (*Turdus merula*). *Science of the Total Environment*, 566–567, 93–101.

Mendenhall, C.D., Archer, H.M., Brenes, F.O., Sekercioglu, C.H., and Sehgal, R.N.M. (2013). Balancing biodiversity with agriculture: land sharing mitigates avian malaria prevalence. *Conservation Letters*, 6(2), 125–31.

Mitra, A., Chatterjee, C., and Mandal, F.B. (2011). Synthetic chemical pesticides and their effects on birds. *Research Journal of Environmental Toxicology*, 5(2), 81–96.

Moyers, S.C., Adelman, J.S., Farine, D.R., Thomason, C.A., and Hawley, D.M. (2018). Feeder density enhances house finch disease transmission in experimental epidemics. *Philosophical Transactions of the Royal Society B: Biological Sciences*, 373(1745).

Murphy, M.T. (2003). Avian population trends within the evolving agricultural landscape of eastern and central United States. *The Auk*, 120(1), 20–34.

Murray, C.G., Kasel, S., Szantyr, E., Barratt, R., and Hamilton, A.J. (2014). Waterbird use of different treatment stages in waste-stabilisation pond systems. *Emu*, 114(1), 30–40.

Murray, M. (2017). Anticoagulant rodenticide exposure and toxicosis in four species of birds of prey in Massachusetts, USA, 2012–2016, in relation to use of rodenticides by pest management professionals. *Ecotoxicology*, 26(8), 1041–50.

Murray, M.H., Becker, D. J., Hall, R.J., and Hernandez, S.M. (2016). Wildlife health and supplemental feeding: a review and management recommendations. *Biological Conservation*, 204, 163–74.

Murray, M.H., Kidd, A.D., Curry, S.E., et al. (2018). From wetland specialist to hand-fed generalist: shifts in diet and condition with provisioning for a recently urbanized wading bird. *Philosophical Transactions of the Royal Society B: Biological Sciences*, 373, 20170100.

Murray, M.H., Sánchez, C., Becker, D.J., Byers, K.A., Worsley-Tonks, K.E.L., and Craft, M.E. (2019). City sicker? A meta-analysis of wildlife health and urbanization. *Frontiers in Ecology and the Environment*, 17(10), 575–83.

Navara, K.J. and Nelson, R.J. (2007). The dark side of light at night: physiological, epidemiological, and ecological consequences. *Journal of Pineal Research*, 43(3), 215–24.

Ogrzewalska, M., Uezu, A., Jenkins, C.N., and Labruna, M.B. (2011). Effect of forest fragmentation on tick infestations of birds and tick infection rates by *Rickettsia* in the Atlantic Forest of Brazil. *EcoHealth*, 8(3), 320–31.

Oke, T.R. (1973). City size and the urban heat island. *Atmospheric Environment*, 7, 769–79.

Ouyang, J.Q., De Jong, M., Hau, M., Visser, M.E., Van Grunsven, R.H.A., and Spoelstra, K. (2015). Stressful colours: corticosterone concentrations in a free-living songbird vary with the spectral composition of experimental illumination. *Biology Letters*, 11(8).

Ouyang, J.Q., de Jong, M., van Grunsven, R.H.A., et al. (2017). Restless roosts: light pollution affects behavior, sleep, and physiology in a free-living songbird. *Global Change Biology*, 23(11), 4987–94.

Owens, I.P.F. and Bennett, P.M. (2000). Ecological basis of extinction risk in birds: habitat loss versus human persecution and introduced predators. *Proceedings of the National Academy of Sciences of the United States of America*, 97(22), 12144–8.

Pain, D.J. (1996). Lead in waterfowl. In *Environmental Contaminants in Wildlife: Interpreting Tissue Concentrations*, SETAC Special Publications Series, pp. 251–64. Lewis Publishers, Boca Raton, FL.

Palacio, R.D., Kattan, G.H., and Pimm, S.L. (2019). Bird extirpations and community dynamics in an Andean cloud forest over 100 years of land-use change. *Conservation Biology*, 34(3), 677–87.

Plummer, K.E., Risely, K., Toms, M.P., and Siriwardena, G.M. (2019). The composition of British bird communities is associated with long-term garden bird feeding. *Nature Communications*, 10(1).

Purple, K.E., Humm, J.M., Kirby, R.B., Saidak, C.G., and Gerhold, R. (2015). *Trichomonas gallinae* persistence in four water treatments. *Journal of Wildlife Diseases*, 51(3), 739–42.

Qiao, X., Schmidt, A.H., Tang, Y., Xu, Y., and Zhang, C. (2014). Demonstrating urban pollution using toxic metals of road dust and roadside soil in Chengdu, southwestern China. *Stochastic Environmental Research and Risk Assessment*, 28(4), 911–19.

Reid, J.M., Arcese, P., Keller, L.F., Elliott, K.H., Sampson, L., and Hasselquist, D. (2007). Inbreeding effects on immune response in free-living song sparrows (*Melospiza melodia*). *Proceedings of the Royal Society B: Biological Sciences*, 274(1610), 697–706.

Rioux, S., Garant, D., Pelletier, F., and Bélisle, M. (2013). Seasonal patterns in tree swallow prey (*Diptera*) abundance are affected by agricultural intensification. *Ecological Applications*, 23(1), 122–33.

Ritchie, H. and Roser, M. (2020). *Land Use*. OurWorldInData. Org.

Rosenberg, K.V., Dokter, A.M., Blancher, P.J., et al. (2019). Decline of the North American avifauna. *Science*, 366, 120–4.

Sagerup, K., Henriksen, E.O., Skorping, A., Skaare, J.U., and Gabrielsen, G.W. (2000). Intensity of parasitic nematodes increases with organochlorine levels in the glaucous gull. *Journal of Applied Ecology*, 37(3), 532–9.

Sánchez-Virosta, P., Espín, S., García-Fernández, A.J., and Eeva, T. (2015). A review on exposure and effects of arsenic in passerine birds. *Science of the Total Environment*, 512–513, 506–25.

Santiago-Alarcon, D., Carbó-Ramírez, P., Macgregor-Fors, I., Chávez-Zichinelli, C.A., and Yeh, P.J. (2020). The prevalence of avian haemosporidian parasites in an invasive bird is lower in urban than in non-urban environments. *Ibis*, 162, 201–14.

Savoca, M.S., Wohlfeil, M.E., Ebeler, S.E., and Nevitt, G.A. (2016). Marine plastic debris emits a keystone infochemical for olfactory foraging seabirds. *Science Advances*, 2(11), 1–9.

Sehgal, R. (2010). Deforestation and avian infectious diseases. *Journal of Experimental Biology*, 213(6), 955–60.

Sehgal, R. (2015). Manifold habitat effects on the prevalence and diversity of avian blood parasites. *International Journal for Parasitology: Parasites and Wildlife*, 4(3), 421–30.

Snoeijs, T., Dauwe, T., Pinxten, R., Vandesande, F., and Eens, M. (2004). Heavy metal exposure affects the humoral immune response in a free-living small songbird, the great tit (*Parus major*). *Archives of Environmental Contamination and Toxicology*, 46(3), 399–404.

Spiller, K.J. and Dettmers, R. (2019). Evidence for multiple drivers of aerial insectivore declines in North America. *Ornithological Applications*, 121, 1–13.

Stephens, R., Johnson, A., Plumb, R., Dickerson, K., McKinstry, M., and Anderson, S. (2005). Secondary lead poisoning in golden eagle and ferruginous hawk chicks consuming shot black-tailed prairie dogs, Thunder Basin National Grassland, Wyoming. Contaminants Report Number R6/720/05. *US Fish & Wildlife Publications*, 230.

Suorsa, P., Helle, H., Koivunen, V., Huhta, E., Nikula, A., and Hakkarainen, H. (2004). Effects of forest patch size on physiological stress and immunocompetence in an area-sensitive passerine, the Eurasian treecreeper (*Certhia familiaris*): an experiment. *Proceedings of the Royal Society B: Biological Sciences*, 271(1537), 435–40.

Taff, C.C., Weis, A.M., Wheeler, S., et al. (2016). Influence of host ecology and behavior on *Campylobacter jejuni* prevalence and environmental contamination risk in a synanthropic wild bird species. *Applied and Environmental Microbiology*, 82(15), 4811–20.

Teuten, E.L., Saquing, J.M., Knappe, D.R.U., et al. (2009). Transport and release of chemicals from plastics to the environment and to wildlife. *Philosophical Transactions of the Royal Society B: Biological Sciences*, 364(1526), 2027–45.

Tizard, I. (2004). Salmonellosis in wild birds. *Seminars in Avian and Exotic Pet Medicine*, 13(2), 50–66.

US Department of the Interior, US Fish and Wildlife Service, US Department of Commerce, and US Census Bureau (2018). *2016 National Survey of Fishing, Hunting, and Wildlife-Associated Recreation*. https://www.fws.gov/wsfrprograms/subpages/nationalsurvey/nat_survey2016.pdf.

van Hoesel, W., Marzal, A., Magallanes, S., Santiago-Alarcon, D., Ibáñez-Bernal, S., and Renner, S.C. (2019). Management of ecosystems alters vector dynamics and haemosporidian infections. *Scientific Reports*, 9(1), 1–11.

Venter, O., Sanderson, E.W., Magrach, A., et al. (2016). Sixteen years of change in the global terrestrial human footprint and implications for biodiversity conservation. *Nature Communications*, 7, 12558.

Vermeulen, A., Müller, W., Matson, K.D., Tieleman, B.I., Bervoets, L., and Eens, M. (2015). Sources of variation in innate immunity in great tit nestlings living along a metal pollution gradient: an individual-based approach. *Science of the Total Environment*, 508(2015), 297–306.

Vittor, A.Y., Gilman, R.H., Tielsch, J., et al. (2006). The effect of deforestation on the human-biting rate of *Anopheles darlingi*, the primary vector of falciparum malaria in the Peruvian Amazon. *American Journal of Tropical Medicine and Hygiene*, 74(1), 3–11.

Weis, A.M., Miller, W.A., Byrne, B.A., et al. (2014). Prevalence and pathogenic potential of *Campylobacter* isolates from free-living, human-commensal American crows. *Applied and Environmental Microbiology*, 80(5), 1639–44.

Whitney, M. and Cristol, D. (2017). Impacts of sublethal mercury exposure on birds: a detailed review. In *Reviews of Environmental Contamination and Toxicology*, pp. 113–63. Springer, Cham.

Wilcox, C., Van Sebille, E., Hardesty, B.D., and Estes, J.A. (2015). Threat of plastic pollution to seabirds is global, pervasive, and increasing. *Proceedings of the National Academy of Sciences of the United States of America*, 112(38), 11899–904.

Wilcoxen, T.E., Horn, D.J., Hogan, B.M., et al. (2015). Effects of bird-feeding activities on the health of wild birds. *Conservation Physiology*, 3, 1–13.

Zhong, D., Wang, X., Xu, T., et al. (2016). Effects of microclimate condition changes due to land use and land cover changes on the survivorship of malaria vectors in China–Myanmar border region. *PLoS ONE*, 11(5).

CHAPTER 10

Climate Change and Avian Disease

Richard J. Hall

10.1 Introduction

Rapid and accelerating changes in global temperature, precipitation patterns, and other abiotic variables are having profound effects on biodiversity. As a diverse, highly visible, and globally distributed taxon, avian responses to ongoing global change have been relatively well documented. The capacity for long-distance flight and dietary or habitat flexibility have allowed some species, such as the blackcap (*Sylvia atricapilla*) and cattle egret (*Bubulcus ibis*), to rapidly colonize novel seasonal environments (Berthold et al. 1992; Maddock and Geering 1994). In spite of this, avian communities struggle to keep pace with geographic and phenological shifts in suitable environmental conditions (Devictor et al. 2008). Recent analyses confirm that birds are experiencing widespread declines, particularly pronounced among range-restricted habitat specialists, such as grassland birds, and long-distance migrants, such as shorebirds (Rosenberg et al. 2019). Together, these declines lend urgency to understanding the abiotic and biotic factors determining whether species will adapt or perish under ongoing and future climate change.

In addition to the direct fitness consequences of changing abiotic conditions on birds, climate change alters the arena of species interactions, including exposure to infectious pathogens. There are several high-profile instances of climate change influencing the dynamics of avian diseases of conservation concern. The parasite responsible for avian malaria, *Plasmodium relictum*, was introduced into the Hawaiian islands by human activity along with its primary mosquito vector, *Culex quinquefasciatus*, devastating immunologically naïve native

songbird communities (Warner 1968). Climate change continues to support the upslope march of *C. quinquefasciatus* by expanding larval habitat availability and temperatures favoring mosquito survival and parasite development (Atkinson et al. 2014). In late 2019, efforts to introduce critically endangered kiwikiu (*Pseudonestor xanthophrys*) to recently restored habitat were thwarted when at least 9 of 13 birds died from avian malaria within a few weeks of translocation (Warren et al. 2020). On Kauai, a lack of high-elevation sites suggests that it is only a matter of time before malaria reaches the last remaining disease-free refugia for several Hawaiian endemics, including the threatened 'i'iwi (Figure 10.1) (Paxton et al. 2016).

Increasingly, pathogens harbored by birds and other wildlife emerge and spill over into humans and domestic animals, motivating studies of how climate change might influence zoonotic spillover risk (see Chapter 12). One high-profile example is highly pathogenic avian influenza viruses, which can cause high human mortality and devastating economic losses to poultry and which are naturally maintained in wild waterbirds (Webster et al. 1992; see Chapter 11). Changes in environmental conditions throughout waterbird annual ranges may promote increased intercontinental mixing in Arctic breeding grounds and stopover sites and increased contact between domestic and wild birds using agricultural lands in the non-breeding season, both of which could provide novel opportunities for viral reassortment and cross-species transmission (Morin et al. 2018; Figure 10.1). Complicating predictions of future risk, warming water temperatures can reduce environmental viral persistence (Brown et al. 2009) and promote increased residency in

Richard J. Hall, *Climate Change and Avian Disease* In: *Infectious Disease Ecology of Wild Birds*. Edited by: Jennifer C. Owen, Dana M. Hawley, and Kathryn P. Huyvaert, Oxford University Press. © Oxford University Press 2021.
DOI: 10.1093/oso/9780198746249.003.0010

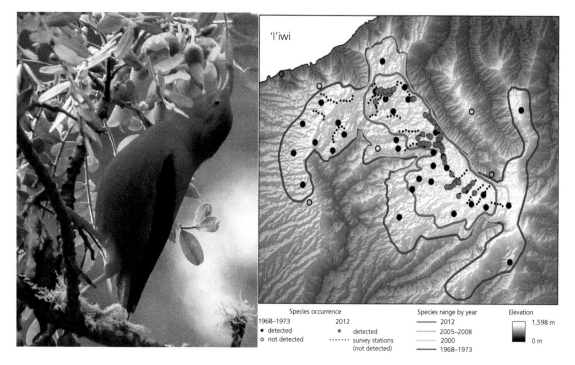

Figure 10.1 Climate warming and altered precipitation is increasing transmission of the parasite *Plasmodium relictum* that causes avian malaria in high-elevation Hawaiian forests, threatening the persistence of native Hawaiian honeycreepers, including the 'i'iwi (*Vestiaria coccinea*). Map shows estimated range contraction of 'i'iwi on Kauai based on count data from 1968 (purple), 2000 (blue), and 2012 (red). (Photo credit: R. Hall. Figure from Paxton et al. 2016.)

some waterfowl (Spivey et al. 2017). Understanding how climate change simultaneously influences host and pathogen ecology is crucial for predicting whether parasitism is likely to increase or decrease in the future.

The goal of this chapter is to survey theoretical predictions and empirical evidence for how climate change will affect avian host–parasite interactions through changes in bird distributions and timing of life history events that influence exposure and transmission (see Section 10.2); through shifts in host physiology and behavior that influence susceptibility and impacts of infection (see Section 10.3); and through effects on the replication and persistence of diverse parasite types outside of avian hosts (i.e., in vectors or the external environment; see Section 10.4). Finally, I will highlight key knowledge gaps and future research directions aimed at improving capacity to anticipate and effectively respond to shifting patterns of avian diseases (see Section 10.5).

10.2 Infection consequences of changes in bird distribution and phenology

Climate change is shifting the distribution and timing of environmental conditions that maximize breeding success and non-breeding survival for birds. Among the commonest and most readily observable responses to climate change across taxa are range shifts and/or shifts in the timing of habitat use and events in the annual cycle (Cotton 2003; La Sorte and Thompson 2007); however, there is much variation among bird species in their propensity to undergo these shifts. Whether birds colonize new environments or continue to occupy their historic seasonal ranges, most are facing altered environmental conditions and novel species assemblages, which could influence the diversity of infectious agents to which they are exposed. Relative to interactions with larger-bodied organisms, however, the effects of shifts in distribution and phenology on host exposure to and transmission of infectious agents have been less thoroughly explored.

10.2.1 Latitudinal and elevational shifts in avian breeding ranges

Many bird species are undergoing climate-induced shifts in their breeding ranges. Increased temperatures underlie range shifts predominantly towards the poles (Hitch and Leberg 2007) or to higher elevations (Freeman et al. 2018); however, changes in precipitation patterns in some montane regions can lead to downslope range shifts (Tingley et al. 2012). Ecological theory suggests that species adapted to harsh conditions at high elevations and latitudes benefit from reduced diversity of natural enemies, including parasites (Greiner et al. 1975), and thus these range shifts can result in novel contact zones between species and the potential for acquiring novel parasites. For example, bidirectional parasite-sharing of avian malaria parasite lineages has been documented in the contact zones of range-expanding melodious warblers (*Hippolais polyglotta*) and historically breeding congeneric icterine warblers (*Hippolais icterina*) (Reullier et al. 2006).

On the other hand, colonizing species are predicted to experience lower levels of parasitism through two mechanisms. First, the enemy release hypothesis suggests that successfully colonizing species may benefit from a lack of coevolved natural enemies in their new range, which may increase their competitive ability (Keane and Crawley 2002). This hypothesis predicts that colonizers leave behind parasites from their historic breeding range, as observed in avian malaria parasite infections in introduced versus native house sparrows (*Passer domesticus*) (Marzal et al. 2011). Second, colonizing species may experience reduced parasite exposure risk, since avian communities at higher latitudes and elevations often harbor lower parasite diversity and prevalence, as seen with avian malaria in the tropical highlands (Zamoria-Vilchis et al. 2012) and at high latitudes in Alaska (Loiseau et al. 2012). However, there are cases in which colonizing species appear to experience neither enemy release nor reduced exposure. For example, range-expanding pied flycatchers (*Ficedula hypoleuca*) maintained

their diversity of specialist avian malaria parasites while also acquiring more new lineages from congeneric collared flycatchers (*Ficedula albicollis*) in their contact zones (Jones et al. 2018).

Of particular concern is the potential for range-expanding species to introduce novel parasites against which recipient host communities have no prior exposure history. Lack of adaptation to these introduced parasites could result in high virulence to endemic species and, in turn, increase the competitive advantage for colonizing species, often referred to as the novel weapons hypothesis (Prenter et al. 2004). Notable examples that were not generated by climate change *per se* include the high mortality and large-scale declines in Hawaiian honeycreepers (Drepanididae spp.) following introduction of *Plasmodium relictum* (Atkinson et al. 1995; Van Riper et al. 1986) and mortality in endemic Mauritius pink pigeons (*Nesoenas mayeri*) resulting from trichomoniasis (Bunbury et al. 2007). Historically, low parasite pressure near the poles could render polar species especially vulnerable to the virulent effects of parasites, as seen in wild Magellanic penguins (*Spheniscus magellanicus*) brought into zoos that subsequently experienced high mortality from avian malaria (Fix et al. 1988). The availability of immunologically naïve but competent hosts could also facilitate transmission of generalist parasites that pose risks to public health, including avian influenza and West Nile virus (Fuller et al. 2012). Thus, importation of generalist pathogens by colonizers into high-latitude, high-elevation refugia may exacerbate extinction risk of many cool-adapted species under warming.

By contrast, birds whose breeding ranges are contracting under climate warming may not face increased exposure risk to parasites. Extant populations in the core breeding range may remain in cool, parasite-free refugia, while populations at the 'warmer,' trailing range edge could decline below densities necessary for sustained transmission (May and Anderson 1979). Trends in parasitism at the trailing range edge remain understudied in avian systems; however, studies of parasite coextinction in seabird ectoparasites suggest that specialized parasites are likely to be extirpated as host populations contract and decline (Koh et al. 2004).

10.2.1.1 Changes in non-breeding range and movement patterns

Climate change could also impact avian disease dynamics by altering migration patterns and non-breeding distributions (Figure 10.2). Bird migration intimately shapes host–parasite interactions and has been proposed as a mechanism for reducing parasite transmission through the avian annual cycle. Escaping high parasite pressure by breeding in high-latitude, parasite-poor habitats has been suggested as a selective driver of migration (Piersma 1997), while seasonal departure from pathogen-contaminated habitats could reduce transmission opportunities ('migratory escape;' Altizer et al. 2011). Since migration can impose extreme energetic demands and physiological changes, migration can also reduce infection prevalence in hosts by reducing the survival probability of infected individuals ('migratory culling;' Altizer et al. 2011). Sedentary dark-eyed juncos (*Junco hyemalis*) exhibit higher prevalence of haemosporidian (blood parasite) infections than long-distance migrants that co-occur on their wintering grounds (Slowinski et al. 2018), providing indirect support for these mechanisms. Further, mathematical models demonstrate that host migration strategy is a crucial determinant of the population impacts of highly transmissible, virulent pathogens (Hall et al. 2014) and that migratory propensity may be shaped by parasite exposure over evolutionary time (Shaw et al. 2019). Therefore, climate-induced changes in migratory propensity could have important consequences for parasite transmission and impacts.

Climate warming is reducing winter harshness in many regions; consequently, the wintering range of many birds is shifting polewards (La Sorte and Thompson 2007) and many migrations are shortening (Visser et al. 2009). Reduced energetic demands of shorter migrations could increase avian population sizes by reducing migratory mortality; mathematical models suggest that these higher host population sizes coupled with reduced efficacy of migratory culling can lead to explosive seasonal outbreaks and very high infection prevalence (Hall et al. 2016). Less harsh winters are enabling snow geese (*Chen caerulescens*) to overwinter at higher latitudes, where agricultural food subsidies are contributing to their rapid population growth (Gauthier et al. 2005); because these geese are an important carrier of the bacterium *Pasteurella multocida* that causes avian cholera (Samuel et al. 2005), their continued population expansion could increase the frequency of outbreaks in other co-occurring waterfowl.

Climate warming is also associated with increases in residency in many partially migratory bird species (Meller et al. 2016; Németh 2017). Relative to migrants, resident populations can support

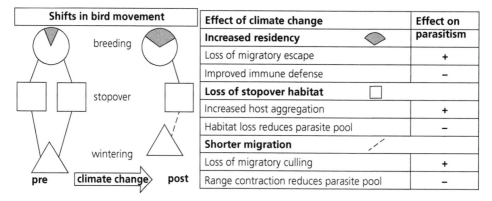

Figure 10.2 Schematic summarizing potential effects of climate change on avian host–parasite dynamics through shifts in bird movement patterns. Hypothetical migration network pre- (left) and post- (right) climate change, where a breeding site (circle) is shared by migrants and residents and lines indicate migration routes connecting to stopover habitats (squares) and an overwintering site (triangle). Climate change influences movement by increasing residency (gray shading), removing stopover habitat and reducing the distance (dashed line) to chosen overwintering site. The table lists some of the proposed mechanisms by which changing movement patterns might increase (+) or decrease (−) parasitism (as measured by infection prevalence, intensity, or diversity) under climate change.

sustained, year-round transmission and avoid migratory culling. In a model of host–parasite dynamics in partial migrants that share breeding grounds, increasing the proportion of residents elevated parasite exposure risk in returning migrants (eroding the benefits of migratory escape); because more migrants became infected, migratory culling resulted in disproportionate mortality of migrants relative to residents, suggesting that climate-induced shifts to residency could threaten the persistence of migration (Brown and Hall 2018). Mallards (*Anas platyrhynchos*) are increasingly overwintering at high latitudes and, in Alaska, maintain transmission of avian influenza virus subtypes throughout the winter (Spivey et al. 2017); thus, they may increasingly be important for the maintenance and transmission of avian influenza viruses to returning migrants (Van Hemert et al. 2014).

Changes in migration distance, routes, and habitats used for stopover and overwintering could positively or negatively influence parasite diversity in hosts. Greater migration distance and migratory propensity are positively associated with hematozoan blood parasite richness of waterfowl (Figuerola and Green 2000) and nematode diversity in shorebirds, thrushes, and hawks (Koprivnikar and Leung 2014). Climate change that results in shortened migrations could therefore reduce parasite diversity in migrants, especially if parasite-rich tropical regions are avoided (Clark et al. 2016). Different habitats used in the non-breeding period may also harbor distinct parasite communities; reduced availability or use of these habitats could therefore alter parasite richness and diversity in migrants. For example, habitat generalism is associated with increased helminth and haemosporidian richness and prevalence in migratory shorebirds relative to those using only saltwater habitats (Clark et al. 2016; Gutierrez et al. 2017). Increased drought frequency that reduces availability of inland stopover sites could reduce infection prevalence if more shorebirds switch to migrating along coastal routes with lower parasite richness. Alternatively, if loss of intertidal mudflats through sea level rise increases use of inland sites with higher parasite diversity, aggregation of migrants in remaining stopover habitats could enhance transmission risk.

Continental and global scale changes to air and ocean currents or to habitats that act as physical barriers to migration could shift migration routes and increase connections between formerly isolated flyways, with implications for the long-distance spread and introduction of novel pathogens, including zoonoses (see Chapter 12). For example, transoceanic movements of Arctic-breeding waterfowl and shorebirds transport avian influenza virus lineages between Eurasia and North America (Winker and Gibson 2010). Longer periods of ice-free water in the Arctic could therefore increase the frequency of parasite transfer and the potential for strain reassortment between flyways and continents. Overall, shifts in migration routes will alter the community of parasites to which migrants are exposed, including the potential for birds to spread pathogens into new geographic regions and new hosts.

10.2.2 Changes in timing of life history events through the annual cycle

Climate change is shifting the timing of suitable environmental conditions and resource availability; associated shifts in the timing of life history events could alter parasite transmission via seasonal changes to host exposure and susceptibility. For example, in some species, climate change is favoring advanced arrival at breeding grounds, advanced breeding, and thus longer breeding seasons, with the potential for multiple broods (Cooper et al. 2011; Smith and Moore 2005). Migrants able to advance their breeding arrival dates could face an extended window for transmission of generalist parasites harbored by resident birds and vector-borne diseases (Brown and Hall 2018; Møller 2010). Birds able to increase the number of broods, or re-nesting probability, in response to an extended breeding season could experience more parasite transmission and impacts through an increase in the number of immunologically naïve young birds. For example, in barn swallows (*Hirundo rustica*), parasitism by nest ectoparasites is thought to increase with the number of broods, which in turn influences the quality and quantity of offspring raised over a season (de Lope and Møller 1993).

Alternatively, bird populations less able to track advances in resource phenology, such as long-distance migrants (Saino et al. 2010), could experience reductions in transmission due to shifts in resource availability. In a mathematical model of vector-borne parasite transmission at a migrant's breeding ground, infection prevalence decreased with the degree of mismatch between host arrival and advancing resource phenology, because mismatch reduced host density and advancing emergence of vectors reduced their overlap with returning migrants (host–parasite mismatch; Hall et al. 2016; Box 10.1). Similarly, a model investigating the effects of mismatch between horseshoe crab spawning and ruddy turnstone (*Arenaria interpres*) arrival on transmission of avian influenza virus in Delaware Bay predicted an increase in the probability of viral extinction with mismatch (Brown and Rohani 2012). In some systems, parasite infection can alter the timing of movement in infected hosts, delaying arrival at breeding sites (Møller et al. 2004) or increasing the duration of stopover (Hegemann et al. 2018). Climate change could exacerbate the inability of infected birds to track resource

Box 10.1 Modeling effects of climate change on bird migration and its consequences for host–parasite interactions

(A,B) In this hypothetical scenario explored by Hall et al. (2016), birds differ in their ability to track climate-mediated advances in resource phenology, potentially leading to population declines through phenological mismatch and altered overlap with reservoir hosts and disease vectors.

(A) Climate change at the migrant's breeding site shifts the peak (black circle) and duration (black horizontal line) of biting vector emergence and optimal arrival date (black vertical arrow) for returning migrants to match breeding resources by one month relative to historical phenology (gray).

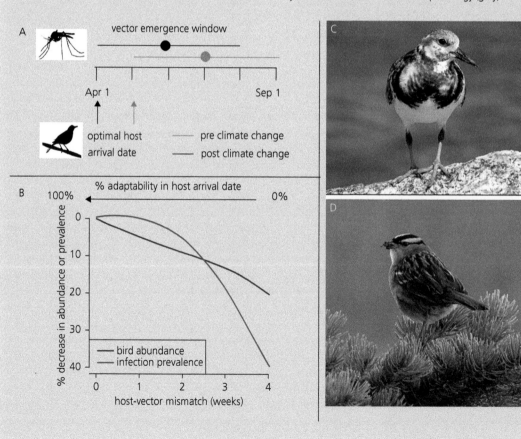

(B) The extent to which migratory insectivores update their arrival date at breeding sites to track resource phenology and vector emergence (host–vector mismatch, x-axis), with 'adaptability' of arrival date decreasing left to right, influences the migrant's peak population size (blue line) and peak infection prevalence (red line) of a vector-borne pathogen over an annual cycle. Changes in population size and infection prevalence are shown as percent declines relative to their respective values assuming no mismatch (i.e., host arrival perfectly tracks resource and vector phenology). Population size and infection prevalence in the migrant both decline with increasing phenological mismatch with invertebrate prey and biting vectors (e.g., decreasing adaptability); however, migrant population declines are more modest than declines in prevalence under high mismatch (low adaptability) because reduced recruitment from resource mismatch is offset by reduced disease-induced mortality. This scenario illustrates the need to consider the diverse effects of climate on fitness-relevant traits to understand the combined implications of climate change and pathogens for avian populations.

(C) In a model of avian influenza virus (AIV) transmission at a spring stopover site of ruddy turnstones (*Arenaria interpres*), Brown and Rohani (2012) found that if turnstones advanced their arrival time to match spawning horseshoe crab phenology, AIV transmission increased due to increased overlap with resident and wintering ducks; alternatively, if only horseshoe crab phenology advanced, resulting declines in ruddy turnstones caused AIV prevalence to drop. Photo credit: R. Hall.

(D) In a model of transmission of the haemosporidian *Leucocytozoon fringillarium* by blackflies (*Simulium* spp.) at high-elevation breeding sites of white-crowned sparrows (*Zonotrichia leucophrys*), Murdock et al. (2013) explored how advancing blackfly (but not sparrow) phenology influenced transmission. They found that a small advance in blackfly phenology increased parasite transmission due to increased overlap of blackflies with relapsing birds at the start of the breeding season; however, further advances in phenology reduced transmission due to reduced overlap between blackflies and parasite-susceptible fledglings. Photo credit: R. Hall

phenology, which could lower infection prevalence through increased mortality of infected individuals and increased temporal separation of susceptible and infected birds that reduces transmission. Thus, climate change has the potential to have opposing effects on disease risk via changes in the timing of life history events.

10.3 Changes in bird physiology, behavior, and response to infection

Although changes in distribution and phenology are frequently observed responses to climate change, physiological and behavioral adaptations to environmental change within a species' current range may be just as important for species' persistence, especially those with restricted ranges and in rapidly warming regions. However, responses such as behavioral thermoregulation often involve trade-offs with other activities such as foraging (du Plessis et al. 2012). These individual-level responses, in turn, influence parasite acquisition and impacts: physiological changes may require reallocation of resources, with consequences for resistance and tolerance to infection; behavioral changes can similarly influence exposure to and onward transmission of infectious agents (see also Chapters 3 and 4).

10.3.1 Changes in immunity and infection impacts

Immune defense in birds varies latitudinally (Martin et al. 2004), potentially reflecting differential immune costs determined by local environmental conditions and parasite exposure (Møller 2010). Climate-driven shifts in these conditions could therefore alter infection dynamics and outcomes by altering within-host processes influencing competence for infection and disease-induced mortality. For example, tree swallows (*Tachycineta bicolor*) demonstrate latitudinal variation in reproductive investment, which is thought to reflect higher investment in self-maintenance over offspring quality in parts of the breeding range where interannual survival is high (Ardia 2005; Figure 10.3A). Experimental augmentation of swallow clutch size resulted in differential changes to parental nest provisioning rates and immune defense in parents and offspring. In Alaska, where adult apparent survival is low, parents increased their provisioning rates and incurred immune costs (Figure 10.3B), while nestling condition (Figure 10.3C) and immunity (Figure 10.3D) were maintained. In Tennessee, where adult swallow survival is higher, parents did not increase provisioning rates and maintained their immune

Figure 10.3 Geographic variation in trade-offs between immune and reproductive investment in tree swallows (*Tachycineta bicolor*; A) suggests that climate change could alter parasite infection outcomes via changes in age-dependent immunity (Ardia 2005). Reducing ('Red.') or Enlarging ('Enl.') brood size of tree swallow nests in Alaska (left column) or Tennessee (right column) relative to controls ('Con.') demonstrated regional differences in effects of brood augmentation on (B) adult cellular immunity, (C) offspring body condition (residual body mass), and (D) nestling immune response. Alaskan females incurred an immune cost under brood augmentation but nestlings maintained similar condition and immune function across treatments, while in Tennessee, females maintained immune function but offspring condition and immunity were reduced under brood augmentation. Photo credit: J. Owen

defense, but their offspring incurred immune costs. Thus, if patterns seen in swallows also occur in other avian species, shifts in adult survival under climate change could differentially influence infection outcomes in different parts of the range; transmission could increase in areas where females prioritize self-maintenance and thus produce immunocompromised offspring and decrease in areas where they produce few high-quality offspring.

Immune defense also varies seasonally (Hegemann et al. 2012) and, in resident birds, costs of immune defense and infection may be higher in winter due to increased metabolic demands (see Chapter 3). Reduced winter harshness could reduce these energetic costs; this could either reduce transmission through improved immune defense or increase prevalence if hosts are better able to tolerate infection. In house finches (*Haemorhous mexicanus*), additive costs of thermoregulation and *Mycoplasma gallisepticum* infection suggest that birds may need to forage more in cold weather,

which could in turn increase their exposure to fomites on feeders (Hawley et al. 2012); milder winters might therefore reduce both parasite exposure and susceptibility. Since mounting an immune defense negatively influences body mass in overwintering house sparrows (Moreno-Rueda 2011), reduced winter harshness could alter infection dynamics by modifying the trade-off between immunity and maintenance; this could reduce transmission if birds are better able to resist infection or increase transmission if infected birds better tolerate infection. By contrast, increased frequency of extreme weather events, such as heatwaves in summer, could increase the negative impact of infections on hosts; for example, the negative effects of enteric *Salmonella* infection on broiler chicken condition were exacerbated when birds were subjected to five weeks of heat stress (temperatures of 31°C; Quinteiro-Filho et al. 2012). The combined effect of heat stress and infection in wild birds remains unknown.

Shifts towards shorter migrations and residency under climate change could confer benefits to host immune defense that reduce parasite transmission or impacts. Migratory species have larger immune defense organs than related resident species (Møller and Erritzøe 1998), probably reflecting exposure to diverse pathogens throughout the migratory range; thus, migratory populations that transition to residency may be able to mount more effective immune responses. Since physiological changes associated with active migration can be immunologically costly (Owen and Moore 2008), foregoing migration could further increase immune defense over the annual cycle. Indirect evidence for this comes from partially migratory common blackbirds (*Turdus merula*), where migrants have lower baseline immune defense than residents (Eikenaar and Hegemann 2016) and are increasingly becoming resident in urban areas (Németh 2017) and where urban populations experience lower prevalence and diversity of hematozoan blood parasite infections than forest-breeding populations (Geue and Partecke 2008). Overall, there are several ways in which climate change may alter the nature of physiological trade-offs between immune defense and other fitness components, but the difficulty in quantifying these trade-offs means that empirical evidence for climate change effects on immune defense is lacking in wild birds.

10.3.2 Changes in behavior

Changes in abiotic conditions such as temperature as well as climate effects on the amount and distribution of food resources could alter seasonal patterns of aggregation in relation to pathogen exposure risk. Since house finch flock sizes tend to peak in the coldest months and increase with latitude, and the prevalence of *M. gallisepticum* in house finch flocks increases with flock size (Altizer et al. 2004), shorter periods of cold temperatures at higher latitudes could reduce infection risk and change the timing of peak infection (Hosseini et al. 2004). Communal roosting can alter pathogen transmission and persistence in opposing ways. For example, some studies of West Nile virus suggest that corvid roosts can maintain persistence of the virus over winter (Montecino-Latorre and

Barker 2018) and that American robin (*Turdus migratorius*) roosts can increase infection prevalence in mosquitoes (Diuk-Wasser et al. 2010); however, other studies on American robins have found a negative effect of roosts on transmission through encounter dilution (Krebs et al. 2014). Communal roosting provides thermoregulatory benefits, and in species such as the sociable weaver (*Philetairus socius*), lower nocturnal temperatures are associated with larger roosting groups (Paquet et al. 2016). Thus, warmer temperatures that reduce the size or duration of non-breeding communal roosts could reduce parasite transmission potential. Conversely, loss of suitable migratory stopover or overwintering habitat is aggregating birds into remaining habitats, potentially increasing transmission risk. For example, extended droughts lead to seasonal reduction of natural wetland availability, increasing concentrations of waterfowl using human-created refuges and agricultural fields; these artificially high densities may promote large outbreaks of avian cholera (Wobeser 1992).

Climate change is causing many species to undergo shifts in foraging behavior and diet, which could influence susceptibility and exposure to parasites. In gulls, decreasing availability of preferred marine prey under ocean warming has increased the proportion of human-derived food in their diet over time (Blight et al. 2015). Foraging at landfills can increase gull exposure to *Clostridium botulinum*, the bacterium responsible for botulism (Ortiz and Smith 1994), and the switch to nutritionally poor or toxin-contaminated anthropogenic food could also result in immunocompromise and susceptibility to infection (Murray et al. 2016; see also Chapter 9). Other dietary shifts could reduce exposure to trophically transmitted, complex life cycle parasites. For example, reduced bill size in red knots associated with a phenological mismatch with resources at Arctic breeding sites has resulted in a shift from foraging on invertebrates to vegetation in the African wintering range (van Gils et al. 2016), potentially reducing exposure to trematode parasites in intermediate hosts.

Climate-related shifts in daily activity budgets could also mediate parasite encounter and susceptibility. Behavioral adjustments to warming temperatures in northern bobwhite (*Colinus virginianus*)

include reduced movement and seeking taller vegetation at the hottest times of day, (Carroll et al. 2015). Increased heat and drought periods appear to be associated with increased helminth infections and bobwhite declines (Blanchard et al. 2019); although the mechanisms underlying this likely reflect multiple climate effects on the helminths and their intermediate hosts, changes in host activity patterns and habitat associations could mediate exposure risk. Direct evidence of shifting activity patterns on parasitism in birds remains undocumented.

Changes in average or extreme temperatures have the potential to modify the effectiveness of antiparasite behaviors. For example, 'sunning,' where birds maximize their surface area exposed to direct sunlight, can reduce ectoparasite burdens through exposure to UV and may induce behavioral fever that kills internal pathogens (Bush and Clayton 2018; see also Chapter 4). Because sunning was more frequent in ectoparasite-infected violet-green swallows (*Tachycineta thalassina*) than those from which parasites were removed by fumigation (Blem and Blem 1993) and sunning birds can exhibit signs of heat stress such as panting, sunning may represent a costly activity in hot climates. In areas where temperatures are predicted to approach the upper thermal limits of birds, a behavioral trade-off between thermoregulation and parasite clearance may result in birds forgoing parasite removal opportunities. Overall, behavioral adaptations to climate change that alter diurnal or seasonal activity patterns and aggregation or that trade off with immune defense could increase or decrease parasite transmission risk.

10.4 Effects of climate change on parasites with external transmission stages

Theory predicts that the abundance and phenology of parasites that live obligately on or in hosts will be less affected by climate change than parasites that spend part of their life cycle off of hosts (Cizauskas et al. 2017). In support of this, the abundance of lice and mites on barn swallows has changed less with climate change than the infection prevalence of blood parasites transmitted by biting insects (Møller 2010). For avian parasites that spend part of their life cycle outside of hosts in ectothermic vectors, intermediate hosts, or the abiotic environment, responses of parasite transmission to climate change may be more strongly regulated by sensitivity of these off-host stages to environmental temperature or relative humidity.

10.4.1 Vector-borne and trophically transmitted parasites

Arthropod vectors and intermediate hosts of avian parasites are often limited in their distribution and phenology by environmental variables. Warming in temperate regions can increase transmission by increasing the number of days of the year in which biting vectors are active. Consistent with this idea, earlier emergence of mosquitoes and blackflies is associated with increasing avian malaria prevalence in barn swallows (Møller 2010). Changes in temperature and relative humidity can also allow vectors to increase their geographic range by increasing both adult survival and larval habitat availability. This has been observed in the upslope expansion of *Plasmodium relictum*-transmitting mosquitoes in Hawaii (Atkinson et al. 2014) and may underlie global increases in transmission of West Nile virus at higher latitudes (Paz 2015). In the US, moderate drought can improve larval mosquito habitat quality and increase contact between vectors and birds visiting water and has been positively associated with West Nile virus transmission (Epstein 2001); these conditions are likely to become more frequent under climate warming projections. Another mechanism by which vectors could expand their range is through transport by migratory birds (Cohen et al. 2015). At high latitudes, historically harsh winters limit the establishment of ticks introduced by returning migrants, but milder winters could allow more ticks to successfully overwinter and transmit infectious agents such as *Borrelia burgdorferi*, the bacterium responsible for Lyme disease (Ogden et al. 2008; see also Chapter 8).

Life history traits of arthropod vectors and the parasites they harbor are also sensitive to climate change (Figure 10.4A). For example, in *Culex* mosquitoes that transmit West Nile virus and *Plasmodium relictum* in birds, reductions in the length of the gonotrophic cycle (the interval between blood meals) with temperature could also increase the intensity of transmission by increasing the number of blood meals taken from birds (Reisen et al. 2014). Warming can also speed up parasite production, as observed in trematode parasites of birds, where infected snails produce more cercariae at warmer temperatures (Poulin 2006). A strong limiting factor on transmission can be the extrinsic incubation period of the parasite (i.e. the time from parasite ingestion by the vector to the vector being infectious), which is often determined by temperature. In *P. relictum*, this extrinsic incubation period in mosquitoes varies from 6–28 days and parasite development slows substantially below 21°C (LaPointe et al. 2010). Therefore, cool, high-elevation (Atkinson et al. 2014) or high latitude (Oakgrove et al. 2014) sites can act as parasite-free refugia because few infectious vectors survive long enough to acquire and transmit the parasite. Climate warming could reduce the extent of these parasite-free refugia by speeding up parasite development and allowing sustained transmission (Figure 10.4B).

While multiple empirically supported mechanisms predict increased transmission of vector-borne pathogens under climate change, changing environmental conditions in some regions could reduce

transmission through reductions in parasite or vector thermal performance (Figure 10.4B) or by altering vector habitat suitability. For example, increased warming in tropical and subtropical lowland regions could move key traits of vectors or the parasites they harbor past their thermal maxima, as predicted for some tropical mosquito-borne diseases of humans (Mordecai et al. 2019). Increases in the frequency of extreme weather events and climatic conditions could also reduce vector populations; for example, heavy rainfall and severe drought are associated with reduced adult and larval mosquito survival, respectively, and models suggest that this could limit avian malaria transmission in Hawai'i during dry years or immediately following extreme rainfall (Ahumada et al. 2004). Finally, warming temperatures could also modify the competitive outcome of co-infecting parasites with differing thermal optima. A study of blood parasite co-infections of Alaskan birds found a negative association between *Haemoproteus* and cold-adapted *Leucocytozoon* co-infections (Oakgrove et al. 2014), suggesting that warming could shift the competitive balance in favor of *Haemoproteus*.

Overall, the non-linear dependence of many traits of parasites and vectors on temperature mean

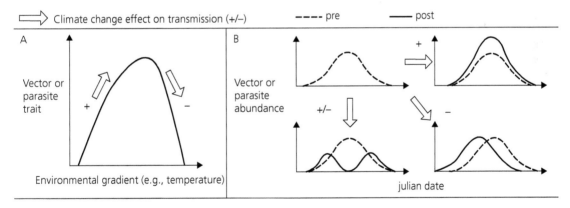

Figure 10.4 Effects of climate change (arrows) on vector or free-living parasite (A) traits and (B) abundance that could increase (+) or decrease (–) transmission in seasonal environments. (A) Many vector or parasite traits show unimodal relationships with an environmental variable, including temperature. Climate change that shifts traits towards their optima tends to increase transmission potential, while shifts away from the environmental optima will decrease transmission. (B) Seasonality in environmental conditions influences multiple traits to determine seasonal abundance of vectors or parasite propagules (top left panel, dashed line). Climate shifts towards trait optima throughout the annual cycle could increase the intensity and duration of transmission (e.g., warming in temperate regions; top right panel). Seasonal changes that move traits past their environmental tolerance (e.g., extreme summer heat in subtropical regions) could result in multiple peaks in annual transmission that increase the duration but lower the intensity of transmission (bottom left panel). Shifting phenology that advances the emergence window of vectors or parasites faster than the response in host phenology could reduce transmission via host–parasite mismatch (bottom right panel).

that moderate warming is generally expected to support increased vector-borne disease transmission in temperate, polar, and montane regions, while effects of extremes in temperature and precipitation could interrupt or prevent transmission where thermal optima are exceeded and larval habitat dries up (Figure 10.4).

10.4.2 Environmentally transmitted pathogens

For avian pathogens with externally reproducing environmental stages, a variety of environmental factors such as temperature, pH, and water depth can influence production or germination of infectious propagules. Several fungal and bacterial pathogens infect birds opportunistically and cause morbidity and mass mortality under environmental conditions that cause birds to encounter unusually high numbers of propagules. Positive associations between temperature and spore production in *Aspergillum* fungi (Seyedmousavi et al. 2015) suggest that warming might increase the number of fatal respiratory tract infections in birds. Mass mortality of waterbirds from botulism results from neurotoxins released during germination of dormant spores of *Clostridium botulinum*, and these events are associated with shallow, stagnant water and high temperatures (Rocke and Samuel 1999). Positive associations between outbreaks and water temperature in the Great Lakes (Lafrancois et al. 2011) suggest that conditions conducive to mass mortality could increase under warming.

Some pathogens have environmental stages that remain infectious for relatively short periods; therefore, changes in temperature and other environmental covariates may be crucial determinants of the seasonal buildup and duration of the transmission window. As one example, the protozoan *Trichomonas gallinae* has caused high mortality in Columbiformes and raptors where it was introduced with non-native pigeons and has emerged as a virulent pathogen of cardueline finches in Europe that co-occur at bird feeders with pigeon hosts. The persistence time of *T. gallinae* in water (up to one hour; Purple et al. 2015) has been associated with higher temperatures (Bunbury et al. 2007) and with increased humidity in bird seed (up to two days; McBurney et al. 2017); thus, regions predicted to

experience warmer and wetter climates may experience increased outbreak risk.

In contrast, some cold-adapted pathogens have persistent environmental stages lasting weeks to more than a year in favorable conditions; in these cases, warming could reduce interannual environmental persistence. For example, Newcastle disease virus shows dramatic, ten-fold declines in persistence times with increasing temperature, from 370 days at 4°C to just 6 days at 37°C (Davis-Fields et al. 2014). Similarly, the duration of infectivity in water of avian influenza viruses is associated with cool temperatures and low salinity (Brown et al. 2009). Disproportionate warming or sea level rise that increases salinity of freshwater in the Arctic breeding and staging grounds of waterfowl could therefore reduce the amount of free-living virus overwintering in Arctic waters. The resulting loss of overwintering environmental viruses could lead to transmission being largely driven by co-occurrence with shedding waterfowl, with implications for viral diversity and evolution (Hill et al. 2016).

10.5 Synthesis and future research directions

Across a diverse array of bird and parasite taxa, theoretical, field, and experimental studies of avian host–parasite interactions suggest that climate change will continue to have important and pervasive effects on parasite transmission and impacts on host populations. This review suggests that some birds, including island endemics, birds inhabiting extreme environments, and long-distance migrants, especially those that congregate in scarce and declining habitats such as wetlands, are particularly likely to face increased parasite pressure. Additionally, environmental influences on host physiology and behavior and parasite traits and transmission modes critically determine the intensity of these impacts.

The majority of empirical evidence for how climate change will influence avian host–parasite interactions examines direct or indirect effects of warming at the 'cooler' limit of host and parasite ranges, in most cases predicting increased parasitism (e.g., Garamszegi 2011). However, mathematical models and physiological theory suggest that parasitism could decline under climate change through multiple mechanisms, including: increased parasite, vector, or infected host mortality at the 'hot' range edge (or under increasing frequency of environmental extremes); improved antiparasite defense in hosts resulting from reduced demands of surviving milder winters and shortening migrations; phenological mismatch of hosts with parasite vectors; and environmental limits to vector or parasite development or environmental pathogen persistence being exceeded (Brown et al. 2009; Eikenaar and Hegemann 2016; Murdock et al. 2013). It is unclear whether the comparative lack of evidence for climate ameliorating avian parasitism reflects true trends or publication bias towards emerging pathogens that threaten wildlife or people or geographic biases in research locations (Ducatez and Lefebre 2014). Given that tropical regions host a high diversity of birds and parasites and historically more stable climates have led to narrower thermal niches, more empirical research documenting changes in parasitism in these regions could yield a more global perspective on climate–disease relationships.

Several studies demonstrate the potential for trade-offs between infection risk and host behavior, physiology, and other components of host fitness, which are likely mediated by the external environment. However, the logistical constraints of quantifying these trade-offs over multiple host generations mean that there is limited direct evidence for how changing climates will alter avian host competence for infection. 'Space-for-time' substitutions, where organisms are sampled from multiple locations spanning latitudinal or elevational gradients, are commonly used to provide indirect evidence for how populations will persist under future climate change (Pickett 1989), but these have less frequently been used to study avian host–parasite associations. Quantifying how host densities, immune performance, and parasite pressure relate to underlying environmental conditions in the core and extremes of current bird ranges will be crucial to identifying how these components interact to shape current and future transmission (Becker et al. 2020).

There are several limitations in relating experimental manipulations of environmental conditions to predictions for real-world avian–parasite interactions under global change. First, the majority of experimental studies that quantify vector performance or environmental parasite persistence use temperature as the only covariate, while (co)variation in rainfall crucially determines bird distributions and movement and habitat suitability for vector breeding and pathogen survival (Boyle et al. 2010; LaPointe et al. 2012). Second, although vectors and parasites frequently exhibit unimodal responses to environmental covariates, experimental constraints mean that the number and range of environmental conditions at which these traits are measured are often too limited to capture this non-linearity. As a consequence, directional predictions of climate change on infection dynamics are often monotonic and region-specific and should only be cautiously extrapolated to environmental extremes and wider geographic areas. Third, extrapolating results from constant-temperature experiments to environments where temperature (and humidity) varies daily, seasonally, and by microhabitat type can underestimate effects of environmental extremes on vector or parasite survival (Carrington et al. 2013). Finally, direct effects of these environmental extremes on bird hosts in relation to infection processes are rarely quantified, with evidence particularly lacking from wild birds. Addressing these knowledge gaps will be crucial for anticipating climate-induced changes to parasite transmission and impacts.

Another major knowledge gap relates to host and parasite adaptation to climate change. The rapid pace of global climate change is probably exerting strong selection pressure on parasites, vectors, and hosts, with consequences for transmission. For example, environmental extremes that reduce expected lifespan of birds could result in selection for increased offspring production over immune defense, while reductions in seasonal harshness could promote tolerance of infection (Ardia 2005). Since parasites may be able to adapt more quickly to environmental change than their hosts, climate influences on out-of-host parasite survival or the diversity of competing parasites could lead to selection on virulence and transmissibility (Koelle et al. 2005). Theoretical and empirical studies investigating the consequences of plastic or evolutionary responses to climate change will be crucial for predicting and managing parasite infections in imperiled avian populations.

Literature cited

Ahumada, J.A., Lapointe, D., and Samuel, M.D. (2004). Modeling the population dynamics of *Culex quinquefasciatus* (Diptera: Culicidae), along an elevational gradient in Hawaii. *Journal of Medical Entomology*, 41, 1157–70.

Altizer, S., Bartel, R., and Han, B.A. (2011). Animal migration and infectious disease risk. *Science*, 331, 296–302.

Altizer, S., Hochachka, W.M., and Dhondt, A.A. (2004). Seasonal dynamics of mycoplasmal conjunctivitis in eastern North American house finches. *Journal of Animal Ecology*, 73, 309–22.

Ardia, D.R. (2005). Tree swallows trade off immune function and reproductive effort differently across their range. *Ecology*, 86, 2040–6.

Atkinson, C.T., Woods, K.L., Dusek, R.J., Sileo, L.S., and Iko, W.M. (1995). Wildlife disease and conservation in Hawaii: pathogenicity of avian malaria (*Plasmodium relictum*) in experimentally infected i'iwi (*Vestiaria coccinea*). *Parasitology*, 111, S59–69.

Atkinson, C.T., Utzurrum, R.B., Lapointe, D.A., et al. (2014). Changing climate and the altitudinal range of avian malaria in the Hawaiian islands—an ongoing conservation crisis on the island of Kaua'i. *Global Change Biology*, 20, 2426–36.

Becker, D.J., Albery, G.F., Kessler, M.K., et al. (2020). Macroimmunology: the drivers and consequences of spatial patterns in wildlife immune defence. *Journal of Animal Ecology*, 89, 972–95.

Berthold, P., Helbig, A.J., Mohr, G., and Querner, U. (1992). Rapid microevolution of migratory behaviour in a wild bird species. *Nature*, 360, 668–70.

Blanchard, K.R., Kalyanasundaram, A., Henry, C., Brym, M.Z., Surles, J.G., and Kendall, R.J. (2019). Predicting seasonal infection of eyeworm (*Oxyspirura petrowi*) and caecal worm (*Aulonocephalus pennula*) in northern bobwhite quail (*Colinus virginianus*) of the Rolling Plains Ecoregion of Texas, USA. *International Journal for Parasitology: Parasites and Wildlife*, 8, 50–5.

Blem, C.R. and Blem, L.B. (1993). Do swallows sunbathe to control ectoparasites? An experimental test. *The Condor*, 95, 728–30.

Blight, L.K., Hobson, K.A., Kyser, T.K., and Arcese, P. (2015). Changing gull diet in a changing world: a 150-year stable isotope (δ13C, δ15N) record from feathers collected in the Pacific Northwest of North America. *Global Change Biology*, 21, 1497–507.

Boyle, W.A., Norris, D.R., and Guglielmo, C.G. (2010). Storms drive altitudinal migration in a tropical bird.

Proceedings of the Royal Society B: Biological Sciences, 277, 2511–19.

Brown, J.D., Goekjian, G., Poulson, R., Valeika, S., and Stallknecht, D. E. (2009). Avian influenza virus in water: infectivity is dependent on pH, salinity and temperature. *Veterinary Microbiology*, 136, 20–6.

Brown, L.M. and Hall, R.J. (2018). Consequences of resource supplementation for disease risk in a partially migratory population. *Philosophical Transactions of the Royal Society B: Biological Sciences*, 373, 20170095.

Brown, V.L. and Rohani, P. (2012). The consequences of climate change at an avian influenza 'hotspot'. *Biology Letters*, 8, 1036–9.

Bunbury, N., Jones, C.G., Greenwood, A.G., and Bell, D.J. (2007). *Trichomonas gallinae* in Mauritian columbids: implications for an endangered endemic. *Journal of Wildlife Diseases*, 43, 399–407.

Bush, S.E. and Clayton, D.H. (2018). Anti-parasite behaviour of birds. *Philosophical Transactions of the Royal Society B: Biological Sciences*, 373, 20170196.

Carrington, L.B., Seifert, S.N., Willits, N.H., et al. (2013). Large diurnal temperature fluctuations negatively influence *Aedes aegypti* (Diptera: Culicidae) life-history traits. *Journal of Medical Entomology*, 50, 43–51.

Carroll, J.M., Davis, C.A., Elmore, R.D., Fuhlendorf, S.D., and Thacker, E.T. (2015). Thermal patterns constrain diurnal behavior of a ground-dwelling bird. *Ecosphere*, 6(11), 1–15.

Cizauskas, C. A., Carlson, C. J., Burgio, K. R., et al. (2017). Parasite vulnerability to climate change: an evidence-based functional trait approach. *Royal Society Open Science*, 4, 160535.

Clark, N J., Clegg, S.M., and Klaassen, M. (2016). Migration strategy and pathogen risk: non-breeding distribution drives malaria prevalence in migratory waders. *Oikos*, 125, 1358–68.

Cohen, E.B., Auckland, L.D., Marra, P.P., and Hamer, S.A. (2015). Avian migrants facilitate invasions of neotropical ticks and tick-borne pathogens into the United States. *Applied Environmental Microbiology*, 81, 8366–78.

Cooper, N.W., Murphy, M.T., Redmond, L.J., et al. (2011). Reproductive correlates of spring arrival date in the eastern kingbird *Tyrannus tyrannus*. *Journal of Ornithology*, 152, 143–52.

Cotton, P.A. (2003). Avian migration phenology and global climate change. *Proceedings of the National Academy of Sciences of the United States of America*, 100, 12219–22.

Davis-Fields, M.K., Allison, A.B., Brown, J.R., Poulson, R.L., and Stallknecht, D.E. (2014). Effects of temperature and pH on the persistence of avian paramyxovirus-1 in water. *Journal of Wildlife Diseases*, 50, 998–1000.

de Lope, F. and Møller, A.P. (1993). Effects of ectoparasites on reproduction of their swallow hosts: a cost of being multi-brooded. *Oikos*, 67, 557–62.

Devictor, V., Julliard, R., Couvet, D., and Jiguet, F. (2008). Birds are tracking climate warming, but not fast enough. *Proceedings of the Royal Society B: Biological Sciences*, 275, 2743–8.

Diuk-Wasser, M.A., Molaei, G., Simpson, J.E., Folsom-O'Keefe, C.M., Armstrong, P.M., and Andreadis, T.G. (2010). Avian communal roosts as amplification foci for West Nile virus in urban areas in northeastern United States. *American Journal of Tropical Medicine and Hygiene*, 82, 337–43.

du Plessis, K.L., Martin, R.O., Hockey, P.A., Cunningham, S.J., and Ridley, A.R. (2012). The costs of keeping cool in a warming world: implications of high temperatures for foraging, thermoregulation and body condition of an arid-zone bird. *Global Change Biology*, 18, 3063–70.

Ducatez, S. and Lefebvre, L. (2014). Patterns of research effort in birds. *PLoS ONE*, 9, e89955.

Eikenaar, C., and Hegemann, A. (2016). Migratory common blackbirds have lower innate immune function during autumn migration than resident conspecifics. *Biology Letters*, 12, 20160078.

Epstein, P.R. (2001). West Nile virus and the climate. *Journal of Urban Health*, 78, 367–71.

Figuerola, J. and Green, A.J. (2000). Haematozoan parasites and migratory behaviour in waterfowl. *Evolutionary Ecology*, 14, 143–53.

Fix, A.S., Waterhouse, C., Greiner, E.C., and Stoskopf, M.K. (1988). *Plasmodium relictum* as a cause of avian malaria in wild-caught Magellanic penguins (*Spheniscus magellanicus*). *Journal of Wildlife Diseases*, 24, 610–19.

Freeman, B.G., Scholer, M.N., Ruiz-Gutierrez, V., and Fitzpatrick, J.W. (2018). Climate change causes upslope shifts and mountaintop extirpations in a tropical bird community. *Proceedings of the National Academy of Sciences of the United States of America*, 115, 11982–7.

Fuller, T., Bensch, S., Müller, I., et al. (2012). The ecology of emerging infectious diseases in migratory birds: an assessment of the role of climate change and priorities for future research. *EcoHealth*, 9, 80–8.

Garamszegi, L.Z. (2011). Climate change increases the risk of malaria in birds. *Global Change Biology*, 17, 1751–9.

Gauthier, G., Giroux, J.F., Reed, A., Béchet, A., and Bélanger, L.U.C. (2005). Interactions between land use, habitat use, and population increase in greater snow geese: what are the consequences for natural wetlands? *Global Change Biology*, 11, 856–68.

Geue, D. and Partecke, J. (2008). Reduced parasite infestation in urban Eurasian blackbirds (*Turdus merula*): a factor favoring urbanization? *Canadian Journal of Zoology*, 86, 1419–25.

Greiner, E.C., Bennett, G.F., White, E.M., and Coombs, R.F. (1975). Distribution of the avian hematozoa of North America. *Canadian Journal of Zoology*, 53, 1762–87.

Gutiérrez, J.S., Rakhimberdiev, E., Piersma, T., and Thieltges, D.W. (2017). Migration and parasitism: habitat use, not migration distance, influences helminth species richness in Charadriiform birds. *Journal of Biogeography*, 44, 1137–47.

Hall, R.J., Altizer, S., and Bartel, R.A. (2014). Greater migratory propensity in hosts lowers pathogen transmission and impacts. *Journal of Animal Ecology*, 83, 1068–77.

Hall, R.J., Brown, L.M., and Altizer, S. (2016). Modeling vector-borne disease risk in migratory animals under climate change. *Integrative and Comparative Biology*, 56, 353–64.

Hawley, D.M., DuRant, S.E., Wilson, A.F., Adelman, J.S., and Hopkins, W.A. (2012). Additive metabolic costs of thermoregulation and pathogen infection. *Functional Ecology*, 26, 701–10.

Hegemann, A., Matson, K.D., Both, C., and Tieleman, B. I. (2012). Immune function in a free-living bird varies over the annual cycle, but seasonal patterns differ between years. *Oecologia*, 170, 605–18.

Hegemann, A., Abril, P.A., Muheim, R., et al. (2018). Immune function and blood parasite infections impact stopover ecology in passerine birds. *Oecologia*, 188, 1011–24.

Hill, N.J., Ma, E.J., Meixell, B.W., Lindberg, M.S., Boyce, W.M., and Runstadler, J.A. (2016). Transmission of influenza reflects seasonality of wild birds across the annual cycle. *Ecology Letters*, 19, 915–25.

Hitch, A.T. and Leberg, P. L. (2007). Breeding distributions of North American bird species moving north as a result of climate change. *Conservation Biology*, 21, 534–9.

Hosseini, P.R., Dhondt, A.A., and Dobson, A. (2004). Seasonality and wildlife disease: how seasonal birth, aggregation and variation in immunity affect the dynamics of *Mycoplasma gallisepticum* in house finches. *Proceedings of the Royal Society B: Biological Sciences*, 271, 2569–77.

Jones, W., Kulma, K., Bensch, S., et al. (2018). Interspecific transfer of parasites following a range-shift in *Ficedula* flycatchers. *Ecology and Evolution*, 8, 12183–92.

Keane, R.M. and Crawley, M.J. (2002). Exotic plant invasions and the enemy release hypothesis. *Trends in Ecology and Evolution*, 17, 164–70.

Koelle, K., Pascual, M., and Yunus, M. (2005). Pathogen adaptation to seasonal forcing and climate change. *Proceedings of the Royal Society B: Biological Sciences*, 272, 971–7.

Koh, L.P., Dunn, R.R., Sodhi, N.S., Colwell, R.K., Proctor, H.C., and Smith, V.S. (2004). Species coextinctions and the biodiversity crisis. *Science*, 305, 1632–4.

Koprivnikar, J. and Leung, T.L. (2015). Flying with diverse passengers: greater richness of parasitic nematodes in migratory birds. *Oikos*, 124, 399–405.

Krebs, B.L., Anderson, T.K., Goldberg, T.L., et al. (2014). Host group formation decreases exposure to vector-borne disease: a field experiment in a 'hotspot' of West Nile virus transmission. *Proceedings of the Royal Society B: Biological Sciences*, 281, 20141586.

La Sorte, F.A. and Thompson III F.R. (2007). Poleward shifts in winter ranges of North American birds. *Ecology*, 88, 1803–12.

Lafrancois, B.M., Riley, S.C., Blehert, D.S., and Ballmann, A.E. (2011). Links between type E botulism outbreaks, lake levels, and surface water temperatures in Lake Michigan, 1963–2008. *Journal of Great Lakes Research*, 37, 86–91.

LaPointe, D.A., Goff, M.L., and Atkinson, C.T. (2010). Thermal constraints to the sporogonic development and altitudinal distribution of avian malaria *Plasmodium relictum* in Hawai'i. *Journal of Parasitology*, 96, 318–24.

LaPointe, D.A., Atkinson, C.T., and Samuel, M.D. (2012). Ecology and conservation biology of avian malaria. *Annals of the New York Academy of Sciences*, 1249, 211–26.

Loiseau, C., Harrigan, R.J., Cornel, A.J., et al. (2012). First evidence and predictions of *Plasmodium* transmission in Alaskan bird populations. *PLoS ONE*, 7, e44729.

Maddock, M. and Geering, D. (1994). Range expansion and migration of the cattle egret. *Ostrich*, 65, 191–203.

Martin II, L.B., Pless, M., Svoboda, J., and Wikelski, M. (2004). Immune activity in temperate and tropical house sparrows: a common-garden experiment. *Ecology*, 85, 2323–31.

Marzal, A., Ricklefs, R.E., Valkiūnas, G., et al. (2011). Diversity, loss, and gain of malaria parasites in a globally invasive bird. *PLoS ONE*, 6, e21905.

May, R.M. and Anderson, R.M. (1979). Population biology of infectious diseases: Part II. *Nature*, 280, 455.

McBurney, S., Kelly-Clark, W.K., Forzán, M.J., Vanderstichel, R., Teather, K., and Greenwood, S.J. (2017). Persistence of *Trichomonas gallinae* in birdseed. *Avian Diseases*, 61, 311–15.

Meller, K., Vähätalo, A.V., Hokkanen, T., Rintala, J., Piha, M., and Lehikoinen, A. (2016). Interannual variation and long-term trends in proportions of resident individuals in partially migratory birds. *Journal of Animal Ecology*, 85, 570–80.

Møller, A.P. (2010). Host–parasite interactions and vectors in the barn swallow in relation to climate change. *Global Change Biology*, 16, 1158–70.

Møller, A.P. and Erritzøe, J. (1998). Host immune defence and migration in birds. *Evolutionary Ecology*, 12, 945–53.

Møller, A.P., de Lope, F., and Saino, N. (2004). Parasitism, immunity, and arrival date in a migratory bird, the barn swallow. *Ecology*, 85, 206–19.

Montecino-Latorre, D. and Barker, C.M. (2018). Overwintering of West Nile virus in a bird community with a communal crow roost. *Scientific Reports* 8, 6088.

Mordecai, E.A., Caldwell, J.M., Grossman, M.K., et al. (2019). Thermal biology of mosquito-borne disease. *Ecology Letters*, 22, 1690–708.

Moreno-Rueda, G. (2011). Trade-off between immune response and body mass in wintering house sparrows (*Passer domesticus*). *Ecological Research*, 26, 943.

Morin, C. W., Stoner-Duncan, B., Winker, K., et al. (2018). Avian influenza virus ecology and evolution through a climatic lens. *Environment international*, 119, 241–49.

Moyer, B.R. and Wagenbach, G.E. (1995). Sunning by black noddies (*Anous minutus*) may kill chewing lice (*Quadraceps hopkinsi*). *The Auk*, 112, 1073–7.

Murdock, C.C., Foufopoulos, J., and Simon, C.P. (2013). A transmission model for the ecology of an avian blood parasite in a temperate ecosystem. *PLoS ONE*, 8, e76126.

Murray, M.H., Becker, D.J., Hall, R.J., and Hernandez, S.M. (2016). Wildlife health and supplemental feeding: a review and management recommendations. *Biological Conservation*, 204, 163–74.

Németh, Z. (2017). Partial migration and decreasing migration distance in the Hungarian population of the common blackbird (*Turdus merula* Linnaeus, 1758): analysis of 85 years of ring recovery data. *Ornis Hungarica*, 25, 101–8.

Oakgrove, K.S., Harrigan, R.J., Loiseau, C., Guers, S., Seppi, B., and Sehgal, R.N. (2014). Distribution, diversity and drivers of blood-borne parasite co-infections in Alaskan bird populations. *International Journal for Parasitology*, 44, 717–27.

Ogden, N.H., St-Onge, L., Barker, I.K., et al. (2008). Risk maps for range expansion of the Lyme disease vector, *Ixodes scapularis*, in Canada now and with climate change. *International Journal of Health Geographics*, 7, 24.

Ortiz, N.E. and Smith, G.R. (1994). Landfill sites, botulism and gulls. *Epidemiology and Infection*, 112, 385–91.

Owen, J.C. and Moore, F.R. (2008). Swainson's thrushes in migratory disposition exhibit reduced immune function. *Journal of Ethology*, 26, 383–8.

Paquet, M., Doutrelant, C., Loubon, M., Theron, F., Rat, M., and Covas, R. (2016). Communal roosting, thermoregulatory benefits and breeding group size predictability in cooperatively breeding sociable weavers. *Journal of Avian Biology*, 47, 749–55.

Paxton, E.H., Camp, R.J., Gorresen, P.M., Crampton, L.H., Leonard, D.L., and VanderWerf, E.A. (2016). Collapsing avian community on a Hawaiian island. *Science Advances*, 2, e1600029.

Paz, S. (2015). Climate change impacts on West Nile virus transmission in a global context. *Philosophical Transactions of the Royal Society B: Biological Sciences*, 370, 20130561.

Pickett, S.T. (1989). Space-for-time substitution as an alternative to long-term studies. In G. Kikens, ed. *Long-Term Studies in Ecology*, pp. 110–35. Springer, New York.

Piersma, T. (1997). Do global patterns of habitat use and migration strategies co-evolve with relative investments in immunocompetence due to spatial variation in parasite pressure? *Oikos*, 80, 623–31.

Poulin, R. (2006). Global warming and temperature-mediated increases in cercarial emergence in trematode parasites. *Parasitology*, 132, 143–51.

Prenter, J., MacNeil, C., Dick, J.T., and Dunn, A.M. (2004). Roles of parasites in animal invasions. *Trends in Ecology and Evolution*, 19, 385–90.

Purple, K.E., Humm, J.M., Kirby, R.B., Saidak, C.G., and Gerhold, R. (2015). *Trichomonas gallinae* persistence in four water treatments. *Journal of Wildlife Diseases*, 51, 739–42.

Quinteiro-Filho, W.M., Gomes, A.V.S., Pinheiro, M.L., et al. (2012). Heat stress impairs performance and induces intestinal inflammation in broiler chickens infected with *Salmonella enteritidis*. *Avian Pathology*, 41, 421–7.

Reisen, W.K., Fang, Y., and Martinez, V.M. (2014). Effects of temperature on the transmission of West Nile virus by *Culex tarsalis* (Diptera: Culicidae). *Journal of Medical Entomology*, 43, 309–17.

Reullier, J., Pérez-Tris, J.A., Bensch, S., and Secondi, J. (2006). Diversity, distribution and exchange of blood parasites meeting at an avian moving contact zone. *Molecular Ecology*, 15, 753–63.

Rocke, T.E. and Samuel, M.D. (1999). Water and sediment characteristics associated with avian botulism outbreaks in wetlands. *Journal of Wildlife Management*, 63, 1249–60.

Rosenberg, K.V., Dokter, A.M., Blancher, P.J., et al. (2019). Decline of the North American avifauna. *Science*, 366, 120–4.

Saino, N., Ambrosini, R., Rubolini, D., et al. (2010). Climate warming, ecological mismatch at arrival and population decline in migratory birds. *Proceedings of the Royal Society B: Biological Sciences*, 278, 835–42.

Samuel, M.D., Shadduck, D.J., Goldberg, D.R., and Johnson, W.P. (2005). Avian cholera in waterfowl: the role of lesser snow and Ross's geese as disease carriers in the Playa Lakes region. *Journal of Wildlife Diseases*, 41, 48–57.

Seyedmousavi, S., Guillot, J., Arné, P., et al. (2015). *Aspergillus* and aspergilloses in wild and domestic animals: a global health concern with parallels to human disease. *Medical Mycology*, 53, 765–97.

Shaw, A.K., Craft, M.E., Zuk, M., and Binning, S.A. (2019). Host migration strategy is shaped by forms of parasite transmission and infection cost. *Journal of Animal Ecology*, 88, 1601–12.

Slowinski, S.P., Fudickar, A.M., Hughes, A.M., et al. (2018). Sedentary songbirds maintain higher prevalence of haemosporidian parasite infections than migratory conspecifics during seasonal sympatry. *PLoS ONE*, 13, e0201563.

Smith, R.J. and Moore, F.R. (2005). Arrival timing and seasonal reproductive performance in a long-distance migratory landbird. *Behavioral Ecology and Sociobiology*, 57, 231–9.

Spivey, T.J., Lindberg, M.S., Meixell, B.W., et al. (2017). Maintenance of influenza A viruses and antibody response in mallards (*Anas platyrhynchos*) sampled during the non-breeding season in Alaska. *PLoS ONE*, 12, e0183505.

Tingley, M.W., Koo, M.S., Moritz, C., Rush, A.C., and Beissinger, S.R. (2012). The push and pull of climate change causes heterogeneous shifts in avian elevational ranges. *Global Change Biology*, 18, 3279–90.

van Gils, J.A., Lisovski, S., Lok, T., et al. (2016). Body shrinkage due to Arctic warming reduces red knot fitness in tropical wintering range. *Science*, 352, 819–21.

Van Hemert, C.R., Pearce, J.M., and Handel, C.M. (2014). Wildlife health in a rapidly changing north: focus on avian disease. *Frontiers in Ecology and the Environment*, 12, 548–56.

Van Riper III, C., Van Riper, S.G., Goff, M.L., and Laird, M. (1986). The epizootiology and ecological significance of malaria in Hawaiian land birds. *Ecological Monographs*, 56, 327–44.

Visser, M.E., Perdeck, A.C., van Balen, J.H., and Both, C. (2009). Climate change leads to decreasing bird migration distances. *Global Change Biology*, 15, 1859–65.

Warner, R.E. (1968). The role of introduced diseases in the extinction of the endemic Hawaiian avifauna. *The Condor*, 70, 101–20.

Warren, C.C., Berthold, L.K., Mounce, H.L., Luscomb, P., Masuda, B, and Berry, L. (2020). Kiwikiu Translocation Report 2019. Internal Report, pp. 1–101.

Webster, R.G., Bean, W.J., Gorman, O.T., Chambers, T.M., and Kawaoka, Y. (1992). Evolution and ecology of influenza A viruses. *Microbiology and Molecular Biology Reviews*, 56, 152–79.

Winker, K. and Gibson, D.D. (2010). The Asia-to-America influx of avian influenza wild bird hosts is large. *Avian Diseases*, 54, 477–82.

Wobeser, G. (1992). Avian cholera and waterfowl biology. *Journal of Wildlife Diseases*, 28, 674–82.

Zamora-Vilchis, I., Williams, S.E., and Johnson, C.N. (2012). Environmental temperature affects prevalence of blood parasites of birds on an elevation gradient: implications for disease in a warming climate. *PLoS ONE*, 7, e39208.

CHAPTER 11

Pathogens from Wild Birds at the Wildlife–Agriculture Interface

Alan B. Franklin, Sarah N. Bevins, and Susan A. Shriner

11.1 Introduction

Wild birds have long been identified as carrying pathogens that can affect both human and agricultural health (Chung et al. 2018; Elmberg et al. 2017; Greig et al. 2015). There is concern that wild birds can transmit these pathogens to crops, livestock (e.g., domestic cattle and swine), and poultry, which can affect animal health but also enter the human food chain and affect human health (Jay-Russell and Doyle 2016). After multiple outbreaks of food-borne pathogens, industry and governmental regulation concerning fecal contamination of agricultural operations by wildlife has intensified (Gennet et al. 2013). For example, fruit and vegetable farmers in the US are regulated by commercial produce buyers that have encouraged the removal of wildlife habitat in agricultural landscapes to reduce the risk of food-borne illness (Gennet et al. 2013; Karp et al. 2015). In addition, global outbreaks of avian influenza A viruses (hereafter, we use IAV/IAVs to encompass the group of influenza A viruses that can also affect non-avian hosts) in poultry have been attributed to waterfowl (Ren et al. 2016; Tian et al. 2015; Verhagen et al. 2015), the natural hosts for these viruses (Webster et al. 1992). Thus, there is a perception that wild birds pose a threat to agricultural and human health because of the pathogens that wild birds carry.

However, the degree to which wild avian species pose risks for contaminating livestock and crops with pathogens is subject to debate (Smith et al. 2020) and depends on a variety of factors, which we explore in this chapter. Here, we first identify the need to define focal avian species that frequently use agricultural operations and couple this with identifying host types and potential pathways for pathogen contamination. We then identify the major pathogens of concern to agriculture, which can also be prevalent in the focal species of wild birds. Finally, we conceptually explore the potential of focal species to traffic pathogens of concern to agricultural operations at different scales. There have been several excellent reviews on pathogens from wild birds at the wildlife–agriculture interface (such as Clark 2014 and Erickson 2016) which we attempted to complement rather than duplicate by focusing on concepts in disease ecology that have applications to the wildlife–agriculture interface.

11.2 Avian use of agricultural operations

Wildlife species that live in and around human-modified landscapes are referred to as **peridomestic** or **synanthropic**. While many avian species can be considered synanthropic, the most common species associated with agriculture are in the orders Charadriiformes (e.g., gulls), Anseriformes (ducks and geese), Columbiformes (pigeons and doves), and Passeriformes (e.g., blackbirds, starlings, and sparrows) (Clark 2014; Table 11.1). In most cases, common synanthropic birds at the wildlife–agriculture interface have broad, continental geographic ranges (e.g., Table 11.1). Often, they are also invasive species; for example, 22.2% of the synanthropic species at the wildlife–agriculture interface in North America are considered invasive (Table 11.1).

Alan B. Franklin, Sarah N. Bevins, and Susan A. Shriner, *Pathogens from Wild Birds at the Wildlife–Agriculture Interface* In: *Infectious Disease Ecology of Wild Birds*. Edited by: Jennifer C. Owen, Dana M. Hawley, and Kathryn P. Huyvaert, Oxford University Press. © Oxford University Press 2021. DOI: 10.1093/oso/9780198746249.003.0011

Table 11.1 Examples of common synanthropic avian species associated with agricultural environments in North America (from Burns et al. 2012; Clark 2014; Fox et al. 2017; Johnston 2001; Tormoehlen et al. 2019). Localized species with limited geographic distributions are not included.

Common Name	Species	Type	Range[1]
Charadriformes			
Laughing gull	*Larus atricilla*	Native	Regional
Franklin's gull	*Larus pipixcan*	Native	Regional
Ring-billed gull	*Larus delawarensis*	Native	Continental
Killdeer	*Charadrius vociferus*	Native	Continental
Columbiformes			
Rock pigeon	*Columbia livia*	Invasive	Continental
Eurasian collared dove	*Streptopilia decaoto*	Invasive	Continental
Mourning dove	*Zenaida macroura*	Native	Continental
Anseriformes			
Snow goose	*Chen caerulescens*	Native	Regional
Canada goose	*Branta canadensis*	Native	Continental
Mallard	*Anas platyrhynchos*	Native	Continental
Passeriformes			
Barn swallow	*Hirundo rustica*	Native	Continental
European starling	*Sturnus vulgaris*	Invasive	Continental
House sparrow	*Passer domesticus*	Invasive	Continental
Savannah sparrow	*Passerculus sandwichensis*	Native	Continental
House finch	*Haemorhous mexicanus*	Native	Continental
Red-winged blackbird	*Agelaius phoeniceus*	Native	Continental
Brown-headed cowbird	*Molothrus ater*	Native	Continental
Horned lark	*Eremophila alpestris*	Native	Continental
American robin	*Turdus migratorius*	Native	Continental
Common grackle	*Quiscalus quiscula*	Native	Regional

[1] Ranges are relative to North America only.

11.2.1 Degree of use of agricultural operations by synanthropic birds

The abundance and diversity of birds using agricultural fields and facilities differs depending on the season, facility type, and resource availability. For example, European starlings (*Sturnus vulgaris*) can have flocks totaling 50,000 individuals in livestock feedlots in winter and fall (Carlson et al. 2011d; Figure 11.1A), but in the breeding season, when pairs are territorial, they may occur in low densities in the same area (Cabe 1993). This pattern in starling abundance is also reflected by the degree of fecal contamination on Ohio dairies, with lower levels of contamination during the breeding season and higher levels during the late fall and winter (Medhanie et al. 2014) (Figure 11.1B). On the other hand, starlings may present a pathogen incursion risk to poultry farms during the breeding season when this cavity-nesting species seeks building openings and crevices for nesting and infected birds may breach the interiors of poultry houses.

Figure 11.1 Example of use of feedlots and dairies by European starlings. Photo (A) illustrates the extent of European starling flocks using these types of facilities. (B) Starlings perched on feed troughs with potential to contaminate cattle feed with pathogens from their feces. (Photographs are from the USDA Wildlife Services photograph archive.)

Contamination of crop fields by avian synanthropes is another avenue for pathogen incursion at the wildlife–agriculture interface. For example, gulls were implicated as a source for contaminating tomato fields with *Salmonella*, especially when fields are near landfills that are also visited by gulls (Gruszynski et al. 2014). Thus, impacts from individual species can vary by spatial and temporal factors such as location, species, and season.

Use of agricultural operations by wild birds is not as well documented in the Americas as it is in other parts of the world, such as Europe (Bergervoet et al. 2019; Butler et al. 2007; Elbers and Gonzales 2020) and Asia (e.g., Fujioka et al. 2010). Data on agricultural contamination by wild birds in the Americas and other parts of the world are also generally lacking. Exceptions include Carlson et al. (2011b) and Carlson et al. (2011c), described above, and Burns et al. (2012), who estimated both abundance and use of poultry facilities by individual species of wild birds. Some species-specific metrics used in these studies were the total number of birds, times observed in immediate barn areas, and times observed entering barns. Kitts-Morgan et al. (2015) found that the most common birds using equine, dairy, and cattle operations in the southeastern US were European starlings, house sparrows (*Passer domesticus*), barn swallows (*Hirundo rustica*), and rock pigeons (*Columbia livia*). In addition, most of the fecal contamination of livestock feed at these facilities was from birds. While several studies of birds at the wildlife–agriculture interface have focused on the effects of agricultural intensity on wild birds (e.g., Belden et al. 2018; McLaughlin and Mineau 1995; Quinn et al. 2017; Stanton et al. 2018) or differences in use and diversity of wild birds between different agricultural methods (e.g., Beecher et al. 2002; Bengtsson et al. 2005), these studies either have focused on the conservation implications at coarser scales or have not provided the data to document species-specific use of agricultural facilities by wild birds. Species-specific studies are needed to better understand wild bird use of a wide variety of agricultural facilities, especially in the context of pathogen introduction.

11.2.2 Identification of focal synanthropic avian species

Identification of focal synanthropic avian species that intensively use agricultural facilities and have the potential to contaminate agricultural products with pathogens of concern is context- and scale-specific. In most cases, focal species will include many of those identified in Table 11.1. However, local or regional conditions may dictate the use of other criteria, such as abundance, as outlined in Table 11.2. For example, Burns et al. (2012) created a high priority list of wild bird species for IAV surveillance on poultry farms in British Columbia and Ontario, Canada, based on the following annual criteria:

• Anseriformes and Charadriiformes species observed in the immediate barn area two or more times (since species in these avian orders were the natural reservoirs for IAVs).

Table 11.2 Examples of criteria to identify focal synanthropic avian species using agricultural operations.

Criteria	Example
Species documented as highly abundant on agricultural facilities	European starlings in flocks of 1,000-50,000 individuals visiting livestock facilities (Carlson et al. 2011c).
Species identified as having positive population trends relative to increasing agricultural intensity	At a regional scale, killdeer abundance increased with agricultural expansion and agricultural biomass produced (Quinn et al. 2017).
Species associated with outbreaks of food-borne pathogens in humans	An outbreak of *Campylobacter jejuni* in humans was traced to contamination of peas by sandhill crane (*Grus canadensis*) feces (Gardner et al. 2011).
Species identified as reservoirs for pathogens of concern	Waterfowl, such as mallards, have been identified as reservoirs of influenza A virus (Webster et al. 1992)
Species identified as causing extensive damage	Red-winged blackbirds cause $17.5 million of annual damage to sunflower crops (Shwiff et al. 2017).

- Any species observed entering poultry barns.
- Any species observed ten or more times in both the immediate barn area and wetland/cropland areas.

Based on one or more of these criteria, focal species for continued scrutiny in this agricultural system would include 19 species (mallards, *Anas platyrhynchos*; trumpeter swan, *Cygnus buccinators*; Canada goose, *Branta canadensis*; glaucous-winged gull, *Larus glaucescens*; mew gull, *Larus canus*; ring-billed gull, *Larus delawarensis*; killdeer, *Charadrius vociferous*; European starlings; rock pigeons; barn swallows; northwestern crows, *Corvus caurinus*; house finch, *Haemorhous mexicanus*; common grackle, *Quiscalus quiscula*; American robin, *Turdus migratorius*; house sparrow; song sparrow, *Melospiza melodia*; dark-eyed junco, *Junco hyemalis*; savannah sparrow, *Passerculus sandwichensis*; and horned lark, *Eremophila alpestris*) out of the 121 species of wild birds observed at these facilities. Three of the species on this list (northwestern crow, glaucous gull, and mew gull) have geographic ranges limited to Alaska and western Canada, illustrating the need for local knowledge in identifying focal avian species. By developing and combining multiple criteria, agricultural operations at different scales can

develop a list of focal avian species for further evaluation, such as increased scrutiny of farm use by wild birds and increased testing for pathogens in the agricultural system. This type of approach can also be made quantitative using cluster analysis in a Bayesian framework, similar to that used to prioritize disease risks (Hwang et al. 2018).

11.3 Host types and pathways for pathogen contamination by birds

Focal synanthropic avian species encompass several different conceptual host types, which are involved in numerous pathways leading to contamination of agricultural operations with pathogens of concern. As with most natural systems, the wildlife aspect of the wildlife–agriculture interface likely involves multihost systems, with different species acting as distinct types of hosts with unique roles in transmission pathways. For example, 77% of livestock pathogens infect multiple hosts, including wildlife (Cleaveland et al. 2001).

11.3.1 Host types

Most pathogens at the wildlife–agriculture interface are generalists with multiple hosts; hence, characterizing transmission cycles and applying successful interventions for disease prevention or control is complex. A clear understanding of the different host types, their functional role in transmission, and the interactions between these populations is critical to understand the mechanisms of transmission between wildlife and agricultural operations, especially when dealing with multihost wildlife communities.

At a basic level, pathogens persist through a chain of infection in which a pathogen replicates in a host, exits the host, and is then transmitted to another host directly or indirectly (see Chapter 2 for in-depth discussion of the chain of infection). Generally, the primary entity or entities in which a pathogen multiplies and persists—humans, animals, and/or the environment—is considered the **reservoir**. Reservoirs can range from a single population of a single species to a multispecies community of interacting populations and the environment. In multihost systems, especially those at the wildlife–agriculture interface which often have a broad host range (i.e., a diversity of potential hosts), refinement of the general reservoir concept is essential for effective disease management (Haydon et al. 2002; Viana et al. 2014) because the host status of a particular species or group of species might differ depending on direction of transmission (wildlife to agriculture or agriculture to wildlife) or the impact of transmission (risk to wildlife or risk to agriculture). Moreover, some strains of a pathogen might largely persist in wildlife populations, while other strains primarily persist in agricultural species. In this scenario, the reservoir status of a host is pathogen strain-specific and may vary depending on contact rates between wild and agricultural species.

One of the most common ways to characterize the role of different host types is by their contribution to pathogen persistence in nature. **Maintenance hosts**, defined as one or more epidemiologically connected populations or environments where a pathogen is permanently maintained (Haydon et al. 2002), transmit pathogens to the defined target hosts. Maintenance hosts contribute to pathogen persistence without external inputs, whereas non-maintenance hosts cannot sustain pathogen persistence over time without external inputs (Viana et al. 2014). Within this framework, a useful refinement for defining host types at the agriculture interface is to identify a **target population** of interest (Haydon et al. 2002). Broadly, agricultural target populations can encompass a variety of agricultural operations such as domestic animal populations (backyard farms, live bird markets, commercial poultry, and livestock feedlots), field crops, or human populations affected through the food chain.

Pathogen transmission between maintenance hosts and target hosts can be uni- or multidirectional, which allows for three main possibilities at the agriculture interface: primary transmission of pathogens from wild birds to agricultural operations (pathogen spillover), primary transmission from livestock to wild birds (pathogen spillback), or transmission in both directions. Unidirectional transmission is most likely to occur when the target population is an incidental or accidental non-maintenance host. For example, when the poultry-associated bacterial pathogen *Mycoplasma gallisepticum* spilled over from domestic chickens to house finches around 1994, it quickly adapted to its new host such that subsequent experiments with the diverged strain in chickens showed significantly reduced susceptibility with no onward transmission (Bonneaud et al. 2019). Multidirectional transmission is relatively common between maintenance and target hosts. For example, IAVs commonly exhibit bidirectional transmission between domestic and wild birds (Bahl et al. 2016).

Because maintenance hosts and target hosts do not always interact directly, the concept of **bridge hosts** was developed (Caron et al. 2015). Bridge hosts provide a link between maintenance hosts or maintenance communities and target hosts (Figure 11.2A). Caron et al. (2015) defined two essential requirements for a host to qualify as a bridge host: (1) it must be replication competent for the pathogen (i.e., pathogens must be able to successfully reproduce within the host) or have the capacity for mechanical transmission where pathogens are transported on body surfaces; and (2) the bridge host must have infectious contacts with the target population. By definition, bridge hosts lack the capacity to maintain pathogen persistence without inputs from maintenance hosts; thus, R_0 should be < 1 for these populations (see Chapters 2 and 7 for more on R_0). Further, bridge hosts need to overlap in time and space with both maintenance hosts and target hosts to effectively link these populations (Figure 11.2B). Thus, understanding both biological (transmission and infection characteristics) and ecological (habitat use, behavior, contact rates) traits is required to assign bridge host status.

Avian IAVs provide a useful case of a host community where bridge hosts may link maintenance and target populations because, in many cases, natural maintenance hosts (waterfowl and aquatic birds) are not frequently found in or around poultry barns, yet these viruses frequently cause outbreaks in commercial poultry. In a study of wild and domestic birds on farms, wetlands, and villages in Zimbabwe, Caron et al. (2014) identified several potential bridge hosts that shared habitats with both domestic and wild birds. They estimated that contacts between maintenance and target hosts were 20 times more likely via bridge hosts compared to direct contacts between maintenance and target populations.

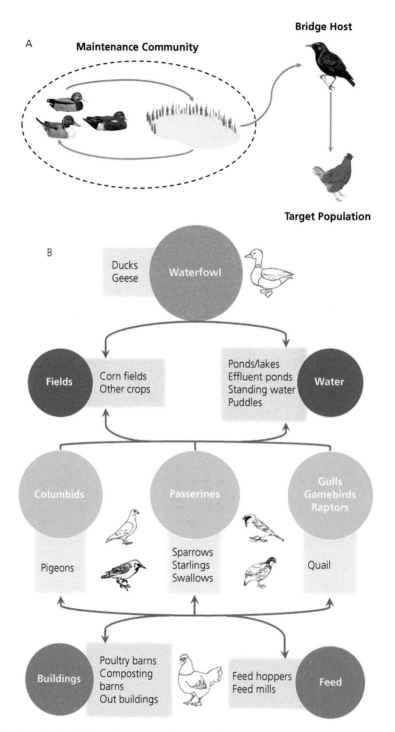

Figure 11.2 Examples of relationships of different host types with target populations at the wildlife–agriculture interface. (A) In this example, the pathogen of concern is an IAV where waterfowl and aquatic habitats are the maintenance hosts (Webster et al. 1992) and European starlings are potential bridge hosts that transmit the virus to poultry, the target population (Shriner et al. 2016). (B) Potential bridge hosts such as pigeons, starlings, sparrows, and gamebirds could bridge IAV between maintenance and target populations via shared resources.

In addition to pathogen persistence, other common tools used to classify species host types are pathogen prevalence, seroprevalence, and genetics (Viana et al. 2014). Phylogenetics is increasingly used to identify networks of species that share pathogens at the agriculture interface and to provide a potential method for confirming pathogen movement among wildlife, bridge, and agricultural species. In many cases, strain matching can be used not only to confirm shared pathogens between maintenance and target hosts but also to assess the direction and timing of pathogen movement between host types. For example, Kwon et al. (2020) used Bayesian phylogenetics to determine that an avian IAV most likely spread from wild waterfowl to domestic ducks and then back to chickens. In this case, waterfowl are known maintenance hosts of IAVs and domestic ducks and chickens are target hosts often requiring intervention to control the spread of the pathogen.

Surveillance systems that provide information on prevalence and/or seroprevalence can be used to identify host species in a given pathogen system. However, these tools often cannot provide information on maintenance or persistence of pathogens within host populations to allow appropriate assignments of host types. Fully understanding pathogen networks at the agriculture interface requires a multidimensional approach that includes both field studies to determine habitat use and contact rates and experimental pen studies to assess intra- and interspecies infection and transmission dynamics.

A final consideration in defining host types is that multihost systems are dynamic and change over time and space as pathogens and hosts evolve. As additional strains and serotypes emerge and re-emerge, new host relationships arise. For example, the well-known Asian H5N1 highly pathogenic avian IAV emerged in China in 1996. The virus was first isolated from a domestic goose in Guangdong province. While wild birds are considered the likely progenitor of the virus and the primary maintenance hosts for IAVs in general, the virus re-emerged in the early 2000s and subsequently spread and was maintained in poultry throughout Southeast Asia (in addition to spilling over into human populations and causing hundreds of deaths; Wan 2012).

In 2005 a new variant of the virus spilled back into wild birds where it caused mass die-offs at Qinghai Lake in China. Thus, while most IAVs are maintained in wild bird hosts, some strains adapt to and are largely maintained in poultry, illustrating that the primary maintenance host of a particular lineage can change.

11.3.2 Types of pathways for contamination

Pathways of contamination from birds to agricultural operations can be either direct or indirect. Most pathogen spread is probably through the fecal–oral route, which can be direct if feces are ingested either intentionally or accidentally or indirect through a secondary source such as water or a bridge host. Direct pathogen spread to livestock is probably rare, whereas direct contamination (e.g., fecal deposition on plants) of agricultural crops is probably more common. For example, fecal contamination of raw peas by sandhill cranes (*Grus canadensis*) was the inferred pathway of *Campylobacter* spillover into human consumers (Gardner et al. 2011). The spread of antimicrobial-resistant bacteria by wild birds is an example of multiple sources of pathogens infecting wild birds that then contaminate different agricultural operations through direct and indirect pathways (Figure 11.3).

Wild birds, in particular synanthropic species, are drawn to agricultural areas for a variety of reasons, often because of resource availability, such as water, food, and shelter. Water attractants include troughs, water tanks, and ponds maintained for livestock as well as aerobic effluent ponds that can provide rich forage associated with high nutrient loads. Crop irrigation also provides plentiful water as an attractant but can also be a source of contamination of agricultural fields if raw water is used for irrigation (Figure 11.3). In addition, crops provide forage for many bird species attracted to seeds, new plant shoots, mature plants, and waste grain. At animal operations, pastures, open feed bunks, spilled feed, compost, and carcasses all provide forage opportunities for wild birds. In addition to these key resources, farms can provide shelter. For example, rock pigeons often nest in and around open barns, American robins nest on building ledges, swallows

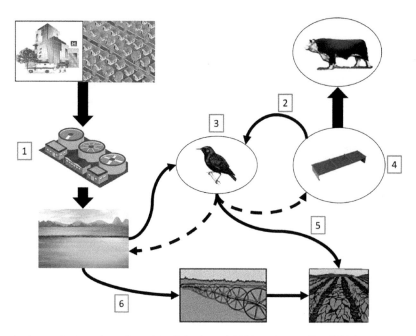

Figure 11.3 Example of contamination pathways for antimicrobial-resistant (AMR) bacteria spread by wild birds such as European starlings (node 3) (Carlson et al. 2015). Sources of contamination include human contributions (e.g., hospitals, residential areas) through wastewater treatment plants (pathway 1) (Pärnänen et al. 2019) and antibiotic-laced feed in livestock feedlots (pathway 2) (Aarestrup 2015). From these sources, wild birds (node 3) can further contaminate livestock facilities through contamination of feed (node 4) or crops through direct contamination with feces (pathway 5) or indirectly through contamination of raw water used to irrigate crops (pathway 6).

build nests on open eaves, and European starlings commonly nest in farm building cavities, often in structural gaps or gashes.

The use of these agricultural resources by synanthropic species presents a range of risk pathways for pathogen transmission and resource contamination. Wild birds can directly or mechanically transmit pathogens to livestock or indirectly contaminate food, water, aerosols, or fomites that humans or livestock are subsequently exposed to. Wild birds on farms have the potential to transport pathogens directly to outdoor livestock or even to domestic animals maintained in closed barns by entering through roof vents, fan vents, structural breaches, or opportunistically through open doors (Shriner et al. 2016). These risk pathways can be mitigated with three general types of action that target synanthropic species: (1) reducing attractants, (2) preventing access, and (3) utilizing deterrents (Shriner et al. 2016).

Additional risk pathways involving synanthropic bird species arise from their association with feed mills and stored feed. While biosecurity is a high priority for these operations and most feed mills operate using closed systems, breaches may occur, resulting in potential pathogen contamination. A US Department of Agriculture risk assessment evaluating the risk of poultry feed contamination with an avian IAV found that the risk of wild birds spreading IAV to feed was low to very low (US Department of Agriculture 2015). However, the risk of contamination by birds in other types of agricultural operations, such as stored feed in livestock operations (Figure 11.4), may pose a significantly higher risk (Daniels et al. 2003). Another potential risk pathway is humans who interact with both wild birds and poultry (e.g., waterfowl hunters). If these individuals do not practice strong biosecurity between handling wild birds and handling poultry, they risk transmitting pathogens between these two groups. Most wild bird strains of IAV have limited zoonotic potential, so the risk of transmission to humans from handling waterfowl is low (Gill et al. 2006; Siembieda et al. 2008); however, a lack of

Figure 11.4 Bonaparte's gulls (*Chroicocephalus philadelphia*) foraging on stored livestock feed at a small-scale cattle feedlot. (Photograph by Nicole Barrett.)

sound biosecurity practice while handling birds could lead to mechanical transmission through contaminated shoes, boots, clothing, or gear.

Beyond introducing or acquiring pathogens from fields or farms, at a broader scale, synanthropic birds exhibit the potential to disperse pathogens between farms and therefore pose a risk pathway for pathogen spread between operations, potentially leading to outbreaks. For example, studies of European starlings have shown that these birds may travel between multiple farms rather than exploiting a single farm (Gaukler et al. 2009; Gaukler et al. 2012). Similarly, radio-tracked rock pigeons have been shown to visit multiple farms (Carlson et al. 2011a).

Cumulatively, between their vagility and their predisposition for exploiting agricultural resources, synanthropic wild birds pose a risk to agricultural operations. The nature of that risk depends not only on an available pathway for introducing a pathogen but also on a complex suite of factors that must be evaluated to quantify risk. For example, actual risk will vary depending on within-host dynamics (individual- and species-level replication competence, immune history, body condition, age, sex, reproductive status), behavior, population size, community composition, season, and environmental conditions.

11.3.3 Spillover versus spillback

Spillover is when a pathogen is transmitted from a reservoir population to another sympatric species

(Daszak et al. 2000; Woodroffe 1999), and so studying avian pathogen transmission at the wildlife–agriculture interface is essentially, at its core, a study of pathogen spillover. Nearly every example mentioned in this chapter, whether bacterial or viral, whether direct transmission or indirect, is an example of spillover either to or from wild birds. In many cases, research is structured to examine spillover as a one-way process. Examples include studies on the transmission of animal pathogens to people (e.g., Islam et al. 2016) or a focus on transmission of domestic animal pathogens to endangered wildlife species (e.g., Deem et al. 2012). However, at the agriculture interface, spillover is bidirectional, with wild birds often infecting domestic birds but pathogens also moving from domestic species to wild birds. A key example are IAVs, where viruses regularly move from wild birds into domestic poultry but poultry-adapted viruses also move from domestic birds to wild species (Figure 11.5). In this latter case, the viruses originate in wild birds, which is an example of spillback where pathogens move from a native host, into an introduced host, and then spread back to the native host again. Spillback is particularly problematic when it involves pathogens that mutate rapidly (such as RNA viruses), because the pathogen may be substantially changed when it spills back to the native host (Novella et al. 1995). Despite this bidirectionality, the majority of research on spillover involving wild birds at the agriculture interface focuses on pathogens moving from wild birds to domestic poultry and humans. Likely this bias is due to the risk these pathogens pose in terms of both the animal health burden and the economic impact to the agricultural sector.

While the risks of spillover to domestic poultry are clear, events associated with pathogens spilling over in the opposite direction, from domestic birds to wild birds, have the potential to be problematic for multiple reasons as well. One is that animal husbandry practices often keep domestic animals at extremely high densities, and these densities can maintain an artificially high prevalence of infected animals that become a continual source of pathogen transmission to wild birds. Another is that spillover is often accompanied by genetic changes to the pathogen. These changes often reflect random

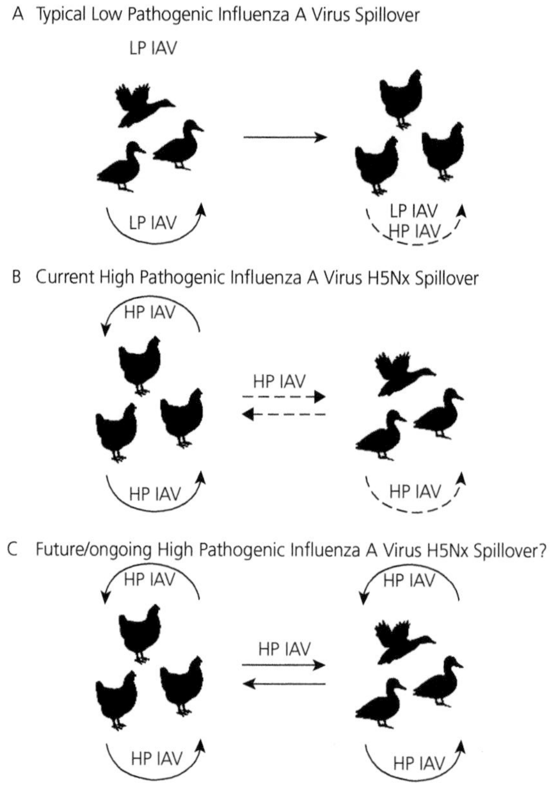

Figure 11.5 Examples of viral spillover between domestic poultry and wild birds in Asia and Europe. Curved solid arrows: continuous circulation among wild waterbirds or poultry. Curved dotted arrows: temporary circulation among wild waterbirds or poultry. Straight solid arrows: regular transmission from poultry to wild waterbirds or vice versa. Straight dotted arrows: occasional transmission from poultry to wild waterbirds or vice versa. (Reprinted from Bodewes and Kuiken, 'Changing role of wild birds in the epidemiology of avian influenza A viruses', *Advances in Virus Research*, Vol 100, 279–307, © 2018, with permission from Elsevier.)

mutation or adaptation to a new host species and these types of spillover events and viral host jumps have led to major disease outbreaks in the past (Pepin et al. 2010). Another example of pathogens spilling over from domestic to wild birds is the increasing number of viral vaccine strains such as Newcastle disease virus and avian coronaviruses that are being detected in wild bird populations (Ayala et al. 2016; Cardenas Garcia et al. 2013; Rohaim et al. 2017; Snoeck et al. 2013). Other pathogens can spill over from domestic to wild birds as well, including various avian astroviruses (Wille et al. 2017); there are also likely many other pathogens that we do not yet know about. Taken together, this body of research paints a picture of spillover that is not a one-way street, even though it is often portrayed that way. Pathogens can move from wild birds to domestic birds, from domestic birds to wild birds, or they can even spillback (at times with new mutations onboard) to the original host. Preventing pathogen spillover at the agriculture interface or implementing successful management strategies during spillover events will require an increased understanding of the risk factors involved and acknowledging the multidirectionality of pathogen transmission

11.4 Pathogen prevalence and transmission in wild birds

11.4.1 Major pathogens of concern

At the agriculture interface, birds have been implicated in contaminating agricultural operations with bacterial, viral, and macroparasitic pathogens (see review in Clark 2014). In terms of bacterial diseases, food-borne pathogens are of critical concern (Painter et al. 2013). These include *Campylobacter*, *Listeria*, *Salmonella*, and Shiga-toxin producing *Escherichia coli* (STEC). Viral pathogens of primary concern at the avian–agriculture interface include IAVs and Newcastle disease virus.

11.4.1.1 Bacterial pathogens

Bacterial contamination by wild birds at agricultural operations is largely focused on whether it ultimately affects human food safety. Food-borne bacteria are often associated with wildlife (Greig et al. 2015) and cause illnesses in human consumers, with negative economic consequences that can cost billions of dollars (Hoffman et al. 2015). The specific bacterial pathogens of concern are STEC, *Salmonella enterica*, *Listeria monocytogenes*, and *Campylobacter jejuni*. While these are not the only wildlife-associated bacterial pathogens of concern, they cause most human food-borne illnesses (Batz et al. 2011; Scallan et al. 2011). The issue of food-borne bacterial contamination has been further compounded in recent decades by the increase of antimicrobial resistance (AMR) in some strains of these pathogens (Caniça et al. 2019; Koluman and Dikici 2013). For example, tetracycline, streptomycin, ampicillin, and sulphonamide resistance of *Salmonella enterica* serovar 4 increased substantially in humans from $\leq 11\%$ to $\geq 50\%$ in the US from 2004 through 2017 (Michael and Schwarz 2016).

In a scoping review of prevalence of *E. coli*, *Salmonella*, and *Campylobacter* in wildlife (Greig et al. 2015), birds represented the largest body of evidence for pathogen transmission to the human food chain. Wild birds have been implicated in transmission of these pathogens to recreational water, domestic animals, and humans, with shared ranges with livestock, particularly cattle, as a significant risk factor (Greig et al. 2015). Migratory bird species were more likely to carry pathogenic bacteria and harbor more bacterial species resistant to antibiotics than non-migratory birds, with rock pigeons having the highest number of antimicrobial resistant bacteria among non-migratory birds (Chung et al. 2018). Interestingly, birds in natural habitats had much greater numbers (n = 137) of pathogenic bacteria species than agricultural environments (n = 81) (Chung et al. 2018).

In general, STEC, such as *E. coli* O157, has low (0 to 5%) prevalence in birds (see review in Langholz and Jay-Russell 2013). However, 20 of 35 (57.1%) different STEC serogroups have been isolated from birds (Pedersen and Clark 2007). This trade-off between low prevalence and multiple carriage of STEC serogroups is exemplified in a study of gulls where prevalence was 2% but two STEC serogroups were detected in a single individual (Makino et al. 2000). Some species, such as rock pigeons in Italy, have been reported to have higher prevalence of STEC (10.7% of 649 sampled) from seven serogroups (Morabito et al. 2001). Thus, seemingly rare pathogens can become prevalent in some avian species depending on the geographic locations examined.

While avian carriers of STEC are generally asymptomatic, birds infected with *S. enterica* can appear healthy or exhibit disease; *Typhimurium* is the serotype that causes disease in some avian species (Benskin et al. 2009). Similar to STEC, *Salmonella* often has a low prevalence in birds (0–52.6%), with most studies (57.1% of 28) reporting estimates of < 3% in avian species (Erickson 2016). However, high prevalences have been sporadically reported in some avian species, such as 17.0% in yellow-legged gulls (*Larus michahellis*) and 52.6% in griffon vultures (*Gyps fulvus*) in Spain (Erickson 2016).

In contrast, *C. jejuni* appears to have higher prevalences in birds (0–69.1%) than other taxa, with prevalences of > 7% in 66.7% of nine studies (Erickson 2016). Rock pigeons and American crows (*Corvus brachyrhynchos*) had especially high prevalences (48.3% of 1800 and 55.1% of 127 samples, respectively). In southern Ontario, Canada, *C. jejuni* was detected in 20.4% of 54 samples from seven species of waterfowl, with seven subtypes identified (Vogt et al. 2019). Two of the identified subtypes were similar to those from human clinical cases, suggesting that prevalence of bacterial species alone does not necessarily reflect the potential impact on human health. In addition, bacterial colonization of avian hosts can differ by species. In a challenge experiment with mallards, colonization of a *C. jejuni* strain from song thrushes (*Turdus philomelos*) was weak and cleared after 10 days, whereas colonization from chicken and mallard strains remained stable over the 18-day experimental period (Atterby et al. 2018).

Of the four bacterial pathogens of concern, the least is known about *L. monocytogenes* in wild birds (Greig et al. 2015). Prior research found relatively high prevalence (15.2%) of *L. monocytogenes* in gulls feeding at sewage disposal sites (Fenlon 1985). More recently, studies have also found high prevalence in three species of gulls (7.9%), in rock pigeons (13%), and in house sparrows (25%) in Finland (Hellström et al. 2008). In contrast to the other bacterial pathogens, *L. monocytogenes* can live and grow as a saprophyte outside of the animal digestive system; this ability to persist in the environment minimizes the relative role of wild birds as contamination sources in food outbreaks (Jay-Russell 2013).

11.4.1.2 Newcastle Disease virus

Newcastle Disease virus (NDV), also referred to as avian paramyxovirus 1, is an avulavirus with multiple strains that circulate in both wild and domestic birds. The World Bank ranked Newcastle disease as the third most significant disease of poultry and number eight on the list of most important wildlife diseases (World Bank 2011). Transmission is through ingestion or inhalation of infectious virus in respiratory secretions or in feces from NDV-positive birds, and while not all NDV strains are virulent (vNDV), those that are can cause up to 100% mortality in naïve birds. The virulent strains are either mesogenic (moderate morbidity) or velogenic (severe morbidity and mortality). The less-virulent forms of NDV (lentogenic strains, also referred to as low-virulent or loNDV) are ubiquitous in both wild waterfowl and poultry around the world (Brown and Bevins 2017; Pedersen et al. 2013; Ramey et al. 2013; Ramey et al. 2017). Velogenic NDV is more restricted and, in wild birds, the only species groups known to maintain endemic, virulent NDV are birds in the Columbiformes and North American cormorant species. Large-scale NDV-related die-offs of wild birds have been repeatedly documented in cormorants and other species, with some outbreaks involving thousands of birds (Wobeser et al. 1993). Despite these and other smaller-scale mortality events, NDV antibodies are routinely seen in wild birds (Pedersen et al. 2013), indicating regular exposure and survival. Routine and widespread detection of NDV antibodies also suggests that wild birds, particularly cormorants, likely contribute to viral maintenance in the environment.

Spillover of virulent NDV from wild birds to domestic birds has been repeatedly documented around the world (Glaser et al. 1999; Kim et al. 2012; Yin et al. 2017). In Scotland, an outbreak at a poultry farm that also housed endangered and rare breeds of significant conservation value was linked with virulent Columbidae-associated NDV genotypes that presumably were associated with pigeons that roosted on site (Irvine et al. 2009). Additional studies in Europe and in the Democratic Republic of Korea have shown NDV moving among wild birds, farms, live-bird markets, and even neighboring countries (Alexander 2011; Kim et al. 2012).

Despite this body of research documenting spillover of vNDV from wild to domestic species, most of these events occur in smaller poultry holdings (i.e., backyard poultry producers or small household flocks) with poor biosecurity or in poultry reared outside where they can come into direct contact with wild birds. Spillover to birds in commercial poultry operations, on the other hand, is less widespread. Typically, large-scale facilities that undergo intensive poultry production have enhanced biosecurity practices. To date, all recorded outbreaks of vNDV in US commercial poultry facilities are linked to infected domestic species, rather than wildlife, showing the efficacy of biosecurity in preventing transmission of this virus at the wildlife–agriculture interface. In addition, there is little evidence that the NDV strains circulating in wild birds are pathogenic to domestic poultry (Ramey et al. 2017). Laboratory experiments have demonstrated that it takes a higher viral dose to infect chickens with pigeon-origin and cormorant-origin viruses than it does to infect with chicken-origin NDV. The chicken-origin viruses also transmitted much more efficiently to contact birds, suggesting again that viral adaptation may need to occur for efficient transmission and successful NDV spillover events (Collins et al. 1994; Ferreira et al. 2019); however, there was a case where an outbreak in turkeys was caused by vNDV that was indistinguishable from virus found in nearby cormorants (Heckert et al. 1996).

The high mortality seen with virulent NDV, combined with the decreased production seen in poultry infected with less-virulent NDV, has led producers to continually vaccinate poultry. The result is billions of live-attenuated NDV vaccines used around the world and what is likely widespread spillover of those vaccine viruses into wild birds. Newcastle vaccine viruses have been detected in wild birds in multiple countries around the globe, including Mexico (Cardenas Garcia et al. 2013), Luxembourg (Snoeck et al. 2013), and Egypt (Rohaim et al. 2017). To date, morbidity and mortality of wild birds infected with these vaccine strains has not been documented, but this spillover raises the possibility of wild birds disseminating these viruses along their migratory corridors and the potential for viral evolution into a more virulent form.

11.4.1.3 Avian influenza viruses

Avian influenza A virus (IAV), commonly known as bird flu to the general public, is a member of the Orthomyxoviridae family. Influenza viruses in this family are grouped into types A, B, C, or D, but only type A viruses infect birds. IAVs are classified into subtypes based on combinations of two surface glycoproteins: hemagglutinin (H) and neuraminidase (N). Sixteen Hs (H1–H16) and 9 Ns (N1–N9) have been isolated in birds. Avian IAVs are further categorized regarding their virulence in chickens as either high pathogenic (HP) or low pathogenic (LP). To date, only viruses of the H5 and H7 subtypes have been characterized as HP viruses. Naturally occurring wild bird IAVs have low pathogenicity in both chickens and wild birds and it is only after the spillover of these viruses to chickens or other poultry that HP strains emerge (Figure 11.5). Wild birds infected with LP viruses exhibit limited clinical symptoms, while strains categorized as HP in chickens may or may not cause significant morbidity or mortality if these strains spillback into wild birds. Some mass die-offs have occurred in wild birds as the result of spillback of HP strains (Kleyheeg et al. 2017), with emerging IAV strains a significant conservation threat to wild birds in addition to the potential for severe impacts to the poultry industry.

Wild aquatic birds in the orders Anseriformes (ducks, geese, swans) and Charadriiformes (gulls, terns, shorebirds) are considered the natural maintenance hosts for IAVs. Mean prevalence in reservoir hosts is commonly over 10% and can be much higher during fall migration peaks in temperate zones (Bevins et al. 2014; Wille et al. 2017). Spillover into poultry is frequent, especially in birds maintained outdoors or at live bird markets. In Asia where live bird markets are more common, several IAV strains are now maintained in poultry, leading researchers to question whether live bird markets might now be reservoirs of IAVs (Cardona et al. 2009; Hassan et al. 2017).

A susceptible individual can become infected through direct contact with an infectious individual or indirect contact with water, soil, feces, or fluids contaminated with infectious virus. Environmental contamination occurs when birds excrete high levels of virus into the environment which supports indirect contact by environmental transmission. In natural hosts, the fecal–oral route is considered the primary route of transmission whereby susceptible hosts ingest virus from contaminated water or food and infectious hosts excrete virus in their feces. Many studies have shown that IAVs can persist for extended periods of time in water, thus promoting the spread of IAVs by extending the infectious period and supporting co-infections that can lead to reassortments (Pepin et al. 2019; Stallknecht et al. 1990). In poultry, and in particular for HP IAVs, viral replication is more common in respiratory tissues, which supports aerosol transmission. Once an IAV is established in a poultry flock, most strains are highly transmissible and no further introductions may be needed for forward transmission within or among flocks, particularly if between-flock transmission is assisted by humans, such as moving infected poultry or IAV-contaminated vehicles or equipment.

IAVs are a true One Health concern given their global distribution and their potential to cause significant harm in humans, livestock, and wildlife. IAVs have a multipartite RNA genome, meaning that the eight genes encoding the core viral proteins are independent entities within the viral particle. A primary consequence of a segmented genome is that when a cell is co-infected with more than one IAV strain, the genes can mix and match in a process referred to as reassortment. Reassortment, combined with the high mutation rate characteristic of IAVs, leads to frequent emergence of new IAV strains that have the potential to evade immune systems, leading to the ever-present possibility that a novel strain will emerge resulting in a human pandemic or bird panzootic.

The World Organization for Animal Health (OIE) is an intergovernmental organization designed to monitor and control animal diseases at a global scale. The OIE maintains a list of notifiable animal diseases that requires any occurrence of vNDV or H5 and H7 IAVs in wild or domestic animals to be reported by law. Intermittent outbreaks of vNDV and continuous outbreaks and emergence of novel IAV strains have led the OIE to designate virulent Newcastle disease and avian influenza as diseases of major interest, requiring international coordination to reduce risks to animals and humans (for IAVs). Much of this coordination relies on national surveillance programs to monitor wild and domestic birds as an early warning system to detect and report disease outbreaks and to identify novel IAV strains as soon as possible to give the world a chance to eliminate outbreaks or respond to emerging strains that have the potential to cause significant harm to humans, livestock, or wildlife. For example, in Great Britain, the Animal and Plant Health Agency (APHA) routinely surveils poultry flock health in the country and conducts investigations in response to production issues (Reid et al. 2019).

11.5 Contamination potential of agricultural operations by birds

To better understand the risk of pathogen contamination, we propose a theoretical construct, **contamination potential**, which we define as the *potential of a given wildlife species to introduce a pathogen to an agricultural product*. Contamination potential can be expressed quantitatively as a function of the characteristics of both the pathogen (e.g., ability to persist in the environment, infectious dose) and its avian hosts (e.g., abundance, visitation rate to agricultural operations, pathogen prevalence, fecal shedding rates of pathogen, pathogen load in feces, fecal amounts). Contamination potential is currently a theoretical construct that we propose to identify

avian species (or other taxa) that have the greatest potential to introduce or spread pathogens at agricultural operations and, therefore, pose the greatest risk for contamination of pathogens at the wildlife–agriculture interface. Despite the potential utility of developing a quantitative estimator for contamination potential, the basic data to populate such an estimator are largely lacking. In addition, complexities, such as temporal and spatial variation in wild host characteristics and carriage of multiple pathogens by host species make it difficult to use a simple estimator to quantify risk. For example, Taff et al. (2016) found that seasonal habitat selection and movements of American crows drastically amplified risks of *C. jejuni* contamination in specific locations, while Gargiulo et al. (2018) found raptor species simultaneously shed *Campylobacter*, *E. coli*, and *Salmonella* in their feces.

A contamination potential index might also need to consider whether transmission pathways are direct or indirect. Of six outbreaks in humans of illness caused by food-borne bacteria that have been associated with wildlife, only one was associated with birds (Langholz and Jay-Russell 2013), where sandhill cranes contaminated pea crops in Alaska with *C. jejuni* (Gardner et al. 2011). However, while proximate causes of outbreaks are often identified as the underlying culprit for pathogen introduction, such as contamination of water used to irrigate crops, the ultimate causes, such as wildlife contaminating the irrigation water, are often elusive. Thus, wildlife epidemiological investigations that consider ultimate causes of pathogen introduction might identify more outbreaks as having been ultimately caused by wild birds through indirect pathways compared to investigations that only consider pathogen introductions based on proximate causes.

11.6 Long-distance movements and pathogen introductions

A primary concern with wild bird-mediated pathogen transmission is the ability of avian species to move disease-causing pathogens across borders and between countries during migration events (see also Chapter 4). There are several well-documented examples of pathogens being introduced by wild birds, including the dispersal of West Nile virus from sub-Saharan Africa to Europe (López et al. 2008). IAV is, once again, the poster child of a bird-associated pathogen that threatens agricultural production and can be moved long-distance by wild birds. The working hypothesis behind the introduction of Eurasian IAV subtypes into North America in 2014 (Bevins et al. 2016; Lee et al. 2015) is that wild birds migrating from Asia to North America in the fall carried the virus across the Pacific Ocean. The virus reassorted with North American viruses (Hill et al. 2017) and spilled over into poultry within weeks or months, leading to an outbreak that caused the loss of nearly 50 million domestic chickens and turkeys at a cost of $3.3 billion dollars to the US economy (Greene 2015). Relatedly, an H6N5 IAV was sequenced several years later from wild birds in Asia and all eight gene segments clustered with known IAV sequences from North America, suggesting the virus is of North American ancestry and was introduced into Eurasia by wild birds (Jeong et al. 2019). This discovery, combined with other detections of North American IAV gene segments in other parts of the world (Lee et al. 2011; Ramey et al. 2018), provides evidence of wild bird-mediated viral dissemination across large spatial extents.

Long-distance pathogen introduction events receive the bulk of the attention on disease emergence, but short-distance migrants and resident wild birds also contribute to pathogen introduction at regional and local scales. Surveillance of both wild birds and poultry in the Netherlands showed that most IAVs in poultry were related to locally circulating wild bird IAVs, suggesting that infection of poultry with wild bird viruses was not uncommon and often went undetected (Bergervoet et al. 2019). Some of these poultry IAVs were more closely related to IAVs detected in other counties or in other regions of the world. It is possible that these IAVs were introduced though long-distance movements combined with circulation at local and regional scales before spillover into poultry, but it is also possible that more incremental movements occurred and that more closely related viruses in local wild bird populations were simply missed during routine surveillance, indicating the importance of a more thorough and intensive approach to

detecting pathogens in wild birds (Bergervoet et al. 2019; Lycett et al. 2016). Mechanical transmission (i.e., transport of pathogens on feet, feathers, or other body surfaces; see Chapter 2) is another way that pathogens can disperse, which is how European starlings introduced *S. enterica* bacteria into livestock operations (Carlson et al. 2015).

While long-distance movements provide the opportunity for avian pathogens to be introduced to new areas, there is considerable debate on whether migration in general provides a pathway or serves as a barrier to pathogen introduction (Altizer et al. 2011; van Dijk et al. 2018). There are clear examples of pathogen introductions tied to migration, such as the 2014 Eurasian IAV introductions to North America already discussed, but there are examples of pathogens that never emerged despite predictions to the contrary. Sustained transmission of HP H5N1 IAV was detected in wild birds for the first time in 2005 (Chen et al. 2005), alerting researchers and governments around the world to possible introductions by migrating waterfowl. Large-scale spread never happened, likely for a variety of reasons (Arsnoe et al. 2011; Dannemiller et al. 2017; Gaidet et al. 2010; Tian et al. 2015), and the virus was never detected in North or South America. Migration may also act as a barrier to transmission because infected hosts cannot survive a strenuous journey (Altizer et al. 2011). Alternatively, migration has been proposed as a way to escape pathogens or heavy parasite infestations during certain periods of the year, which again would limit transmission rather than enhance it (Loehle 1995).

The final way that wild bird pathogens are introduced into new regions is through legal and illegal import of live birds or bird products. Most countries have regulatory pathways in place to verify the disease-free status of legal imports; however, illegal imports are a different matter. Data from the US show that nearly 400,000 wild avian specimens were illegally imported or confiscated from 2006 through 2016 and it is likely that only a small percentage of illegal products are intercepted (Brown and Bevins 2017). A well-publicized seizure of wild birds smuggled into Belgium that were infected with HP Asian H5N1 IAV (Van Borm et al. 2005) along with the regular interception of illegal poultry imports (Davis et al. 2010) demonstrate this introduction risk. While these human-mediated long-distance pathogen introductions are difficult to detect, there is no disputing that dispersal of many wild bird pathogens combined with intensive poultry production practices, live-bird markets, and the international trade of birds, bird products, and contaminated equipment have been associated with multiple avian disease outbreaks in the recent past.

11.7 Future directions

Prioritizing future research directions for pathogen transmission that involves wild birds at the agriculture interface is challenging. Lines of inquiry regarding wild bird pathogens can range from the impacts on wild bird populations, the economic losses faced by agricultural producers, to public health risks. These questions are embedded in ecologically complex, multihost pathogen systems and answering them will require the assembly of expertise on the different hosts, pathogens, and ecological systems involved.

Given the many scales at which wild birds operate, solving problems at the wildlife–agriculture interface will require novel approaches, such as forming large collaborative scientific enterprises that make data collection at large scales feasible and working with up-and-coming analytical tools that allow inferences to be scaled up. Anticipating what types of data are needed and moving from a reactive to a proactive response are also priorities. Outbreaks and pandemics often require ecologists and modelers to produce results rapidly, often when the information and data needed to produce those results are severely limited (Getz et al. 2018). Harnessing the power of up-and-coming analytics and genetic tools will hopefully allow us to move past detecting pathogens transmitted by wild birds and towards predicting which pathogens will be transmitted by wild birds.

Literature cited

Aarestrup, F.M. (2015). The livestock reservoir for antimicrobial resistance: a personal view on changing patterns of risks, effects of interventions and the way forward. *Philosophical Transactions of the Royal Society B: Biological Sciences*, 370, 20140085.

Alexander, D.J. (2011). Newcastle disease in the European Union 2000 to 2009. *Avian Pathology*, 40, 547–58.

Altizer, S., Bartel, R., and Han, B.A. (2011). Animal migration and infectious disease risk. *Science*, 331, 296–302.

Arsnoe, D.M., Ip, H.S., and Owen, J.C. (2011). Influence of body condition on influenza A virus infection in mallard ducks: Experimental infection data. *PLoS ONE*, 6, e22633.

Atterby, C., Mourkas, E., Méric, G., et al. (2018). The potential of isolation source to predict colonization in avian hosts: a case study in *Campylobacter jejuni* strains from three bird species. *Frontiers in Microbiology*, 9, 591.

Ayala, A.J., Dimitrov, K.M., Becker, C.R., et al. (2016). Presence of vaccine-derived Newcastle disease viruses in wild birds. *PLoS ONE*, 11, e0162484.

Bahl, J., Pham, T.T., Hill, N.J., et al. (2016). Ecosystem interactions underlie the spread of avian influenza A viruses with pandemic potential. *PLoS Pathogens*, 12, e1005620.

Batz, M.B., Hoffmann, S., and Morris, J.G., Jr. (2011). *Ranking the Risks: The 10 Pathogen–Food Combinations with the Greatest Burden on Public Health*. University of Florida Emerging Pathogens Institute, Gainesville.

Beecher, N.A., Johnson, R.J., Brandle, J.R., Case, R.M., and Young, L.J. (2002). Agroecology of birds in organic and nonorganic farmland. *Conservation Biology*, 16, 1620–31.

Belden, J.B., McMurry, S.T., Maul, J.D., Brain, R.A., and Ghebremichael, L.T. (2018). Relative abundance trends of bird populations in high intensity croplands in the central United States. *Integrated Environmental Assessment and Management*, 14, 692–702.

Bengtsson, J., Ahnström, J., and Weibull, A.-C. (2005). The effects of organic agriculture on biodiversity and abundance: a meta-analysis. *Journal of Applied Ecology*, 42, 261–9.

Benskin, C.M.H., Wilson, K., Jones, K., and Hartley, I.R. (2009). Bacterial pathogens in wild birds: a review of the frequency and effects of infection. *Biological Reviews*, 84, 349–73.

Bergervoet, S.A., Pritz-Verschuren, S.B.E., Gonzales, J.L., et al. (2019). Circulation of low pathogenic avian influenza (LPAI) viruses in wild birds and poultry in the Netherlands, 2006–2016. *Scientific Reports*, 9, 13681.

Bevins, S.N., Pedersen, K., Lutman, M.W., et al. (2014). Large-scale avian influenza surveillance in wild birds throughout the United States. *PLoS ONE*, 9, e104360.

Bevins, S.N., Dusek, R.J., White, C.L., et al. (2016). Widespread detection of highly pathogenic H5 influenza viruses in wild birds from the Pacific Flyway of the United States. *Scientific Reports*, 6, 28980.

Bonneaud, C., Weinert, L.A., and Kuijper, B. (2019). Understanding the emergence of bacterial pathogens in novel hosts. *Philosophical Transactions of the Royal Society B: Biological Sciences*, 374, 20180328.

Brown, V.R. and Bevins, S.N. (2017). A review of virulent Newcastle disease viruses in the United States and the role of wild birds in viral persistence and spread. *Veterinary Research*, 48, 68.

Burns, T.E., Ribble, C., Stephen, C., et al. (2012). Use of observed wild bird activity on poultry farms and a literature review to target species as high priority for avian influenza testing in 2 regions of Canada. *Canadian Veterinary Journal*, 53, 158–66.

Butler, S.J., Vickery, J.A., and Norris, K. (2007). Farmland biodiversity and the footprint of agriculture. *Science*, 315, 381–4.

Cabe, P.R. (1993). European starling (*Sturnus vulgaris*). The Cornell Lab or Ornithology Birds of the World. https://birdsoftheworld.org/bow/species/eursta/cur/introduction.

Caniça, M., Manageiro, V., Abriouel, H., Moran-Gilad, J., and Franz, C.M.A.P. (2019). Antibiotic resistance in foodborne bacteria. *Trends in Food Science & Technology*, 84, 41–4.

Cardenas Garcia, S., Navarro Lopez, R., Morales, R., et al. (2013). Molecular epidemiology of Newcastle disease in Mexico and the potential spillover of viruses from poultry into wild bird species. *Applied and Environmental Microbiology*, 79, 4985–92.

Cardona, C., Yee, K., and Carpenter, T. (2009). Are live bird markets reservoirs of avian influenza? *Poultry Science*, 88, 856–9.

Carlson, J.C., Clark, L., Antolin, M.F., and Salman, M.D. (2011a). Rock pigeon use of livestock facilities in northern Colorado: implications for improving farm biosecurity. *Human–Wildlife Interactions*, 5, 112–22.

Carlson, J.C., Engeman, R.M., Hyatt, D.R., et al. (2011b). Efficacy of European starling control to reduce *Salmonella enterica* contamination in a concentrated animal feeding operation in the Texas panhandle. *BMC Veterinary Research*, 7, 9.

Carlson, J.C., Franklin, A.B., Hyatt, D.R., Pettit, S.E., and Linz, G.M. (2011c). The role of starlings in the spread of *Salmonella* within concentrated animal feeding operations. *Journal of Applied Ecology*, 48, 479–86.

Carlson, J.C., Linz, G.M., Ballweber, L.R., Elmore, S.A., Pettit, S.E., and Franklin, A.B. (2011d). The role of European starlings in the spread of coccidia within concentrated animal feeding operations. *Veterinary Parasitology*, 180, 340–3.

Carlson, J.C., Hyatt, D.R., Ellis, J.W., et al. (2015). Mechanisms of antimicrobial resistant *Salmonella enterica* transmission associated with starling–livestock interactions. *Veterinary Microbiology*, 179, 60–8.

Caron, A., Grosbois, V., Etter, E., Gaidet, N., and de Garine-Wichatitsky, M. (2014). Bridge hosts for avian influenza viruses at the wildlife/domestic interface: an eco-epidemiological framework implemented in southern Africa. *Preventive Veterinary Medicine*, 117, 590–600.

Caron, A., Cappelle, J., Cumming, G.S., de Garine-Wichatitsky, M., and Gaidet, N. (2015). Bridge hosts, a missing link for disease ecology in multi-host systems. *Veterinary Research*, 46, 1–11.

Chen, H., Smith, G.J.D., Zhang, S.Y., et al. (2005). Avian flu H5N1 virus outbreak in migratory waterfowl. *Nature*, 436, 191–2.

Chung, D.M., Ferree, E., Simon, D.M., and Yeh, P.J. (2018). Patterns of bird–bacteria associations. *EcoHealth*, 15, 627–41.

Clark, L. (2014). Disease risks posed by wild birds associated with agricultural landscapes. In K.R. Matthews, G.M. Sapers, and C.P. Gerba, eds. *The Produce Contamination Problem: Causes and Solutions*, 2nd ed. Food Science and Technology, International Series, pp. 139–65. Academic Press, San Diego, CA.

Cleaveland, S., Laurenson, M.K., and Taylor, L.H. (2001). Diseases of humans and their domestic mammals: pathogen characteristics, host range and the risk of emergence. *Philosophical Transactions of the Royal Society B: Biological Sciences*, 356, 991–9.

Collins, M.S., Strong, I., and Alexander, D.J. (1994). Evaluation of the molecular basis of pathogenicity of the variant Newcastle disease viruses termed 'pigeon PMV-1 viruses'. *Archives of Virology*, 134, 403–11.

Daniels, M.J., Hutchings, M.R., and Greig, A. (2003). The risk of disease transmission to livestock posed by contamination of farm stored feed by wildlife excreta. *Epidemiology & Infection*, 130, 561–8.

Dannemiller, N.G., Webb, C.T., Wilson, K.R., et al. (2017). Impact of body condition on influenza A virus infection dynamics in mallards following a secondary exposure. *PLoS ONE*, 12, e0175757.

Daszak, P., Cunningham, A.A., and Hyatt, A.D. (2000). Emerging infectious diseases of wildlife—threats to biodiversity and human health. *Science*, 287, 443–9.

Davis, C.T., Balish, A.L., O'Neill, E., et al. (2010). Detection and characterization of clade 7 high pathogenicity avian influenza H5N1 viruses in chickens seized at ports of entry and live poultry markets in Vietnam. *Avian Diseases*, 54, 307–12.

Deem, S.L., Cruz, M.B., Higashiguchi, J.M., and Parker, P.G. (2012). Diseases of poultry and endemic birds in Galapagos: implications for the reintroduction of native species. *Animal Conservation*, 15, 73–82.

Elbers, A.R.W. and Gonzales, J.L. (2020). Quantification of visits of wild fauna to a commercial free-range layer farm in the Netherlands located in an avian influenza hot-spot area assessed by video-camera monitoring. *Transboundary and Emerging Diseases*, 67, 661–77.

Elmberg, J., Berg, C., Lerner, H., Waldenström, J., and Hessel, R. (2017). Potential disease transmission from wild geese and swans to livestock, poultry and humans: a review of the scientific literature from a One Health perspective. *Infection Ecology & Epidemiology*, 7, 1300450.

Erickson, M.C. (2016). Overview: foodborne pathogens in wildlife populations. In M. Jay-Russell and M.P. Doyle, eds. *Food Safety Risks from Wildlife: Challenges in Agriculture, Conservation, and Public Health*, pp. 1–30. Springer, Cham.

Fenlon, D.R. (1985). Wild birds and silage as reservoirs of *Listeria* in the agricultural environment. *Journal of Applied Bacteriology*, 59, 537–43.

Ferreira, H.L., Taylor, T.L., Dimitrov, K.M., Sabra, M., Afonso, C.L., and Suarez, D.L. (2019). Virulent Newcastle disease viruses from chicken origin are more pathogenic and transmissible to chickens than viruses normally maintained in wild birds. *Veterinary Microbiology*, 235, 25–34.

Fujioka, M., Lee, S.D., Kurechi, M., and Yoshida, H. (2010). Bird use of rice fields in Korea and Japan. *Waterbirds*, 33, 8–29.

Gaidet, N., Cappelle, J., Takekawa, J.Y., et al. (2010). Potential spread of highly pathogenic avian influenza H5N1 by wildfowl: dispersal ranges and rates determined from large-scale satellite telemetry. *Journal of Applied Ecology*, 47, 1147–57.

Gardner, T.J., Fitzgerald, C., Xavier, C., et al. (2011). Outbreak of campylobacteriosis associated with consumption of raw peas. *Clinical Infectious Diseases*, 53, 26–32.

Gargiulo, A., Fioretti, A., Russo, T.P., et al. (2018). Occurrence of enteropathogenic bacteria in birds of prey in Italy. *Letters in Applied Microbiology*, 66, 202–6.

Gaukler, S.M., Linz, G.M., Sherwood, J.S., et al. (2009). *Escherichia coli*, *Salmonella*, and *Mycobacterium avium* subsp. *paratuberculosis* in wild European starlings at a Kansas cattle feedlot. *Avian Diseases*, 53, 544–51.

Gaukler, S.M., Homan, H.J., Linz, G.M., and Bleier, W.J. (2012). Using radio-telemetry to assess the risk European starlings pose in pathogen transmission among feedlots. *Human–Wildlife Interactions*, 6, 30–7.

Gennet, S., Howard, J., Langholz, J., Andrews, K., Reynolds, M.D., and Morrison, S.A. (2013). Farm practices for food safety: an emerging threat to floodplain and riparian ecosystems. *Frontiers in Ecology and the Environment*, 11, 236–42.

Getz, W.M., Marshall, C.R., Carlson, C.J., et al. (2018). Making ecological models adequate. *Ecology Letters*, 21, 153–66.

Gill, J.S., Webby, R., Gilchrist, M.J.R., and Gray, G.C. (2006). Avian influenza among waterfowl hunters and wildlife professionals. *Emerging Infectious Diseases*, 12, 1284–6.

Glaser, L.C., Barker, I.K., Weseloh, D.V.C., et al. (1999). The 1992 epizootic of Newcastle disease in double-crested cormorants in North America. *Journal of Wildlife Diseases*, 35, 319–30.

Greene, J.L. (2015). *Update on the Highly-Pathogenic Avian Influenza Outbreak of 2014–2015*. Congressional Research Service, Washington, DC.

Greig, J., Rajić, A., Young, I., Mascarenhas, M., Waddell, L., and LeJeune, J. (2015). A scoping review of the role of wildlife in the transmission of bacterial pathogens and antimicrobial resistance to the food chain. *Zoonoses and Public Health*, 62, 269–84.

Gruszynski, K., Pao, S., Kim, C., et al. (2014). Evaluating gulls as potential vehicles of *Salmonella enterica* serotype Newport (JJPX01.0061) contamination of tomatoes grown on the eastern shore of Virginia. *Applied and Environmental Microbiology*, 80, 235–8.

Hassan, M.M., Hoque, M.A., Debnath, N.C., Yamage, M., and Klaassen, M. (2017). Are poultry or wild birds the main reservoirs for avian influenza in Bangladesh? *EcoHealth*, 14, 490–500.

Haydon, D.T., Cleaveland, S., Taylor, L.H., and Laurenson, M.K. (2002). Identifying reservoirs of infection: a conceptual and practical challenge. *Emerging Infectious Diseases*, 8, 1468–73.

Heckert, R.A., Collins, M.S., Manvell, R.J., Strong, I., Pearson, J.E., and Alexander, D.J. (1996). Comparison of Newcastle disease viruses isolated from cormorants in Canada and the USA in 1975, 1990 and 1992. *Canadian Journal of Veterinary Research*, 60, 50–4.

Hellström, S., Kiviniemi, K., Autio, T., and Korkeala, H. (2008). *Listeria monocytogenes* is common in wild birds in Helsinki region and genotypes are frequently similar with those found along the food chain. *Journal of Applied Microbiology*, 104, 883–8.

Hill, N.J., Hussein, I.T.M., Davis, K.R., et al. (2017). Reassortment of influenza A viruses in wild birds in Alaska before H5 clade 2.3.4.4 outbreaks. *Emerging Infectious Diseases*, 23, 654–7.

Hoffman, S., Maculloch, B., and Batz, M. (2015). *Economic Burden of Major Foodborne Illnesses Acquired in the United States*. United States Department of Agriculture Economic Research Service, Washington, DC.

Hwang, J., Lee, K., Walsh, D., Kim, S.W., Sleeman, J.M., and Lee, H. (2018). Semi-quantitative assessment of disease risks at the human, livestock, wildlife interface for the Republic of Korea using a nationwide survey of experts: a model for other countries. *Transboundary and Emerging Diseases*, 65, e155–64.

Irvine, R.M., Aldous, E.W., Manvell, R.J., et al. (2009). Outbreak of Newcastle disease due to pigeon paramyxovirus type 1 in grey partridges (*Perdix perdix*) in Scotland in October 2006. *Veterinary Record*, 165, 531–5.

Islam, M.S., Sazzad, o.M.S., Satter, S.M., et al. (2016). Nipah virus transmission from bats to humans associated with drinking traditional liquor made from date palm sap, Bangladesh, 2011–2014. *Emerging Infectious Diseases*, 22, 664.

Jay-Russell, M.T. (2013). What is the risk from wild animals in food-borne pathogen contamination of plants? *CAB Reviews*, 8, 1–16.

Jay-Russell, M. and Doyle, M.P. (2016). *Food Safety Risks from Wildlife: Challenges in Agriculture, Conservation, and Public Health.* Springer, Cham.

Jeong, S., Lee, D.-H., Kim, Y.-J., et al. (2019). Introduction of avian influenza A (H6N5) virus into Asia from North America by wild birds. *Emerging Infectious Diseases*, 25, 2138–40.

Johnston R.F. (2001) Synanthropic birds of North America. In J.M. Marzluff, R. Bowman, R. Donnelly, eds. *Avian Ecology and Conservation in an Urbanizing World.* Springer, Boston, MA. https://doi.org/10.1007/978-1-4615-1531-9_3

Karp, D.S., Gennet, S., Kilonzo, C., et al. (2015). Comanaging fresh produce for nature conservation and food safety. *Proceedings of the National Academy of Sciences of the United States of America*, 112, 11126–31.

Kim, B.-Y., Lee, D.-H., Kim, M.-S., et al. (2012). Exchange of Newcastle disease viruses in Korea: the relatedness of isolates between wild birds, live bird markets, poultry farms and neighboring countries. *Infection, Genetics and Evolution*, 12, 478–82.

Kitts-Morgan, S.E., Carleton, R.E., Barrow, S.L., Hilburn, K.A., and Kyle, A.K. (2015). Wildlife visitation on a multi-unit educational livestock facility in northwestern Georgia. *Southeastern Naturalist*, 14, 267–80.

Kleyheeg, E., Slaterus, R., Bodewes, R., et al. (2017). Deaths among wild birds during highly pathogenic avian influenza A (H5N8) virus outbreak, the Netherlands. *Emerging Infectious Diseases*, 23, 2050–4.

Koluman, A. and Dikici, A. (2013). Antimicrobial resistance of emerging foodborne pathogens: status quo and global trends. *Critical Reviews in Microbiology*, 39, 57–69.

Kwon, J.-H., Bahl, J., Swayne, D.E., et al. (2020). Domestic ducks play a major role in the maintenance and spread of H5N8 highly pathogenic avian influenza viruses in South Korea. *Transboundary and Emerging Diseases*, 67, 844–51.

Langholz, J.A. and Jay-Russell, M.T. (2013). Potential role of wildlife in pathogenic contamination of fresh produce. *Human–Wildlife Interactions*, 7, 140–57.

Lee, D.-H., Lee, H.-J., Lee, Y.-N., et al. (2011). Evidence of intercontinental transfer of North American lineage avian influenza virus into Korea. *Infection, Genetics and Evolution*, 11, 232–6.

Lee, D.-H., Torchetti, M.K., Winker, K., Ip, H.S., Song, C.-S., and Swayne, D.E. (2015). Intercontinental spread of Asian-origin H5N8 to North America through Beringia by migratory birds. *Journal of Virology*, 89, 6521–4.

Loehle, C. (1995). Social barriers to pathogen transmission in wild animal populations. *Ecology*, 76, 326–35.

López, G., Jiménez-Clavero, M.Á., Tejedor, C.G., Soriguer, R., and Figuerola, J. (2008). Prevalence of West Nile virus neutralizing antibodies in Spain is related to the behavior of migratory birds. *Vector-Borne and Zoonotic Diseases*, 8, 615–22.

Lycett, S.J., Bodewes, R., Pohlmann, A., et al. (2016). Role for migratory wild birds in the global spread of avian influenza H5N8. *Science*, 354, 213–17.

Makino, S., Kobori, H., Asakura, H., et al. (2000). Detection and characterization of Shiga toxin-producing *Escherichia coli* from seagulls. *Epidemiology and Infection*, 125, 55–61.

McLaughlin, A. and Mineau, P. (1995). The impact of agricultural practices on biodiversity. *Agriculture, Ecosystems & Environment*, 55, 201–12.

Medhanie, G.A., Pearl, D.L., McEwen, S.A., et al. (2014). A longitudinal study of feed contamination by European starling excreta in Ohio dairy farms (2007–2008). *Journal of Dairy Science*, 97, 5230–8.

Michael, G.B. and Schwarz, S. (2016). Antimicrobial resistance in zoonotic nontyphoidal *Salmonella*: an alarming trend? *Clinical Microbiology and Infection*, 22, 968–74.

Morabito, S., Dell'Omo, G., Agrimi, U., et al. (2001). Detection and characterization of Shiga toxin-producing *Escherichia coli* in feral pigeons. *Veterinary Microbiology*, 82, 275–83.

Novella, I.S., Duarte, E.A., Elena, S.F., Moya, A., Domingo, E., and Holland, J.J. (1995). Exponential increases of RNA virus fitness during large population transmissions. *Proceedings of the National Academy of Sciences of the United States of America*, 92, 5841–4.

Painter, J.A., Hoekstra, R.M., Ayers, T., et al. (2013). Attribution of foodborne illnesses, hospitalizations, and deaths to food commodities by using outbreak data, United States, 1998–2008. *Emerging Infectious Diseases*, 19, 407–15.

Pärnänen, K.M., Narciso-da-Rocha, C., Kneis, D., et al. (2019). Antibiotic resistance in European wastewater treatment plants mirrors the pattern of clinical antibiotic resistance prevalence. *Science Advances*, 5, eaau9124.

Pedersen, K. and Clark, L. (2007). A review of Shiga toxin *Escherichia coli* and *Salmonella enterica* in cattle and free-ranging birds: potential association and epidemiological links. *Human–Wildlife Conflicts*, 1, 68–77.

Pedersen, K., Marks, D.R., Arsnoe, D.M., et al. (2013). Avian Paramyxovirus serotype 1 (Newcastle disease virus), avian influenza virus, and *Salmonella* spp. in mute swans (*Cygnus olor*) in the Great Lakes region and Atlantic coast of the United States. *Avian Diseases*, 58, 129–36.

Pepin, K.M., Lass, S., Pulliam, J.R., Read, A.F., and Lloyd-Smith, J.O. (2010). Identifying genetic markers of adaptation for surveillance of viral host jumps. *Nature Reviews Microbiology*, 8, 802.

Pepin, K.M., Hopken, M.W., Shriner, S.A., et al. (2019). Improving risk assessment of the emergence of novel influenza A viruses by incorporating environmental surveillance. *Philosophical Transactions of the Royal Society B: Biological Sciences*, 374(1782).

Quinn, J.E., Awada, T., Trindade, F., Fulginiti, L., and Perrin, R. (2017). Combining habitat loss and agricultural intensification improves our understanding of drivers of change in avian abundance in a North American cropland anthrome. *Ecology and Evolution*, 7, 803–14.

Ramey, A.M., Reeves, A.B., Ogawa, H., et al. (2013). Genetic diversity and mutation of avian paramyxovirus serotype 1 (Newcastle disease virus) in wild birds and evidence for intercontinental spread. *Archives of Virology*, 158, 2495–503.

Ramey, A.M., Goraichuk, I.V., Hicks, J.T., et al. (2017). Assessment of contemporary genetic diversity and inter-taxa/inter-region exchange of avian paramyxovirus serotype 1 in wild birds sampled in North America. *Virology Journal*, 14, 43.

Ramey, A.M., Reeves, A.B., Donnelly, T., Poulson, R.L., and Stallknecht, D.E. (2018). Introduction of Eurasian-origin influenza A (H8N4) virus into North America by migratory birds. *Emerging Infectious Diseases*, 24, 1950–3.

Reid, S.M., Manvell, R., Seekings, J.M., et al. (2019). Surveillance and investigative diagnosis of a poultry flock in Great Britain co-infected with an influenza A virus and an avirulent avian avulavirus type 1. *Transboundary and Emerging Diseases*, 66, 696–704.

Ren, H., Jin, Y., Hu, M., et al. (2016). Ecological dynamics of influenza A viruses: cross-species transmission and global migration. *Scientific Reports*, 6, 36839.

Rohaim, M.A., El Naggar, R.F., Helal, A.M., Hussein, H.A., and Munir, M. (2017). Reverse spillover of avian viral vaccine strains from domesticated poultry to wild birds. *Vaccine*, 35, 3523–7.

Scallan, E., Hoekstra, R.M., Angulo, F.J., et al. (2011). Foodborne illness acquired in the United States—major pathogens. *Emerging Infectious Diseases*, 17, 7–15.

Shriner, S.A., Root, J.J., Lutman, M.W., et al. (2016). Surveillance for highly pathogenic H5 avian influenza virus in synanthropic wildlife associated with poultry farms during an acute outbreak. *Scientific Reports*, 6, 36237.

Shwiff, S.A., Ernest, K.L., Degroot, S.L., Anderson, A.M., and Shwiff, S.S. (2017). The economic impact of blackbird damage to crops. In G.M. Linz, M.L. Avery, and R.A. Dolbeer. eds. Ecology and Management of Blackbirds (Icteridae) in North America, CRC Press, Boca Raton, FL.

Siembieda, J., Johnson, C.K., Boyce, W., Sandrock, C., and Cardona, C. (2008). Risk for avian influenza virus exposure at human–wildlife interface. *Emerging Infectious Diseases*, 14, 1151–3.

Smith, O.M., Snyder, W.E., and Owen, J.P. (2020). Are we overestimating risk of enteric pathogen spillover from wild birds to humans? *Biological Reviews*, 95, 652–79.

Snoeck, C.J., Marinelli, M., Charpentier, E., et al. (2013). Characterization of Newcastle disease viruses in wild and domestic birds in Luxembourg from 2006 to 2008. *Applied and Environmental Microbiology*, 79, 639–45.

Stallknecht, D.E., Shane, S.M., Kearney, M.T., and Zwank, P.J. (1990). Persistence of avian influenza viruses in water. *Avian Diseases*, 34, 406–11.

Stanton, R.L., Morrissey, C.A., and Clark, R.G. (2018). Analysis of trends and agricultural drivers of farmland bird declines in North America: a review. *Agriculture, Ecosystems & Environment*, 254, 244–54.

Taff, C.C., Weis, A.M., Wheeler, S., et al. (2016). Influence of host ecology and behavior on *Campylobacter jejuni* prevalence and environmental contamination risk in a synanthropic wild bird species. *Applied and Environmental Microbiology*, 82, 4811–20.

Tian, H., Zhou, S., Dong, L., et al. (2015). Avian influenza H5N1 viral and bird migration networks in Asia. *Proceedings of the National Academy of Sciences of the United States of America*, 112, 172–7.

Tormoehlen K, Johnson-Walker Y.J., Lankau E.W., Myint M.S., Herrmann J.A. (2019). Considerations for studying transmission of antimicrobial resistant enteric bacteria between wild birds and the environment on intensive dairy and beef cattle operations. *PeerJ*, 7:e6460 https://doi.org/10.7717/peerj.6460

US Department of Agriculture (2015). *Risk that Poultry Feed Made with Corn—Potentially Contaminated with Eurasian North American Lineage H5N2 HPAI Virus from Wild Migratory Birds—Results in Exposure of Susceptible Commercial Poultry*. Animal and Plant Health Inspection Service, Veterinary Services, Science, Technology, and Analysis Services, Center for Epidemiology and Animal Health, Fort Collins, CO.

Van Borm, S., Thomas, I., Hanquet, G., et al. (2005). Highly pathogenic H5N1 influenza virus in smuggled Thai eagles, Belgium. *Emerging Infectious Disease*, 11, 702–5.

van Dijk, J.G.B., Verhagen, J.H., Wille, M., and Waldenström, J. (2018). Host and virus ecology as determinants of influenza A virus transmission in wild birds. *Current Opinion in Virology*, 28, 26–36.

Verhagen, J.H., Herfst, S., and Fouchier, R.A.M. (2015). How a virus travels the world. *Science*, 347, 616–17.

Viana, M., Mancy, R., Biek, R., et al. (2014). Assembling evidence for identifying reservoirs of infection. *Trends in Ecology & Evolution*, 29, 270–9.

Vogt, N.A., Pearl, D.L., Taboada, E.N., et al. (2019). Carriage of *Campylobacter*, *Salmonella*, and antimicrobial-resistant, nonspecific *Escherichia coli* by waterfowl species collected from three sources in southern Ontario, Canada. *Journal of Wildlife Diseases*, 55, 917–22.

Wan, X.F. (2012). Lessons from emergence of A/Goose/Guangdong/1996-like H5N1 highly pathogenic avian influenza viruses and recent influenza surveillance efforts in southern China. *Zoonoses and Public Health*, 59, 32–42.

Webster, R., Bean, W.J., Gorman, O.T., Chambers, T.M., and Kawaoka, Y. (1992). Evolution and ecology of influenza A viruses. *Microbiological Reviews*, 56, 152–79.

Wille, M., Latorre-Margalef, N., and Waldenström, J. (2017). Of ducks and men: ecology and evolution of a zoonotic pathogen in a wild reservoir host. In C.J. Hurst, ed. *Modeling the Transmission and Prevention of Infectious Disease*, pp. 247–86. Springer, Cham.

Wobeser, G., Leighton, F.A., Norman, R., et al. (1993). Newcastle disease in wild water birds in western Canada, 1990. *Canadian Veterinary Journal*, 34, 353–9.

Woodroffe, R. (1999). Managing disease threats to wild mammals. *Animal Conservation*, 2, 185–93.

World Bank (2011). *World Livestock Disease Atlas: A Quantitative Analysis of Global Animal Health Data (2006–2009)*. World Bank, Washington, DC.

Yin, R., Zhang, P., Liu, X., et al. (2017). Dispersal and transmission of avian paramyxovirus serotype 4 among wild birds and domestic poultry. *Frontiers in Cellular and Infection Microbiology*, 7, 212.

CHAPTER 12

Pathogen Transmission at the Expanding Bird–Human Interface

Sarah A. Hamer and Gabriel L. Hamer

12.1 Introduction

Birds and humans have been sharing infectious disease agents for centuries. One of the first historical records linking avian disease to human deaths involved Alexander the Great who died in 323 BCE presumably of West Nile encephalitis, shortly following the death of a flock of sick ravens that may have been infected with the same agent (Marr and Calisher 2003). As another example, the H1N1 virus responsible for the influenza pandemic of 1918 resulted in about 50 million human deaths worldwide and is believed to be of avian origin (Taubenberger and Morens 2006). Most of the emerging infectious diseases (EIDs) in humans today are zoonotic, and wildlife, including birds, play key roles as reservoirs (Jones et al. 2008). This chapter covers the interface between avian health and human health, with an emphasis on the underlying biology and circumstances in which pathogens circulating in wild avian populations spill over into human populations.

12.2 Examples of zoonoses at the intersection of wild bird, domestic bird, and human populations

Birds and humans often have close associations, resulting in key opportunities for pathogen exchange at the expanding interface between them. These opportunities occur across three primary avian interfaces: wild birds, pet birds, and agricultural birds. First, human interactions with wild birds occur largely when game birds are recreationally hunted, during backyard bird feeding, or via shared insect vectors that feed on wild birds and humans, as often occurs in urban areas. Second, close associations with humans can occur with pet birds within households or veterinary clinics. Finally, human interactions with agricultural birds occur in farm or slaughter settings (Figure 12.1). Pathogen transmission from birds to humans, defined here as **spillover**, across these interfaces can occur via direct contact, contamination of water or feces, or vectors that transmit pathogens across species. In this section, we review some of the key parasite and pathogen groups that are shared between humans and birds across these three interfaces.

Enteropathogens—including *Salmonella* spp., *Campylobacter* spp., and *Escherichia coli*—are commonly transmitted from birds to humans, either directly or indirectly from avian contamination of the environment (see also Chapter 11). For example, garden birds act as the primary reservoir of the pathogens that cause human salmonellosis in England and Wales, where most passerine salmonellosis outbreaks were identified at and around feeding stations that serve as sites of public exposure to sick birds and their feces (Lawson et al. 2014). Game birds commonly carry enteropathogens, which infect the intestinal tract, and may serve as sources of these pathogens to hunters (Luechtefeld et al. 1980; Nebola et al. 2007). For example, in Spain, free-living waterfowl harbored zoonotic *Campylobacter* spp., underscoring the importance of waterfowl as potential sources of infection for domestic animals and humans and the need for hunters to use personal protective practices when

Sarah A. Hamer and Gabriel L. Hamer, *Pathogen Transmission at the Expanding Bird–Human Interface* In: *Infectious Disease Ecology of Wild Birds.*
Edited by: Jennifer C. Owen, Dana M. Hawley, and Kathryn P. Huyvaert, Oxford University Press. © Oxford University Press 2021.
DOI: 10.1093/oso/9780198746249.003.0012

Figure 12.1 Pathogen transmission occurs at the expanding bird–human interface, involving agriculture or domestic birds in production settings; wild birds in natural and urban environments; and pet birds in and around the home. We illustrate some of the likely modes of transmission at each interface including arthropod vectors, fecal–oral, water-borne, and direct, though we note that all transmission modes are possible at all three interfaces. (Created with BioRender.com.)

in contact with hunted waterfowl (Antilles et al. 2015). However, because spillover of enteric pathogens from birds to humans is a complex process that depends upon pathogen acquisition, reservoir competence and bacterial shedding, contact with people and their food, and pathogen survival in the environment (Smith et al. 2020), measures of avian infection alone should not be equated to measures of human risk of infection.

Avian schistosomes (trematodes, also known as blood flukes) are responsible for a majority of human cases of cercarial dermatitis ('swimmer's itch') around the world, a disease considered to be re-emerging globally (Horak et al. 2015). Transmission occurs when the cercaria, the last larval stage developing in an aquatic snail intermediate host, leave the snail and invade the skin of a warm-blooded vertebrate, including humans. The diversity of avian schistosomes found around the world reflects the thousands of migratory birds

that carry these parasites across space, exposing snails in different habitats (Horak et al. 2015). A survey of over 350 birds (of 46 species) and 10,000 snails (of 10 species) across North America showed that approximately 25% of birds and 1% of snails were infected with at least six *Trichobilharzia* schistosome species detected. Identifications of parasites from hatch-year birds allowed for linking the infections to the local environments where the birds were captured (Brant and Loker 2009). Knowing the background distribution of different schistosome species that infect different avian hosts can be powerful in source-tracking human outbreaks of cercarial dermatitis. For example, human outbreaks in British Columbia were linked to swimming at a recreational lake (Leighton et al. 2000). As a public health management technique, snail habitat was mechanically disturbed using boat-mounted rototillers or a tractor and rakes to reduce snail populations.

Pet birds carry a suite of zoonotic agents that pose a risk for human exposure, including bacteria, viruses, and macroparasites (Boseret et al. 2013). Psittacosis—also known as chlamydiosis, parrot fever, or ornithosis—is caused by infection with the zoonotic bacterium *Chlamydophila psittaci*, and infections have been documented in 465 avian species of 30 avian orders, with at least 153 species in the order Psittaciformes (parrots, cockatoos, and relatives; Kaleta and Taday 2003). In birds, the bacteria can cause respiratory signs, nasal and conjunctival discharge, diarrhea, polyuria, and dullness (Billington 2005). However, some birds can have latent infections with no clinical signs and serve as silent carriers with intermittent shedding of bacteria over long periods of time. In humans, infections may be asymptomatic or cause fevers, severe pneumonia, and associated complications. Most human exposures result from contact with feces from infected pet birds (Yung and Grayson 1988), but there is increasing recognition of outbreaks linked to wild birds. For example, direct contact with wild birds was a significant risk factor in a case control study of patients hospitalized from a psittacosis outbreak in Australia (Telfer et al. 2005). Similarly, after an increase in incidence of psittacosis in Sweden, a retrospective study showed that infected-human cases were more likely than case controls to have cleaned wild bird feeders or been exposed to wild bird droppings in other ways (Rehn et al. 2013). Studies like these suggest that personal protective equipment, including air filter face masks, are recommended for preventing disease, yet such practices are very rarely implemented when handling birds or feeders recreationally, in pet ownership, or in veterinary medicine.

Several arthropod-borne viruses (arboviruses) that use wild birds as reservoir hosts are emerging globally due to increasing human-facilitated movements of infected vectors and climate changes (see Chapter 10) that allow vector spread into and survival in new areas (Gould et al. 2017). These viral agents are amplified in bird–arthropod cycles and are occasionally capable of spillover transmission to humans. An improved understanding of the avian host involvement in the transmission of emerging arboviruses is essential for management programs. In many cases, the geographic patterns of the viruses have been strongly influenced by the migration patterns of their avian reservoirs (Esser et al. 2019; see Section 12.4). Among the emerging bird-associated arboviruses are West Nile virus (WNV), St. Louis encephalitis virus, the equine encephalitis viruses (eastern, western, and Venezuelan), Usutu virus (USUV), Mayoro virus, and Japanese encephalitis virus (Vasilakis and Gubler 2016).

The basic transmission cycles are similar, in which an ornithophilic mosquito infects the bird during blood-feeding, birds amplify the virus and infect new mosquitoes, and occasionally the viruses spill over into humans or other animals that are incidental or dead-end hosts. In some cases, the birds may develop disease and die from infection; in other cases, birds may be asymptomatic carriers. Because predominantly bird-feeding mosquito species transmit these arboviruses among birds in nature, there is minimal human risk due to the narrow host range of the vectors. However, when mosquito species with broader host ranges enter the transmission cycle (e.g., mosquitoes that feed on birds and mammals), they may serve as a bridge to allow the pathogen to move from the bird–mosquito cycle to a bird–mosquito–human cycle (Figure 12.1). For example, across freshwater hardwood swamps of the eastern US, eastern equine encephalitis (EEE) virus is maintained in bird–mosquito transmission cycles predominantly by the mosquito species *Culiseta melanura*. *Culiseta melanura* rarely bites mammals; instead, outbreaks of EEE in horses and humans are attributed to bridge transmission whereby *Aedes*, *Coquillettidia*, and *Culex* species mosquitoes first feed on birds and subsequently feed on and transmit the pathogen to mammals, which may result in febrile illness or fatal neurologic disease (Lindsey et al. 2018). Additionally, blood-meal analysis studies of vectors show that even mosquito species that feed almost exclusively on birds will occasionally feed on humans, with these incidental meals serving as high risk events for transmission of the virus to humans (Hamer et al. 2008).

12.2.1 Human surveillance for avian zoonoses—a call for more

Given that wild, domestic, and pet birds can be infected with zoonotic agents, several studies have investigated whether high-risk human individuals—those with high levels of contact with birds through recreational or occupational exposures—show elevated levels of exposure to avian zoonotic pathogens. Studies that assess risk of human exposure to avian influenza viruses suggest that the transmission of avian influenza viruses from migratory birds to US-based bird handlers is rare yet can have potentially significant public health consequences (Shafir et al. 2012). For example, a small number of duck hunters and wildlife professionals with extensive histories of direct interaction with waterfowl and game birds in Iowa, US, showed evidence of avian influenza virus exposure (Gill et al. 2006). This suggests that handling wild waterfowl, especially ducks, is a risk factor for influenza virus transmission to humans. In contrast, a nationwide survey of US bird banders who, on average, placed bands on 300 birds per year for 15 years, showed sparse evidence of prior infection with avian influenza viruses (Gray et al. 2011). Thus, while contact with wild birds can result in exposure to avian pathogens, the type of wild birds handled, and perhaps the level of contact with bird excrement or secretions, will ultimately determine the degree of exposure to pathogens such as avian influenza viruses.

In contrast to wild or natural settings, agricultural and household pet settings may generally provide even greater opportunity for human exposure to avian pathogens, owing in part to high densities of birds and/or indoor settings that may facilitate transmission. For example, surveys of turkey farmers in the midwestern US showed that they had 3.9–15.3 greater odds of having antibodies against several avian influenza virus strains when compared to case controls with no exposure to turkeys, raising concern for zoonotic transmission to agricultural workers (Kayali et al. 2010). There have been few studies examining zoonotic risk from avian pets, but, to better understand zoonotic risk of psittacosis, sample collection kits were sent to breeding facilities for Psittaciformes (cockatoos, parrots, parakeets, lories) across Belgium, to allow for pharyngeal samples to be collected from the birds and humans that handle pet birds (Vanrompay et al. 2007). Overall, 20% of the birds and 13% of the bird owners plus a veterinary student were positive for *Chlamydophila psittaci*. All infected humans had

rhinitis and a cough but no severe symptoms, likely owing to a history of regular exposure to the pathogen, thereby protecting against severe disease.

Overall, epidemiological surveillance of humans for zoonotic agents in relation to their level of avian contact has the potential to inform the level of transmission risk and direct personal protective strategies that are needed. However, avian-associated zoonoses have received relatively less attention than mammal-associated zoonoses. In order to assess zoonoses from wild mammals, human surveillance of high-risk populations has provided novel insights (e.g., survey of wildlife rehabilitators for racoon roundworm [Sapp et al. 2016]; survey of field biologists attending the annual mammalogy meeting for hantavirus [Fulhorst et al. 2007]). We call for new studies of targeted human populations for avian-associated zoonoses to provide comparable data on spillover risk.

12.3 Ecological and evolutionary factors important for zoonotic emergence at the wild bird–human interface

The degree to which parasites and pathogens transmit across the bird–human interface (Figure 12.1) will depend in part on the ability of pathogens to successfully infect both human and avian hosts. Pathogen **host shifts**—whereby a pathogen 'leaps' from infecting one species to infecting another species—are common in nature and have resulted in the emergence of high-profile human infectious diseases such as human immunodeficiency virus, Ebola virus, and SARS-CoV-2 (Cleaveland et al. 2001; Zhou et al. 2020). Multiple studies have investigated pathogen characteristics that influence the rate of host shifts among taxa (Kamiya et al. 2014) and have found that high mutation rates, theoretically, increase adaptation to new hosts (Gandon and Michalakis 2002). This corroborates empirical data showing that RNA viruses—with short genomes, rapid mutation rates, and high replication rates—are most likely to be zoonotic emerging infectious agents of disease in humans (Taylor et al. 2001).

The process of pathogen spillover from other animals to humans is complex, with many barriers that result in a small subset of animal pathogens successfully being established in human transmission cycles. Plowright et al. (2017) presented a synthetic framework to identify the mechanisms and steps involved in the process for pathogen spillover from animals into humans which results in emerging infectious diseases. This process is broken down into the phases of **pathogen pressure** (i.e., the total amount of pathogen leaving animal hosts and present in the environment), pathogen exposure (human and vector behavior that determines likelihood, route, and dose of exposure), and probability and severity of infection (influenced by dose and route of exposure along with genetic, physiological, and immunological attributes of the recipient human host). Each phase represents a barrier to pathogen spillover, such that the majority of pathogens are filtered out and do not successfully infect humans. For example, avian influenza viruses are common in wild and domestic poultry but only rarely do these viruses bridge into human populations (Kuiken et al. 2006; Lipsitch et al. 2016). Reasons for the lack of spillover include that many of these viruses lack the ability to replicate in human cells and human innate immune responses block infections. Nonetheless, although many RNA virus mutations will be deleterious, some avian viruses could possess a suite of mutations allowing for cell infection and replication in humans (Kuiken et al. 2006).

Knowledge of the traits of viruses that are associated with spillover from animals to humans could be useful in the prediction and management of human emerging infectious diseases. Olival et al. (2017) conducted an analysis using a database of 2,805 mammal–virus associations and ranked the viral traits that increase the likelihood of a virus being zoonotic and capable of infecting humans. They found that the proportion of zoonotic viruses per species increases with host phylogenetic proximity to humans and is driven by the large number of viruses shared among non-human primates and humans. In addition, they found that the number of viruses that infect humans scales positively with the size of the **zoonotic pool**—a measure of the abundance of viruses documented in a wild mammal species. Bats have emerged as animals with a large zoonotic pool and are associated with an increased chance that some of these viruses will spill over into human populations (Olival et al.

2017). A newer analysis, however, suggests that variation in the frequency of zoonoses among major bird and mammal reservoir groups is attributed to variation in host and virus species richness and not intrinsic or ecological differences among animal groups (Mollentze and Streicker 2020). While the phylogenetic distance between birds and humans suggests that birds may be less likely to share pathogens with humans than non-human mammalian taxa, further analyses of the predictive factors associated with zoonotic bird-associated pathogens would be useful.

12.4 Birds as vehicles for the movement of arthropod vectors and pathogens of human health concern

Wild birds are unique in their ability to move long distances over short periods of time (e.g., migration; post-fledgling dispersal), thereby affording the opportunity for pathogens or ectoparasites of birds also to be dispersed. When bird-associated pathogens or parasites are introduced into new environments that are not suitable for their survival or establishment, their introduction may go unnoticed and is likely to be inconsequential for human, animal, and ecological health. However, when birds disperse pathogens or parasites into new environments where the available hosts, climate, and other factors that allow for their establishment and spread to occur are present, then important human health impacts may follow.

Most ticks move very little when they are not on vertebrate hosts (Crooks and Randolph 2006), such that emergence of tick populations in new regions often reflects the patterns of movement of infested hosts. Birds are increasingly recognized as having a role in sculpting patterns of ticks and tick-borne disease across the landscape (Loss et al. 2016). Wild birds have long imported tick species into new regions (Hasle 2013); most commonly, spring migrants transport ticks northward beyond the recognized distributional limits of the ticks. These bird-enabled tick dispersals are increasingly being understood to have public health consequences. For example, inspection of tens of thousands of northward-migrating birds arriving to banding stations in eastern Canada showed an overall low tick infestation prevalence (0.35–2.2% of birds harbored ticks) and overall low burden of ticks on infested birds (mean of 1.66 ticks per bird), yet the large number of migrants to the region every spring suggests that millions of *Ixodes scapularis* ticks are dispersed into Canada each spring (Ogden et al. 2008). While these bird-imported ticks ('adventitious ticks') were long thought to reach an unsuitable climate and die-out, a warming climate and other factors now support the northward range expansion of *I. scapularis* ticks in southeastern Canada, with a concomitant dramatic increase in the incidence of human Lyme disease (Bouchard et al. 2015).

These migratory movements facilitate the importation of exotic species—including human pathogens—from the Neotropics into the US. For example, of the eight species of tick encountered on birds arriving to south Texas during spring migration, only one tick species is known to be established in the US, whereas all others were neotropical species from Central or South America and Mexico (Cohen et al. 2015). Further, some of these ticks were infected with spotted fever group *Rickettsia* species of human health concern. It is likely that biotic or abiotic barriers currently prevent these exotic ticks from establishing in spite of the large annual **propagule pressure**, or number of individuals arriving to a new region; for example, the climate may be unsuitable for ticks to survive, molt, and encounter another host, or the host species available in the US may not include the hosts that are commonly used in the tick's native range. However, environmental changes may allow for these bird-imported ticks to encounter habitats suitable for their survival and spread in the future, with unknown human health consequences. Indeed, because birds typically import the immature life stages of ticks (larvae and nymphs), it is noteworthy that an adult stage of a neotropical tick (*Amblyomma longirostre*) was recently found questing in the US (Noden et al. 2015). Observations like this may suggest progress along the pathway of invasion, given that the most likely mechanism of importation of this species would be an immature tick arriving on a migratory bird. Encountering an adult in the wild suggests that a bird-imported immature tick survived and molted to emerge as an adult and is likely to feed on local wild or domestic animals. The presence of neotropical ticks on humans in the US is noteworthy, raising a call for rigorous inspections of imported livestock and pets at ports of entry (Molaei et al. 2019). Nonetheless, the annual importation of exotic ticks by migrating wild birds will proceed as a natural and unchecked process, circumventing any biosecurity practices.

Studies in other regions have also documented the movement of tick-associated pathogens facilitated by birds. For example, Crimean Congo hemorrhagic fever (CCHF) virus is a bunyavirus spread by ticks (mainly *Hyalomma* spp.) that causes fever and hemorrhage in humans and is endemic to Africa, the Balkans, the Middle East, and Asia. Wild bird migration has been proposed as the mechanism of virus spread accounting for the increase in human incidence of CCHF virus in Turkey, as virus-infected ticks were collected from two migratory bird species: great reed warbler (*Acrocephalus arundinaceus*) and European robin (*Erithacus rubecula*) (Leblebicioglu et al. 2014). Similarly, wild birds have been implicated in moving CCHF virus to Italy and Greece from Africa, with examples that include infected ticks carried by the woodchat shrike (*Lanius senator senator*) during spring migration; this species winters in a belt from Senegal to Somalia and breeds in southern Europe and northern Africa (Lindeborg et al. 2012).

Avian movements are also recognized as important in sculpting the geographic patterns of distribution of the Lyme disease bacterium *Borrelia burgdorferi*. A recent study reconstructed the evolutionary history of *B. burgdorferi* through sequencing of genomes from across the US and southern Canada. Unprecedented levels of gene flow and evidence of long-distance migration events between the three major geographic regions sampled were attributed to long-distance, bird-mediated dispersal (Walter et al. 2017). Similarly, in an analysis of the Lyme disease spirochete species that circulate in Europe, the bird-associated species (*B. garinii* and *B. valaisiana*) showed limited geographic structuring when compared to the mammal-associated species (*B. afzelii*), owing to high rates of migration and mixing facilitated by avian hosts (Vollmer et al. 2011).

Finally, while patterns of bird-associated movement are likely strongest for ticks and their pathogens (given the limited mobility of ticks), birds have also been implicated in the movement of mosquito-borne pathogens, such as WNV. WNV is maintained in nature by several avian bird species, with vast variation in the degree to which infected birds can serve as a source of viral infection to feeding mosquitoes (Komar et al. 2003). Phylogenetic approaches have shown that the patterns of migration of WNV are consistent with the looped flight paths of many terrestrial migratory birds in the eastern flyway and central flyway (Swetnam et al. 2018).

Overall, there are now numerous examples of arthropod vector and pathogen movement by birds. However, the importation of the vector or agent alone is only significant when it arrives to a receptive environment with supportive climate and suitable vertebrate hosts to become established. Records of avian importation of pathogens or vectors of medical importance should be understood as a warning for future establishment of new human disease foci, which is likely to be augmented in many regions by climate change (see Chapter 10).

12.5 Reverse zoonoses (anthroponoses): examples of pathogen transmission from humans to birds

Whereas the sharing of pathogens between humans and other animals is most commonly thought of as a directional process with transmission going from non-human animal reservoirs to human hosts, the reverse directionality (i.e., from human to non-human animals) must also be considered. Among the most common examples of **reverse zoonoses** (also known as anthroponoses or 'spillback') are human respiratory viruses that cause disease in non-human primates, often with implications for conservation efforts (Dunay et al. 2018). Infectious agents from humans likely spill over into birds frequently, yet there are few examples of reverse zoonoses in birds, perhaps owing to challenges in their detection (Messenger et al. 2014; see Chapter 2). Identification of the definitive agents of disease in a wild or even captive bird is rarely done, especially relative to differential diagnosis of disease in human patients.

One unique geographic area that provides opportunity for inquiry into reverse zoonoses in birds,

however, is Antarctica, where diverse fauna have remained isolated from humans until recent encroachment. Cerda-Cuellara et al. (2019) sampled 24 seabird species over a broad geographical range in Antarctica and characterized the bacterial species, including identification of serovars, genotypes, and resistance profiles to antibiotics commonly used in human and veterinary medicine. They then compared these bird-associated bacteria to those found in humans in order to document reverse zoonoses. They detected *Salmonella* serovars and *Campylobacter* genotypes in seabirds reported almost exclusively from humans and domestic animals from developed countries, including some with resistance to antibiotics used in human and veterinary medicine. Similarly, a study in New Zealand documented *Campylobacter jejuni* in captive brown kiwi (*Apteryx mantelli*), which is interpreted as a reverse zoonosis (On et al. 2019). These studies point to the need for stricter biosecurity measures in order to protect birds (wild or captive) from human pathogens.

Wildlife management plans may include specific guidance to reduce the transmission of human pathogens to animals. For example, best practices for human tourists that track and observe great apes in the wild include: self-reporting of illnesses or illness screenings; refraining from tracking while sick; hand washing; shoe disinfection; maintaining 7–10 m physical distance from the animals; use of disposable facemasks; and burying human excrement (Muehlenbein and Ancrenaz 2009). Among birds, the avifauna of the Galápagos islands are threatened by several infectious diseases, including not only invasive poultry pathogens but also pathogens that may be introduced by humans, especially tourists. There is continual concern about avian diseases including *Toxoplasma gondii*, *Salmonella* spp., cholera, botulism, and WNV (Wikelski et al. 2004), some of which may be introduced by humans. Predictive models have been used to quantify the risk of WNV introduction to the islands via different routes, including the arrival of infected humans, but this pathway of introduction was determined to present a much lower risk than others, including the arrival of infected mosquitoes by airplane and migratory birds (Kilpatrick et al. 2006b). Preventive measures aimed at halting the introduction of new pathogens to the islands include spraying of airplanes for vector control, control of domestic animal entry, and a quarantine program to prevent the introduction of organisms via shipments (Wikelski et al. 2004). Risk posed to wildlife directly by humans—including their excrement and respiratory secretions—may be harder to manage (Curry et al. 2002).

Close associations between humans and poultry offer several opportunities for zoonotic pathogen exchange from birds to humans (see Chapter 11); this ecological setting has also allowed for reverse zoonoses. For example, while spillover transmission of avian influenza viruses from birds to humans commonly occurs, influenza viruses may also be transmitted from humans to birds. A study in Canada documents H1N1 viruses causing disease in breeder turkeys to be of close homology to viruses found in humans and swine; when turkeys were experimentally infected with human pH1N1, viral shedding and seroconversion occurred (Berhane et al. 2010). As another example, phylogenetic analysis of whole-genome sequences of *Staphylococcus aureus* from diverse vertebrates revealed that the bacteria made a single host-jump from humans to broiler chickens, likely in the 1970s in or near Poland (Lowder et al. 2009). Considerations to enhance biosecurity in poultry settings to reduce human-to-bird transmission may be warranted.

12.6 Parasite interactions in birds—implications for human health

Birds are infected with a large suite of parasites and pathogens, many of which are capable of infecting humans, and many receive attention and funding in medical and research communities simply because of their impact on humans. In contrast, the non-zoonotic parasites and pathogens that birds harbor, and that may influence co-infecting zoonotic pathogens, receive less research attention. Nonetheless, the field of community ecology has grown to appreciate the complexities inherent to vertebrate co-infection with multiple pathogens (Cox 2001; Ezenwa 2016; Graham 2008; Pedersen and Fenton 2007; Tompkins et al. 2011; see also Chapter 8). The interactions among parasites can

be synergistic or antagonistic and can occur directly, such as via competition for resources, or indirectly, as modulated through the host immune system. Importantly, these interactions within the host can shape the distribution of pathogens in populations at the ecosystem level (Ezenwa and Jolles 2011; Lloyd-Smith et al. 2008; Pathak et al. 2012).

Parasite co-infection has received considerable attention in humans, such as the investigation of how helminth infections facilitate the transmission of human immunodeficiency virus (HIV) (Kamal and Khalifa 2006; Noblick et al. 2011). Within-host immune-mediated interactions are particularly important between extracellular macroparasites (e.g., helminths) and intracellular microparasites (e.g., viruses and bacteria) due to the different immune responses produced during infections (Cox 2001). In mammals, macroparasite exposure typically induces CD4+ T-helper type 2 cells to produce cytokines such as IL-4 (see also Chapter 3). Microparasites induce T-helper type 1 cells which produce cytokines such as interferon γ. The polarization of Th1/Th2 immune responses implies that they are produced as a trade-off: the upregulation of one results in a downregulation of the other (Graham 2008). This trade-off has been implicated as being involved in many co-infection systems, where the infection of a macroparasite increases susceptibility and transmissibility for subsequent microparasite infections (Fenton 2008).

Although most progress in understanding parasite community ecology has been made in mammals, many similar parallels have been observed in birds. For example, Schwarz et al. (2011) provide support that the Th1/Th2 polarization also occurs for birds, where gastrointestinal nematode infections upregulate a Th2 immune response in chickens and downregulate a Th1 immune response, while Newcastle disease virus has the opposite effect (Degen et al. 2005). There is also empirical evidence to suggest that helminth infection in free-ranging chickens results in an immunosuppressive effect for subsequent viral infections (Hørning et al. 2003). In wild cormorants (e.g., European shag, *Phalacrocorax aristotelis*) in Europe, Newbold et al. (2017) found that gastrointestinal helminth infection resulted in distinct changes to the gastrointestinal microbiome. A more recent study in song sparrows (*Melospiza melodia*) identified that gastrointestinal helminth infections reduced the immune response to a simulated bacterial infection (Vaziri et al. 2019). In addition, wild song sparrows with higher helminth infections had a higher prevalence of avian malaria.

Although infection of wild birds with multiple parasites or pathogens is likely ubiquitous, the way in which these infections increase zoonotic risk of spillover to humans has received minimal attention. The primary system receiving this focus of inquiry has been zoonotic mosquito-borne viruses that amplify in the bird–mosquito cycle and are transmitted to humans through a bridge mosquito vector that feeds on birds and then humans. These viruses include WNV, western equine encephalitis virus (WEEV), and USUV. Helminth infection is predicted to have a net positive effect on WNV transmission based on the way in which helminth infection would modify parameters informing the WNV reproductive number, R_0 (Ezenwa and Jolles 2011), although minimal empirical data exist for avian helminth–arbovirus systems. Other co-infecting parasites, such as blood protozoa, may instead suppress arbovirus infection within hosts. This can occur if, for example, infection with a blood protozoan stimulates aspects of host immunity that are also effective against viruses. In the 1940s in California, the results of detailed studies of WEEV (Hammon et al. 1945) and avian malaria parasites (*Plasmodium* spp.) (Herman et al. 1954) noted that the same birds and same vectors were involved in *Plasmodium* and WEEV transmission. This led to investigations of the potential for avian malaria parasites to influence the transmission of WEEV, which was pursued in controlled laboratory co-infection experiments using canaries (*Serinus canaria domestica*). These experiments found suppression of titers of WEEV in the blood of birds that were co-infected with *Plasmodium* (Barnett 1956), though the mechanisms generating this interaction remain unknown.

More recently, research from suburban Chicago has shown that the same birds responsible for amplifying WNV are also infected with helminths (Hamer and Muzzall 2013; Hamer et al. 2013),

trypanosomes (Hamer et al. 2013), and *Plasmodium* or other blood parasites that cause avian malaria (Medeiros et al. 2014; Figure 12.2). Additional studies from the same region show that the *Culex* mosquitoes responsible for WNV transmission are also infected with these avian parasites (Boothe et al. 2015), and at the population level, WNV and *Plasmodium* spp. amplify at the same time and the same location (Medeiros et al. 2016). In Europe, researchers studying the emergence of USUV have noticed high levels of co-infection of this virus and *Plasmodium* spp. in birds (Rijks et al. 2016; Rouffaer et al. 2018). Taken together, these studies suggest that multiple parasites simultaneously infect birds and vectors, and the potential health consequences of these co-infections are only beginning to be understood.

In addition to direct or immune-mediated interactions within birds, additional opportunities exist for parasites to impact zoonotic pathogen transmission within mosquito vectors. In some cases, the same avian parasites (e.g., *Plasmodium* spp.) that suppress viral titers within hosts (as occurs for canaries and WEEV; see above) can facilitate virus transmission within vectors. For example, *Plasmodium* infection of mosquitoes removes the salivary gland barrier and allows for the enhanced transmission of Rift Valley fever virus (Vaughan and Turell 1996b). Similarly, when mosquitoes ingest microfilarial nematodes concurrent with viruses, it shortens the time needed for the virus to reach the salivary glands and be transmitted to a new host (Turell et al. 1984; Vaughan and Turell 1996a). *Plasmodium*-infected birds were found to be more attractive to mosquitoes (Cornet et al. 2013) and *Plasmodium*-infected mosquitoes obtained blood-meals later (and smaller blood-meals) than uninfected mosquitoes (Cornet et al. 2019). Collectively, these studies suggest that co-infection is likely to modify a zoonotic pathogen's vectorial capacity, which is the combined effect of extrinsic and intrinsic factors on the mosquito population's capacity to transmit an infectious agent to a susceptible population (Kramer and Ciota 2015; Macdonald 1961). Thus, strictly avian parasites can have important consequences for human health by modifying the vector-borne transmission of zoonotic agents.

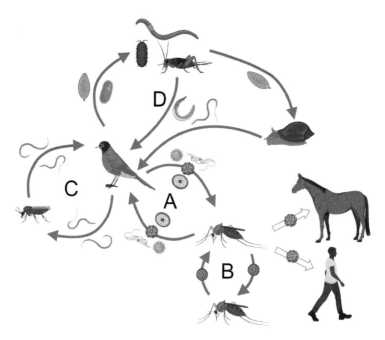

Figure 12.2 Pathogens co-circulating in the WNV system. At the core of the enzootic cycle are *Culex* spp. mosquitoes transmitting WNV among wild passerine birds with occasional bridge transmission to dead-end mammalian hosts. These same *Culex* mosquitoes and avian hosts also circulate other viruses, Haemosporida, and trypanosomes (A). The mosquitoes that vector WNV are also infected with insect-specific viruses which are maintained vertically (B). *Culicoides* sp. midges transmit filarial worms to the same avian hosts which amplify WNV (C). Birds are the definitive host for a suite of helminths which use arthropods and gastropods as intermediate hosts (D). (Created with BioRender.com.)

12.7 Management strategies to minimize avian zoonoses

Given the diverse infectious agents that are shared among birds and humans, management to minimize avian zoonoses must integrate approaches with specific attention on the environments in which transmission occurs. There are three broad contexts in which these avian zoonoses are shared: the domestic environment in which pet birds interact with humans; agricultural environments where humans and poultry or other farmed birds share pathogens; and the natural environment in which wild birds share pathogens with humans in a number of direct and indirect ways (Figure 12.1). As mentioned, these transmission contexts include hunters processing wild game birds, humans coming into contact with wild bird feces, and through vector-borne transmission in which arthropods bridge infectious agents from birds to humans.

Approaches for mitigating zoonoses in a captive bird setting have been reviewed (Boseret et al. 2013). Management activities for pet birds include the categories of household hygiene (e.g., clean clothes, wash hands, change water frequently or only provide access to water for limited times each day, wash cages weekly, preserve food in sealed containers), importation regulation (e.g., identify accurate traceability of birds, quarantine new birds, restrict access, isolate birds), and veterinary care (e.g., notice disease and promptly care for sick birds or consult a veterinarian). For agricultural contexts, poultry operations need to maintain high levels of biosecurity in order to minimize economically costly impacts from disease-causing organisms, such as Newcastle disease virus and avian influenza virus (Cattoli et al. 2011; USDA 2014; for more, see Chapter 11). These practices include keeping visitors to a minimum, limiting visitation of poultry growers among farms, keeping all wild animals out of poultry houses, practicing pest control, avoiding poultry contact with wild birds, performing routine flock inspections, and maintaining a hygienic environment with low humidity.

Relative to management of zoonotic risk in captive avian environments, there are far more challenges to the management of zoonotic risk from wild bird populations. Some of the hygiene-based strategies used with captive birds can also apply to interactions with wild birds. For example, supplemental feeding of birds at feeding stations can contribute to the transmission of *Salmonella enterica*, the causative agent of human salmonellosis. Routine cleaning of bird feeders with a 10% bleach solution is one way to reduce *Salmonella* persistence (Feliciano et al. 2018). *Histoplasma capsulatum* is a fungus that grows in accumulated bird feces (Benedict and Mody 2016) and often presents problems in areas where pigeons, starlings, and other birds roost or nest near buildings where humans are at risk of histoplasmosis via inhalation of fungal spores. The best management strategy in these cases is to prevent birds from spending excessive time in these sensitive areas through hazing techniques.

The management of infectious agents at the population level in wild birds is difficult and there are thus few examples of these activities. One good system providing a case study for this concept is that of WNV. Wild birds are the only amplification hosts that drive the transmission of WNV, which results in a high burden of human morbidity and mortality in the US and is an arbovirus with one of the largest global distributions (Kramer et al. 2008). WNV is often associated with the urban environment where *Culex pipiens* complex mosquitoes are the primary vector. Standardized mosquito trapping and dead bird surveillance programs have provided data that can be used to provide an early warning and predict regions and times of elevated human disease risk, given that the virus first amplifies between birds and mosquitoes prior to spillover to humans (Guptill et al. 2003). Similarly, caged birds can be placed in areas and sampled regularly to learn the local temporal dynamics of infection, as these birds will be exposed to local vectors (e.g., sentinel chickens for WNV detection) (Morales-Betoulle et al. 2013).

A few studies have evaluated the role of different bird species in the urban community and have identified a limited number of bird species contributing to the majority of transmission (Hamer et al. 2011; Kilpatrick et al. 2006a), such as American robin (*Turdus migratorius*), house sparrow (*Passer domesticus*), and house finch (*Haemorhous mexicanus*). The targeted population suppression of these bird species would likely reduce WNV transmission, although this management option is not ethically, socially, or legally feasible. This is because many of these avian amplification reservoirs are protected by the Migratory Bird Treaty Act and communities impacted by WNV would likely object to the area-wide removal of these birds. Alternative approaches could be the vaccination of wild birds for WNV, which has shown promise in laboratory settings (Bunning et al. 2007; Kilpatrick et al. 2010) but has not been deployed in the field, given significant logistical challenges including the injection of large numbers of wild birds. This approach could be similar to the successful programs for vaccinating wild raccoons and other mesopredators in the US (Maki et al. 2017), based on oral vaccine-laden baits. A new approach for mitigating human risk of WNV involves treatment of wild birds to kill the mosquito vectors that feed upon them. In California, seeds impregnated with ivermectin are delivered to birds; this drug has broad insecticidal properties once ingested by wild birds (Nguyen et al. 2019). Such approaches may have the potential to simultaneously protect both human and bird health.

Disease management approaches have also included biological control, in which the introduction of a competitor, predator, or parasite into a system is intended to have a detrimental effect on a key reservoir or vector to suppress pathogen transmission. In the context of zoonoses associated with birds, a government program in Turkey introduced thousands of exotic helmeted guineafowl (*Numida meleagris*) to the landscape to control the *Hyalomma marginatum* ticks carrying CCHF virus, given that the guineafowl could eat ticks. However, guineafowl are associated with low frequency of tick consumption and, instead, can increase tick populations by serving as blood-meal hosts (Sekercioglu 2013). While some unintended consequences of management programs may be unavoidable, the management of zoonoses—given their complex ecologies—necessitates communication among multiple stakeholders to represent diverse elements of the disease systems that must be considered.

12.8 Summary

Owing to their often close associations with humans, many types of birds—especially urban and game birds, pet birds, and agricultural birds—have the potential to share pathogens with humans directly or indirectly (Figure 12.1). Although many barriers exist for pathogen spread from birds to humans, spillover events do occur and they are capable of resulting in pandemics. While most examples of avian–human zoonoses document spread from birds to humans, we present evidence that reverse zoonoses also occur whereby human pathogens spread to birds. Elucidating the diverse ways in which birds impact human health must consider not only avian infection with zoonotic agents but also the roles that co-circulating, non-zoonotic pathogens may play in modulating spillover risk. Further, the unique ability of wild birds to migrate—often over long distances in short periods of time—allows birds to sculpt broad-scale patterns of distribution of vectors and pathogens, because birds migrate while infested by ticks and infected by pathogens that can be deposited into new landscapes. Given that some of these vectors and pathogens can infect humans, these bird movements can have important human health consequences. However, the suitability of the environment into which bird-dispersed vectors and pathogens are imported must also be considered (see Chapter 10), because human health consequences may arise only after the successful establishment and spread of these agents.

The management of avian zoonotic pathogen transmission must take the form of integrated approaches to target high-risk birds and their environments and may include hygiene, drug treatments, and vaccination campaigns coupled with standardized surveillance programs. While studies of avian infection with zoonotic agents are common, we suggest an expansion of human serosurveillance for bird zoonoses, focusing on groups with high exposures to different types of birds. In this respect, epidemiological studies of high-risk human sentinels could provide insight not only on spillover events that could impact human health but also on areas where avian zoonoses could pose a risk to birds and other animals.

Literature cited

Antilles, N., Sanglas, A., and Cerda-Cuellar, M. (2015). Free-living waterfowl as a source of zoonotic bacteria in a dense wild bird population area in northeastern Spain. *Transboundary and Emerging Diseases*, 62, 516–21.

Barnett, H.C. (1956). Experimental studies of concurrent infection of canaries and of the mosquito *Culex tarsalis* with *Plasmodium relictum* and western equine encephalitis virus. *American Journal of Tropical Medicine and Hygiene*, 5, 99–109.

Benedict, K. and Mody, R.K. (2016). Epidemiology of histoplasmosis outbreaks, United States, 1938–2013. *Emerging Infectious Diseases*, 22(3), 370–8.

Berhane, Y., Ojkic, D., Neufeld, J., et al. (2010). Molecular characterization of pandemic H1N1 influenza viruses isolated from turkeys and pathogenicity of a human pH1N1 isolate in turkeys. *Avian Diseases*, 54, 1275–85.

Billington, S. (2005). Clinical and zoonotic aspects of psittacosis. *In Practice*, 27(5), 256–63.

Boothe, E., Medeiros, M.C.I., Kitron, U.D., et al. (2015). Identification of avian and hemoparasite DNA in blood-engorged abdomens of *Culex pipiens* (Diptera; Culicidae) from a West Nile virus epidemic region in suburban Chicago, Illinois. *Journal of Medical Entomology*, 52, 461–8.

Boseret, G., Losson, B., Mainil, J.G., Thiry, E., and Saegerman, C. (2013). Zoonoses in pet birds: review and perspectives. *Veterinary Research*, 44(36).

Bouchard, C., Leonard, E.K., Koffi, J.K., et al. (2015). The increasing risk of Lyme disease in Canada. *Canadian Veterinary Journal*, 56, 693–9.

Brant, S.V. and Loker, E.S. (2009). Molecular systematics of the avian schistosome genus *Trichobilharzia* (Trematoda: Schistosomatidae) in North America. *Journal of Parasitology*, 95, 941–63.

Bunning, M.L., Fox, P.E., Bowen, R.A., et al. (2007). DNA vaccination of the American crow (*Corvus brachyrhynchos*) provides partial protection against lethal challenge with West Nile virus. *Avian Diseases*, 51, 573–7.

Cattoli, G., Susta, L., Terregino, C., and Brown, C. (2011). Newcastle disease: a review of field recognition and current methods of laboratory detection. *Journal of Veterinary Diagnostic Investigation*, 23, 637–56.

Cerda-Cuellar, M., Moré, E., Ayats, T., et al. (2019). Do humans spread zoonotic enteric bacteria in Antarctica? *Science of the Total Environment*, 654, 190–6.

Cleaveland, S., Laurenson, M.K., and Taylor, L.H. (2001). Diseases of humans and their domestic mammals: pathogen characteristics, host range and the risk of emergence. *Philosophical Transactions of the Royal Society B: Biological Sciences*, 356, 991–9.

Cohen, E.B., Auckland, L.D. Marra, P.P., and Hamer, S.A. (2015). Avian migrants facilitate invasions of neotropical ticks and tick-borne pathogens into the United States. *Applied and Environmental Microbiology*, 81, 8366–78.

Cornet, S., Nicot, A., Rivero, A., and Gandon, S. (2013). Malaria infection increases bird attractiveness to uninfected mosquitoes. *Ecology Letters*, 16, 323–9.

Cornet, S., Nicot, A., Rivero, A., and Gandon, S. (2019). Avian malaria alters the dynamics of blood feeding in *Culex pipiens* mosquitoes. *Malaria Journal*, 18, 82.

Cox, F.E.G. (2001). Concomitant infections, parasites and immune responses. *Parasitology*, 122, S23–38.

Crooks, E. and Randolph, S.E. (2006). Walking by *Ixodes ricinus* ticks: intrinsic and extrinsic factors determine the attraction of moisture or host odour. *Journal of Experimental Biology*, 209, 2138–42.

Curry, C.H., McCarthy, J.S., Darragh, H.M., Wake, R.A., Todhunter, R., and Terris, J. (2002). Could tourist boots act as vectors for disease transmission in Antarctica? *Journal of Travel Medicine*, 9, 190–3.

Degen, W.G., van Daal, N., Rothwell, L., Kaiser, P., and Schijns, V.E.J.C. (2005). Th1/Th2 polarization by viral and helminth infection in birds. *Veterinary Microbiology*, 105, 163–7.

Dunay, E., Apakupakul, K., Leard, S., Palmer, J.L., and Deem, S.L. (2018). Pathogen transmission from humans to great apes is a growing threat to primate conservation. *Ecohealth*, 15, 148–62.

Esser, H.J., Mögling, R., Cleton, N.B., et al. (2019). Risk factors associated with sustained circulation of six zoonotic arboviruses: a systematic review for selection of surveillance sites in non-endemic areas. *Parasites & Vectors*, 12, 265.

Ezenwa, V.O. (2016). Helminth–microparasite co-infection in wildlife: lessons from ruminants, rodents and rabbits. *Parasite Immunology*, 38, 527–34.

Ezenwa, V.O. and Jolles, A.E. (2011). From host immunity to pathogen invasion: the effects of helminth coinfection on the dynamics of microparasites. *Integrative and Comparative Biology*, 51, 540–51.

Feliciano, L.M., Underwood, T.J., and Aruscavage, D.F. (2018). The effectiveness of bird feeder cleaning methods with and without debris. *The Wilson Journal of Ornithology*, 130, 313–20.

Fenton, A. (2008). Worms and germs: the population dynamic consequences of microparasite–macroparasite co-infection. *Parasitology*, 135, 1545–60.

Fulhorst, C.F., Milazzo, M.L., Armstrong, L.R., et al. (2007). Hantavirus and arenavirus antibodies in persons with occupational rodent exposure, North America. *Emerging Infectious Diseases*, 13, 532–8.

Gandon, S. and Michalakis, Y. (2002). Local adaptation, evolutionary potential and host–parasite coevolution: interactions between migration, mutation, population size and generation time. *Journal of Evolutionary Biology*, 15, 451–62.

Gill, J.S., Webby, R., Gilchrist, M.J.R., and Gray, G.C. (2006). Avian influenza among waterfowl hunters and wildlife professionals. *Emerging Infectious Diseases*, 12, 1284–6.

Gould, E., Pettersson, J., Higgs, S., Charrel, R., and de Lamballerie, X. (2017). Emerging arboviruses: Why today? *One Health*, 4, 1–13.

Graham, A.L. (2008). Ecological rules governing helminth–microparasite coinfection. *Proceedings of the National Academy of Sciences of the United States of America*, 105, 566–70.

Gray, G.C., Ferguson, D.D., Lowther, P.E., Heil, G.L., and Friary, J.A. (2011). A national study of US bird banders for evidence of avian influenza virus infections. *Journal of Clinical Virology*, 51, 132–5.

Guptill, S.C., Julian, K.G., Campbell, G.L., Price, S.D., and Marfin, A.A. (2003). Early-season avian deaths from West Nile virus as warnings of human infection. *Emerging Infectious Diseases*, 9, 483–4.

Hamer, G.L. and Muzzall, P.M. (2013). Helminths of American robins, *Turdus migratorius* and house sparrows, *Passer domesticus* (Order: Passiformes) from suburban Chicago, Illinois, USA. *Comparative Parasitology*, 80, 287–91.

Hamer, G.L., Kitron, U.D., Brawn, J.D., et al. (2008). *Culex pipiens* (Diptera: Culicidae): a bridge vector of West Nile virus to humans. *Journal of Medical Entomology*, 45, 125–8.

Hamer, G.L., Kitron, U.D., Brawn, J.D., et al. (2011). Fine-scale variation in vector host use and force of infection drive localized patterns of West Nile virus transmission. *PLoS ONE*, 6, e23767.

Hamer, G.L., Anderson, T.K., Berry, G.E., et al. (2013). Prevalence of filarioid nematodes and trypanosomes in American robins and house sparrows, Chicago USA. *International Journal of Parasitology: Parasites and Wildlife*, 2, 42–9.

Hammon, W., Reeves, W., and Galindo, P. (1945). Epidemiological studies of encephalitis in the San Joaquin Valley of California, 1943, with the isolation of virus from mosquitoes. *American Journal of Hygiene*, 42, 299–306.

Hasle, G. (2013). Transport of ixodid ticks and tick-borne pathogens by migratory birds. *Frontiers in Cellular and Infection Microbiology*, 3, 48.

Herman, C.M., Reeves, W.C., Moclure, H.E., et al. (1954). Studies on avian malaria in vectors and hosts of encephalitis in Kern county, California. I. Infections in avian hosts. *American Journal of Tropical Medicine and Hygiene*, 3, 676–95.

Horak, P., Mikes, L., Lichtenbergova, L., Skala, V., Solanova, M., and Brant, S.V. (2015). Avian schistosomes and outbreaks of cercarial dermatitis. *Clinical Microbiology Reviews*, 28, 165–90.

Hørning, G., Rasmussen, S., Permin, A., and Bisgaard, M. (2003). Investigations on the influence of helminth parasites on vaccination of chickens against Newcastle Disease Virus under village conditions. *Tropical Animal Health and Production*, 35, 415–24.

Jones, K.E., Patel, N.G., Levy, M.A., et al. (2008). Global trends in emerging infectious diseases. *Nature*, 451, 990–3.

Kaleta, E.F. and Taday, E.M.A. (2003). Avian host range of *Chlamydophila* spp. based on isolation, antigen detection and serology. *Avian Pathology*, 32, 435–62.

Kamal, S.M. and Khalifa, K.E. (2006). Immune modulation by helminthic infections: worms and viral infections. *Parasite Immunology*, 28, 483–96.

Kamiya, T., O'Dwyer, K., Nakagawa, S., and Poulin, R. (2014). Host diversity drives parasite diversity: meta-analytical insights into patterns and causal mechanisms. *Ecography*, 37, 689–97.

Kayali, G., Ortiz, E.J., Chorazy, M.L., and Gray, G.C. (2010). Evidence of previous avian influenza infection among US turkey workers. *Zoonoses and Public Health*, 57, 265–72.

Kilpatrick, A.M., Daszak, P., Jones, M.J., Marra, P.P., and Kramer, L.D. (2006a). Host heterogeneity dominates West Nile virus transmission. *Proceedings of the Royal Society B: Biological Sciences*, 273, 2327–33.

Kilpatrick, A.M., Daszak, P., Goodman, S.J., et al. (2006b). Predicting pathogen introduction: West Nile virus spread to Galápagos. *Conservation Biology*, 20, 1224–31.

Kilpatrick, A.M. Dupuis, II, A.P., Chang, G.-J.J., and Kramer, L.D. (2010). DNA vaccination of American robins (*Turdus migratorius*) against West Nile virus. *Vector Borne and Zoonotic Diseases*, 10, 377–80.

Komar, N., Langevin, S., Hinten, S., et al. (2003). Experimental infection of North American birds with the New York 1999 strain of West Nile virus. *Emerging Infectious Diseases*, 9, 311–22.

Kramer, L.D. and Ciota, A.T. (2015). Dissecting vectorial capacity for mosquito-borne viruses. *Current Opinion in Virology*, 15, 112–18.

Kramer, L.D., Styer, L.M., and Ebel, G.D. (2008). A global perspective on the epidemiology of West Nile virus. *Annual Review of Entomology*, 53, 61–81.

Kuiken, T., Holmes, E.C., McCauley, J., Rimmelzwaan, G.F., Williams, C.S., and Grenfell, B.T. (2006). Host species barriers to influenza virus infections. *Science*, 312, 394–7.

Lawson, B., de Pinna, E., Horton, R.A., et al. (2014). Epidemiological evidence that garden birds are a source of human salmonellosis in England and Wales. *PLoS ONE*, 9, e88968.

Leblebicioglu, H., Eroglu, C., Erciyas-Yavuz, K., Hokelek, M., Acici, M., and Yilmaz, H. (2014). Role of migratory birds in spreading Crimean-Congo hemorrhagic fever, Turkey. *Emerging Infectious Diseases*, 20, 1331–4.

Leighton, B.J., Zervos, S., and Webster, J.M. (2000). Ecological factors in schistosome transmission, and an environmentally benign method for controlling snails in a recreational lake with a record of schistosome dermatitis. *Parasitology International*, 49, 9–17.

Lindeborg, M., Barboutis, C., Ehrenborg, C., et al. (2012). Migratory birds, ticks, and Crimean-Congo hemorrhagic fever virus. *Emerging Infectious Diseases*, 18, 2095–7.

Lindsey, N.P., Staples, J. E., and Fischer, M. (2018). Eastern equine encephalitis virus in the United States, 2003–2016. *American Journal of Tropical Medicine and Hygiene*, 98, 1472–7.

Lipsitch, M., Barclay, W., Raman, R., et al. (2016). Viral factors in influenza pandemic risk assessment. *Elife*, 5, e18491.

Lloyd-Smith, J.O., Poss, M., and Grenfell, B.T. (2008). HIV-1/parasite co-infection and the emergence of new parasite strains. *Parasitology*, 135, 795–806.

Loss, S.R., Noden, B.H., Hamer, G.L., and Hamer, S.A. (2016). A quantitative synthesis of the role of birds in carrying ticks and tick-borne pathogens in North America. *Oecologia*, 182, 947–59.

Lowder, B.V., Guinane, C.M., Ben Zakour, N.L., et al. (2009). Recent human-to-poultry host jump, adaptation, and pandemic spread of *Staphylococcus aureus*. *Proceedings of the National Academy of Sciences of the United States of America*, 106, 19545–50.

Luechtefeld, N.A., Blaser, M.J., Reller, L.B., and Wang, W.L. (1980). Isolation of *Campylobacter fetus* subsp. *jejuni* from migratory waterfowl. *Journal of Clinical Microbiology*, 12, 406–8.

Macdonald, G. (1961). Epidemiologic models in studies of vector-borne diseases. *Public Health Reports*, 76, 753–64.

Maki, J., Guiot, A.-L., Aubert, M., et al. (2017). Oral vaccination of wildlife using a vaccinia–rabies–glycoprotein recombinant virus vaccine (RABORAL V-RG(R)): a global review. *Veterinary Research*, 48, 57.

Marr, J.S. and Calisher, C.H. (2003). Alexander the Great and West Nile virus encephalitis. *Emerging Infectious Diseases*, 9, 1599–603.

Medeiros, M.C.I., Anderson, T.K., Higashiguchi, J.M., et al. (2014). An inverse association between West Nile virus serostatus and avian malaria infection status. *Parasites & Vectors*, 7, 415.

Medeiros, M.C.I., Ricklefs, R.E., Brawn, J.D., Ruiz, M.O., Goldberg, T.L., and Hamer, G.L. (2016). Overlap in the seasonal infection patterns of avian malaria parasites and West Nile virus in vectors and hosts. *American Journal of Tropical Medicine and Hygiene*, 95, 1121–9.

Messenger, A.M., Barnes, A.N., and Gray, G.C. (2014). Reverse zoonotic disease transmission (zooanthroponosis): a systematic review of seldom-documented human biological threats to animals. *PLoS ONE*, 9, e89055.

Molaei, G., Karpathy, S.E., and Andreadis, T.G. (2019). First report of the introduction of an exotic tick, *Amblyomma coelebs* (Acari: Ixodidae), feeding on a human traveler returning to the United States from Central America. *Journal of Parasitology*, 105, 571–5.

Mollentze, N. and Streicker, D.G. (2020). Viral zoonotic risk is homogenous among taxonomic orders of mammalian and avian reservoir hosts. *Proceedings of the National Academy of Sciences of the United States of America*, 117, 9423–30.

Morales-Betoulle, M.E., Komar, N., Panella, N.A., et al. (2013). West Nile virus ecology in a tropical ecosystem in Guatemala. *American Journal of Tropical Medicine and Hygiene*, 88, 116–26.

Muehlenbein, M.P. and Ancrenaz, M. (2009). Minimizing pathogen transmission at primate ecotourism destinations: the need for input from travel medicine. *Journal of Travel Medicine*, 16, 229–32.

Nebola, M., Borilova, G., and Steinhauserova, I. (2007). Prevalence of *Campylobacter* subtypes in pheasants (*Phasianus colchicus* spp. *torquatus*) in the Czech Republic. *Veterinarni Medicina*, 52, 496–501.

Newbold, L.K., Burthe, S.J., Oliver, A.E., et al. (2017). Helminth burden and ecological factors associated with alterations in wild host gastrointestinal microbiota. *ISME Journal*, 11, 663–75.

Nguyen, C., Gray, M., Burton, T.A., et al. (2019). Evaluation of a novel West Nile virus transmission control strategy that targets *Culex tarsalis* with endectocide-containing blood meals. *PLoS Neglected Tropical Diseases*, 13, e0007210.

Noblick, J., Skolnik, R., and Hotez, P.J. (2011). Linking global HIV/AIDS treatments with national programs for the control and elimination of the neglected tropical diseases. *PLoS Neglected Tropical Diseases*, 5, e1022.

Noden, B.H., Arnold, D., and Grantham, R. (2015). First report of adult *Amblyomma longirostre* (Acari: Ixodidae) in Oklahoma. *Systematic and Applied Acarology*, 20, 468–70.

Ogden, N.H., Lindsay, L.R., Hanincová, K., et al. (2008). Role of migratory birds in introduction and range expansion of *Ixodes scapularis* ticks and of *Borrelia burgdorferi* and *Anaplasma phagocytophilum* in Canada. *Applied and Environmental Microbiology*, 74, 1780–90.

Olival, K.J., Hosseini, P.R., Zambrana-Torrelio, C., Ross, N., Bogich, T.L., and Daszak, P. (2017). Host and viral traits predict zoonotic spillover from mammals. *Nature*, 546, 646–50.

On, S., Brett, B., Horan, S., Erskine, H., Lin, S., and Cornelius, A. (2019). Isolation and genotyping of *Campylobacter* species from kiwi (*Apteryx* spp.) in captivity: implications for transmission to and from humans. *New Zealand Veterinary Journal*, 67, 134–7.

Pathak, A.K., Pelensky, C., Boag, B., and Cattadori, I.M. (2012). Immuno-epidemiology of chronic bacterial and helminth co-infections: observations from the field and evidence from the laboratory. *International Journal for Parasitology*, 42, 647–55.

Pedersen, A.B. and Fenton, A. (2007). Emphasizing the ecology in parasite community ecology. *Trends in Ecology & Evolution*, 22, 133–9.

Plowright, R.K., Parrish, C.R., McCallum, H., et al. (2017). Pathways to zoonotic spillover. *Nature Reviews Microbiology*, 15, 502–10.

Rehn, M., Ringberg, H., Runehagen, A., et al. (2013). Unusual increase of psittacosis in southern Sweden linked to wild bird exposure, January to April 2013. *Eurosurveillance*, 18, 13–20.

Rijks, J.M., Kik, M.L., Slaterus, R., et al. (2016). Widespread Usutu virus outbreak in birds in the Netherlands, 2016. *Eurosurveillance*, 21, 45.

Rouffaer, L.O., Steensels, M., Verlinden, M., et al. (2018). Usutu virus epizootic and *Plasmodium* coinfection in Eurasian blackbirds (*Turdus merula*) in Flanders, Belgium. *Journal of Wildlife Diseases*, 54, 859–62.

Sapp, S.G., Rascoe, L.N., Wilkins, P.P., et al. (2016). *Baylisascaris procyonis* roundworm seroprevalence among wildlife rehabilitators, United States and Canada, 2012–2015. *Emerging Infectious Diseases*, 22, 2128–31.

Schwarz, A., Gauly, M., Abel, H., et al. (2011). Immunopathogenesis of *Ascaridia galli* infection in layer chicken. *Developmental & Comparative Immunology*, 35, 774–84.

Şekercioğlu, Ç.H. (2013). Guineafowl, ticks and Crimean-Congo hemorrhagic fever in Turkey: the perfect storm? *Trends in Parasitology*, 29, 1–2.

Shafir, S.C., Fuller, T., Smith, T.B., and Rimoin, A.W. (2012). A national study of individuals who handle migratory birds for evidence of avian and swine-origin influenza virus infections. *Journal of Clinical Virology*, 54, 364–7.

Smith, O.M., Snyder, W.E., and Owen, J.P. (2020). Are we overestimating risk of enteric pathogen spillover from wild birds to humans? *Biological Reviews*, 95, 652–79.

Swetnam, D., Widen, S.G., Wood, T.G., et al. (2018). Terrestrial bird migration and West Nile virus circulation, United States. *Emerging Infectious Diseases*, 24, 2184–94.

Taubenberger, J.K. and Morens, D.M. (2006). 1918 influenza: the mother of all pandemics. *Emerging Infectious Diseases*, 12, 15–22.

Taylor, L.H., Latham, S.M., and Woolhouse, M.E. (2001). Risk factors for human disease emergence. *Philosophical*

Transactions of the Royal Society B: Biological Sciences, 356, 983–9.

Telfer, B.L., Moberley, S.A., Hort, K.P., et al. (2005). Probable psittacosis outbreak linked to wild birds. *Emerging Infectious Diseases*, 11, 391–7.

Tompkins, D.M., Dunn, A.M., Smith, M.J., and Telfer, S. (2011). Wildlife diseases: from individuals to ecosystems. *Journal of Animal Ecology*, 80, 19–38.

Turell, M., Rossignol, P., Spielman, A., Rossi, C., and Bailey, C. (1984). Enhanced arboviral transmission by mosquitoes that concurrently ingested microfilariae. *Science*, 225, 1039–41.

USDA (2014). *Biosecurity Guide for Poultry and Bird Owners.* Program Aid No. 1885. Animal and Plant Health Inspection Service, United States Department of Agriculture, Washington, DC.

Vanrompay, D., Harkinezhad, T., Van de Walle, M., et al. (2007). *Chlamydophila psittaci* transmission from pet birds to humans. *Emerging Infectious Diseases*, 13, 1108–10.

Vasilakis, N. and Gubler, D.J. (2016). *Arboviruses: Molecular Biology, Evolution and Control.* Caister Academic Press, Wymondham, Norfolk.

Vaughan, J.A. and Turell, M.J. (1996a). Dual host infections: enhanced infectivity of eastern equine encephalitis virus to *Aedes* mosquitoes mediated by *Brugia* microfilariae. *American Journal of Tropical Medicine and Hygiene*, 54, 105–9.

Vaughan, J.A. and Turell, M.J. (1996b). Facilitation of Rift Valley fever virus transmission by *Plasmodium berghei* sporozoites in *Anopheles stephensi* mosquitoes. *American Journal of Tropical Medicine and Hygiene*, 55, 407–9.

Vaziri, G.J., Muñoz, S.A., Martinsen, E.S., and Adelman, J.S. (2019). Gut parasite levels are associated with severity of response to immune challenge in a wild songbird. *Journal of Wildlife Diseases*, 55, 64–73.

Vollmer, S.A., Bormane, A., Dinnis, R.E., et al. (2011). Host migration impacts on the phylogeography of Lyme borreliosis spirochaete species in Europe. *Environmental Microbiology*, 13, 184–92.

Walter, K.S., Carpi, G., Caccone, A., and Diuk-Wasser, M.A. (2017). Genomic insights into the ancient spread of Lyme disease across North America. *Nature Ecology & Evolution*, 1, 1569–76.

Wikelski, M., Foufopoulos, J., Vargas, H., and Snell, H. (2004). Galápagos birds and diseases: invasive pathogens as threats for island species. *Ecology and Society*, 9(1), 5.

Yung, A.P. and Grayson, M.L. (1988). Psittacosis—a review of 135 cases. *Medical Journal of Australia*, 148, 228–33.

Zhou, H., Chen, X., Hu, T., et al. (2020). A novel bat coronavirus closely related to SARS-CoV-2 contains natural insertions at the S1/S2 cleavage site of the spike protein. *Current Biology*, 30, 2196–203.

CHAPTER 13

A Flight Path Forward for Avian Infectious Disease Ecology

Dana M. Hawley, Kathryn P. Huyvaert, and Jennifer C. Owen

13.1 Introduction

The field of avian infectious disease ecology is at a key precipice, poised for exciting new 'flight paths' in the coming decades. The chapters throughout this book illustrate the explosion of interest and work in the field in recent years, with much more to come. This rapid growth reflects, at least to some degree, the specialness of birds as a study taxon for understanding infectious disease ecology more broadly (Clayton and Moore 1997). Birds are arguably unparalleled in their combination of charisma and human interest, scientific tractability, and economic and ecological importance (Cox and Gaston 2018; Whelan et al. 2015). While most of the ecosystem services that birds provide (e.g., pollination, seed dispersal) are considered positive in nature, birds also serve as important sources of zoonotic and agriculturally important pathogens (Whelan et al. 2015). Further, birds know no geographic boundaries; their parasites and pathogens, whether zoonotic or not, connect continents and hemispheres (Newman et al. 2009; Ogden et al. 2008; Reed et al. 2003; Weber and Stilianakis 2007). Studying and managing infectious diseases in birds is thus an unprecedented exercise in crossing scales, requiring a flight path that goes from understanding processes acting within individual hosts up to factors influencing movements across ecosystems and around the globe (Figure 13.1).

Bird populations may also, in and of themselves, be at a key precipice. Recent analyses of long-term monitoring data indicate that bird populations in some parts of the world are declining precipitously (Rosenberg et al. 2019), potentially for reasons relevant to questions addressed in this book. What is the role of parasites and pathogens in avian host fitness and population declines (Chapters 6 and 7)? How might bird populations evolve to persist in the face of pathogens—whether using immunity or behavior—and how might avian evolution lead to reciprocal evolution of parasites and pathogens (Chapters 3–5)? How do the changing biological communities in which birds reside alter avian infectious disease dynamics (Chapter 8)? Will land use and climate change exacerbate negative impacts of parasites and pathogens on bird populations and spill over into agricultural animals and humans (Chapters 9–12)?

The chapters in this book have begun to address these questions and many others, but, given the complexity inherent in infectious disease ecology (Chapter 2), studies to date have often generated more questions than answers. Here, we briefly lay out a flight path forward for avian infectious disease ecology—one that bridges silos and crosses scales—and one that innovates while also leveraging historical strengths of birds as a focal study group.

13.2 A flight path for avian disease ecology that crosses scales and disciplines

'Not so long ago, the chasm between the study of disease and the study of ecology was so great that no one seemed able to cross it' **(Moore 2012).**

Dana M. Hawley, Kathryn P. Huyvaert, and Jennifer C. Owen, *A Flight Path Forward for Avian Infectious Disease Ecology* In: *Infectious Disease Ecology of Wild Birds*. Edited by: Jennifer C. Owen, Dana M. Hawley, and Kathryn P. Huyvaert, Oxford University Press. © Oxford University Press 2021.
DOI: 10.1093/oso/9780198746249.003.0013

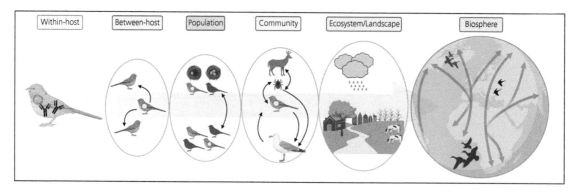

Figure 13.1 Effectively studying infectious disease ecology in birds requires understanding interactions and processes that act, and in some cases emerge, across biological levels of organization. These include (from left to right) host–pathogen interactions within individual hosts; between-host transmission; population-level dynamics of hosts and pathogens; interactions among species of host, pathogen, or vector within complex biological communities; ecosystem or landscape-level processes that link biological communities to the abiotic environment; and biosphere-scale processes that link across continents or hemispheres. (Created with BioRender.com).

One of the key challenges in infectious disease ecology, regardless of study taxon, is linking across scales. Crossing scales in infectious disease ecology can take many different forms and can often be critical for identifying emergent patterns and insights (e.g., Clay et al. 2019). One key type of scale-crossing in infectious disease ecology lies in understanding how within-host infection dynamics (Chapter 2), driven by a host's immune system, pathogen characteristics, and ecological context, scale up to influence the between-host processes core to a pathogen's fitness (e.g., Handel and Rohani 2015). Crossing scales in infectious disease ecology can also encompass the ability to bridge to higher levels of biological organization, building from the factors that drive individual host susceptibility (e.g., Becker et al. 2019), up to the community context in which host–pathogen interactions occur (e.g., Johnson et al. 2015). In many cases, the integration of diverse approaches and disciplines (Martínez-Padilla et al. 2019) is necessary to effectively cross biological levels of organization and glean the deepest understanding of the system. Thus, the strongest inferences in infectious disease ecology often come from multipronged, integrative research strategies that merge methods and fields such as molecular or phenotypic studies of host and or pathogen, experimental infections or controlled transmission experiments (where possible), field-based studies that are experimental or longitudinal in nature, and mathematical models of dynamic between-host

processes (e.g., red grouse and their nematode parasites: Martínez-Padilla et al. 2019; Hawaiian honeycreepers and avian blood parasites: McClure et al. 2020).

Crossing scales in infectious disease ecology can also mean expanding and bridging spatial or geographic units of analysis, which is particularly critical in birds due to their seemingly unconstrained geographic boundaries (e.g., Reed et al. 2003). Studies that bridge from small forest patches or backyard gardens up to continents or even hemispheres can help elucidate global dynamics and emergent spatial properties of infectious disease. For example, localized effects of temperature or pollution within small habitat patches, such as the impacts of artificial light at night on the within-host dynamics of infection relapse (e.g., Becker et al. 2020), are likely to interact with both seasonal physiology and migratory movements (e.g., Gylfe et al. 2000) to influence pathogen emergence and avian distributions on global scales (see also Chapter 10). Further, supplemental feeding of wild birds in individual backyards can alter pathogen persistence and bird densities (see Chapter 9) in ways that ultimately scale up to impact population-level declines in some avian taxa (e.g., Lawson et al. 2018; see Chapter 7).

Crossing temporal scales is also critical for infectious disease ecology and sometimes goes hand in hand with bridging spatial scales, particularly for migratory birds that spend parts of their annual cycle in distinct hemispheres. Work outside of an infectious disease context has illustrated the importance of making connections across the full annual cycle in birds, whereby non-lethal events occurring during overwintering, for example, can carry over to influence individual fitness on the breeding grounds (e.g., O'Connor et al. 2014). While carry-over effects specifically caused by parasites are not well studied in birds, evidence that parasite status influences migratory stopover behavior (Garvin et al. 2006; Hegemann et al. 2018) points to the potential for the fitness effects of parasite infection to bridge distinct components of the annual cycle in birds. In addition to carry-over effects with respect to host fitness, migratory birds can cross temporal and spatial scales simultaneously when they literally 'carry' and thus disperse parasites over time and space via their annual cycles (Hasle 2013; Liu et al. 2005). For example, migratory birds were likely involved in the emergence and spread of one highly pathogenic avian influenza virus (HPAIV) belonging to the H5N8 group in Europe and North America in 2014 (Lee et al. 2015). Furthermore, evidence of *Toxoplasma gondii* infection in Arctic fox in northern systems implicates migrating geese in the movement of the parasite from temperate wintering grounds to Arctic nesting areas, where felid definitive hosts are absent or rare (Prestrud et al. 2007). Thus, studies that cross spatial and temporal scales can reveal new hosts and new transmission routes for *T. gondii* (Elmore et al. 2012) that would not have been fully realized if the role of birds had been ignored.

In addition to multiscale approaches, diverse disciplinary perspectives are critical for making effective progress in infectious disease ecology (Scheiner and Rosenthal 2006). Parasites and pathogens do not follow traditional disciplinary boundaries (Borer et al. 2011), nor do the birds that carry them. Host–pathogen interactions are inherently interdisciplinary and integrative, often requiring the merging of fields like genomics and physiology with mathematics to understand parasite consequences for hosts at population scales (e.g., Martínez-Padilla et al. 2019). Further, the role of wild birds and agricultural birds in the maintenance and spread of pathogens of human health concern and the interconnectedness between anthropogenic activities (e.g., land use, climate change, supplemental feed-

ing) and avian health together underscore the need to approach avian infectious disease ecology from a One Health perspective (Chapters 9–12). The growing field of One Health, which recognizes this interconnectedness of the health of humans, wild and domestic animals, and the environment, is inherently interdisciplinary (Figure 13.2). Yet, despite significant growth in copublication between scientists in different disciplines working on shared One Health questions, the majority of One Health research still occurs largely in separate silos (Manlove et al. 2016). Further, any measurable progress to date has largely been interdisciplinary, bringing together veterinarians and ecologists who both study infectious diseases of animals from different perspectives, rather than transdisciplinary—an approach that goes a step further by truly integrating collective knowledge (Manlove et al. 2016).

Given the unique human interest in birds and the economic and ecological interdependence of humans and birds (e.g., Whelan et al. 2015), avian infectious disease ecology is uniquely suited to cross traditional boundaries of expertise and perhaps to generate effective transdisciplinary research studies (e.g., Parkes et al. 2005). The strongest innovations in the field of avian infectious disease ecology will involve work that bridges and transcends disciplines

Figure 13.2 One Health Venn diagram illustrating the interconnectedness of avian health with that of humans and the environment. (Created with BioRender.com).

as diverse as parasitology, statistics, anthropology, sociology, microbiology, human psychology and behavioral sciences, computational sciences, economics, veterinary medicine, public health, biosecurity, geography, mathematics, behavioral ecology, evolutionary biology, and ornithology. Thus, the flight path forward for a One Health approach to avian infectious disease ecology requires a convergence of all of these paths and ways of knowing.

13.3 Leveraging existing strengths to promote innovation

A flight path forward for avian infectious disease ecology should leverage existing strengths of the field to move research and conservation in innovative directions. Work in avian infectious disease ecology to date is notable in several key ways. First, the strong ornithological history of banding (ringing) and tracking birds has led to numerous longitudinal studies, some of which have been leveraged to examine ecological factors that influence the fitness consequences of endemic infections (e.g., Lachish et al. 2011) or the degree of effects of outbreaks on populations where longitudinal data collection began prior to pathogen emergence, such as avian poxvirus in juvenile waved albatross (*Phoebastria irrorata*) (Tompkins et al. 2017) and great tits (*Parus major*) (Lachish et al. 2012). The ease of recapturing many birds (such as nest-box breeding birds; e.g., family Paridae) has also facilitated the use of experimental manipulations in longitudinal studies to causally elucidate the effects of pathogens on these hosts. Such approaches have been critical for revealing some of the subtle, non-lethal fitness effects that chronic infection with many parasites and pathogens have on their hosts that would otherwise remain undetected (e.g., Knowles et al. 2009). Finally, many avian taxa are also amenable to experimental infection and/or transmission studies that allow direct measurement of within-host and between-host processes, which is key to crossing scales. For example, such studies in house finches have allowed researchers to identify characteristics of the bacterial pathogen *Mycoplasma gallisepticum* that are associated with distinct host immune responses, transmission dynamics, and evolutionary trajectories (e.g.,

Bonneaud et al. 2018; Fleming-Davies et al. 2018; Williams et al. 2014).

One of the most unique historical strengths of avian infectious disease ecology derives from the incredible fascination and commitment that humans have with respect to wild birds (Jones 2018). Researchers in avian infectious disease ecology have leveraged this human fascination by engaging vast numbers of volunteer birdwatchers to act as community scientists, collecting key data (often in their own backyards) that documented the spatial spread of novel avian pathogens (Dhondt et al. 2005) and the degree to which emerging pathogens affected avian population dynamics (Hochachka and Dhondt 2000; LaDeau et al. 2007; Lawson et al. 2012). Although each community scientist collects data at a small spatial scale, the vast number of birdwatchers willing to engage in these studies generates datasets with spatial scales that are unmatched by those produced by more traditional research approaches. Given the spatial scales across which birds move and interact with their environment (Figure 13.1), the flight path for avian infectious disease ecology should continue to leverage the enthusiasm and commitment of community scientists for large-scale data collection, working in transdisciplinary contexts with computational and social scientists to take the critical questions and emerging insights in innovative directions.

The flight path forward for avian infectious disease ecology also relies on our continued leveraging of concepts and quantitative tools that have strong roots in population and community ecology (see Chapter 7). Surveillance and monitoring efforts that account for heterogeneity in detection by using, for example, occupancy or multistate mark-recapture models, offer such a set of tools for 'keeping an eye out' for effects of pathogens on birds. A key component of disease surveillance in wildlife is the initial and subsequent detection of disease or the pathogens/parasites that cause it, including having 'early warning systems'—sentinels—to indicate that an outbreak is starting in the population (Halliday et al. 2007; Ryser-Degiorgis 2013). Birds have long served as important wildlife sentinels for human health risks due to their physiological sensitivity to environmental changes ('canaries in a coal mine;' Rabinowitz et al. 2010). Further, some groups of birds (e.g., aquatic, pelagic birds) make especially good sentinels because the spatial scale of their movements across the landscape mean that they provide a broad sample of the ecosystem and can alert us to the presence of a pathogen (Newman et al. 2007) well before a focused study might reveal an issue. At the same time, these characteristics that make birds ideal sentinels of pathogen presence or spread (i.e., physiological sensitivity and exposure to broad portions of the landscape) also mean that bird populations are particularly vulnerable to global anthropogenic changes (Rosenberg et al. 2019). Thus, the responses of bird populations to environmental stressors can serve as an important early warning system for pathogens, as it did for pesticide use in the 1960s (Carson 2002, 1962). Overall, these diverse characteristics and strengths of birds as a study taxon, and of ornithology as a field, must continue to be leveraged along the flight path forward for the study of infectious diseases in birds.

13.4 A flight path forward for avian infectious disease ecology, management, and conservation

'... we can claim that ornithology has entered the era of big data' **(López-López 2016).**

A flight path for avian infectious disease ecology most relevant for management and conservation requires both forward-looking and innovative approaches. Effective monitoring and surveillance for pathogen threats, as well as a basic understanding of host–pathogen interactions and their consequences for avian ecology and evolution, requires the ability to accurately detect parasites and pathogens and the communities of microbes within the host or vector with which they interact. While detection methods for specific avian parasites and pathogens are increasingly sensitive and cost-effective (Table 2.1), these traditional methods are designed to detect parasite and pathogen taxa that are actively under study and for which assays already exist. In order to detect novel pathogens as well as emergent properties of co-infections, studies of avian infectious disease must harness novel

technologies to characterize microbial communities in an unbiased way (e.g., next-generation sequencing and -omics; Table 2.1; Zylberberg 2019). For example, the causative agent of avian keratin disorder, an emerging disease of Alaskan birds characterized by debilitating beak deformities, was identified as a novel picornavirus using non-targeted metagenomic sequencing approaches (Zylberberg et al. 2016). In addition to assays that allow for discovery of novel pathogens in the absence of a suspected agent, sampling and monitoring methods that are themselves not dependent on live captures of birds, which can be logistically challenging and biased with respect to infection status (see Chapter 7), can help to improve detection of pathogens that affect wild bird populations. Thus, the flight path forward for avian infectious disease ecology should also include innovative approaches to detect pathogens in the absence of infected hosts (e.g., clams as biological detectors of avian influenza viruses in water, e.g., Huyvaert et al. 2012).

Effective monitoring and surveillance for infectious disease ecology also requires innovative and forward-thinking ways to keep an eye on bird populations. Community science efforts are a key tool in the avian disease surveillance toolbox because they provide a cost-effective way of keeping many eyes out for signs of infection or disease in bird populations over the long term (Lawson et al. 2015). These efforts can be targeted to particular clinical signs such as swollen eyes associated with mycoplasmal conjunctivitis in birds visiting feeders (Hochachka and Dhondt 2000), and these reports can then be used to indicate an issue of potential conservation concern in a wild bird population. In other cases, human reporting of dead bird carcasses can help to sound the alarm on avian infectious diseases with potential population-level or even human health effects. For example, retrospective analyses of dead bird reports showed that community-based data collection was useful as an early warning for the presence of West Nile virus in wild birds (i.e., crows in Chicago, IL) two weeks in advance of its emergence in humans (e.g., Watson et al. 2004). Finally, even in the absence of known or visible disease threats such as clinical signs or carcasses, community science efforts provide important baseline population monitoring that can be used to identify long-term trends for particular taxa or guilds and ultimately to understand population-level effects of diseases that emerge in particular avian hosts that are already actively being monitored (LaDeau et al. 2007).

Better surveillance—reaching broader spatial and longer temporal scales—will come from leveraging the strengths of community science as well as the strengths of the science community. Broadening participation and incentives for community science activities such as backyard bird monitoring programs, where participants are currently unrepresentative of the available human community (e.g., Martin and Greig 2019), can help engage a broader proportion of birdwatchers in avian conservation and simultaneously extend the spatial scale of surveillance to reach more diverse environments such as strictly urban areas. A particularly strong component of community-based science is the genuine spirit of collaboration between researchers and the community whose energy and passion for birds are being harnessed (Bonney et al. 2016). In some cases, community scientists that are observing birds are also actively engaging in on-the-ground management to reduce impacts of disease: for example, a survey of participants in the Cornell Lab of Ornithology's backyard monitoring program Project FeederWatch found that 70% of respondents report that they have or would take actions (typically cleaning their feeders) in response to observing birds with disease at their feeders (Dayer et al. 2019). Nonetheless, only 6% of respondents indicated that their feeding is likely to contribute to birds experiencing more disease, suggesting a potential disconnect between the goals and outcomes of some public interactions with wild birds (Cox and Gaston 2018; Dayer et al. 2019). Effective management and conservation of backyard birds in the face of disease should be done in direct collaboration with engaged public stakeholders such as bird feeders and bird watchers.

Similarly, the future of avian infectious disease ecology and management requires a substantial broadening of engagement and perspectives, with respect to both individual backgrounds and institutions. Much like the umbrella disciplines of ecology and evolutionary biology, avian infectious disease ecology must address the paucity of racial and ethnic diversity in the field by working to broaden participation and inclusivity of Black, Indigenous, and Latinx individuals in particular (Tseng et al. 2020). Diverse representation is also needed with respect to institutional roles and boundaries, particularly for managing and conserving birds at the interfaces between human-dominated landscapes and wildlands (see Chapters 11 and 12). Researchers and managers based at diverse types of organizations, including universities, local and regional governmental organizations, and non-governmental organizations, address diverse aspects of infectious disease in avian species around the globe. Effective collaborations within and across these types of organizations require calling on agencies and researchers to step out of their persistent silos (e.g., Manlove et al. 2016) and to better leverage their disciplinary expertise to bridge the 'chasm' between disciplines and to solve common problems. Further, given the ability of birds and their pathogens to cross countries and continents, effective management and conservation of birds in the face of disease will, in many cases, require effective international collaboration and engagement. Although not focused on disease threats in particular, the North American Bird Conservation Initiative serves as one example of an integrated and visionary approach for bird conservation that crosses institutional, stakeholder, and political boundaries (NABCI 2013). The field of avian infectious disease ecology should better leverage the geographic reach of international bodies dedicated to the health of wild and domestic birds, including, for example, the Food and Agriculture Organization of the United Nations (FAO), the World Organization for Animal Health (OIE), and the Wildlife Conservation Society (WCS).

Finally, effective surveillance that crosses spatial and temporal scales requires methodological innovations for tracking bird populations over space and time. The last few decades have brought a technological revolution in our ability to both process and collect large-scale scientific data on birds (e.g., López-López 2016). Many of these innovations have been computational in nature and have, for example, allowed for better validation (Bonter and Cooper 2012) and analysis of large-scale, community science datasets (Kelling et al. 2009) or more streamlined management of large-scale datasets on avian movements (reviewed in López-López 2016). However, technological innovations are also needed to improve our ability to collect particular kinds of large-scale data, such as movements of individual birds within and across hemispheres and parts of their annual cycles. The ability to track individual bird movements and, in some cases, aspects of their physiology relevant for infection status (Adelman et al. 2014) allows studies to link within- and between-host processes, facilitating the study of emergent spatial and temporal dynamics such as those associated with migration and carry-over effects (reviewed in López-López 2016). Currently, technologies for effectively tracking individual movements or physiology either are limited for use only in large-sized birds (López-López 2016) or require physical recapture (e.g., geolocators) or existing automated telemetry towers (e.g., Cooper and Marra 2020), which limits sample size and their utility for infectious disease studies. Future technical innovations that allow miniaturization of tracking devices and further integration of remote biomonitoring of physiological information (Adelman et al. 2014) will promote data collection relevant for avian infectious disease ecology at unprecedented spatial and temporal resolutions. Further, innovations such as artificial intelligence (AI) may allow tracking of individuals that are not individually marked (Ferreira et al. 2019); and perhaps could even be used to non-invasively quantify the presence of clinical signs of disease that are present in some systems (e.g., pox lesions, conjunctivitis, beak deformities). Notably, both the design and implementation of any new technologies to address avian disease management and conservation will require bridging disciplinary boundaries effectively and innovatively (Pimm et al. 2015)

13.5 Conclusions

'It's not just what they [birds] do for the environment, it's what they do for our souls'   **(Franzen 2018).**

The connections that people have with birds range from the spiritual and psychological, whereby backyard birds have been linked with stress relief and a sense of connection to nature (Cox and Gaston 2016; Cox et al. 2017), to the intimate biological connections between human and avian health highlighted in several chapters of this book (Chapters 9–12). A flight path forward for avian infectious disease ecology needs to recognize and embrace all of these connections. Management of avian parasites and pathogens should also consider the inherent ecological and conservation value of parasites and pathogens themselves (Gómez and Nichols 2013; Wood and Johnson 2015). Although often overlooked, parasites, much like birds, serve important roles in communities and ecosystems, such that, in some cases, the presence and diversity of parasites with complex life cycles can be used as an indicator of bird diversity (Hechinger and Lafferty 2005) and healthy, intact ecosystems (Hudson et al. 2006). Thus, a flight path forward for avian conservation and management in the face of disease should also include a path forward for the conservation of parasite biodiversity (Carlson et al. 2020).

While parasites of birds can be effective indicators of local ecosystem health, the spatial and temporal movements of birds themselves and the vast numbers of people watching them recreationally (Cox and Gaston 2018) make them particularly unique indicators of ecosystems around the globe (Figure 13.3). Avian sentinels like the endemic Hawaiian honeycreepers are often our first indication

Figure 13.3 Birds like this waved albatross (*Phoebastria irrorata*) have flight paths that readily cross broad spatial and temporal scales, allowing birds to serve as important ecosystem indicators. The flight path forward for avian infectious disease ecology should similarly know no boundaries. (Photo by K.P. Huyvaert.)

of the irreversible damage that introduced infectious diseases can have for Earth's unique biodiversity. Further, the multiple interfaces across which human and bird health are intimately connected (Chapter 12) mean that birds can alert us to impending threats of emerging or reemerging pathogens of humans. But birds are far more than good indicators of broader environmental problems or the presence of infectious diseases of relevance for avian conservation or human health. Birds are fascinating in and of themselves for their diversity, their beauty, and their sophisticated biology, all of which allow them to navigate the globe and withstand some of the harshest environments on Earth. And most importantly, birds are accessible to each and every one of us; no matter where one is on the planet, one can witness the marvel of birds. As scientists, the unique biology and accessibility of birds is one of many reasons that birds provide us with the inspiration to make innovative and broad insights regarding the ecology and evolution of infectious diseases. These insights cross scales from host–pathogen interactions occurring within individual birds (Chapters 2 and 3) up to movements of pathogens among individuals and through populations (Chapters 4–7), communities, and landscapes (Chapters 8 and 9) and across hemispheres (Chapters 10–12). With effective community engagement, transdisciplinary collaboration, and technological innovation, the flight path forward for avian infectious disease ecology can, just like birds themselves, know no boundaries.

Literature cited

Adelman, J.S., Moyers, S.C., and Hawley, D.M. (2014). Using remote biomonitoring to understand heterogeneity in immune-responses and disease-dynamics in small, free-living animals. *Integrative and Comparative Biology*, 54, 377–86.

Becker, D.J., Downs, C.J., and Martin, L.B. (2019). Multi-scale drivers of immunological variation and consequences for infectious disease dynamics. *Integrative and Comparative Biology*, 59, 1129–37.

Becker, D.J., Singh, D., Pan, Q., et al. (2020). Artificial light at night amplifies seasonal relapse of haemosporidian parasites in a widespread songbird. *Proceedings of the Royal Society of London B*, 287, 20201831.

Bonneaud, C., Giraudeau, M., Tardy, L., Staley, M., Hill, G.E., and McGraw, K.J. (2018). Rapid antagonistic coevolution in an emerging pathogen and its vertebrate host. *Current Biology*, 28, 2978–83.

Bonney, R., Phillips, T.B., Ballard, H.L., and Enck, J.W. (2016). Can citizen science enhance public understanding of science? *Public Understanding of Science*, 25, 2–16.

Bonter, D.N. and Cooper, C.B. (2012). Data validation in citizen science: a case study from Project FeederWatch. *Frontiers in Ecology and the Environment*, 10, 305–7.

Borer, E.T., Antonovics, J., Kinkel, L.L., et al. (2011). Bridging taxonomic and disciplinary divides in infectious disease. *Ecohealth*, 8, 261–7.

Carlson, C.J., Hopkins, S., Bell, K.C., et al. (2020). A global parasite conservation plan. *Biological Conservation*, 250(108596).

Carson, R. (2002, 1962). *Silent Spring*. Houghton Mifflin Harcourt, Boston.

Clay, P.A., Cortez, M.H., Duffy, M.A., and Rudolf, V.H. (2019). Priority effects within coinfected hosts can drive unexpected population-scale patterns of parasite prevalence. *Oikos*, 128, 571–83.

Clayton, D.H. and Moore, J. (1997). *Host–Parasite Evolution: General Principles and Avian Models*. Oxford University Press, Oxford.

Cooper, N.W. and Marra, P.P. (2020). Hidden long-distance movements by a migratory bird. *Current Biology*, 30(20), 4056–62.

Cox, D.T. and Gaston, K.J. (2016). Urban bird feeding: connecting people with nature. *PLoS ONE*, 11, e0158717.

Cox, D.T. and Gaston, K.J. (2018). Human–nature interactions and the consequences and drivers of provisioning wildlife. *Philosophical Transactions of the Royal Society B: Biological Sciences*, 373, 20170092.

Cox, D.T., Shanahan, D.F., Hudson, H.L., et al. (2017). Doses of neighborhood nature: the benefits for mental health of living with nature. *Bioscience*, 67, 147–55.

Dayer, A.A., Rosenblatt, C., Bonter, D.N., et al. (2019). Observations at backyard bird feeders influence the emotions and actions of people that feed birds. *People and Nature*, 1, 138–51.

Dhondt, A.A., Altizer, S., Cooch, E.G., et al. (2005). Dynamics of a novel pathogen in an avian host: mycoplasmal conjunctivitis in house finches. *Acta Tropica*, 94, 77–93.

Elmore, S.A., Jenkins, E.J., Huyvaert, K.P., Polley, L., Root, J.J., and Moore, C.G. (2012). *Toxoplasma gondii* in circumpolar people and wildlife. *Vector-Borne and Zoonotic Diseases*, 12, 1–9.

Ferreira, A.C., Silva, L.R., Renna, F., et al. (2019). Deep learning-based methods for individual recognition in small birds. *bioRxiv*, 862557. https://www.biorxiv.org/content/10.1101/862557v1.

Fleming-Davies, A.E., Williams, P.D., Dhondt, A.A., et al. (2018). Incomplete host immunity favors the evolution of virulence in an emergent pathogen. *Science*, 359, 1030–3.

Franzen, J. (2018). Why birds matter. *The National Geographic Magazine,* January.

Garvin, M.C., Szell, C.C., and Moore, F.R. (2006). Blood parasites of nearctic–neotropical migrant passerine birds during spring trans-gulf migration: impact on host body condition. *Journal of Parasitology,* 92, 990–6.

Gómez, A. and Nichols, E. (2013). Neglected wild life: parasitic biodiversity as a conservation target. *International Journal for Parasitology: Parasites and Wildlife,* 2, 222–7.

Gylfe, A., Bergstrom, S., Lundstrom, J., and Olsen, B. (2000). Reactivation of Borrelia infection in birds. *Nature,* 403, 724–25.

Halliday, J.E., Meredith, A.L., Knobel, D.L., Shaw, D.J., Bronsvoort, B.M.d.C., and Cleaveland, S. (2007). A framework for evaluating animals as sentinels for infectious disease surveillance. *Journal of the Royal Society Interface,* 4, 973–84.

Handel A, and Rohani P. (2015). Crossing the scale from within-host infection dynamics to between-host transmission fitness: a discussion of current assumptions and knowledge. *Philosophical Transactions of the Royal Society of London B,* 370, 20140302.

Hasle, G. (2013). Transport of ixodid ticks and tick-borne pathogens by migratory birds. *Frontiers in Cellular and Infection Microbiology,* 3, 48.

Hechinger, R.F. and Lafferty, K.D. (2005). Host diversity begets parasite diversity: bird final hosts and trematodes in snail intermediate hosts. *Proceedings of the Royal Society B: Biological Sciences,* 272, 1059–66.

Hegemann, A., Abril, P.A., Muheim, R., et al. (2018). Immune function and blood parasite infections impact stopover ecology in passerine birds. *Oecologia,* 188, 1011–24.

Hochachka, W.M. and Dhondt, A.A. (2000). Density-dependent decline of host abundance resulting from a new infectious disease. *Proceedings of the National Academy of Sciences of the United States of America,* 97, 5303–6.

Hudson, P.J., Dobson, A.P., and Lafferty, K.D. (2006). Is a healthy ecosystem one that is rich in parasites? *Trends in Ecology & Evolution,* 21, 381–5.

Huyvaert, K.P., Carlson, J.S., Bentler, K.T., Cobble, K.R., Nolte, D.L., and Franklin, A.B. (2012). Freshwater clams as bioconcentrators of avian influenza virus in water. *Vector-Borne and Zoonotic Diseases,* 12, 904–6.

Johnson, P.T., De Roode, J.C., and Fenton, A. (2015). Why infectious disease research needs community ecology. *Science,* 349(6252), 1259504.

Jones, D. (2018). *The Birds at my Table: Why we Feed Wild Birds and Why it Matters.* Cornell University Press, Ithaca, NY.

Kelling, S., Hochachka, W.M., Fink, D., et al. (2009). Data-intensive science: a new paradigm for biodiversity studies. *Bioscience,* 59, 613–20.

Knowles, S.C.L., Nakagawa, S., and Sheldon, B.C. (2009). Elevated reproductive effort increases blood parasitaemia and decreases immune function in birds: a meta-regression approach. *Functional Ecology,* 23, 405–15.

Lachish, S., Knowles, S.C.L., Alves, R., Wood, M.J., and Sheldon, B.C. (2011). Fitness effects of endemic malaria infections in a wild bird population: the importance of ecological structure: Fitness effects of endemic malaria infections in a wild bird population. *Journal of Animal Ecology,* 80, 1196–206.

Lachish, S., Lawson, B., Cunningham, A.A., and Sheldon, B.C. (2012). Epidemiology of the emergent disease paridae pox in an intensively studied wild bird population. *PLoS ONE,* 7, e38316.

LaDeau, S.L., Kilpatrick, A.M., and Marra, P.P. (2007). West Nile virus emergence and large-scale declines of North American bird populations. *Nature,* 447, 710–13.

Lawson, B., Robinson, R.A., Colvile, K.M., et al. (2012). The emergence and spread of finch trichomonosis in the British Isles. *Philosophical Transactions of the Royal Society B: Biological Sciences,* 367, 2852–63.

Lawson, B., Petrovan, S.O., and Cunningham, A.A. (2015). Citizen science and wildlife disease surveillance. *Ecohealth,* 12, 693–702.

Lawson, B., Robinson, R.A., Toms, M.P., Risely, K., MacDonald, S., and Cunningham, A.A. (2018). Health hazards to wild birds and risk factors associated with anthropogenic food provisioning. *Philosophical Transactions of the Royal Society B: Biological Sciences,* 373, 20170091.

Lee, D.-H., Torchetti, M.K., Winker, K., Ip, H.S., Song, C.-S., and Swayne, D.E. (2015). Intercontinental spread of Asian-origin H5N8 to North America through Beringia by migratory birds. *Journal of Virology,* 89, 6521–4.

Liu, J., Xiao, H., Lei, F., et al. (2005). Highly pathogenic H5N1 influenza virus infection in migratory birds. *Science,* 309, 1206.

López-López, P. (2016). Individual-based tracking systems in ornithology: welcome to the era of big data. *Ardeola,* 63, 103–36.

Manlove, K.R., Walker, J.G., Craft, M.E., et al. (2016). 'One Health' or three? Publication silos among the One Health disciplines. *PLoS Biology,* 14, e1002448.

Martin, V.Y. and Greig, E.I. (2019). Young adults' motivations to feed wild birds and influences on their potential participation in citizen science: an exploratory study. *Biological Conservation,* 235, 295–307.

Martínez-Padilla, J., Wenzel, F.M., and Lorenzo Perez-Rodriguez, S. (2019). Parasite-mediated selection in red grouse—consequences for population dynamics and mate choice. In K. Wilson, A. Fenton, and D. Tompkins, eds. *Wildlife Disease Ecology: Linking Theory to Data and Application,* Chapter 10, pp. 296–320. Cambridge University Press, Cambridge.

McClure, K.M., Fleischer, R.C., and Kilpatrick, A.M. (2020). The role of native and introduced birds in transmission of avian malaria in Hawaii. *Ecology*, 101(7), e03038.

Moore, J. (2012). A history of parasites and hosts, science and fashion. In D.P. Hughes, J. Brodeur, and F. Thomas, eds. *Host Manipulation by Parasites*. Oxford University Press, Oxford.

NABCI (2013). *North American Bird Conservation Initiative in the United States: A Vision of American Bird Conservation*. Division of Bird Habitat Conservation, US Fish and Wildlife Service, Department of the Interior, Arlington, VA.

Newman, S.H., Chmura, A., Converse, K., et al. (2007). Aquatic bird disease and mortality as an indicator of changing ecosystem health. *Marine Ecology Progress Series*, 352, 299.

Newman, S.H., Iverson, S.A., Takekawa, J.Y., et al. (2009). Migration of whooper swans and outbreaks of highly pathogenic avian influenza H5N1 virus in eastern Asia. *PLoS ONE*, 4, e5729.

O'Connor, C.M., Norris, D.R., Crossin, G.T., and Cooke, S.J. (2014). Biological carryover effects: linking common concepts and mechanisms in ecology and evolution. *Ecosphere*, 5, 1–11.

Ogden, N.H., Lindsay, L.R., Hanincova, K., et al. (2008). Role of migratory birds in introduction and range expansion of *Ixodes scapularis* ticks and of *Borrelia burgdorferi* and *Anaplasma phagocytophilum* in Canada. *Applied and Environmental Microbiology*, 74, 1780–90.

Parkes, M.W., Bienen, L., Breilh, J., et al. (2005). All hands on deck: transdisciplinary approaches to emerging infectious disease. *Ecohealth*, 2, 258–72.

Pimm, S.L., Alibhai, S., Bergl, R., et al. (2015). Emerging technologies to conserve biodiversity. *Trends in Ecology & Evolution*, 30, 685–96.

Prestrud, K.W., Åsbakk, K., Fuglei, E., et al. (2007). Serosurvey for *Toxoplasma gondii* in arctic foxes and possible sources of infection in the high Arctic of Svalbard. *Veterinary Parasitology*, 150, 6–12.

Rabinowitz, P.M., Scotch, M.L., and Conti, L.A. (2010). Animals as sentinels: using comparative medicine to move beyond the laboratory. *ILAR Journal*, 51, 262–7.

Reed, K.D., Meece, J.K., Henkel, J.S., and Shukla, S.K. (2003). Birds, migration and emerging zoonoses: West Nile virus, Lyme disease, influenza A and enteropathogens. *Clinical Medicine and Research*, 1, 5–12.

Rosenberg, K.V., Dokter, A.M., Blancher, P.J., et al. (2019). Decline of the North American avifauna. *Science*, 366, 120–4.

Ryser-Degiorgis, M.-P. (2013). Wildlife health investigations: needs, challenges and recommendations. *BMC Veterinary Research*, 9, 223.

Scheiner, S.M. and Rosenthal, J.P. (2006). Ecology of infectious disease: forging an alliance. *Ecohealth*, 3, 204–8.

Tompkins, E.M., Anderson, D.J., Pabilonia, K.L., and Huyvaert, K.P. (2017). Avian pox discovered in the critically endangered waved albatross (*Phoebastria irrorata*) from the Galápagos islands, Ecuador. *Journal of Wildlife Diseases*, 53, 891.

Tseng, M., El-Sabaawi, R.W., Kantar, M.B., Pantel, J.H., Srivastava, D.S., and Ware, J.L. (2020). Strategies and support for Black, Indigenous, and people of colour in ecology and evolutionary biology. *Nature Ecology & Evolution*, 4(10), 1288–90.

Watson, J.T., Jones, R.C., Gibbs, K., and Paul, W. (2004). Dead crow reports and location of human West Nile virus cases, Chicago, 2002. *Emerging Infectious Diseases*, 10, 938.

Weber, T.P. and Stilianakis, N.I. (2007). Ecologic immunology of avian influenza (H5N1) in migratory birds. *Emerging Infectious Diseases*, 13, 1139–43.

Whelan, C.J., Şekercioğlu, Ç.H., and Wenny, D.G. (2015). Why birds matter: from economic ornithology to ecosystem services. *Journal of Ornithology*, 156, 227–38.

Williams, P.D., Dobson, A.P., Dhondt, K.V., Hawley, D.M., and Dhondt, A.A. (2014). Evidence of trade-offs shaping virulence evolution in an emerging wildlife pathogen. *Journal of Evolutionary Biology*, 27, 1271–8.

Wood, C.L. and Johnson, P.T. (2015). A world without parasites: exploring the hidden ecology of infection. *Frontiers in Ecology and the Environment*, 13, 425–34.

Zylberberg, M. (2019). Next-generation ecological immunology. *Physiological and Biochemical Zoology*, 92, 177–88.

Zylberberg, M., Van Hemert, C., Dumbacher, J.P., Handel, C.M., Tihan, T., and DeRisi, J.L. (2016). Novel picornavirus associated with avian keratin disorder in Alaskan birds. *mBio*, 7(4), e00874-16.

Index

Note: Tables, figures, and boxes are indicated by an italic *t*, *f*, and *b* following the page number.

A

acute phase responses 31, 57, 58
Adelina tribolii 150
Aedes mosquitoes 11, 233
  Aedes aegypti 183
aerial parasites 56
age, and immune responses 34, 35*f*
agent-based models 134–5
agriculture 173, 231, 242, 249
  animal feedlots 181
  dairy farms 181
AIVs *see* avian influenza viruses
  (AIVs)
albatross
  Indian yellow-nosed albatross
    (*Thalassarche carteri*) 104
  waved albatross (*Phoebastria
    irrorata*) 108, 123, 250
    flight paths 251*f*
    pox lesions on the bill of 124*f*
allopreening (mutual preening) 55–6,
  67–8
Amblyomma loculosum 159*f*
American robin (*Turdus
  migratorius*) 12, 66, 86–7, 152,
  151*f*, 199, 215
amphibian hosts, transmission of
  Ribeiroia ondatrae and diversity
  of 154
ampicillin 219
Animal and Plant Health Agency
  (APHA), Great Britain 222
Anisakis nematode infection, in
  European shags (*Phalacrocorax
  aristotelis*) 114
Anopheles mosquitoes 175
Anseriformes (ducks, geese,
  swans) 209, 210*t*, 211, 221
anthropogenic resources 115,
  179–81
  and avian immune systems 37
anthroponoses *see* reverse zoonoses
  (anthroponoses)

antibodies 16
  maternal transfer to chicks 17–18
  natural 30
  persistence of 22
antigens 15–16, 31, 32, 43
antimicrobial peptides 15
antimicrobial resistance 219
antimicrobial-resistant (AMR)
    bacteria, contamination
    pathways for 216*f*
antiparasite behaviors 54–6, 200
  anting 55*t*
  deterrence of aerial parasites 56
  dust-bathing 55*t*
  nest 'self-medication' 56
  preening 55–6
'apapane (*Himatione sanguinea*) 83, 140
  effect of avian malaria on
    movement and foraging of
    juvenile 61
aquatic warbler (*Acrocephalus
  paludicola*) 109
Arctic fox, *Toxoplasma gondii* infection
    in 249
Arctic geese, seroprevalence of
    Toxoplasma gondii in 129
arsenic 177
arthropod vectors 10, 11–12, 14, 200,
    235–7
  birds as vehicles for movement
    of 235–7
arthropod-borne viruses *see* viruses,
    arboviruses
aspergillosis 180
Aspergillum fungi 202
Atlantic puffin (*Fratercula arctica*)
    160, 163
avian behaviors *see* behavior(s)
avian blood parasites *see* blood
    parasites
avian cholera 99, 100, 135, 194,
    199, 237 see also *Pasteurella
    multocida*

in common eider (*Somateria
    mollissima*) 134
avian hosts *see* hosts
avian infectious disease ecology 247–56
    see also disease ecology
  conceptual framework for 3*f*
  crossing scales in 248
  innovation in 250–1
  management, and
    conservation 251–3
avian influenza viruses (AIVs) 59, 81,
    125, 191, 209, 215, 221–2, 235
    see also influenza type A
    viruses (IAV)
  duration of infectivity in water
    of 202
  highly pathogenic (HP) or low
    pathogenic (LP) 221
  matrix gene 125
  reassortment 221
  spillover transmission from birds
    to humans 238
avian keratin disorder 251
avian malaria 21, 61, 139–40, 142,
    155, 175
  in endemic Hawaiian birds 82, 189
  implications for human health 239
  parasites (*Plasmodium* spp.) 239
    see also *Plasmodium* spp.
  in tawny pipits (*Anthus
    campestris*) 33
avian mange 180
avian paramyxovirus 1 *see* viruses,
    Newcastle disease virus (NDV)
avian populations *see* population, of
    wild birds
avian poxvirus 10, 11, 101*t*, 108, 115,
    180
  in great tit (*Parus major*) 106
  in ground finch (*Geospiza
    fuliginosa*) 109
  in juvenile waved albatross
    (*Phoebastria irrorata*) 250

avian schistosomes (trematodes/ blood flukes) 232
avian zoonoses *see* zoonoses

B

Bacillus licheniformis 63
bacteria and bacterial pathogens 4, 5, 219–20
  antibiotic resistant 219
  bacterial killing ability of house sparrows (*Passer domesticus*) 33
  food-borne bacterial contamination 219
  survival of infected and uninfected wild birds 101*t*
banding (ringing) 4, 235, 250
bark feeding 59
barn swallow (*Hirundo rustica*) 34, 61
  avian malaria in 200
  lice and mites on 200
  nest ectoparasites 195
  using agricultural operations 211
basic reproductive number (R_0) 23, 133, 139 *see also* reproductive number
Bbsl spirochaetes see *Borrelia* spp.
B-cells 16, 32
beak and feather disease, in Cape parrots (*Poicephalus robustus*) 115
behavior(s) 3
  antiparasitic 54–6
  avoidance 68
  climate change and changes in 199–200
  and disease ecology 53–4
  feather maintenance 55*t*
  foraging and movement 58–61
  foraging behavior
    effects of disease on 60–1
    and exposure risk 58–60
  important to parasite exposure, resistance, and spread 54–68
  and infectious disease dynamics in birds 53–75
  mate choice and sexual selection 63–5
  personality 61–3
  sickness behaviors 56–8
  social 65–8
β term (transmission parameter) 23, 45, 131, 133
Bewick's swan (*Cygnus columbianus*) 59, 60
BIDE model 138
biodiversity 153–5
biological control 241

biosphere 246*f*
bird banders, infection with avian influenza viruses 234
bird droppings, in urbanized areas 157
bird feeders 180–1
  garden birds as reservoir of pathogens that cause human salmonellosis 231
  parasites and high densities of birds at 157
  transmission of parasites 180
bird flu *see* avian influenza viruses (AIVs)
bird movements, tracking 253
bird–human interface
  pathogen transmission at 231–46
  zoonotic emergence at 234–5
biting midges (*Culicoides* spp.) 11
blackbirds *see also* Passeriformes (blackbirds, starlings, and sparrows)
  red-winged blackbird (*Agelaius phoeniceus*) 16, 210*t*
    infested with *Ixodes* ticks 184
  European blackbird (*Turdus merula*) 175, 182, 199
    immune function during the migratory season compared with resident 35, 36*f*
blackcap (*Sylvia atricapilla*) 191
black-capped chickadee (*Poecile atricapillus*) 67
blackflies (*Simulium* spp.) 195*b*
black-legged kittiwake (*Rissa tridactyla*) 150, 160, 162
black-tailed prairie dog (*Cynomys ludovicianus*) 177
black-throated blue warbler (*Setophaga caerulescens*) 139
blister beetles, canthardidin-containing 60
blood parasites (haemosporidian parasites) 89, 91, 107–8, 109, 115, 139–40, 182, 194, 201
  see also *Haemoproteus* spp., *Leucocytozoon* spp., and *Plasmodium* spp. parasites
  molecular detection of 20
  phylogenetic tree of partial cytochrome b gene of 90*f*
  survival of infected and uninfected groups of wild birds 102*t*, 103*t*
  western bluebirds (*Sialia mexicana*) infected with 43*f*
blue tit (*Cyanistes caeruleus*) 82
  antibody responses against diphtheria vaccine 43

clutch size after blood parasites infection 109
fledging success after infections in 113, 114
Haemoproteus blood parasites in 108
parental provisioning rates of 114
Plasmodium infection in 113, 129
blue-footed booby (*Sula nebouxii*) 65
Borrelia spp.
  Borrelia burgdorferi 21, 59, 101*t*, 135, 200, 236
  Borrelia burgdorferi sensu lato (Bbsl) complex 159, 162, 163, 164
  Borrelia garinii 160, 163, 236
botulism 199, 202, 237
breeding
  breeding areas
    annual return rates to 104
    latitudinal and elevational shifts in 193–5
  breeding phenology 109
  breeding season 22, 33, 34, 35, 36, 37, 58, 66, 108, 114, 158, 160, 193, 195*b*, 208
bridge hosts 213
brood size 113
brown kiwi (*Apteryx mantelli*) 237
Buggy Creek virus 60, 108, 154
bunyavirus 236

C

cadmium 177
California condor (*Gymnogyps californianus*) 177
Campylobacter spp. 215, 231, 237
  Campylobacter jejuni 58, 181, 210*t*, 219, 222
Canada goose (*Branta canadensis*) 157
canary (*Serinus canaria domestica*) 239
  Plasmodium relictum in 67
Cape parrot (*Poicephalus robustus*), beak and feather disease in 115
capture-recapture models, open and closed 127–8
cardueline finches 202
cattle egret (*Bubulcus ibis*) 191
cercariae, free-swimming 154
cercarial dermatitis ('swimmer's itch') 232
chaffinch (*Fringilla coelebs*) 23, 108, 137
Charadriiformes (gulls, terns, shorebirds) 209, 210*t*, 211, 221
chickens (*Gallus gallus domesticus*) 18, 22 *see also* domestic poultry

helminths infection in
free-ranging 238
Chlamydophila psittaci 233, 234
chlamydiosis 233
cholera *see* avian cholera
chromium 177
cliff swallow (*Petrochelidon
pyrrhonota*) 59, 60, 150, 154
climate change 37
and avian disease 191–207
and changes in bird physiology,
behavior, and response to
infection 197–200
and changes in non-breeding range
and movement patterns 194–5
and changes in timing of life
history events 195–7
climate warming 194–5
effects on bird–parasite
interactions 4–5
effects on parasites with external
transmission stages 200–2
future research 202–4
host and parasite adaptation to 203
infections and changes in bird
distribution and
phenology 192–7
clorophacinone 178
Clostridium botulinum 181, 199, 202
clutch size, infection and 109–13
coccidial parasites (coccidia) 20, 58,
66, 178, 182
coevolution *see also* evolution
of avian hosts and their
parasites 3–4, 89–92
requirements for
demonstrating 77, 78b
co-infection 107, 239
collared flycatcher (*Ficedula
albicollis*) 62, 193
Columbiformes (pigeons and
doves) 23, 175, 202, 209, 210t
common eider (*Somateria
mollissima*) 100, 134, 154
common murre (*Uria aalge*) 162
communities
assembly of 148
circulation of vector-borne
parasites within host
communities 161f
community ecology, and the place
of birds 147–9
community-level interactions 4
diversity of communities and
parasite circulation 151–7
dynamics of 148
interactions among species of host,
parasite, or vector within

complex biological
communities 246f
metacommunities 149, 156
modified 156–7
stability of 148
structure of natural 148f
compartmental models 13–14, 131–4
see also SIR epidemiological
models
competition 81, 149
apparent 150
interspecific 150
nestling 39f
seabird communities 158
conjunctivitis 84–5
mycoplasmal conjunctivitis 101t,
251
mycoplasmosis 180
contamination potential 222
Cooper's hawk (*Accipiter cooperii*) 178
Coquillettidia 233
Cormack–Jolly–Seber (CJS)
models 128, 140
cormorants
double-crested cormorant
(*Phalacrocorax auritus*) 22, 34,
108
gastrointestinal helminth infection
in 238–9
great cormorant (*Phalacrocorax
carbo*) 130, 158
seropositivity for NDV
antibodies 22
coronaviruses 218
gamma and delta 2
Corsican blue tit (*Parus caeruleus
ogliastrae*), deterrence of *Culex*
mosquitoes 56
corticosterone 16, 38, 179
co-speciation 89
Crimean Congo hemorrhagic fever
(CCHF) virus 236, 241
crows
American crow (*Corvus
brachyrhynchos*) 10, 64, 86–7,
108, 219
allopreening (mutual preening)
in 67–8
foraging sites 181
with poxviral dermatitis
lesions 59f
risks of *C. jejuni* contamination
and movements of 222
WNV in 137
northwestern crow (*Corvus
caurinus*) 128
Cryptosporidium spp. 103t, 113, 129
Cryptosporidium baileyi 108

Cryptosporidium parvum 8
Culex mosquitoes *see* mosquitoes
Culicoides midges 238f
Culiseta melanura 9, 233
cytokines 238

D
dabbling ducks, risk for AIVs in 59
dark-eyed junco (*Junco hyemalis*) 194
Darwin's finches 55
dead bird carcasses, reporting of 251
deforestation 173, 175
delayed type hypersensitivity 32,
37, 38
density-dependent transmission 134
diazinon 178
dichlorodiphenyltrichloroethane
(DDT) 178
dilution effect 151, 150b, 154, 165
diphtheria vaccine, antibody
responses of blue tits to 43
Diptera 176
disease
causation of 7–12
effects on movement and
foraging 60–1
disease ecology 2 *see also* avian
infectious disease ecology
imperfect detection in 122b
dispersal 148
domestic poultry *see also* chickens
(*Gallus gallus domesticus*)
age and immune defenses 34
vaccinations to protect from
disease outbreaks 17
viral spillover between wild birds
and 216f
dominance behavior 66–7

E
early warning systems 222, 250–1
eastern equine encephalitis (EEE)
virus 9, 126, 233
ecoimmunology 2, 29–52
ecological communities 147 *see also*
communities
ectoparasites 21, 54
and cigarette butts in nests 57f
detection of 18, 129
feather 61
feather lice 129
ticks. *see* ticks
effective reproduction number
(R_e) 23, 134
emerging infectious diseases
(EIDs) 68, 231, 235
encephalitis viruses 11 *see also*
viruses

'encounter-dilution effect' 66
enemy release hypothesis 193
enteropathogens 231
environment, abiotic and biotic
        features 12
environmental stressors
    anthropogenic factors 37
    food availability 37
    and immune responses 38
    and physiological stress
        responses 36–8
enzyme-linked immunosorbent
        assays (ELISAs) 21
epidemics 23
epidemiological modeling 130–5
    compartmental models 131–4
    resources for 136
epidemiology 2
epizootic 23
epizootiology 124
epornitic 23
equine encephalitis viruses 233
    see also viruses
Escherichia coli 231
estrogen, and immune responses 33
Euhaplorchis californiensis 149
Eurasian blue tit (*Cyanistes
        caeruleus*) 106, 107f
Eurasian IAV subtypes, introduction
        into North America 223
Eurasian treecreeper (*Certhia
        familiaris*) 176
European goldfinch (*Carduelis
        carduelis*) 108
European greenfinch (*Chloris
        chloris*) 108, 137, 136f
European house sparrow (*Passer
        domesticus*) *see* house sparrow
        (*Passer domesticus*)
European rabbit (*Oryctolagus
        cuniculus*) 151
European robin (*Erithacus
        rubecula*) 236
European shag (*Phalacrocorax
        aristotelis*), Anisakis infection
        in 114
European starling (*Sturnus
        vulgaris*) 56, 157, 210t,
        216, 217
    as bridge host that transmits IAV to
        poultry 214f
    introduction of *S. enterica* bacteria
        into livestock operations 223
    nestling weight and development
        of immune system in 39f
    sickness behaviors in 58
    use of agricultural operations
        by 210, 211

Eustrongylides ignotus 181
evolution *see also* coevolution
    adaptive pathogen virulence
        evolution in birds 88–9
    evolutionary responses of
        pathogens to their avian
        hosts 85–9
    host–pathogen coevolution in
        avian systems 77–97
    phenotypic 83–5
    requirements for
        demonstrating 78b
experimental manipulations, to detect
        avian parasites 19t
exposed state 13, 23
extinctions
    from habitat loss 175
    of host populations 142
extra-pair copulations (EPCs) 63–4
extreme weather events 198
    and vector populations 201

F
fecundity 114–15
finch trichomoniasis 137
fitness effects
    of parasite infection in birds 99–121
    of hosts 4, 149
flaviviral encephalitis 85
flea (*Ceratophyllus celsus*) 150
fledging success 113–14
flocoumafen 178
flour beetles (*Tribolium* spp.) 150
food availability, and immune
        defense 37
food-borne bacterial
        contamination 219
force of infection (λ) 45, 133, 134

G
Galápagos hawk (*Buteo
        galapagoensis*) 43
Galápagos islands 108, 237
Galápagos mockingbird (*Mimus
        parvulus*) 108
game birds 231
gamma γ (recovery or removal
        rate) 131, 132
geese
    seroprevalence in 81
    Toxoplasma gondii in 130
genetic drift 148
genetics 215
global change, impacts on natural
        communities and their
        dynamics 164
good genes hypothesis 64
gray catbird (*Dumatella carolensis*) 61

great bustard (*Otis tarda*) 60, 64, 178–9
Great Island virus (Orbivirus) 163
great reed warbler (*Acrocephalus
        arundinaceus*) 82, 236
great tit (*Parus major*) 59, 61, 177
    artificial light at night and
        corticosterone in 179
    avian poxvirus in 106, 115, 250
    co-infections with types of
        haemosporidian parasites 107
    immune defense of females while
        breeding 34
    immune responses of nestlings 34
    mortality rate of juveniles infected
        with avian poxvirus 108
    Plasmodium parasites 91
    studies of *Mhc* genes in 82
    trypanosome infection of 113
greenfinch (*Chloris chloris*) 23, 62, 64,
        108
    infection with Sindbis virus 67
    trichomoniasis in 137
grey partridge (*Perdix perdix*) 150
griffon vulture (*Gyps fulvus*) 219
ground finch (*Geospiza fuliginosa*) 109
ground-feeding birds, and exposure
        risk 58–9
guineafowl 241
gulls (*Larus* spp.) 154 *see also*
        Charadriiformes (gulls, terns,
        shorebirds)
    Bonaparte's gull (*Chroicocephalus
        philadelphia*) 215f
    climate change and foraging
        behavior of 199
    contaminating tomato fields with
        Salmonella 211
    glaucous gull (*Larus
        hyperboreus*) 178
    helminths and ring-billed gull
        (*Larus delawarensis*) 156
    Listeria monocytogenes in 220
    red-billed gull (*Chroicocephalus
        novaehollandiae scopulinus*),
        Plasmodium infection in 113
    yellow-legged gull (*Larus
        michahellis*) 157, 219

H
habitat loss and degradation 174–6
    fragmentation of habitats 174–6
Haemoproteus parasites 89, 91, 106,
        108, 139, 201 *see also* blood
        parasites
    and cold-adapted *Leucocytozoon*
        co-infections 201
    infection in male yellowhammers
        (*Emberiza citrinella*) 114

phylogenetic tree of partial
cytochrome b gene of 90*f*
survival of infected and
uninfected groups of 102*t*, 103*t*
western bluebirds (*Sialia mexicana*)
infected with 43*f*
Haemoproteus beckeri 102*t*, 111*t*, 115
Haemoproteus columbae 102*t*
Haemoproteus majoris 82
Haemoproteus prognei 113*f*
haemosporidian parasites *see* blood
parasites
hatching success 113
Hawai'i, introduction of non-native
birds to 154–5
Hawai'i 'amakihi (*Chlorodrepanis
virens/Hemignathus virens*) 83,
139, 140
clutch sizes 114
pox lesions on the leg of 126*f*
tolerance and resistance to avian
malaria 83
Hawaiian birds
avian malaria in 83
avian poxvirus in 107
Hawaiian honeycreepers 1, 99, 139,
192*f*
altitudinal migrations in response
to avian malaria, 68
avian malaria in 140
deforestation and malaria
prevalence for 175
loss due to avian malaria 142
mortality following introduction of
Plasmodium relictum 193
roosting observations 68–9
susceptibility to *P. relictum* 108
hawks 195
heat island effect 173, 183
heavy metals 177
helmeted guineafowl (*Numida
meleagris*) 241
helminths 103*t*, 105, 239, 238*f*
infection in free-ranging
chickens 238
and ring-billed gulls (*Larus
delawarensis*) 156
helper T-cells 32
hemagglutination inhibition
assays 21–2
hemagglutinin 15*b*, 221
heptachlor 178
herd immunity 23, 134
Heterakis gallinarum 150
heterophils 31
lymphocytes ratios 176, 177
Histoplasma capsulatum 241
homozygosity, and parasites 64

horseshoe crab 196
host competence 44
host immune response
adaptive 15, 16
innate 15
host shifting 89, 90, 158, 234
host–pathogen evolution and
coevolution 77–97
coevolution of avian hosts and
their pathogens 89–92
evolutionary responses of avian
hosts to their pathogens 80–5
host responses to infection 79*f*, 87
life history changes to reduce costs
of infections 81
resistance and tolerance 81–5
host–pathogen interactions 2, 7–28
environment 12
infection status of hosts 18–22
infection timeline 12–17
microparasites and
macroparasites 9–10
parasite-specific immunity 17–18
pathogens and parasites 9
spatial variation in 90–1
vectors 11–12
virulence and pathogenicity 10–11
within individual hosts 246*f*
within-host dynamics
(pathogenesis) 14–17
hosts 8–9
avian host competence for
infection 203
barriers to prevent establishment of
parasites 14–15
biodiversity 153–5
'dead-end' or 'incidental' 9
definitive 8
dilution 151
estimation of population size
of 126–7
host types and pathways for
parasite contamination by
birds 212–18
imperfect detection 122*b*
infection status/state 18–22
intermediate 8
interspecific host
heterogeneity 151–3
local extinctions of 142
mechanical, transport, or
paratenic 8
progression of host's infectious
state 13
reservoir (maintenance) 8–9
susceptible 8
visual detection of infection
status 18–20

house finch (*Haemorhous
mexicanus*) 12, 45, 62, 122*b*,
125*b*, 152
cigarette butts in nests of 56
co-infections with *Plasmodium* and
M. gallisepticum 107
cold temperatures at higher
latitudes and the prevalence
of *M. gallisepticum* 199
defense behaviours of 68
exhibiting sickness behaviors 57*f*
immune responses to
M. gallisepticum 88
infected with *M. gallisepticum*
12, 44, 59, 88, 91–2, 107,
131, 157
antibody persistence in 22
dominance behavior and severity
of 67
epidemics of 69
infectiousness of 14
survival after 106
infection avoidance behavior
of 80–1
inflammation responses to
phytohemagglutinin 67
mycoplasmal conjunctivitis in 61,
84–5, 141
parasite prevalence in urban 182
prevalence of *M. gallisepticum* 65
protection against *M. gallisepticum*
infection 22
sickness behaviors in 57
thermoregulation and *M.
gallisepticum* infection in 198
transmission dynamics within
groups of 89
use of bird feeders and
M. gallisepticum 181
house martin (*Delichon urbicum*),
clutch sizes 109, 113*f*
house sparrow (*Passer domesticus*)
33, 56, 86, 136*f*, 154, 157, 175
see also sparrows
with acute coccidiosis 67
avian malaria parasite infections
in 193
effects of parasites on sparrow
survival and abundance 140
immune responses 34
injected with Newcastle disease
virus vaccine 65
lead concentrations in feathers
of 178
studies of MHC genes in 82
using agricultural operations 211
West Nile virus (WNV) in 44, 45,
107, 179

human health, implications of parasite interactions in birds for 238–9
human immunodeficiency virus (HIV) 238
human Lyme disease 159–60
humans, interactions with birds 231

I

icterine warbler (*Hippolais icterina*) 193
'i'iwi (*Vestiaria coccinea*) 140, 192*f*
immune (recovered) state 13
immune defense *see also* immune system
  age and 34
  drivers of variation in 33–8
  food availability and 37
  life history stage and 34–6
  linking directly to individual fitness and parasite burdens 45
  nestling weight and development of 39*f*
  overview of techniques 40*t*
  and population-level epidemics 45
  primary and secondary 16*f*
  quantifying variation in 38–9
  sex differences and variation in 33
immune phenotypes
  impact on parasite transmission 45
  prediction of infection status or host fitness 39–44
immune responses *see* immune defense
immune system 2, 29–33
  active 17
  adaptive 31–2
  constitutive vs. inducible and non-specific vs. specific 30
  and infection impacts 197–9
  innate 30–1
  parasite-specific 17–18
  passive acquisition of 17
immunity *see* immune system
immunocompetence, and heavy metals 177
immunocompetence handicap hypothesis 64–5
immunological assays 21–2
imperfect detection
  in avian disease ecology 122*b*
  of hosts 124
import of live birds or bird products 223
incubation period 12
individual reproductive number (*V*) 44, 45

individual-based models (IBMs) 134–5
infection status/state
  experimental manipulations of 21
  imperfect detection 122*b*
infection timeline
  compartmental models of states of infectiousness 13–14
  disease state of host 12
  infectious state of host 13
infection(s)
  definition of 7
  integrated response to 32–3
  prevention of 80–1
  reducing costs of 81–5
infectious agents 3 *see also* pathogens/parasites
infectious disease(s)
  definition of 7
  interactions between host, parasite, environment, and vector 8
infectious state 23
infectiousness 13, 45
infectivity 14
infestations 9
influenza type A viruses (IAV) 14, 15*b*, 213, 214*f*, 217
  H1N1 virus 15*b*, 231, 238
  H5N1 virus 60, 215, 223
  H6N5 virus 223
  highly pathogenic avian influenza virus (HPAIV) 249
  low pathogenic avian influenza viruses (LPAIV) 101*t*, 221
  spillover between domestic poultry and wild birds 216*f*
  survival effects in hosts 106
insect vectors, habitat loss and contact between birds and 175
intestinal nematodes 67
  Heterakis gallinarum 150
ivermectin 241
Ixodes spp. ticks 159, 160, 184 *see also* ticks
  Ixodes scapularis 236
  Ixodes uriae 159*f*, 160–2, 158*f*

J

Jolly–Seber models 128

K

Kauai 191
keystone species 149
killer T-cells 32
killdeer (*Charadrius vociferus*) 210*t*
kiwikiu (*Pseudonestor xanthophrys*) 191

L

lambda λ (risk or force of infection) 133, 134
land use change 4
  and avian disease dynamics 173–89
  influences on disease dynamics 174–81
  major drivers of 173–4
  prevalence and intensity of a vector-borne parasite influenced by 172*f*
latent period 13
Laysan finch (*Telespyza cantans*) 83
LD$_{50}$ (lethal dose) 11
Leach's storm-petrel (*Oceanodroma leucorhoa*) 34
lead 37, 176, 177, 178
lesser noddy (*Anous tenuirostris*) 159*f*
lesser snow goose (*Anser caerulescens*) 105
lethal dose (LD$_{50}$) 11
Leucocytozoon parasites *see also* blood parasites
  and cold-adapted *Leucocytozoon* co-infections 201
  infection in male yellowhammers (*Emberiza citrinella*) 114
  Leucocytozoon 106, 139, 201
  Leucocytozoon fringillarium 195*b*
  Leucocytozoon marchouxi 102*t*
  Leucocytozoon ziemanni 115
light pollution 179
Lincoln–Petersen estimator 127
linking across scales 248
lipopolysaccharide 30, 34, 57, 58
Listeria monocytogenes 219, 220
live bird markets 221
louping ill disease 157
Lyme borreliosis *see* Lyme disease
Lyme disease 21, 159–60, 200, 236
  epidemiology of 164
  geographic patterns of distribution of *Borrelia burgdorferi* 236
  in seabird communities 158–64

M

macroparasites 8, 9–10, 105
macrophages 31
maintenance hosts 213
major histocompatibility complex (MHC) 31, 62, 64, 82
malaria *see* avian malaria
mallard (*Anas platyrhynchos*) 195, 219
  infected with avian influenza virus 34, 36*f*, 45
  lead levels in 37

migratory 60
  as reservoirs of influenza A
    virus 210*t*
mammals, exposure to
    macroparasites 238
mangrove finch (*Cactospiza
    heliobates*) 136*f*, 142
mark-recapture methods 104–5,
    127–8, 250
mate choice, and sexual selection 63–5
melodious warbler (*Hippolais
    polyglotta*) 193
memory cells 16, 22, 32
mercury 177
microbial associated molecular
    patterns (MAMPS) 30
microevolutionary studies
  spatial adaptation experiments
    90–1
  temporal adaptation
    experiments 91–2
Microfilaria 103*t*, 239
microparasites 9, 10, 105
microplastics 179
microscopy, to detect avian
    parasites 19*t*, 20
migration 248–9
  and disease dynamics 60
  immune defenses and 34, 36*f*
  introduction of parasites by 223
  migratory culling 60
  patterns 194–5
Migratory Bird Treaty Act 241
migratory birds *see also* migration
  migration timing and tick
    dispersal 135
  parasitic bacteria 219
  transport of vectors by 200
  West Nile virus in 60
molecular detection, of avian
    parasites 19*t*, 20–1
monitoring and surveillance, for
    infectious disease ecology 251
monogamous systems, sex-based
    immune differences in 33
mosquitoes 9, 11, 56, 152, 233, 237,
    239, 238*f*
  climate change and 200
  Culex pipiens 241
  Culex quinquefasciatus 139, 155,
    183, 191
  Plasmodium relictum-transmitting
    mosquitoes 200
  as WNV vector for 11, 12
movement, effects of disease on 60–1
multihost systems 8, 142, 180,
    212, 215

multistate mark-recapture (MSMR)
    models 128
multistate models 129
Mycoplasma gallisepticum 44, 56, 89,
    91–2, 106, 180
  on bird feeders 61
  characteristics of 250
  and conjunctivitis in house
    finches 84–5
  flock epidemics of 69
  and house finch population case
    study 141
  in house finches (*Haemorhous
    mexicanus*) *see* house finch
    (*Haemorhous mexicanus*)
  mycoplasmal conjunctivitis 101*t*
  prevalence of 65
  spillover from domestic chickens to
    house finches 213
  time spent on bird feeders by
    house finches and infection
    with 59
  in urban areas 157

N

natural killer cells 31
natural selection 78*b*
natural wetlands 174, 199
nest ectoparasites 21, 66, 150
nest 'self-medication' 56
network models 135
neuraminidase 15*b*, 221
next-generation sequencing
    techniques 45
noise pollution 179
non-migratory birds, antibiotic
    resistant bacteria in 219
non-passerine birds, and the
    prevalence and density of
    infected *Culex* mosquitoes 152
North American Bird Conservation
    Initiative 252
North American thrushes 60
northern bobwhite (*Colinus
    virginianus*) 199–200
northern cardinal (*Cardinalis
    cardinalis*) 12, 105
northern mockingbird (*Mimus
    polyglottos*) 12, 152
northern spotted owl (*Strix
    occidentalis caurina*) 175
northwestern song sparrow
    (*Melospiza melodia morphna*) 58
  see also song sparrow
notifiable animal diseases 222
nutritional stress, impact on parasites
    reproductive success 115

O

Oahu 'amakihi (*Hemignathus
    chloris*) 83
occupancy models 136, 142
  single-season 128
One Health 5, 45, 184, 249–50
organochlorine compounds 178
Ornithodoros capensis 159*f*
ornithosis 233
outbreaks 23
ovenbird (*Seiurus aurocapillus*) 135
oxidative stress, and immune
    defense 37
oystercatcher (*Haematopus
    ostralegus*) 59

P

Pacific killifish (*Fundulus
    parvipinnis*) 149
pairing success 109
parasite load 20, 21, 33, 129–30
  age and 34
  immune responses and 44
parasites *see* pathogens/parasites
parentage 109
parental care 65, 114
parrot fever 233
partridge 150–1
Passeriformes (blackbirds, starlings,
    and sparrows) 9, 152, 209,
    210*t*, 238*f*
  passerine salmonellosis 231 *see also*
    salmonellosis
Pasteurella multocida (avian
    cholera) 99, 101*t*, 105, 135, 194
  see also avian cholera
pathogenicity 10–11
pathogens *see* pathogens/parasites
pathogens/parasites
  adaptive phenotypic responses
    of 87–9
  adherence to host tissues 14
  aggregation 130
  birds as vehicles for movement
    of 235–7
  characteristics that impact effects
    on host survival 105–7
  coevolution of avian hosts
    and 89–92
  co-infection 238
  in communities 149–51
  and declines in avian host
    population size 136–7
  definition of 9
  deliberate penetration by an
    arthropod vector 14
  detection of 19*t*

pathogens/parasites (cont.)
DNA or RNA 20
effect on reproductive success
109–15
effects on avian host populations 4,
136–42
effects on breeding phenology,
pairing success, and
parentage 109
effects on fledging success 113–14
effects on hatching success 113
effects on host fitness 100f
effects on host survival 99–109
environmentally transmitted 202
evolutionary responses to their
avian hosts 85–9
experimental approaches for
testing 77b, 78f
exposure 235
fecundity 114–15
host characteristics that impact
effects on survival 108–9
host population regulation and
fluctuation linked to 140–2
influence on individual survival in
birds 105–9
long-distance movements and
introduction of 222–4
management of 253
mechanical transmission 223
molecular evolutionary response to
their avian hosts 85–7
mosquito-borne 237
population size 129–30
pressure 235
prevalence and transmission in
wild birds 218–22
prevalence of 215
reproductive strategies and
spread 65
and reproductive success of their
hosts 149
role in dynamics of population of
wild birds 123–46
transgenerational effects 114
transmission between birds and
humans 5, 231–46
transmission cycle 53, 54t
transmission routes 10
vector-borne and trophically
transmitted 200–2
from wild birds at the wildlife–
agriculture interface 209–29
within avian communities
155–6
pattern recognition receptors
(PRRs) 30

penguins
allopreening (mutual preening)
in 55–6
king penguin (Apternodytes
patagonicus) 55
Magellanic penguin (Spheniscus
magellanicus) 193
mortality rates from infection 108
peridomestic species 209
persistent organic pollutants
(POPs) 178
personality 61–3
boldness 62, 63
and exposure 61–2
and infection spread 63
and resistance 62–3
risk-taking behaviors 62
sociability 62
and stress physiology 62
pesticides 176–7, 178
pet birds 231
zoonotic agents in 233, 240
phagocytes 31
pheasants, intestinal nematode
(Heterakis gallinarum) in 150–1
Philornis downsi (parasitic fly) 142
phylogenetics 215
physiological stressors
endocrine-mediated stress
response 37–8
oxidative stress 37
phytohemagglutinin 32, 67
pied flycatcher (Ficedula
hypoleuca) 193
pigeon (Columba livia), prevalence of
haemosporidian parasites
in 182
pine siskin (Spinus pinus) 180
pink pigeon (Nesoenas mayeri) 114, 193
plaque-reduction neutralization tests
(PRNTs) 21
Plasmodium spp. parasites 11, 82, 89,
90f, 102t, 103t, 106–7, 139, 175,
239 see also blood parasites or
avian malaria
in blue tits (Cyanistes caeruleus) 113,
129
local host adaptation 91
Plasmodium circumflexum 82
Plasmodium relictum 82, 155, 165,
178, 191, 193
canary (Serinus canaria
domestica) 67
climate change and 92, 192f, 200
in endemic Hawaiian birds 83
Eurasian blue tits (Cyanistes
caeruleus) infected by 107, 107f

in Hawai'i 'amakihi
(Chlorodrepanis virens) 113
incubation period 201
population-level effects of 139
in socially dominant canaries
(Serinus canaria domestica) 67
susceptibility of honeycreepers
to 109
transmission of 140
in red-billed gulls (Chroicocephalus
novaehollandiae scopulinus) 113
song sparrow (Melospiza melodia)
and local adaptation 91
plastics, effects on bird health 179
pollution 4, 37
chemical (toxicants) 176–9
light and noise 179
localized effects of 248
polygynous systems, sex-based
immune differences in 33
polymerase chain reaction (PCR) 20
polymorphism 64
population
definition of 123
of parasites 129–30
population data 142
population growth 138–40
population-level disease
dynamics 22–4
population-level dynamics of hosts
and parasites 246f
population-level transmission
dynamics 44–5
of wild birds 247
describing infection status
in 123–9
estimates of host population
size 126–7
mark-recapture methods to
estimate 127–8
regulation of 140–2
role of parasites in dynamics
of 123–46
poultry, spillover of NDV from wild
to domestic species 220
poultry farms
minimizing avian zoonoses in 240
poultry feed, contamination with
avian IAV 216
wild bird species for IAV
surveillance on 211–12
Pradel models 139
pre-disease state 12
preening 55–6, 55t, 67–8
pre-infectious state 13
prevalence 18
estimation of 124–6

occupancy methods to
estimate 128–9
point and period 125
sample/naïve 125
primaquine 109, 113*f*
progesterone, and immune
responses 33
propagule pressure 236
proteins, acute phase 31
proteomics 45
protozoal infections 102*t*, 105, 106
Psittaciformes (parrots, cockatoos,
and relatives) 233
psittacosis 233, 234
purple martin (*Progne subis*) 106
fledging success and female
Haemoproteus infection 114

R

R_0 (basic reproductive number) 23,
133, 139 *see also* reproductive
number
raptors 202
infected with avian poxvirus 149
prevalence of *Trichomonas gallinae*
in 175
shedding *Campylobacter*, *E. coli*, and
Salmonella in their feces 222
R_e (effective reproduction
number) 23, 134
real-time quantitative reverse
transcription PCR (qRT-
PCR) 20 *see also* polymerase
chain reaction
recovered state 13, 131
recovery or removal rate (γ) 131, 132
red grouse (*Lagopus lagopus
scotica*) 105, 136*f*, 141, 157
brood sizes when infected with
T. tenuis and with
Cryptosporidium spp 113
clutch sizes 113
effects of *T. tenuis* on winter
survival in 106
respiratory cryptosporidiosis
in 108
red junglefowl (*Gallus gallus*)
comb lengths in male 64–5
infections with intestinal
nematodes 67
red knot (*Calidris canutus*) 36, 37, 199
red-naped sapsucker (*Sphyrapicus
nuchalis*) 149
redwing thrush (*Turdus iliacus*),
Borrelia infection in 60
reproduction
effect of parasites on 109–15

exposure and reproductive
strategies 63–4
immune defenses and 34
parasite spread and reproductive
strategies 65
resistance and reproductive
strategies 64–5
reproductive number
basic reproductive number (R_0) 23,
133–4, 139
effective reproduction number
(R_e) 23, 134
individual reproductive number
(*V*) 44, 45
net reproduction number (R_n) 134
reservoir hosts 151, 212
resistance and tolerance 29, 81–5
avian immune genes 81–3
respiratory cryptosporidiosis 108
reverse zoonoses
(anthroponoses) 237–8
rhesus macaque (*Macaca mulatta*) 67
ring-necked pheasant (*Phasianus
colchicus*) 150
risk or force of infection (λ) 133, 134
R-naught (R_0) (basic reproduction
number) 23, 133–4, 139
rock dove/pigeon (*Columba livia*) 55,
175, 215, 219
antibody persistence in 22
immune priming 68
prevalence of haemosporidian
parasites in 182
using agricultural operations in the
southeastern US 211
rodenticides 178–9
rRNA sequencing 21
ruddy turnstone (*Arenaria interpres*) 196

S

Salmonella spp. 184, 211, 231, 237
in avian feces on feeders 180
infection on broiler chicken 198
Salmonella enterica 219, 241
antimicrobial resistance (AMR)
in 219
salmonellosis 180
garden birds, as reservoir of
pathogens that cause human
salmonellosis 231
sampling populations 124
distance sampling methods 127
resources for 136
sampling units 124
sandhill crane (*Grus canadensis*),
contamination of peas
C. jejuni 210*t*, 215, 222

schistosomes (trematodes/blood
flukes) 232
seabird communities 158–9
breeding colonies and foraging
hotspots at sea 158
community-level interactions
between ectoparasites
and 159*f*
population structure of *Ixodes uriae*
in 160–2
ticks and the circulation of Lyme
borreliosis-causing
agents 159–60
second-generation anticoagulant
rodenticides (SGARS) 178
selection 148
'selective sieve' hypothesis 86
self-vaccination 68
serin (*Serinus serinus*), with avian pox
lesions 122*b*
seropositivity 126
seroprevalence 21, 215
estimates of 126
in geese 81
serosurveillance 126
seropositive hosts 21
sewage treatment plants 174
sex differences, in immune response
and parasite load 33
Shiga-toxin producing *Escherichia coli*
(STEC) 219
shore-birds 195
Siberian chipmunk (*Eutamias
sibiricus*) 61
sickness behaviors 56–8
context-specific 57–8
manifestation and putative
function of 57
during migration 58
and transmission 58
SIR epidemiological models 13,
131–2, 130*b*, 131*f see also*
compartmental models
SEI epidemiological model 13, 23,
130*b*
SIS epidemiological model 130*b*
skylark (*Alauda arvensis*) 35, 36
snow goose (*Chen caerulescens*) 45,
194
sociable weaver (*Philetairus
socius*) 199
social behavior 65–8
collective defense behaviors/social
immunity 67–8
dominance behavior 66–7
of healthy and sick birds 69
transmission routes and 65–6

social network analysis 135
social status 67
song sparrow (*Melospiza melodia*) 151*f*
   endocrine-mediated stress response in 38
   helminth infections 239
   local adaptation to *Plasmodium* 91
song thrush (*Turdus philomelos*) 219
sooty tern (*Onychoprion fuscatus*) 159*f*
Spanish Imperial eagle (*Aquila adalberti*) 151
'Spanish' influenza virus 1918 (H1N1) *see* influenza type A viruses (IAV)
sparrows *see also* house sparrow (*Passer domesticus*)
   mountain white-crowned sparrow (*Zonotrichia leucophrys*) 114
   song sparrow (*Melospiza melodia*) 176 *see also* song sparrow (*Melospiza melodia*)
   white-crowned sparrow (*Zonotrichia leucophrys*) 195*b*
speciation 148
species coexistence 151
spillback 217–18 *see also* reverse zoonoses (anthroponoses)
spillover *see also* zoonoses
   avian influenza A viruses (AIVs) 221–2
   Newcastle disease virus (NDV) 220–1
   of parasites from birds to humans 5, 24, 151, 154, 157, 165, 191, 215, 217–18, 231, 232, 233
spotted fever group (*Rickettsia* spp.) 236
Staphylococcus aureus 238
steroid hormones 33
streptomycin 219
stress, and habitat loss 176
stress hormone levels 177
stress response, and immune defense 37–8
sulphonamide 219
sunning 55*t*, 200
superb fairy wren (*Malurus cyaneus*) 62
superspreaders 63, 69, 151
supplemental feeding 4, 180–1, 241, 248
survival rates of free-living birds 104–5
susceptible state 14, 23, 45, 129, 130*b*
Swainson's thrush (*Catharus ustulatus*) 61

swallow bug (*Oeciacus vicarius*) 150, 154
swallows, nests at agricultural operations 215–16
synanthropic birds 209
   identification of focal synanthropic avian species 211–12
   species associated with agricultural environments in North America 210*t*
   use of agricultural operations by 210–11

T
target hosts 213
tawny pipit (*Anthus campestris*) 33
T-cells 16, 32
   T-helper cells 238
temperature, localized effects of 248
Tengmalm's owl (*Aegolius funereus*), *Leucocytozoon ziemanni* infection in 115
terminal investment 65, 115
testosterone
   and immune responses 33
   and male mating strategies 64
tetracycline 219
Th1/Th2 polarization 238
thallium 177
thermoregulation 31
thrushes, nematode diversity in 195
ticks 58–9
   and the circulation of Lyme borreliosis-causing agents in seabird communities 159–60
   deer tick (*Ixodes scapularis*) 135
   hard (Ixodidae) 20, 159
   hard ticks (Ixodidae) 20
   Hyalomma marginatum ticks 241
   Hyalomma ticks 236
   infected with *Borrelia* spp. 182
   Ixodes ticks 159, 160, 184
   and louping ill disease in wild birds 157
   migratory birds and tick dispersal 135
   movement of 235–6
   neotropical tick (*Amblyomma longirostre*) 236
   seabird tick (*Ixodes uriae*) 150
   soft (Argasidae) 159
   visual detection of 18
tolerance 44
   resistance and 81–5
toll-like receptors (TLRs) 30, 82

tools, for assessing effects of parasites on avian populations 123–9
toxicants 176–9
Toxoplasma spp. 129
   Toxoplasma gondii 237, 249
   in geese 129, 130
tracking birds 250
traffic noise 179
transgenerational effects, of parasites 114
transmission 133, 154
   by arthropod vectors 11–12
   biological 11
   frequency-dependent 134
   between-host 246*f*
   mechanical 11
   routes 10
   sickness behaviors and 58
tree swallow (*Tachycineta bicolor*) 123, 177, 197–8
trematode parasites 201
   Ribeiroia ondatrae 154
Trichobilharzia schistosome species 232
Trichomonas spp. 129
   Trichomonas gallinae 23, 103*t*, 106, 108, 137, 180, 202
trichomoniasis 137, 175, 180, 193
Trichostrongylus tenuis 103*t*, 105, 106, 113, 141
trypanosome, infection of female great tits 113
turkey farmers, antibodies against avian influenza virus strains 234
Typhimurium 219

U
urban birds, lead concentrations in 176
urbanization 156, 173–4
   consequences for avian parasitism 181–4
   density of birds and 157
   effects on avian health 181*f*
   effects on immune responses of wild birds 37
   and wildlife health 182

V
V (individual reproductive number) 44, 45
vectors 8, 11–12
   biological transmission 11
   densities of infected vectors (DIV) 152

mechanical transmission 11
vectorial capacity 11
violet-green swallow (*Tachycineta thalassina*) 200
virulence 10–11
virulence-transmission trade-off hypothesis 87, 89
viruses 9 *see also* influenza type A viruses (IAV); West Nile virus (WNV)
  arboviruses 1, 233, 239, 241
  astroviruses 218
  differences in survival between infected and uninfected groups of free-living wild birds 101*t*
  Japanese encephalitis virus 233
  Marek's disease virus 88
  Mayoro virus 233
  Newcastle disease virus (NDV) 22, 34, 108, 202, 218, 220–1, 238
  picornavirus 251
  rabbit hemorrhagic disease virus (RHDV) 151
  Rift Valley fever virus 239
  RNA viruses 234
  St. Louis encephalitis virus 233
  Sindbis virus 64, 67
  Usutu virus (USUV) 233, 239
  western equine encephalitis virus (WEEV) 239
visual detection, of avian parasites 18–20

W

wading birds, intestinal nematode *Eustrongylides ignotus* in 181
wastewater treatment sites 181
water baths 180
waterbirds
  mortality from botulism 202
  as paratenic hosts of *Cryptosporidium parvum* 8
waterfowl 59, 128, 191–2, 216, 231–2
  at agricultural operations 215
  Arctic-breeding 195, 202
  hematozoan blood parasite richness of 195
  ingestion of lead pellets 177
  as reservoirs of influenza A virus 210*t*, 214*f*
  risk factor for influenza virus transmission to humans 234
  zoonotic *Campylobacter* spp. in 231
weather, extreme 198, 201

West Nile virus (WNV) 1–2, 10, 14, 44, 126, 233
  antibody levels peak during infection 43
  avian species richness and human incidence of 154
  differences in survival between infected and uninfected groups of free-living wild birds 102*t*
  dispersal from sub-Saharan Africa to Europe 223
  early warning for the presence of 251
  in house sparrows (*Passer domesticus*) 56, 107, 179
  infectious load 45
  interspecific differences in susceptibility to infection with 81
  and interspecific host heterogeneity within avian communities 151
  management strategies to minimize 241
  molecular evolutionary response to their avian hosts 85–7
  mortality in northern cardinals (*Cardinalis cardinalis*) 105
  movement of 237
  in North American bird populations 137
  parasites co-circulating in the WNV system 238*f*
  persistence over winter 199
  relationship between avian community composition, mosquito feeding preference, and amplification fraction of 151*f*
  risk in American robins 66
  risk of introduction to the Galápagos islands 237
  survival of infections from 108
  transmission at higher latitudes 200
  transmitted to humans 239
  in urban areas 183
  use of agent-based models (ABMs) to understand dynamics of 135
  vaccination of wild birds for 241
  vectorial competence for 11–12
  West Nile encephalitis 231
  WNV-neutralizing antibodies in chickens 18
western bluebird (*Sialia mexicana*) 43*f*

western sandpiper (*Calidris mauri*), sickness behaviors 58
wetlands, natural 174, 199
white ibis (*Eudocimus albus*) 184
white-browed sparrow weaver (*Plocepasser mahali*) 37
white-crowned sparrow (*Zonotrichia leucophrys*) 91
white-fronted goose (*Anser albifrons albifrons*) 106
wild birds
  differences in survival between infected and uninfected groups of free-living 101*t*
  human interest in 1, 251–2, 253
  minimizing avian zoonoses in 241
  transmission of parasites to domestic livestock 5
  viral spillover between domestic poultry and 216*f*
wild turkey (*Meleagris gallopavo*), snood lengths 66
wildlife management plans 237
wildlife–agriculture interface
  avian use of agricultural operations 209–12
  contamination of agricultural operations by birds 222
  contamination of crop fields by avian synanthropes 211
  host types for parasite contamination by birds 212–15
  migration and parasite introductions 222–4
  pathogen prevalence and transmission in wild birds 218–22
  pathways of contamination from birds to agricultural operations 215–17
  relationship of different host types with target populations at the 214*f*
willow grouse (*Lagopus lagopus*) 130, 141
willow tit (*Poecile montanus*) 176
winter harshness 198
wintering conditions, immune defenses and 35
within-host dynamics 4–17
  attachment and invasion 14–15
  exposure and colonization 14
  pathogen replication and activation of host immune response 15–16
  pathogen shedding/exit 16–17
  resistance and tolerance 16

wood thrush (*Hylocichla mustelina*) 135
woodchat shrike (*Lanius senator senator*) 236
World Organization for Animal Health (OIE) 222

Y

yellowhammer (*Emberiza citrinella*), *Haemoproteus* infection in 114

Z

zebra finch (*Taeniopygia guttata*) 63
exposure to methylmercury 178

sickness behaviors 57, 58
zoonoses 1, 150*b*, 195, 231–4, 235, 242
minimizing avian zoonoses 240–1
reverse (anthroponoses) 237–8, 242
surveillance for avian zoonoses 233–4